T0135327

# Lecture Notes in Networks and Systems

## Volume 381

The series "Lecture Notes in Networks and Systems" publishes the latest developments in Networks and Systems—quickly, informally and with high quality. Original research reported in proceedings and post-proceedings represents the core of LNNS.

Volumes published in LNNS embrace all aspects and subfields of, as well as new challenges in, Networks and Systems.

The series contains proceedings and edited volumes in systems and networks, spanning the areas of Cyber-Physical Systems, Autonomous Systems, Sensor Networks, Control Systems, Energy Systems, Automotive Systems, Biological Systems, Vehicular Networking and Connected Vehicles, Aerospace Systems, Automation, Manufacturing, Smart Grids, Nonlinear Systems, Power Systems, Robotics, Social Systems, Economic Systems and other. Of particular value to both the contributors and the readership are the short publication timeframe and the world-wide distribution and exposure which enable both a wide and rapid dissemination of research output.

The series covers the theory, applications, and perspectives on the state of the art and future developments relevant to systems and networks, decision making, control, complex processes and related areas, as embedded in the fields of interdisciplinary and applied sciences, engineering, computer science, physics, economics, social, and life sciences, as well as the paradigms and methodologies behind them.

Indexed by SCOPUS, INSPEC, WTI Frankfurt eG, zbMATH, SCImago.

All books published in the series are submitted for consideration in Web of Science.

More information about this series at https://link.springer.com/bookseries/15179

Tatiana Antipova
Editor

# Digital Science

## DSIC 2021

 Springer

*Editor*
Tatiana Antipova
Institute of Certified Specialists
Perm, Russia

ISSN 2367-3370               ISSN 2367-3389   (electronic)
Lecture Notes in Networks and Systems
ISBN 978-3-030-93676-1        ISBN 978-3-030-93677-8   (eBook)
https://doi.org/10.1007/978-3-030-93677-8

This Springer imprint is published by the registered company Springer Nature Switzerland AG
The registered company address is: Gewerbestrasse 11, 6330 Cham, Switzerland

# Preface

This book is based on selected papers accepted for presentation and discussion at the 2021 International Conference on Digital Science (DSIC 2021) that was held during October 15–17, 2021. This conference attracted the practitioners and researchers from above 15 countries around the world, providing a value opportunity for interaction between the two communities. DSIC 2021 community is a network of scholars interested in natural and social sciences original research results. The network was established to promote cooperation between scholars of different countries. An important characteristic feature of this conference should be the short publication time and worldwide distribution.

Using computer software continues to be adopted widely, and the submissions to DSIC 2021 reflected a diversity of concerns that can be addressed to broaden the digital science. Alongside challenges that have traditionally been the subject of discussion and research such as Digital Economics; Digital Education; Digital Engineering; Digital Environmental Sciences; Digital Finance, Business & Banking; Digital Health Care, Hospitals & Rehabilitation; Digital Media; Digital Medicine, Pharma & Public Health; Digital Public Administration; Digital Technology & Applied Sciences.

A total of 99 research papers were submitted; each was reviewed by two Program Committee members or invited reviewers, and 52 were accepted (acceptance rate 52%). Collectively, the papers presented here constitute a collection of contributions to the flexible research and experience literature on a wide range of contemporary topics in digital science. The conference program featured a rich set of session topics and session types. Sessions were complemented by ad hoc lightning talks and a vibrant Google Meet tracks. We would like to thank everyone who contributed to this work, including paper authors, session speakers, course chairs, Program Committee members, volunteers and sponsors. Without their support, the event would not have been as successful.

This publication is unique as it captures modern trends in digital technology with a balance of theoretical and experimental work. This book is useful for private and professional research and teaching and for improvement of authors' scientific career by citing and grant applications.

Tatiana Antipova

# Contents

Digital Engineering

Digital Environmental Sciences

## Digital Medicine, Pharma and Public Health

## Digital Public Administration

## Digital Technology and Applied Sciences

# Digital Economics

# The Hexadecimal Factory - Product and Service Design Work in the Digital Economy

John Knight$^{(\boxtimes)}$

Aalto University of Arts, Architecture and Design, Helsinki, Finland
john.knight@aalto.fi

**Abstract.** This article explores digital product and service production. Digital production is a sizeable part of the global economy and growing area of employment. People working in the digital economy are employed in many sectors including e-commerce. Their roles are diverse, working conditions varied and in most cases their occupation blends traditional skills such as management with new ones such as software development. This article focuses on creative workers. Working closely with developers, their tasks span envisioning, designing, developing, delivering and maintaining digital deliverables. Preparatory studies were carried out to understand this work and then to examine the felt nature of creative work in often intense, agile working conditions. Agile was found to have both positive and negative effects on wellbeing. A four-week diary study was then conducted to explore this topic in more detail. The findings helped define how the creative work of designers is integrated within production, the kind of knowledge employed in collaborative, multidisciplinary work and the occupational strains they encounter. These findings are discussed and underpin the article's climatic conclusion on evolving Wilcock's framework for occupational health [9] to account for the specific characteristics of creative labour in the current period.

**Keywords:** Design · Employment · Occupation · Work · Grounded theory

## 1 Introduction

McKinlay and Smith's critical collection of articles focusing on work and employment in creative industries (2010) is a notable exception to a dearth of studies on creative work in the digital economy. Even then it is primarily concerned with film, theatre and television, rather than the broader digital economy that now spans across almost every nation, market, product and service domain and job. Since 2010, this area of economic activity has grown exponentially. For example, only three years ago around 5.9 million jobs [1] could be attributed to the sector in the US. Bukht and Heeks, [4] contend that generally the digital economy contributes at least 5% of GDP in most developed economies. Research by these economists [5] extends into identifying three important sub-categories of economic activity within the whole.

© The Author(s), under exclusive license to Springer Nature Switzerland AG 2022
T. Antipova (Ed.): DSIC 2021, LNNS 381, pp. 3–14, 2022.
https://doi.org/10.1007/978-3-030-93677-8_1

Firstly, they define a *core digital* sector, that includes a broad category of software development driven value creation. This core sector spans infrastructural platforms, and products and services that underpin a digitalised economy. The second, *digital economy* sector includes innovative products, services and inventions, such as social media, artificial intelligence or remote work practices that are only possible within today's technologically entrenched society. Finally, the *digitalised economy* encompasses traditional areas of economic activity that have evolved by technological transformation such as e-commerce. Creative workers are employed in all three of these categories. Software development consumes creative labour through the design of user interfaces that manifests technology as useful and usable interactive products and services for consumers. These interfaces in turn provide the means through which consumers take part in their own economic activity.

## 2   Formative Research

Having identified a sizeable area of creative labour within the digital economy, research was undertaken to understand the nature of occupation in this domain more fully. This was done in two stages. First, a number of small studies (see Table 1) were conducted to map out the domain. Findings helped shape a deeper and more targeted diary study that followed and explored emergent themes elicited in the first set of studies. Both stages of research aimed to address three foundational questions as follows:

What is design work?
What is the impact of this work?
What is the broader context for this work?

**Table 1.** Formative research studies

| Study id | Sample | Method | Theme |
| --- | --- | --- | --- |
| 1 | 19 | Workshop | Designers current project practices |
| 2 | 40 | Survey | Senior designers working practices |
| 3 | 5 | Workshop | In-house design team workflows |
| 4 | 121 | Survey | Designers' attitudes to agile |
| 5 | 52 | Survey | Designers' attitudes to agile |
| 6 | 14 | Survey | Designers' conflicts & beliefs |
| 7 | 59 | Survey | Developers' attitudes to design |
| 8 | 6 | Workshop | Stakeholder's experiences of agile |
| 9 | 24 | Workshop | Design policy stakeholders |
| 10 | 102 | Survey | Stakeholders' attitudes to design |
| Total | 442 | N/A | N/A |

Data from the first stage of studies was analysed by thematic coding of transcripts, interview notes and verbatims. First, textual data was tabulated and open coded. Overarching themes were then identified across the study data and in some cases, further coding was used to identify facets and nodes. However, in the final analysis, the main focus was to define key contextual themes within the broad topic of creative work in order to home in on specific areas for further investigation in the follow up study.

## 2.1 Theme 1 – Designers' Work Encompasses Production, Innovation and Facilitation

A consistent finding across the studies is that designers' primary role is in digital production. This means creating assets for development and working with developers to build digital product and services such as creating wireframes, graphics – even copy and content or page layouts as deliverables. Next, designer's role maps to the more traditional orientation of supporting innovation through planning out a new or improved digital product or service through blueprint type deliverables. Lastly, designers' role in facilitating collaboration is a growing part of their daily working practices. The relative growth of facilitation and co-design activities (against the relative decline of individual design work) has been accelerated by the predominance of agile and design thinking. Naturally, ancillary design activities are also part of everyday work and there is general tendency for this kind of activity to increase proportionally with seniority, especially in seasoned designer's role in 'shaping' projects (estimation and resource planning), 'client-facing' activities and sales support including the ubiquitous proposal pitch.

## 2.2 Theme 2 – Agile Permeates Digital Production and Shapes Creative Work

Scrum is ubiquitous and pervades and conditions how designers work and when and what they produce. Here, designers work on 'missions' in 'pods' through 'sprints' of two weeks duration that comprise time-boxed tasks described in 'tickets' and 'stories', through a 'backlog'. Teams' work is constantly refined via 'grooming' and 'sprint planning' sessions where teams estimate and prioritise tasks through 'story points' to deliver an 'increment' of working software. Kanban boards are used to progress work during 'daily Stand-ups' where a 'scrum master' 'removes impediments' that might slow team's 'velocity'. A number of rituals reinforce agile values and project vision including 'show and tells' and 'hackathons'. Teamwork is enabled by many digital tools and planning aides that help teams to collaborate, progress, produce and manage.

## 2.3 Theme 3 – Agile Has Negative and Positive Impacts on Creative Work

Designers' multifaceted role is often challenging in the fast pace of agile delivery as they need to 'feed' developers' assets and sequence their output to parallel production. Creative workers need to continually (re)align to a fixed project vision despite frequent changes ('pivots'), 'scope creep', evolving stakeholder needs, undocumented requirements and constraints that only emerge 'in sprint'. As a result, rework is often and creatives' role is regularly limited to low-value 'production' work. This is most evident in the tendency for them to be primarily occupied with small, incremental work over more rewarding strategic missions. Time pressure often means that 'cut down' methods are used instead of comprehensive requirements gathering and design thinking time. Agile can have a negative effect on production value and waste as well as eroding well-being through stress, interpersonal conflict and alienation.

Positive aspects of agile should not be underestimated. These, again, include the benefits of developing agency and comradeship in team work as well as positive flow work experiences [6] of collaborative working at pace. Anecdotally the kind of 'buzz' felt in agile pods can be infectious and when work sequencing, vision and resourcing are congruent the progressive aspects of agile are evident and in some ways at odds with more formal and hierarchical ways of working. Björklund & der Marel [3] echo this positive aspect of agile as helping foster competency, relatedness and meaning.

## 2.4    Theme 4 – Agile Missions Deliver Micro, Macro and Transformative Value

Agile development delivers three distinctive levels of value by delivering business and consumer innovation that ranges from (transformative) strategic transformations, (macro) middling product or service improvements through to (micro) small tweaks. These three incremental levels of innovation provide scaled levels of value that can be calculated as 'return on investment' and 'business cases' through to prioritization criteria in sprint planning. In a broader sense, value contribution is also an aspect of the occupational experience of creative workers. Designers gain kudos and satisfaction from high value adding missions but also flourish on a mix of small to large scale work. Middle work which sits between the two extremes and is often the most productively wasteful and least rewarding. This is possibly due to the often relatively unstructured character of this kind of work compared small discrete tasks or extensive, well-resourced and structured missions necessary for transformational outcomes.

## 2.5    Theme 5 – Designers Are Expected to Humanise Technology

Clients and collaborators (developers) expect and cherish designers, for their influence on innovation as much as their productive value. Designers' work including a per-ceived proclivity to humanise technology and an ability engender empathy with cus-tomers into technology driven missions is seen as a key differentiating advantage. Designers' (attributed) progressive prerogative and influence may either be the result or cause of their often-reported dilemmas in regularly encountering ethical challenges. In other words, the value-based role designers are expected to play is a source of conflict for them. The kinds of difficulties faced by designers range from nonmaleficence for consumers, benefiance in applying the right design process, justice in assuring acces-sibility and autonomy to counterbalance prevailing social, technical and commercial values.

## 2.6   Formative Research Conclusions

The first phase of research mapped out the structure of creative work at a general level. The predominance of agile, three levels of innovation spanning tweaks to transformation and designers' expected role in humanizing technology were strong themes in the data that generally had a negative impact on individual experiences of occupation. The negative effects of working on agile missions derives from the often relentless pace of sprints, the pressure of sequencing and producing deliverables across multiple workstreams, the commonplace marginalisation of designers within highly technically focused teams, waste through rework and a prevalence of unrewarding tasks were some of the reported issues found across the studies. Nevertheless, participants reported positive experiences of work too. They derive satisfaction from collaborative work, develop skill and knowledge-based agency and enjoy the social value of comradeship that underpins scrum. This formative study found that designers' sense of belonging, autonomy and engagement in meaningful activity to be in contradiction to their actual (often limited) role in production. Further and more focused research was undertaken to focus in on the experience of creative work at a personal level.

# 3   Summative Research

Participants were recruited through social media platforms. This involved contacting people whose described themselves on their profile as working 'designer's. A variety of tenure and sectors was sought, and participants took an initial online screening questionnaire to check profile matching, gain their informed consent and to give background information about the study. Forty-two subjects were selected down from the initial cohort to complete diary tasks over four weeks. Participants were mostly aged between twenty-five and forty-four (82%), most were female (62%) and the majority were based in the UK (59%). Digital diaries were used to collect data. Each participant was given a unique and secure workspace that provided activity descriptions and space for data entry in the form of interactive tables, pages and forms, depending on the task at hand. Diary entries were prompted through a series of written tasks, exercises and activities given by the moderator to participants via email and described in the weekly task descriptions that follow. The studies produced a large amount of data. Analysis involved tabulation and coding individual diary entries into a consolidated data set. Then open coding was applied by scoring related content to the codes. As coding progressed, themes developed that were then named and through a recursive process a taxonomy emerged, structured around primary themes that contained facets and nodes.

## 3.1   Narrative Inquiry Research and Reflexive Diary Study

In the first activity, participants co-authored a five-act narrative about a typical mission which provided transcript data for qualitative coding into underlying themes, facets and nodes within the data set. Participants online, collaborative story-writing task developed into a 'Practitioners' Tale'. This described their everyday working week in a typical client project as a group of designers. This activity involved participants creating a set of characters on the first day. The moderator provided scaffolding around this activity that included mandating that wherever possible, the language, terms and colloquial language of work was used. Lastly, within the writing activity, participants were asked to describe episodes of harmony and friction within the action in order to draw out occupational experience and balance factors. The structure and sequencing supported the activity well and prompts ensured participants were engaged over the whole week and that emergent findings could be probed within the study. The collaborative story telling provided rich data on the felt experience of creative work. Here, the cohort could express their experiences as a group, reflect on the past, share them with the cohort and perhaps therapeutically use the activity to communicate the daily tribulations of work in a safe environment with the others.

The result was a co-authored narrative containing 584 entries totalling 6052 words including native phrases such as 'Sheep dip (Client speak/jargon for initial meeting/workshop)'. Three main characters (Designer, Engineer and Manager) were described in some detail as well as brief descriptions of the supporting and minor characters. The acts provided structure to a typical mission starting with *'Act 1 - Counting down…'* which detailed the preparatory work needed to ensure sprints run to full velocity and the high expectations and hope felt at the beginning of missions. The narrative then moved to the second act ('Time is tight') where the pressure of working at pace, a lack of structure creates tension, compromise and stress. The story then moves to *'Act 3 - Fight for Power'* where conflict erupts and the overall goal of the mission is called into question through the commonplace changes of direction, rework and wasted production effort. In *'Touchdown'* a critical path emerges and alignment occurs briefly to deliver the 'increment'. Finally, in *'Act 5 - Wrapping up…'* team members recover for the next mission, reconvene with their peers and reflect on the work done. In this closing piece, there is a strong sense of achievement and the team overcoming impossible odds to produce value over and above what was deemed possible. There a complimentary and emphatic sense of the importance of comradeship and celebrating the success of the team.

The narrative provided a corpus within which thematic coding could be conducted. Provisional themes, codes and facets were developed and then tested until a stable set of themes were identified. The themes were iteratively developed from open coding the data and then closed coding to Wilcock's [10] occupational framework of being, doing, belonging and becoming. The framework was used to guide the direction of analysis as it provided a good fit to the emergent themes and also related to the domain level topic of occupation. The full corpus was then copied to three spreadsheet tabs that corresponded to each of the themes and framework constructs. Each tabs' content was then reviewed and extraneous data (e.g., relating to product owners or clients that are outside of the scope of this research) removed in order to make coding easier and accurate. Each theme was then recursively coded to identify facets and nodes and where possible these were quantified by frequency in order to show the relative weighting of each element in the narrative.

## 3.2    Narrative Thematic Findings to Wilcock's Framework

Generally, there was a good fit between all of the data and the framework's themes of being, doing, belonging and becoming. However, the weighting (the frequency of matching data to each framework element) was surprisingly heterogeneous. Data relating to the felt nature of occupation (being) made up just half of all the coded data (51%). The second element of doing scored also relatively highly at 36%. This meant that the two primary themes made up nearly ninety percent of the coded data. The dynamic nature of occupation, described in the third facet of the framework, was significantly less frequent at nine percent. Belonging, similarly had low frequency (4%) within the overall set of four coded themes. These overarching finding suggest a general imbalance in designers' occupational experience. This could be understood positively in those designers are highly engaged in their work and that their occupation is deeply connected with who they are as a person in ways that go beyond instrumental motivations.

## 3.3    Narrative Thematic Findings on Doing

Facets and nodes within the doing theme were relatively easily identifiable as they pertained to the subject doing something or acting on another subject or object. The structural elements of work, identified in the first stage of research, were also found in the second data set, albeit with additional insights into the ratio of different kinds of missions. There was a prevalence of middle work (74%) compared with the distinctively lower frequency of tweak (12%) related and strategic missions (14%) respectively. Occupational balance factor data (see Fig. 1) included participants' conception of missions and the negative and positive aspects of work. Facets and nodes were, again, relatively easily identified as they generally related to the outcome of an action or the relationship with another subject. Content relating to the mission theme itself was the highest frequency facet within this theme. Nodes included the type of challenge (unstructured vs. structured), the power relations in teams, the important issue of production waste including resource issues (for example, frequent personnel changes limit knowledge transfer) and touched on the critical topic of (non)convergence on a shared vision within missions. Pace was the second most frequently reported topic. Awareness, expectations and the level of respect other disciplines afforded to designers made up a third role aligned facet within this theme. Content mapping to relatedness (15%) echoed the strong sense of camaraderie seen in the previous research, as well as the more functional collaboration needed in production tasks and the development of team agency. Relatedness also included some surprisingly stringent content including attribution of 'abuse' in pods, as well as similar conditional nodes (good or bad) such as pride. Conflict was reported within relatedness too. Data relating to control and disempowerment (86%) was also an emphatic finding in the data. Nodes within the occupational balance theme shed further light on the felt nature of occupation. This data included insights into the pressure of work itself, lack of structure and certainty, the prevalence of mundane tasks and uneven skills and knowledge across team members. 'Exhaustion', 'burnout', 'panic', 'juggling' and 'rework' are some of the words used in relation to missions in the diary entries. The language and prevailing negative experiences reported by participants give a flavour of the kinds of stresses encountered in creative work in agile development missions within the wider digital economy.

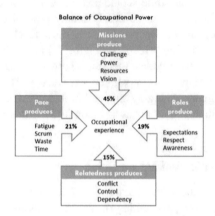

**Fig. 1.** Occupational balance factors.

## 3.4    Narrative Thematic Findings on Knowing

The final theme identified in the data pertained to knowledge. The amount of data relating to this topic was notably higher than to the other themes but was less easily identifiable. The data relating to knowing was relatively homogeneous and well-defined within four discrete facets. Ease of coding was helped by most attributed content containing the word 'know' or 'knowing' or where it could be accurately inferred through phrase that involved the subject [knowing] about something or [knowledge] that was needed to act. Unlike the other themes, while the facets and nodes were well-defined there was no related framework, such as Wilcock or even relevant words that captured the essence of the nodes and was a common term in language. The closest match found was Baumard [2] framework that draws on ancient Greek concepts including Phronesis. However, both the arcane language and the abstract nature of the facets within Baumard's framework, led to a shorthand labeling schema being developed that used colloquial English, including ken and knack. The four elements of this schema were then used to code out facets and nodes, described in Table 2.

Ken connotes a totalising abstract to concrete knowledge capability, attributable through external sources in books and the concretised knowledge and covering the broad body of design knowing and theory. The second, facet 'Chi' can be defined as knowledge capability built on context, contingent on interaction with others and previous experience. Facets within this theme, include reference to methodologies and nodes relating to cases, patterns, principles and structuring work to standard ways of working. Knowing and learning through doing and reflecting on the work was a strong element too. The third node within the contextual knowing facet concerned knowing about external factors influencing work such as costs and dependencies. This kind of contextual knowing correlates strongly with tenure as experience is gained worldly knowledge increases. However, anecdotally, this can lead to an orthodoxy of working that inhibits doing things differently. Knack: physical and mental prowess combined into useful practical knowledge capability was the third facet within the theme, although the one with the lowest frequency across all of the nodes. This theme relates to the technical skills needed to do the work and the competency related knowing that employers are most focused on in most cases. Nous, was a strong node in the data. Not only is this kind of cunning 'practical intelligence' useful in everyday work but it also correlates to career success independently from educational background or verbal intelligence [8]. The data suggested that this situational, quick-witted kind of knowledge capability is emergent in doing and unlike chi is highly personal and indeed correlates to tenure and developing individual strategies and ways of working. Nodes within this facet included knowing and applying knowledge relating to personal values and ethics, the ability to improvise and break with orthodoxy to get the best out of situation – almost at a level of abstraction that flies in the face of convention, individual drive and guile and the ability to influence others are also strongly attributable to this theme. This node also mapped to scrutiny, judgement and empathy.

**Table 2.** Design knowledge framework – theme (T), facets (F) and nodes (Nx)

| T | F | N1 | N2 | N3 | N4 | N5 |
|---|---|---|---|---|---|---|
| Nous | Personality | Curiosity | Pragmatic | Emotion | Empirical | Reflect |
| | Drive | Achieve | Guts | Risk | Own | Juggle |
| | Strategy | Control | Defer | Plan | Change | Plot |
| 71% | Influence | Empathise | Listen | Steer | Rapport | Talk |
| | Intellect | Learn | Diagnose | Explain | | |
| | Approach | Discover | Balance | Scrutinise | | |
| Chi | Practices | Patterns | Principles | Process | Structure | |
| | Context | Quality | Optimise | Value | Estimation | |
| 21% | Experience | Cases | Education | | | |
| | Connect | Collaborate | Facilitate | Consult | Contribute | |
| Ken 6% | | Theory | Abstract | Causality | Clarity | Tradition |
| Knack 2% | Capability | Expertise | Hands on | Hone | Finish | Document |

### 3.5    Thematic Coding Validation

In order to validate the construct further responders kept a reflexive diary for a week and the resulting data helped to assess the knowledge framework. The results also gave good insights into the daily tribulations of design work and importantly enabled mapping coping strategies to the design knowledge capabilities. The results indicated that Nous was the most frequently reported, Chi was second, Ken and Knack being the lowest. These findings, and the narrative together suggest a number of knowledge-based strategies for building resilience. Firstly, designers learned to navigate difficult situations using their own capacity and resources. Secondly, that using cases, methods and standards is a strategy to circumvent marginalisation by legitimising their activities to common practices. The low prevalence of theoretical and practical knowledge is notable too. A factor in the low level of tool-related practical knowledge is perhaps contingent on the kind of work designers do today. This tends to be on computers and that, anecdotally the software they use to render visuals, create flows and produce developer ready assets is becoming less complex and more usable with a decreasing need to learn complicated, bespoke design software. Low levels of theoretical knowledge is also not so surprising perhaps, as the other data points to the pressure of work and often low-level production task designers are occupied by. The validation findings, however, suggest a more fundamental and surprising conclusion in connection to Wilcock's framework. In surfacing the importance of knowledge, as both work resource, capability and enabler in occupational balance, the findings suggest that being, doing, becoming and belonging are underpinned by knowledge. This may not be the case in all occupations and perhaps as information workers, designers are particularly dependent on their knowledge capabilities compared to more traditional occupations.

### 3.6    Occupational Diary

The diary study research progressed beyond the narrative inquiry that identified the importance of knowledge to occupation and included a further two weeks of fieldwork. This study produced data on daily design tasks, participants occupational experiences and emotional states. Participants kept a daily diary and were prompted to answer questions on what they were doing and how they felt at specific times, by adding notes and scoring against a set of descriptors. The descriptors were developed from the findings of the first stage of research and were iteratively refined before the occupational diary study in order to gain a good potential fit with agile working experiences. The sixteen emotional descriptors were scored by diary keepers using a ten-point scale (0 being low) during their working day over one week. They could also add their own descriptors if the predefined set was lacking or ill-fitting. This resulted in tabular data that could be quantitatively analysed to measure frequency and fit and was used to refine the set of descriptors for a second week of diary keeping. The second weeks fieldwork followed a similar procedure to the first, except participants were only given the refined set of descriptors to match against in their diaries as the first set had already been refined.

## 3.7    Thematic Findings on Being

The occupational diary data was coded, and the descriptors paired into eight binary sets of low state to high state such as stressed to calm within an occupational experience framework (see Table 3). Each set was then coded back from the data to derive nine occupational emotional binaries ranging from congruence to pace. These provide insights to the felt nature of creative work with each binary containing a low state stressor and a positive satisficing high point. The data shows that the occupational experience fluctuates between polarities and are highly personal in how they manifest as positive or negative experiences. In other words, a challenge can be both good and bad for different individuals and situations. The findings are surprising in the relative homogeneity of the data with pace being the lowest and arousal the highest. How each of the binaries work together and the valence within each set is an interesting topic for further research. Anecdotally, it would seem that individual difference and personal capacity for work determines their occupational experience. In other words, someone can be positively engaged in challenging task even if they are rushed if they find the activity stimulating. Someone else might find being fully aligned in the purpose of their task with their peers a good thing, while others might strive for more autonomy. However, individuals operate within this framework, it seems to accurately account for designers' occupational experience in the study with little variation from the first set.

**Table 3.** Occupational emotional descriptors and occupational experience themes

| Low-state frequency | Low-state descriptors | High-state descriptors | High-state frequency | Occupational experience theme nodes |
|---|---|---|---|---|
| 4.34% | Alienated | Aligned | 5.41% | Congruence |
| 3.56% | Bored | Excited | 5.49% | Stimulation |
| 7.12% | Calm | Stressed | 6.26% | Arousal |
| 6.04% | Constrained | Unconstrained | 4.97% | Autonomy |
| 5.67% | Disempowered | Empowered | 6.71% | Control |
| 4.56% | Disorded | Ordered | 5.60% | Structure |
| 6.12% | Distracted | Engaged | 7.67% | Engagement |
| 5.00% | Overstretched | Underused | 5.63% | Utilisation |
| 4.97% | Leisurely | Rushed | 4.89% | Pace |

# 4    Conclusion

This research has helped to outline the structural foundations of creative work in the digital economy Two stages of fieldwork were undertaken within an emergent occupational perspective. Formative research helped to identify key constructs with design work in an agile environment. These primarily structural constructs included three levels of design work ranging from the small change tweak type missions through to the strategic. However, this tranche of research gave limited insights into the felt nature of this kind of work. In the sumative phase of research, diary studies were used to gain insights into this topic and to evolve Wilcock's occupational framework of being, doing, knowing and belonging. The occupational perspective not only compliments the proximal fields of social practice theory and design research, but also provides methodological tools and insights into the felt experience of work and perhaps, most importantly the connection between work and well-being [10]. Future work may build on the possibility of using these insights for targeted therapeutic interventions that engender improving generalised and persistent occupational health issues of achieving balance, resilience and empowerment in work. Extending the health potential of design itself and augmenting resilience building with tried and tested design thinking activities is another potential benefit of aligning these two disciplines further. It might be that the very tools and methods used in design practice might compliment the now common-place use of mental health support in work and help build resilience in all kinds of digital work.

# References

1. Barefoot, K., Curtis, D., Joliff, W., Nicholson, J.R., Omohundro, R.: Defining and measuring the digital economy. US Department of Commerce Bureau of Economic Analysis, Washington, DC, 15 March 2018 (2018). https://www.bea.gov/sites/default/files/papers/defining-and-measuring-the-digital-economy.pdf. Accessed 2 April 2021
2. Baumard, P.: The intelligence deed end: how you present it! Comp. Int. Rev. **5**, 53–55 (1994). https://doi.org/10.1002/cir.3880050216
3. Björklund, T., van der Marel, F.: Meaningful moments at work: frames evoked by in-house and consultancy designers. Design J. **22**(6), 753–774 (2019)
4. Bukht, R., Heeks, R.: Development implications of digital economies. University of Manchester, Manchester (2018)
5. Bukht, R., Heeks, R.: Defining, conceptualising and measuring the digital economy. GDI Development Informatics Working Paper 68. Global Development Institute, University of Manchester, Manchester (2017)
6. Knight, J.: Go with the flow: accelerated digital design in the age of post-agility. Design J. **20** (sup1), S2700–S2715 (2017). https://doi.org/10.1080/14606925.2017.1352781
7. Mayer-Ahuja, A., Wolf, H.: Beyond the hype: working in the German internet industry. In: McKinlay, A., Smith, S. (eds.) Creative Labour-Working in the Creative Industries. British Journal of Industrial Relations, London School of Economics, vol. 48, no. 3, pp. 210–233 (2010)
8. Wagner, R.K., Sternberg, R.J.: Practical intelligence in real-world pursuits: the role of tacit knowledge. J. Pers. Soc. Psychol. **49**(2), 436–458 (1985). https://doi.org/10.1037//0022-3514.49.2.436
9. Wilcock, A.A., et al.: The relationship between occupational balance and health: a pilot study. Occup. Ther. Int. **4**(1), 17–30 (1997)
10. Wilcock, A.A.: Reflections on doing, being and becoming. Aust. Occup. Ther. J. **46**, 1–11 (1997). https://doi.org/10.1046/j.1440-1630.1999.00174.x

# The Role of Controlling in Increasing Labor Productivity During the Digital Transformation

Ludmila Popova[1]([envelope]) [iD], Irina Maslova[1] [iD], Yulia Kotlova[2] [iD],
and Zoya Mkrtchyan[1] [iD]

[1] Orel State University, Naugorskoe Highway, 40,
Orel 302020, Russian Federation
[2] Khabarovsk State University of Economics and Law, Tikhookeanskaya Street,
134, Khabarovsk 680034, Russian Federation

**Abstract.** He mode of studying the controlling system, and its implementation in Russian economic entities, and the need to increase labor productivity, both at the state, and corporate levels, leads to the need for a comprehensive study of the possibilities of tools, and methods of the controlling system in the development of mechanisms for their adaptation for increasing labor productivity. The intensity of this discussion reflects the challenges facing the controlling system in the context of its role in increasing the labor productivity of individual business entities, and, as a result, the economic development of the Russian economy as a whole. The importance of studying the controlling system in this direction lies in the spatial aspect of economic development. Approaches to the scientific understanding of the definitions of "controlling," and "labor productivity" and their regulation require effective interaction of methods, and techniques.

This article substantiates the capabilities of the controlling system provided with digital tools in increasing the productivity indicators of an economic entity.

Methodological basis: a set of general scientific (analysis, synthesis, system-structural, and others), and private scientific (specifically sociological, logical, and legal) methods.

Results: A study of the possibilities of a controlling system based on digital transformation in increasing labor productivity by influencing business processes, developing employee competencies, and forming a special operational model of collaboration, and culture of a business entity was conducted.

**Keywords:** Controlling · Labor productivity · Digitalization · Added value · Resistance · Cost approach · Accounting, and analytical system · Stability

## 1 Introduction

In the current economic, and political situation, characterized by insufficient rates of economic growth of our country, due, among other things, to foreign policy, and factors, there is an obvious need to switch to an innovative path of development, and the need for systematic improvement of tools, and procedures for managing business

T. Antipova (Ed.): DSIC 2021, LNNS 381, pp. 15–25, 2022.
https://doi.org/10.1007/978-3-030-93677-8_2

entities in various fields of activity. In his address to the Federal Assembly on February 20, 2019, the President identified "an outstripping rate of labor productivity growth, the formation of competitive industries, and an increase in non-resource exports" among the priority areas of development until 2024. In this regard, a trend in the development of the country's economy today is the national project "Labor Productivity", approved by the Ministry of Economic Development of the Russian Federation in March 2019.

The labor productivity indicator is most important in assessing the efficiency of the economy since it reflects how rationally a business entity uses its labor resources, and indirectly signals how modern the equipment fleet it has, what technologies it uses, how intelligently it conducts business. Russian economic entities have a huge potential in this area. Additionally, the government has adopted information support and active implementation of digitalization processes in all spheres of society as the basis for the innovative development path. Considering recent trends, economists also pay considerable attention to the digitalization of production, and the justification of the direct relationship between digitalization, and labor productivity growth.

The intensity of socio-political processes and the turbulence of the external environment of functioning require economic entities to respond quickly to changing conditions, the ability to quickly process, and analyze large amounts of data. Simultaneously, the requirements for the quality of such information increase, and, as a result, there is a need for special management methods.

Accordingly, the introduction and development of a controlling system based on digital transformation become relevant. Our country can achieve technological, and political independence only by introducing innovative management methods. Under the influence of these processes, the information system of each economic entity, being the main source of information exchange, is also actively transformed. A promising direction of management innovations is associated with the wide use of controlling procedures.

This article substantiates the capabilities of the controlling system provided with digital tools in increasing the productivity indicators of an economic entity.

## 2  Materials, and Methods

Productivity is the most important indicator of the quality of the organization of the production system, its management, and economics as a whole.

Foreign and Russian scientists are interested in the issues of labor productivity management, and the search for mechanisms for its growth.

The first research in this area was aimed at finding opportunities to increase productivity by improving the efficiency of operations (F. W. Taylor, H. Gantt, H. Emerson). More recent studies have begun to focus on finding ways to increase productivity by improving management methods (H. Fayol, W. E. Deming, M. Hammer, and others).

In modern interpretations of productivity, the authors distinguish the construction of effective business processes [5] and the harmonization of production [8], the competitiveness of production [10], which is focused on the high value of the product

[15, 21], meets the expectations of consumers [4] and, as a result, can provide high financial productivity [14], including through the management of knowledge and consciousness of employees [12, 16].

Simultaneously, it is controlling that can become an effective, and comprehensive tool that creates conditions for labor productivity growth. The main task of controlling is to implement effective management of organizational changes in practice and to implement the ideas, and principles of managerial economics in the enterprise.

The business structure of the enterprise, which uses the ideas of budgeting, is harmoniously combined with controlling, in which profit, cost centers, service centers, investment centers are allocated, as well as the structure of products, and services are established, the classification of customers for controlling purposes is performed, and an integrated planning system is used, which allows you to allocate a specialized subsystem of performance controlling.

The systematic information approach used in this article allowed the authors to assess key factors, develop appropriate conclusions and obtain results. The authors used general scientific methods as the basis of the research methodology: analysis, and synthesis, inductive, and deductive methods, historical, logical, and systematic approaches.

Based on the logical generalization of the materials obtained during the study, the authors developed scientific and methodological provisions within the framework of the study. An approach to the formation of a controlling subsystem focused on productivity growth, implemented in the conditions of digital transformation of the economy, is described. The article can be of interest to economists, managers, and anyone interested in management, and controlling issues.

## 3   Theoretical Review

At this stage of socio-economic development, economic entities are objectively forced to consider not only the parameters of their internal environment but also external factors, as well as to form their management system in such a way that it is resistant to the challenges of modern economic realities. This means that there is an objective need to study the possibilities of controlling as the main mechanism that ensures the achievement of management goals in increasing the key parameters of functioning and in particular, labor productivity. Fundamental studies of the nature and role of controlling in the management of modern economic processes are based on solving not only tactical but also strategic tasks, using applied tools and mechanisms for their application. As you know, the origin of the idea of controlling occurred at the end of the XIX century, but its scientific justification was formed later, in the middle of the XX century. Moreover, in scientific circles, its founder is deemed German economist D. Hahn, who was the first to present the interpretation of the term "controlling", indicating that it is "information support focused on the results of enterprise management" [7].

Additionally, he identified two key functions of controlling planning, and control, and for the first time typologies it into two types: strategic, and operational. The first definition of controlling was proposed by D. Khan. Its essence was to perform tasks only of an internal nature, without considering the influence of external factors. Simultaneously, a large role in this process was assigned to the person making managerial decisions [7].

In the future, the ideas of controlling were developed, and transformed by various foreign scientists, bringing their arguments to the interpretation of this scientific category. In Russia, the interest in controlling, and the need for it arose much later abroad. It was due to the rejection of a planned economy, and the development of market relations (in the 1990s). The opening of borders has expanded access to information about foreign trends in management in general, and control in particular. The first introduction of controlling procedures occurred in the banking sector, then it began to spread to other economic entities [20].

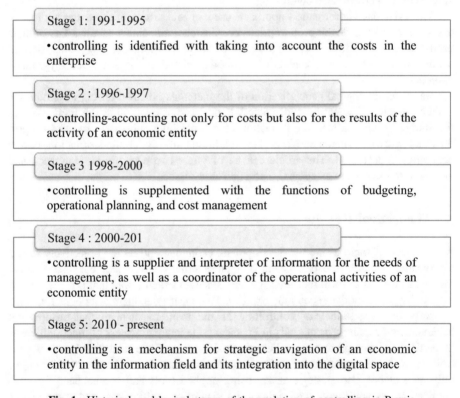

**Fig. 1.** Historical, and logical stages of the evolution of controlling in Russia

The founder of the Russian concept of controlling in companies of various branches of the national economy, and forms of ownership was S. G. Falko. On his initiative, in 2000, the international symposium "Controlling as a philosophy, and methodology of effective management of state organizations, and private business based on Bauman Moscow State Technical University was held for the first time in Russia [9]. Within the framework of this event, the Association of Controllers of Russia was formed, which was headed by S. G. Falko.

The formation of general scientific theoretical approaches to the concept and functional content of controlling is based on several historical, and logical stages (Fig. 1).

Later, other Russian economists also contributed to the development of the domestic concept of controlling.

Over time, scientific knowledge in this area has grown with the results of new research. Scientists and economists continue to pay attention to the study of this category.

Questions about the essence of the categorical definition of "controlling" including its constituent elements, principles, tasks, information support, and its relationship with management, and accounting systems are still debatable. In the current period, further evolution of controlling is already Toward "strategic navigation" [9] of an economic entity, and its integration in the digital information space, which is consistent with the priorities of national development.

# 4   Results

The impact of financial, and economic indicators on the economy fully depends on the level of productivity of the economy as a whole, provided by the labor productivity of individual business entities.

As already noted, the issues of labor productivity growth are among the priority national projects today, therefore, it is necessary to search for effective tools for productivity growth.

Today, labor productivity in certain economic sectors, and subjects are critically low. The main reasons for the low productivity of business entities (Fig. 2) become a consequence of a significant number of negative factors [1].

Among the main reasons for the low productivity, we will highlight the imperfection of the business entity's controlling system and low-effective management. Note that management, as a rule, blames low labor productivity on insufficient automation of processes, not given that the optimization of processes (within the framework of controlling capabilities) should be primary, and then their digitalization, since the digitalization of ordered processes reduces labor intensity, otherwise the opposite effect can be achieved. In this regard, the controlling system has significant potential in achieving the goal of increasing labor productivity.

Among the key tasks of controlling, it is reasonable to single out the practical implementation of effective management of organizational changes. In this regard, based on the complexity, and consistency of the controlling category, it can be argued that it is controlling that can act as an effective, and comprehensive tool that creates conditions for labor productivity growth. Simultaneously, it is impossible to deny the fact that the solution of problems of increasing productivity should be to perform comprehensively, considering not only economic, and managerial but also technical, and technological factors.

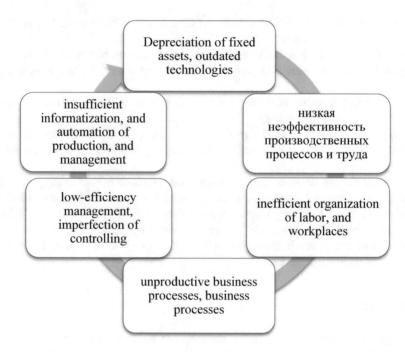

**Fig. 2.** Key reasons for low labor productivity of business entities

Controlling operates with the results of information flows of accounting, and analytical systems, which include, as is known, subsystems of financial, tax, and management accounting, as well as complex economic analysis [22].

The accounting and analytical systems of each economic entity are unique, and its structure determined by the peculiarities of the information needs of the management, and the current regulatory framework (Fig. 3).

As known, the controlling system includes management accounting, planning, control and analysis. By doing it ensures the coordination of the management system as a whole, reducing the time for making a management decision and increasing the efficiency of managing an economic entity. The role of controlling in improving the efficiency of labor productivity management processes is displayed as follows (Table 1).

By embedding an element of labor productivity management in the model of the controlling system, it is important, relying on the information base of accounting, and analytical system of an economic entity, to ensure its expansion by introducing progressive tools due to digital transformation.

To increase productivity, the management system should be focused on development, in which the enterprise can create and produce goods and services with sufficient costs to create a reserve of price competitiveness and profit [2].

The intellectualization of labor, the predominant importance of knowledge, innovation, and information corresponds to the current stage of the development of the world economy. In this regard, the integration of the controlling system with the digital

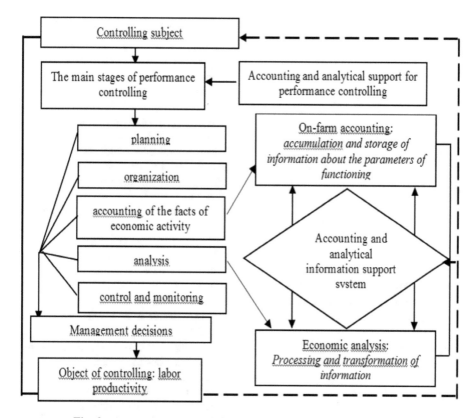

**Fig. 3.** Accounting, and analytical support of performance controlling

potential of an economic entity allows you to bring the process of managing it to a significantly new level.

In July 2021, the Ministry of Industry, and Trade of the Russian Federation adopted the "strategy for the digital transformation of manufacturing industries to achieve their "digital maturity" until 2024, and for the period up to 2030" [17]. It confirms that the digital transformation of industry is a priority direction for developing the domestic economy, providing high adaptability in the formation of business models and management of production processes through the integration of end-to-end digital technologies.

At the heart of the introduction of digital technologies in the industry is the desire to comprehensively improve efficiency, and create conditions for the successful operation of industries. The digital transformation of the sectors of the economy that are strategically important for the state will lead to the functional, and technological independence of not only the industrial sector but also increase the level of security and autonomy of the information technology infrastructure/.In turn, the digital transformation of business processes will objectively decrease costs and increase labor productivity, and product quality. In the end, will decrease the time for the release of a

**Table 1.** The role of controlling in improving the efficiency of management processes

| Criteria for the effectiveness of management processes | The role of controlling in improving the efficiency of management processes |
|---|---|
| The level of manageability of processes | – coordination, analysis, control over the formulation, and maintenance of decision-making, and implementation processes<br>– improving the information, the integrity of the system, individual subsystems |
| Focus on the set goal | – Strategic coordination of decisions made, and the activities of departments, personnel<br>– ensuring the interconnection of external sources of strategic goals, and personnel<br>– ensuring compliance of operational plans with strategic<br>– ensuring the information integrity of departments |
| The duration of the cycle, and the level of the direct flow of processes in the control | – saving time for making a management decision integration of departments |
| A specific mechanism for implementing the process | – making adjustments to management processes because of monitoring the production system |
| Performance | – reducing the risks of making a management decision that does not meet the goal<br>– reducing the time of making a management decision at the stages of strategic planning, the risks of adopting unjustified strategic goals, and guidelines |

business entity's product to the market, and ensure flexible (quickly adaptable to external changes) functioning.

The key task of the digital transformation of industry, by the adopted strategy, is to modernize the management of production processes, which, ultimately, should lead to a significant increase in labor productivity. This approach applies also to individual business entities [17].

Note that among the factors affecting the speed of the introduction of digital technologies in economic entities, there are internal (human resources, technological level of production, etc.) and external (the level of competition, the availability of technologies and capital, as well as the development of legislation, etc.)

Another factor hindering the processes of digital transformation of a business entity is the insufficient digital maturity of current business processes, low level of automation, lack of competencies, and low level of IT literacy of employees [1].

An integrated approach to information, and analytical support of the controlling system based on the introduction of digitalization systems, such as ERP, which digitalizes the accounting system, finance, logistics, procurement, personnel, PLM, which covers the product lifecycle, business intelligence systems (MBA), allows a business entity, even in difficult conditions, to be able to quickly rebuild its trajectories (Fig. 4).

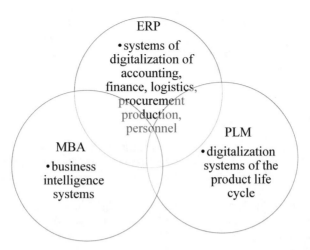

**Fig. 4.** An integrated approach to information, and analytical support of the labor productivity-controlling system

Thus, economic entities with a well-established controlling system focused on achieving the specified parameters of functioning, and productivity built based on digital transformation has a significant potential for rapid adaptation to changes in production conditions, and, as a result, an increase in labor productivity, and added value.

Among their main advantages is the integrated digitalization of all business processes, including production processes, which allow you to quickly respond to changes, allowing you to quickly restructure the system of management.

## 5  Conclusion

From the standpoint of modern market conditions of management, there is no doubt that labor productivity can be defined as one of the most important indicators of the labor results of each organization.

The growth of labor productivity is directly related to the financial, economic, economic, and managerial activities of an economic entity and depends on it while ensuring the cost-benefit ratio in the allocation of resources. The interconnectedness, and interdependence of the financial, and economic parameters of functioning, and effective management methods, which, first, include controlling, to achieve the specified parameters of labor productivity, and its growth, are obvious.

The intellectualization of labor, the predominant importance of knowledge, innovation, and information corresponds to the current stage of the development of the world economy. Currently, the Russian economy is eliminating the consequences of the economic downturn caused by forced isolation due to the pandemic. Today, there is an obvious need for methods that can improve productivity. Controlling has great potential. The results of the study can be useful in eliminating the consequences of the financial and economic global crisis. Thus, the integration of the controlling system with the digital transformation process will allow us to obtain an innovative management model that promotes productivity growth with a flexible response to changes in both internal, and external factors.

# References

1. Abdrakhmanova, G.I., Bykhovsky, K.B., Veselitskaya, N.N., Vishnevsky, K.O., Gokhberg, L.M.: Digital transformation of industries: starting conditions, and priorities. In: XXII International Scientific Conference on Problems of Economic, and Social Development, Moscow (2021)
2. Altukhov, A.I., Dudin, M.N., Anishchenko, A.N.: Digital transformation as a technological breakthrough, and transition to a new level of development of the agro-industrial sector of Russia. Food Policy Secur. **7**, 81–96 (2020)
3. Avdeev, M.Yu.: Approaches to creating a system for controlling labor productivity in industrial enterprises. Eur. Sci. **4**(46) (2019)
4. Bragin, Yu.V.: The way of QFD: design, and production of products based on the expectations of consumers, 240 p. Quality Center, Yaroslavl (2003)
5. Eliferov, V.G., Repin, V.V.: Business processes: regulation, and management: Textbook, INFRA-M, 319 p. (2009)
6. Falco, S.G.: Controlling in Russia: the current state, and prospects of development. In: Controlling in Small, and Medium-Sized Enterprises: A Collection of Scientific Papers of the IV International Congress on Controlling, Moscow, 80 p. (2014)
7. Khan, D.: Planning, and control: the concept of controlling, 800 p. Finance, and Statistics, Moscow (1997)
8. Klevlin, A.I., Moiseeva, N.K.: Organization of harmonious production: theory, and practice, 360 p. Omega-L (2003)
9. Kotlova, Yu.A., Popova, L.V.: Information potential of accounting, and controlling in the conditions of digitalization of the economy. Manag. Acc. **4**, 63–68 (2021)
10. Mishin, Yu.V.: Economic foundations of the organization of competitive production. Moscow, 212 p. (2004)
11. Necheukhina, N.S., Polozova, N.A., Buyanova, T.I.: Controlling as a mechanism for increasing the efficiency of an industrial enterprise in the conditions of using digital technologies. Sci. Tech. Bull. St. Petersburg State Polytech. Univ. Econ. Sci. **10**, 82–85 (2017)
12. Oskin, V.: Management of knowledge workers. Pers. Manag **6**, 51–54 (2004)
13. Popova, L.V., Isakova, R.E., Golovin, T.A.: Controlling. Delo I Service (2003)
14. Rudnitskaya, L.B.: Financial attractiveness, and financial productivity of the company. Finance **11**, 74–77 (2008)
15. Serbinovsky, B.Yu., Yemets, E.V., Timofeeva, M.S.: Labor productivity management at an industrial enterprise using estimates of the value of products. Finance, monetary circulation, and credit. In: Organization of Financial Systems: Materials of the II International Scientific, and Practical Conference, pp. 240–258. YURSTU, Novocherkassk (2007)
16. Shamir, B.: Strategic leadership in the management of consciousness. KubGAU Sci. J. **67** (03), 82–85 (2011)
17. The strategy of digital transformation of manufacturing industries to achieve their "digital maturity" until 2024, and for the period up to 2030. Ministry of Industry, and Trade of the Russian Federation (2021)

18. Shigaev, A.I.: Controlling the strategy of enterprise development, 351 p. UNITIDANA (2008)
19. Vinogradova, S.A.: Genesis of conceptual approaches to controlling. Vest. Univ. **11**, 71–77 (2014)
20. Volkov, D.L.: The theory of value-oriented management: financial, and accounting aspects, 320 p. Petersburg State University, Higher School of Management (2008)
21. Zhidkova, E.A.: Development of accounting, and analytical concept of controlling. In: Theory, and Methodology, 228 p. (2017)

# Implementation of Innovative Technologies and Directions of Digitalization of the Gas Transportation Sector

Zhanna Mingaleva$^{(\boxtimes)}$ ⓘ, Elena Lobova ⓘ, Galina Timofeeva ⓘ,
and Igor Poroshin ⓘ

Perm National Research Polytechnic University, Perm 614990,
Russian Federation
mingall@pstu.ru

**Abstract.** The expansion of digital technologies is increasingly active in all sectors and industries. The process of digitalization of activities is of particular interest in the oil and gas sector. The article is devoted to the study of the level of digital maturity of a particular Russian company that transports gas through main gas pipelines. The most effective directions for the development of further digitalization of the gas transmission company were also identified and analyzed. Structural-logical and content analysis, economic and mathematical calculations were used as research methods. As a result of the study, it was found that at present Russian gas transportation companies widely and actively use various digital technologies. This allows them to provide solutions to many production and commercial problems with a high degree of efficiency. Also, currently applied innovative and digital technologies allow gas transmission companies to successfully solve problems in the field of achieving sustainable development and combating negative climate change. Prospects for digitalization in the gas transportation sector include the introduction of advanced composite materials for pipes, as well as the use of UAVs, drones and other robots for air monitoring of methane levels over pipelines.

**Keywords:** Digitalization · Gas transportation sector · Gas pipeline monitoring · Methane emission · Digital technologies · DOT model

## 1 Introduction

The current stage of economic development is characterized by the expansion of areas of innovations' application in the field of technology and the growing possibilities of interaction of various digital devices to form an innovative and digital economy [1–3]. Currently wireless communication technologies for data collection, various software systems, systems for operational data transmission, control and monitoring of the state of regulatory values of various technological processes are used the most widely in production activities [4].

Companies in the oil and gas sector, like enterprises in other sectors of the economy, are inevitably affected by the need of digital transformation of their activities and are forced to use digital technologies for the normal implementation of economic and

T. Antipova (Ed.): DSIC 2021, LNNS 381, pp. 26–37, 2022.
https://doi.org/10.1007/978-3-030-93677-8_3

trade operations. First of all, this applies to such business processes as document management (including data storage), sales and logistics, IoT technologies, information security, monitoring of harmful emissions into the environment and monitoring the state of the environmental situation in the places of companies' operations. It is in these areas that oil and gas companies are most actively adopting advanced digital technologies. In particular, we are talking about such digital technologies as Blockchain, smart contracts, IoT tags, distribution ledgers, etc. [5–8]. We have already noted earlier that "The introduction of digital technologies in the oil and gas complex of Russia is directly connected with artificial intelligence systems. This is machine learning and in-depth machine learning; botosphere, including: robotization, bots, drones; and virtual reality of objects: improved reality, digital twin, mixed reality" [9, p. 28].

However, it should be noted that the most active digitalization of certain business processes and production activities in the oil and gas sector is carried out in such sub-sectors as hydrocarbon production and oil and gas refining. At present, the concept of a digital well (project "Rig of the Future") is being introduced more and more [10, 11], and various automation and computerization systems are used at refineries [12, 13]. Specifically, the senior upstream advisor of Chevron Trond Unneland noted that "the game-changing innovations in oil and gas in the past few decades include deepwater/subsea applications, real-time reservoir management, 4D seismic technology, horizontal drilling and fracturing, and data analytics. He added that emerging technologies in oil and gas include mobile computing, data science, cloud computing, cognitive computing and the Industrial Internet of Things" [4].

At the same time, such an important field of activity as the transportation of oil and gas through trunk pipelines is still practically outside the processes of digitalization and the application of innovative technologies in general. Wherein large gas producing companies have a very high dependence of their sales activities on the state of the pipeline infrastructure. For example, revenues of the largest gas producer in Russia PJSC Gazprom from the direct sale of natural gas and the provision of gas transportation services account for more than 50% of the company's total revenue [14]. This determines the high importance of the issues of introducing innovative technologies and digitalization of gas transportation through main gas pipelines. So the processes of innovative development and digitalization should be carried out in two directions.

Firstly, it is the prevention of the threat of harm to the environment by reducing gas emissions into the atmosphere.

Secondly, increasing the efficiency of pipeline transport by saving a commercial product (gas transported to consumers) also by reducing gas emissions into the atmosphere.

According to experts, gas emissions into the atmosphere do not only pollute the environment, but also lead to the loss of a commercial product: "...leaks in pipeline networks are one of the major causes of innumerable losses in pipeline operators and nature" [15, p. 1]. Therefore "The safety of the gas transmission infrastructure is one of the main concerns for infrastructure operating companies. Common gas pipelines' tightness control is tedious and time-consuming" [16, p. 1].

Thus, the development and implementation of innovative and digital technologies to reduce the loss of a commercial product when transporting gas through main gas pipelines is extremely important.

## 2    Theoretical Background

The importance of the introduction of innovative technologies and digitalization of gas transportation through main gas pipelines is determined by the special nature of the impact of pipeline transport on the commercial and financial activities of enterprises in the oil and gas sector. In particular, various types of gas leaks from pipelines lead to a reduction in the revenues of such companies.

The financial losses of gas transportation infrastructure enterprises associated with gas leaks from trunk pipelines mainly consist of two key sources.

1. Losses of a commercial product as a result of methane ($NH_4$) emissions into the atmosphere during technological maintenance of gas pipelines (repair work) and gas distribution stations (GDS). As a rule, these losses are known in advance (fixed in the repair plan) and included in the gas sales price.
2. Losses of a commercial product as a result of emissions of methane ($NH_4$) into the atmosphere in emergency situations or unauthorized gas leaks in case of damage to gas pipelines. These losses are not planned and result in direct business damage.

The presence of fundamentally different causes of gas loss from pipelines, which are fundamentally different in nature and basis, requires the development of a set of actions to prevent or reduce them. The main differences in the causes of gas leaks from main gas pipelines are shown in Fig. 1. Accordingly, the set of measures to eliminate them or reduce the damage caused also differs.

An assessment of scientific research depth in these two areas showed that at present the scientific literature mainly discusses the application of various digital technologies for the rapid detection and elimination of unauthorized gas emissions into the atmosphere in emergency situations or unexpected leaks. "Incidents of pipeline failure can result in serious ecological disasters, human casualties and financial loss. In order to avoid such menace and maintain safe and reliable pipeline infrastructure, substantial research efforts have been devoted to implementing pipeline leak detection and localisation using different approaches" [15, p. 1]. The main research here is carried out in the following areas.

Firstly, a large block of work consists of studies related to conducting field experiments to determine the possibility of using UAVs to detect natural gas leaks from underground gas pipelines, introducing digital technologies for wireless data transmission from UAVs, with recognition and operational data transmission [17, 18]. The results of such experiments are reflected in numerous scientific publications. They showed that aerial measurements based on the use of UAVs are a valuable source of information for identifying gas leaks, including from underground sources. Therefore, these innovative technologies are recognized as promising for everyday use by companies operating gas infrastructure.

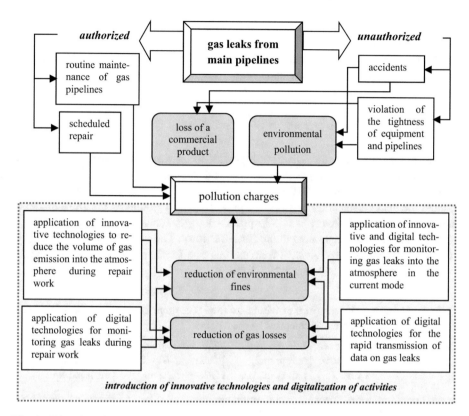

**Fig. 1.** Directions for the introduction of innovative and digital technologies to reduce gas leaks from main pipelines. Source: compiled by the authors

Secondly, the analysis of the possibilities of using lasers to detect methane leaks from gas pipelines is an important area that has been actively developing in recent years [19–21]. These technologies are proposed to be developed and implemented in addition (and in the long term as a complete replacement) to the currently used ground mobile equipment for sampling to check the quality of atmospheric air [22].

Thirdly, new digital technologies continue to develop to improve atmospheric air quality control systems using stationary sensors and measuring instruments fixed at various sections of the main gas pipelines and transmitting information to a central control device - computer stations [23, 24].

In general, the results of the analysis of the main directions of research on the possibilities of digitalization processes in the gas transportation system through main pipelines showed that all most of the studies are aimed at solving the issues of creating and implementing innovative and digital technologies, methods for solving the problem of timely detection of gas leaks from underground gas pipelines and monitoring of external environment indicators. At the same time, there are no comprehensive studies to reduce the permitted emissions of methane into the atmosphere. It is believed that the purchase of emission permits by pipeline infrastructure enterprises can compensate for the environmental damage caused by ongoing activities.

However, modern approaches within the framework of the concept of sustainable development presuppose the widest possible use of innovative technologies, primarily for the physical reduction of emissions of harmful substances into the atmosphere in order to minimize harm to nature and society. The search for such solutions is the goal of this work.

## 3   Research Methodology

In this study, we used the Deloitte approach to assessing the maturity of digital transformation of oil and gas exploration and production activities - the Digital Operations Transformation (DOT) model (The DOT model) [25, p. 5]. We have applied the general methodological principles and logic of building the DOT model to the gas transportation sector. LLC "Gazprom Transgaz Tchaikovsky" is the largest gas transmission company operating in the Perm Territory. The production data of the LLC "Gazprom Transgaz Tchaikovsky" were used as data for the study. The structure of the data taken for analysis is presented in Table 1.

**Table 1.** Main characteristics of the divisions of LLC "Gazprom Transgaz Tchaikovsky" in the Perm Territory (2019). Source: compiled by the authors

| Division | Length of pipelines, km | Volume of gas transported, billion m3 per year | Number of pumping stations (CS* GPU** GRS ***), units | Gross emissions of harmful substances, thousand tons |
|---|---|---|---|---|
| Gremyachinskoe LPUMG | 1056 | 182,5 | 1 CS, 26 GPU | 23,71 |
| Kungurskoye LPUMG | 1126 | 255,5 | 2 CS, 44 GPU | 27,11 |
| Chaikovskoye LPUMG | 736 | 182,5 | 1 CS, 30 GPU | 12,90 |
| Gornozavodskoe LPUMG | 1257 | 182,5 | 1 CS, 42 GPU | 22,43 |
| Bardymskoye LPUMG | 920 | 182,5 | 1 CS | 17,59 |
| Ochersky LPUMG | 330 | 59,86 | 1 CS, 8 GPU | 6,13 |
| Diamond LPUMG | 600 | 80,3 | 1 CS, 30 GPU | 7,68 |
| Permskoe LPUMG | 513 | 91,25 | 2 CS, 21 GPU | 1,36 |
| Bereznikovskoye LPUMG | 353 | 5,4 | 1 GRS | 0,68 |
| Total in the Perm Kray | 6891 | 1222,31 | 10 CS, 201 GPU, 1 GRS | 119,59 |

Designations:
* CS - compressor stations
** GPU - gas pumping units
*** GRS - gas distribution station

In total, about 20 oil and gas companies operate on the territory of the Perm Territory, including the largest ones - LLC "Gazprom Transgaz Tchaikovsky" and LLC "NOVATEK-Perm". LLC "Gazprom Transgaz Tchaikovsky" transports gas through the largest gas pipelines. The length of the gas transmission system operated by the enterprise, together with the branch pipelines, is more than 10 thousand km [26], and on the territory of the Perm Territory - 7 thousand km. (see Table 1). LLC "Gazprom Transgaz Tchaikovsky" is the undisputed leader in terms of the length of pipelines and the scope of work and services for the delivery of gas to consumers.

LLC "NOVATEK-Perm" supplies gas to the largest industrial enterprises of the Perm Territory (PJSC Unipro, PJSC Metafrax, PJSC Uralkali, JSC Sibur-Khimprom, etc., but has a shorter length of gas pipelines.

As a source of statistical data, we used the annual reports on the production and economic activities of LLC "Gazprom Transgaz Tchaikovsky" for 2014–2019, Gazprom's methodological materials on determining the effect of energy saving of fuel and energy resources spent on own technological needs of the main gas transportation [27]. The methodology assumes the following assessment procedure:

$$V = \frac{V_i * P_{av.} * 293}{Z_{com.} * T_{av.} * 0,1013} \tag{1}$$

Where,

V – volume of bleed gas during repair works, thousand m$^3$
Vi – gas pipeline section volume, thousand m$^3$
Pav. – average pressure in the pipe, MPa
Zcom. – gas compressibility factor
Tav. – average gas temperature, K

$$V_i = \frac{\left(\frac{Dps-2*16}{1000}\right)^2 * 3,14}{4} * l * 10^{-3} \tag{2}$$

$D_{ps}$ – diameter of the gas pipeline section, mm
$l$ – length of the bleed-off area, m

$$P_{av.} = \frac{Pr_{av.} + 1}{10,2} \tag{3}$$

$Pr_{av.}$ – average pressure in the pipe, kgf/cm$^2$

$$Z_{av.} = 1 - \left((10,2 * P_{av.} - 6) * \left(\frac{0,345 * 10^{-2} * \rho}{1,2} - 0,446 * 10^{-3}\right) + 0,015\right) \\ * (1,3 - 0,0144 * (T_{av.} - 283,2)) \tag{4}$$

$\rho$ – gas density, kg/m$^3$

$$T_{av.}(K) = T_{av.}(°C) + 273 \tag{5}$$

Tcp. (°C) – average temperature in Celsius.

## 4  Research and Results

Based on the application of the methodological principles and logic of building the Deloitte DOT model, an assessment was made of the maturity of the digital transformation of the LLC "Gazprom Transgaz Tchaikovsky" enterprise in the main area of activity - gas transportation. The assessment is carried out on all 10 stages of digital transformation, grouped into 3 digital realms proposed by Deloitte [25, p. 19]. These stages and realms are summarized in Fig. 2.

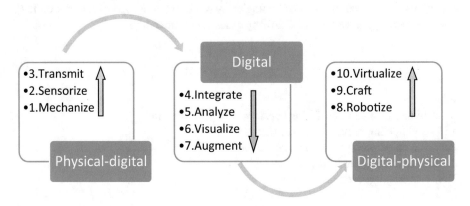

**Fig. 2.**  10 stages of digital transformation according to the Deloitte model. Source: adapted by the authors on [25].

An in-depth analysis of the automated controls and management of production activities used in LLC "Gazprom Transgaz Tchaikovsky" showed that the main stages of digital transformation within the framework of Physical-digital were passed back in 2013–2014.

Firstly, an automated stationary system for monitoring the exhaust gases of gas-pumping units (ASKVG) has been created at the compressor stations of LLC "Gazprom Transgaz Tchaikovsky" [28, p. 33]. ASKVG includes means of interaction with the automated control system of the GPU to obtain the operating characteristics of the unit in the current time mode. This automated system carries out direct measurements of the monitored components of exhaust gases in accordance with the current regulatory documentation. This system provides automatic continuous measurement of the current gas concentrations, determines the current methane emission rate and the total mass of gross pollutant emissions. A total of 51 automated exhaust gas control systems have been installed.

Secondly, at the workplaces of environmental engineers at the Almazny and Tchaikovsky LPUMGs, automated workstations for operators are equipped with a real-time display of all information and data on emissions. Monitoring results make it possible to accurately assess the technical condition of the units, increase the energy efficiency of the GPU, reduce the cost of repairing the GPU by timely diagnostics of its technical condition, and reduce the emissions of pollutants into the air.

Thirdly, innovative technologies for reducing the volume of gas emissions into the atmosphere during scheduled maintenance and emergency repairs on gas pipelines are being tested.

As we noted earlier (see Fig. 1), in the activities of gas transmission companies, one of the key sources of gas loss from trunk pipelines is its release into the atmosphere during various types of repair work. These can be both scheduled repairs of pipelines and hydraulic stations, and emergency repairs in cases of receiving information about a violation of the tightness of pipes, damage to gas pipelines, and emergency gas emissions.

A specific feature of carrying out repair work on certain sections of gas pipelines is the mandatory requirement for complete removal ("bleeding") of gas from the pipe throughout the entire section of the gas pipeline being repaired. Until recently, such repairs have been carried out by releasing ("bleeding") methane into the atmosphere. However, as we have already noted, this leads to the loss of a commercial product, as well as to environmental pollution, for which gas transportation companies pay environmental taxes (in the amount of the standard) or fines (in case of excess emissions).

Currently, an innovative technology for reducing gas emissions during repairs has been developed, but so far little is applied. This technology consists in pumping gas into a parallel gas pipeline using special mobile compressor units (stations). A mobile compressor station is a complex of main and auxiliary equipment. Major equipment includes two mobile compressor units using a 695 kW LMF high pressure compressor, a 750 kW Caterpillar gas engine and an air cooler. The autonomy and mobility of the compressor unit is provided by auxiliary equipment. PJSC "Gazprom" has created a subsidiary, "Gazprom MKS" [29], which develops and manufactures this equipment, but the level of distribution of such stations is still insufficient to significantly increase the efficiency of the gas transmission divisions.

In order to determine the economic feasibility of widespread inclusion of this technology in the digital transformation roadmap of the activities of LLC "Gazprom Transgaz Tchaikovsky", a calculation of the environmental and economic efficiency of the use of MCU was carried out. The calculation was carried out using the example of a unit of planned repair work corresponding to a section of a gas pipeline with a length of 3 812.87 km with a pipe diameter of 1200 mm, which had to be freed of gas for repair work (the repair of this section was actually carried out in 2019). Calculation of the volume of vented gas was carried out according to the officially approved methodology of PJSC "Gazprom" [27].

In the process of assessing the economic and environmental efficiency of the innovative technology, the calculation of three options for the size of the gas release into the atmosphere during repair work was carried out:

- carrying out repairs without the use of the ISS and with complete release of gas into the atmosphere (for example, in the event of an accident and a strong release of gas into the atmosphere) - option 1;
- carrying out repair work using an innovative technique for pumping gas into a parallel gas pipeline (using the ISS) - option 2;
- an intermediate version (actually implemented in 2109 with a partial use of the ISS) - option 3.

For the calculations, the following constants were used:
Gas density ($\rho$) - 0.85 kg/m$^3$
Tav. - 282 K
Pipe length 3 812.87 km
Pipe diameter - 1200 mm
Vi - 4088.263 thousand m$^3$.
The calculation results for the three options are shown in Table 2.

**Table 2.** Comparative characteristics of the effectiveness of the application of innovative technology for pumping gas. Source: compiled by the authors

| Indicator name | Option 1 | Option 2 | Option 3 |
|---|---|---|---|
| Bleed gas volume, thousand m$^3$ | 270 938.062 | 73 375.444 | 140 700.000* |
| Savings, thousand m$^3$ ** | 0 | 197 562.618 | 130 238.062 |
| Additional income from gas sales***, $ | 0 | 117 836.54 | 77 680.70 |
| Reduction of emissions****, thousand m$^3$ | −69 802.962 | 127 759.656 | 60 435.100 |
| Reduction of emissions****, thousand tons | −46 847.625 | 85 744.736 | 40 560.470 |

* According to the annual report, 140.7 million m$^3$ were vented at the considered section of the gas pipeline;
** The amount of savings is calculated as the difference between option 1 and option 2/3;
*** Additional income from gas sales is calculated based on the actual cost of 1,000 m$^3$ of gas 1 as of 2019 ($ 0.596);
**** Emission reduction (thousand tons) calculated relative to the standard values for 2019 (201.135 million m$^3$ of Methane or 134.990 thousand tons of methane) [p. 38]

As for compliance with environmental standards, the actual amount of emissions (option 3) did not exceed the planned standards and was fully included in the gross methane elections within the limits for 2019.

As can be seen from Table 2, when using an innovative technology for pumping gas into a parallel gas pipeline during repair work, the economic and environmental effect is high. This makes it expedient to include this technology in the roadmap for digitalization of gas transportation enterprises, even despite the high investment cost. The most profitable for the company is the use of this technology in the formation of a repair plan. However, even in case of accidents, it can help well to reduce the volume of emissions into the atmosphere. The combination of this technology with stationary or mobile (drones, unmanned aerial vehicles) systems for monitoring gas leaks from main pipelines will be especially useful and effective.

# 5  Conclusions

The assessment of the maturity of the digital transformation of the Gazprom Transgaz Tchaikovsky enterprise in the main area of activity (gas transportation) showed that by now the enterprise has passed the main stages of digital transformation within the digital realms "Physics-digital" and "Digital":

- within the digital realms "Physics-digital" the company uses automatic work processes through electricity, hydraulics, pneumatics, etc. (stage "Mechanize"); detecting changes in the environment and sending information to the computer processor (stage "Sensorize") and digital data transmission over the communication medium to various networks, systems and devices (stage "Transmit");
- within the digital realms "Digital" the company is currently implementing standardization, aggregation and integration of technologies and data (stage "Integrate"); processing and studying large data sets (stage "Analyze"); improvement of information display for better interpretation and usability of data (stage "Visualize").

In the future, the company needs to implement artificial intelligence and intelligent data analysis systems to predict situations, use "clouds", high-performance computing, streaming analytics methods, etc. (stage "Augment"). These digital technologies already exist and can be easily adapted to the gas transportation sector.

As for the last (higher) digital realms "Digital-Physics" of digitalization, here the enterprise also has reserves and opportunities for the implementation of the last three stages (stages: "Robotize", "Craft" and "Virtualize"). In particular, these are the possibilities of using drones to monitor the air condition over main pipelines and other (stage "Robotize"). At the Craft stage, the most promising is the use of new materials (for example, polymer composite materials) in the production of pipes for main gas pipelines. Moreover, modern technologies for creating products from polymer composite materials make it possible to "sew" special sensor sensors directly into the structure of the pipe material, creating systems for monitoring the state of the structure. The most promising for such an implementation are fiber-optic sensors, which have a number of advantages over other sensors [30, p. 19].

In general, the study showed that Russian gas transmission companies have great potential and ample opportunities to digitize their activities, despite the various difficulties and obstacles that are encountered on the path of digitalization in Russia [31].

**Acknowledgment.** The work is carried out based on the task on fulfilment of government contractual work in the field of scientific activities as a part of base portion of the state task of the Ministry of Education and Science of the Russian Federation to Perm National Research Polytechnic University (topic # *FSNM-2020-0026*).

# References

1. Bikmetova, Z.M., Degtyareva, V.V., Makkaeva, R.S.-A.: Innovative development of the digital economy: a view of sustainability. In: Ragulina, J.V., Khachaturyan, A.A., Abdulkadyrov, A.S., Babaeva, Z.S. (eds.) Sustainable Development of Modern Digital Economy. RD, pp. 285–294. Springer, Cham (2021). https://doi.org/10.1007/978-3-030-70194-9_28

2. Bulavko, O.A., Tuktarova, L.R.: Information space concept of interaction between digital and innovative economy. In: Ashmarina, S.I., Mantulenko, V.V. (eds.) Current Achievements, Challenges and Digital Chances of Knowledge Based Economy. LNNS, vol. 133, pp. 11–17. Springer, Cham (2021). https://doi.org/10.1007/978-3-030-47458-4_2

3. Mingaleva, Z., Mirskikh, I.: On innovation and knowledge economy in Russia. World Acad. Sci. Eng. Technol. **42**, 1018–1027 (2010)

4. Montague, J.: Bright oil and gas future will rely on innovation, efficiency. Control Global, 16 March 2017 (2017). https://www.controlglobal.com/articles/2017/abb-customer-world-2017-article-16/. Accessed 15 Feb 2021

5. GE: GE Oil & Gas starts strong in 2017 with innovative digital customer agreements, 31 January 2017 (2017). Accessed 05 July 2021

6. BlockChain—An Opportunity for Energy Producers and Consumers? PwC Global Power Utilities, London, U.K. (2017). https://www.pwc.com/gx/en/industries/assets/pwc-blockchainopportunity-for-energy-producers-and-consumers.pdf. Accessed 08 July 2021

7. Mingaleva, Z., Shironina, E., Buzmakov, D.: Implementation of digitization and blockchain methods in the oil and gas sector. In: Antipova, T. (ed.) ICIS 2020. LNNS, vol. 136, pp. 144–153. Springer, Cham (2021). https://doi.org/10.1007/978-3-030-49264-9_13

8. Mussomeli, A., Gish, D., Laaper, S.: The Rise of the Digital Supply Network. Deloitte University Press (2016). Accessed 05 July 2021

9. Mingaleva, Z., Sevidova, E.: State regulation of the introduction of digital technologies in the oil and gas complex of Russia. J. Digit. Sci. **1**(1), 25–33 (2019). https://doi.org/10.33847/2686-8296.1.1_3

10. Antipova, T.: Streamline management of arctic shelf industry. In: Antipova, T., Rocha, Á. (eds.) Information Technology Science. MOSITS 2017. Advances in Intelligent Systems and Computing, vol. 724, pp. 114–121. Springer, Cham (2018). https://doi.org/10.1007/978-3-319-74980-8_11

11. Opeyemi, B., Holzmann, J., Yaqoob, T.: Application of artificial intelligence methods in drilling system design and operations: a review of the state of the art. J. Artif. Intell. Soft Comput. Res. **5**(2), 121–139 (2015). https://doi.org/10.1515/jaiscr-2015-0024

12. Shell joins digital twin JIP, 19 July 2017. https://www.oedigital.com/news/446079-shell-joins-digital-twin-jip. Accessed 05 July 2021

13. Beaubouef, B.: Industry continues to advance digitization, Offshore Magazine, 5 August 2017. https://www.offshore-mag.com/production/article/16756065/industry-continues-to-advance-digitization. Accessed 05 July 2021

14. Annual report of PJSC Gazprom for 2019. https://www.gazprom.ru/f/posts/77/885487/gazprom-annual-report-2019-ru.pdf. Accessed 15 July 2021

15. Adegboye, M.A., Fung, W.K., Karnik, A.: Recent advances in pipeline monitoring and oil leakage detection 422 technologies: principles and approaches. Sensors **19**, paper # 2548 (2019). https://doi.org/10.3390/s19112548

16. Iwaszenko, S., Kalisz, P., Słota, M., Rudzki, A.: Detection of natural gas leakages using a laser-based methane sensor and UAV. Remote Sens. **13**(3), 510, 1–16 (2021)

17. Golston, L., et al.: Natural gas fugitive leak detection using an unmanned aerial vehicle: localization and quantification of emission rate. Atmosphere **9**, 333 (2018)
18. Yang, S., et al.: Natural gas fugitive leak detection using an unmanned aerial vehicle: measurement system description and mass balance approach. Atmosphere **9**, 383 (2018)
19. Tannant, D., Zheng, W., Smith, K., Cahill, A.: Evaluation of a drone and laser-based methane sensor for detection of a surface release of methane. University of British Columbia (2018)
20. Tannant, D., et al.: Evaluation of a drone and laser-based methane sensor for detection of fugitive methane emissions. British Columbia Oil and Gas Research and Innovation Society, Vancouver (2018)
21. Riurean, S.: Design and evaluation of visible light wireless data communication models. J. Digit. Sci. **2**(2), 3–13 (2020). https://doi.org/10.33847/2686-8296.2.2_1
22. Li, H.Z., Mundia-Howe, M., Reeder, M.D., Pekney, N.J.: Gathering pipeline methane emissions in Utica shale using an unmanned aerial vehicle and ground-based mobile sampling. Atmosphere **11**, 716 (2020)
23. Murvay, P., Silea, I.: A survey on gas leak detection and localization techniques. J. Loss Prev. Process Ind. **25**, 966–973 (2012)
24. Sun, J., Peng, Z., Wen, J.: Leakage aperture recognition based on ensemble local mean decomposition and sparse representation for classification of natural gas pipeline. Measurement **108**, 91–100 (2017)
25. Mittal, A., Slaughter, A., Bansal, V.: From bytes to barrels. The digital transformation in upstream oil and gas. A report by the Deloitte Center for Energy Solutions, 28 p. (2017). https://www2.deloitte.com/insights/us/en/industry/oil-and-gas/digital-transformation-upstream-oil-and-gas.html. Accessed 06 July 2021
26. Annual report of production and economic activity of LLC "Gazprom Transgaz Tchaikovsky" for 2019. https://tchaikovsky-tr.gazprom.ru/. Accessed 15 July 2021
27. STO Gazprom 2-1.20-601-2011 "Methodology for calculating the effect of energy saving of fuel and energy resources spent on own technological needs of main gas transportation" http://ws.gpei.info/product_info.php/products_id/7613?osCsid=3a2162d8734289baae637d0d84e6baca. Accessed 15 July 2021
28. Environmental Report of LLC "Gazprom Transgaz Tchaikovsky" for 2020. 88 p. https://tchaikovsky-tr.gazprom.ru/d/textpage/48/72/ehkologicheskij-otchet_2020_web.pdf. Accessed 15 July 2021
29. LLC "Gazprom MKS". https://mks.gazprom.ru/about/working/. Accessed 01 Aug 2021
30. Pankov, A.A., Bratsun, D.A., Kameneva, A.L., Mingaleva, Zh.A., Oglezneva, S.A., Shipunov, G.S.: Development of theoretical and technological foundations and digital technologies for the design of functional composite materials, multifunctional nanocoatings and diagnostic information systems for monitoring highly loaded elements of aircraft structures. Research report (interim). Perm, 251 p. (2020)
31. Mingaleva, Z., Mirskikh, I.: The problems of digital economy development in Russia. In: Antipova, T., Rocha, A. (eds.) DSIC18 2018. AISC, vol. 850, pp. 48–55. Springer, Cham (2019). https://doi.org/10.1007/978-3-030-02351-5_7

# Assessment of the Level of Digital Maturity of Sectors of the Russian Economy

Nikita Lukovnikov[1]([✉]) [iD], Zhanna Mingaleva[1] [iD], Olga Zakirova[2] [iD], and Yrii Starkov[1] [iD]

[1] Perm National Research Polytechnic University, Perm 614990, Russian Federation
[2] Volga State Technological University, Yoshkar-Ola 424000, Russian Federation

**Abstract.** Successful implementation of a digital enterprise transformation strategy and program depends on a number of factors. One of these factors is the correct assessment of the degree of readiness of the enterprise for the implementation of various digital technologies. The aim of the paper is to assess the level of digital maturity of the main operating segments and types of activities of Russian enterprises. The research methods are economic and mathematical, statistical, dynamic, and composite analysis. In the course of the research, the main directions and areas of application of digital technologies in the Russian economy are highlighted. As a result of the study, it was found that for successful digitalization, it is necessary to assess the level of digital maturity of each operating segment of the enterprise and identify advanced digital technologies to achieve specific business goals.

**Keywords:** Digitalization · Digital technologies · Digital maturity · DOT model

## 1 Introduction

The successful digitalization of the economy and production involves assessing the level of digital maturity of each operating segment of the enterprise and the business model of the enterprise [1–3]. Currently, there is an intensive expansion of the areas of application of innovation in the field of technology and the definition of future advanced digital technologies to achieve specific business goals of enterprises [4–6]. The growing opportunities of various digital devices in the production activities provides a solid foundation for accelerating digitalization.

Digital upgrading of production means revolutionary changes in business models based on the use of digital platforms in order to ensure a significant increase in market volumes by increasing the competitiveness of business activity [7, p. 27]. The most widely used technologies at present in production are wireless communication technologies for data collection, various software systems for monitoring the early detection of early signs of mechanical equipment malfunctions, and operational data transmission systems [8]. The robotization of certain technological processes and business processes is being actively carried out [9, 10].

T. Antipova (Ed.): DSIC 2021, LNNS 381, pp. 38–47, 2022.
https://doi.org/10.1007/978-3-030-93677-8_4

As for the implementation of projects for comprehensive modernization and equipping production equipment with sensors, the decision-making on their implementation depends on the economic efficiency of such projects [11, 12]. In turn, this is influenced by the amount of costs for the implementation of specific digital technologies, as well as the amount of the economic effect from the creation of such a system. In some countries of the world, large expenditures are included in the budgets for the implementation of national programs for the digitalization of society [13].

The researchers note the importance of such areas of digitalization as the formation of supply chains [14], as well as industrial Internet playground [15].

A certain block of research is devoted to the creation of digital twins and their use in various areas of enterprise activity [16–18]. The scientific works devoted to the development and application of various innovative technologies for digitizing complex engineering drawings are also available in modern literature [19].

The issues of prospects and problems of digitalization of production processes at enterprises are also investigated within the framework of the broader phenomenon of the formation of Industry 4.0 and the increase on this basis of the competitiveness of industrial production and individual industries [20–22].

## 2  Theoretical and Methodological Background

Various maturity models have been developed for understanding the diffusion and implementation of new technologies/approaches [21, p. 1]. Today in the scientific literature there are various models for assessing the digital maturity of various economic entities, including small and medium enterprises [23]. A group of researchers proposed a model for strategic management of the digital maturity of an industrial enterprise in the global economic space of the ecosystem economy [24]. A model for assessing the level of digital maturity of enterprises in the countries of Central and Eastern Europe has been developed [25]. Models for assessing the digital maturity of territories during digital transformation complement and expand the models for assessing the economic security of territories [26].

The Digital Operations Transformation (DOT) model (The DOT model), developed and proposed by Deloitte for oil and gas companies, was taken as a theoretical basis for the study [27, p. 5]. Ontological testing of the model has shown that it is universal and can be applied to any sectors and branches of the economy. Therefore, we used The DOT model from Deloitte, as a theoretical and methodological basis for the study.

According to the digital transformation model proposed by Deloitte, there are 10 stages in the formation of a company's digital maturity. Checking the success of companies through all digitalization processes at each stage can be carried out by calculating the main indicators characterizing the level of maturity of each stage. The list of the main stages and the corresponding digital technologies is presented in Fig. 1 (left by Deloitte) [27, p. 5].

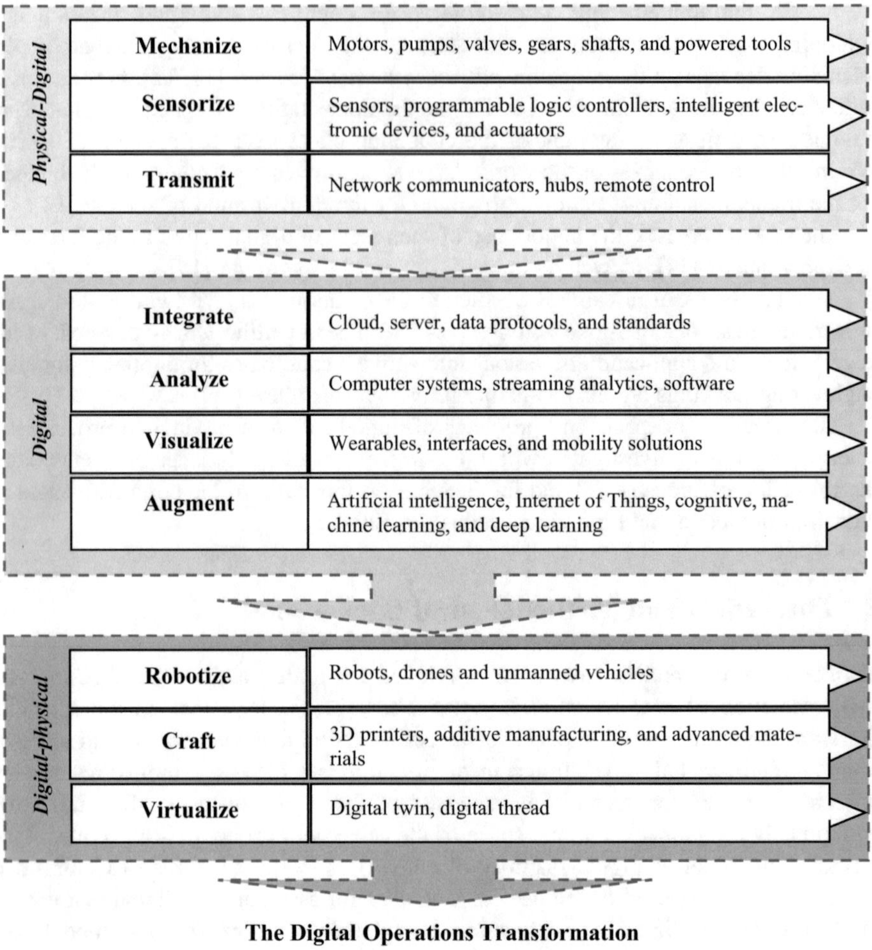

| *Physical-Digital* | **Mechanize** | Motors, pumps, valves, gears, shafts, and powered tools |
| | **Sensorize** | Sensors, programmable logic controllers, intelligent electronic devices, and actuators |
| | **Transmit** | Network communicators, hubs, remote control |

| *Digital* | **Integrate** | Cloud, server, data protocols, and standards |
| | **Analyze** | Computer systems, streaming analytics, software |
| | **Visualize** | Wearables, interfaces, and mobility solutions |
| | **Augment** | Artificial intelligence, Internet of Things, cognitive, machine learning, and deep learning |

| *Digital-physical* | **Robotize** | Robots, drones and unmanned vehicles |
| | **Craft** | 3D printers, additive manufacturing, and advanced materials |
| | **Virtualize** | Digital twin, digital thread |

**The Digital Operations Transformation**

**Fig. 1.** The digital operations transformation model of Deloitte. Source: adapted by the authors on [27].

A quantitative assessment of the availability of advanced production technologies corresponding to each stage and the level of their application in the Russian economy was carried out on the basis of official statistics collected by Rosstat [28]. Also, a comparative assessment of the level of digital maturity of individual sectors of the economy was carried out in accordance with the indicators of the use of basic digital technologies in them, providing the digital operations transformation.

The basic research methods used are economic and statistical, structural and logical, comparative analysis. As data sources, we used Rosstat information on the level and dynamics of the spread of basic digital technologies in the Russian industry.

## 3   Research and Results

The determination of the level of compliance of Russian enterprises and organizations with the main stages of digital transformation of activities will be carried out on the basis of a statistical analysis of the level of technological development of sectors of the Russian economy.

The first step of the study is to determine the general trend of technological development of sectors of the Russian economy based on the introduction of new developments of advanced production technologies. Evaluation of the dynamics of this indicator showed the presence of a steady trend of continuous growth in the number of developed advanced production technologies during the 21st century (see Fig. 2).

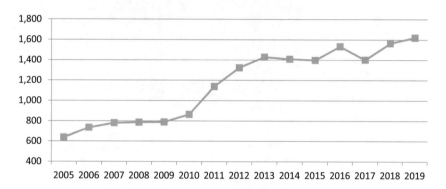

**Fig. 2.** Dynamics of the total number of developed advanced production technologies in Russia in 2005–2019. Source: compiled by the authors

The data in Fig. 2 indicate a steady increase in the number of newly created advanced production technologies in Russia, especially in the period from 2009 to 2016. The small total number of such technologies (just over 1620 in 2019) is offset by the possibility of their scalability and widespread use in all sectors and industries of the economy. For example, digital banking technologies created and applied in the financial, monetary, banking sectors ensure the functioning of the entire economic system of the country. Additive technologies, 3D are successfully used for the production of a wide variety of products for both industrial and household purposes. Drones, robots, unmanned aerial vehicles and other robotic devices are also widely used in all areas of activity, from agriculture (planting control) to the sphere of state and municipal administration (urban management) (for example, traffic control on city streets) [29–31] etc.

Fiber-optic sensors embedded in polymer composite materials make it possible to digitize many procedures for controlling and monitoring the condition of equipment, building and industrial structures, and complex products during their operation [8].

Digital technologies for collecting, transmitting, processing and storing information, which are also widely represented in all areas of the life of Russian society, also have multifunctional applications.

An analysis of the distribution of advanced manufacturing technologies by type of activity and management function showed that almost 1/3 of new technologies (31%) accounted for activities in the production, processing and assembly of finished products (see Fig. 3).

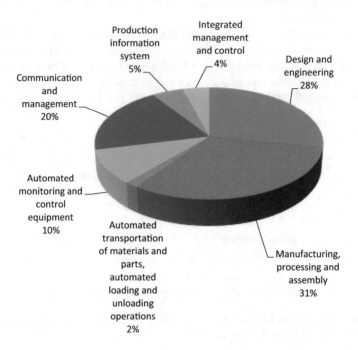

**Fig. 3.** The structure of the distribution of the total number of developed advanced production technologies in Russia in 2019 by areas of application. Source: compiled by the authors

Next in terms of prevalence is the sphere of design and engineering (28%). The third place in terms of the level of renewal and digitalization is steadily occupied by the functions of communication and general management (20%). The rest of the activities account for from 10% (Automated monitoring and/or control equipment) to 2% (advanced technologies to ensure automated transportation of materials and parts, as well as to carry out automated loading and unloading operations) of the total number of new technologies created annually.

Analysis of the structure of distribution of advanced production technologies by type of activity showed that two purely "digital" sectors such as "production information system" and "integrated management and control" to the greatest extent corresponding to the 3rd stage of digitalization occupy in total only 9% of all annually developed advanced production technologies. This fact raises certain concerns, since it indicates a weak readiness of Russian enterprises in the manufacturing sector for the most advanced digital technologies at the digital realms "Digital-physical".

The third important step in assessing the level of digitalization of the Russian economy is a detailed analysis of the key areas of application of special software in the main areas and management functions. Figure 4 shows the structure of the main areas of activity and management functions for the digitalization of which Russian enterprises used special software in 2019.

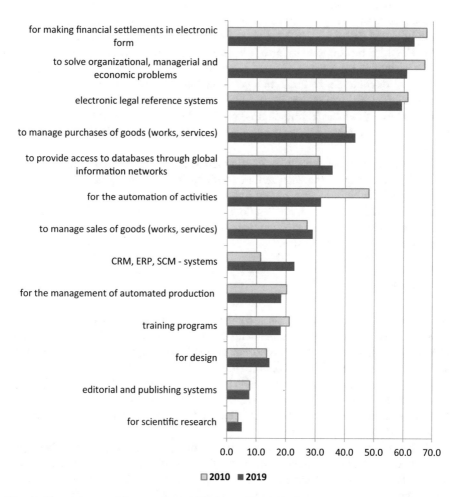

**Fig. 4.** The structure of the use of special software by areas of activity in 2019 (in% of the total number of enterprises and organizations that used such special software). Source: compiled by the authors

As an additional comment to Fig. 4, we note that the area of activity "for the automation of activities" includes systems for the automation of banking activities, automation of trade organizations, ordering, automated library systems, translation programs, dictionaries and other special software tools. Also in the area of activity "for the management of automated production" includes systems for the management of automated production and/or individual technical means and technological processes.

As can be seen from Fig. 4, among the areas of activity where special software is most actively used, the most significant is the area of financial settlements in electronic form - they are carried out by almost 64% of all enterprises and organizations that implement digital technologies. Another 61% of economic entities actively use digital technologies to solve organizational, managerial and economic problems. Moreover, it should be noted that since 2010 the share of such entities has decreased by 6% (from 67%). The third most important area is electronic legal reference systems (59%).

The second group consists of areas of activity with an interval from 30% to 45% of the level of distribution. In 2019, this group included the special software to manage purchases of goods (works, services) - 43%; the special software to provide access to databases through global information networks - 35%, 6; the special software for the automation of activities - 31.7%; the special software to manage sales of goods (works, services) - 29.4%.

The use of the special software in other areas and spheres is less than 30%.

However, among all the analyzed areas of application of the special software, one should especially highlight the area of implementation of CRM, ERP, SCM - systems, the prevalence of which has more than doubled over the past decade. This indicates that some enterprises and organizations in their digital transformation have entered the third digital realms – "Digital-physical".

## 4  Conclusions

In the course of the study, an assessment was made of the level of compliance of Russian enterprises and organizations with the main stages of digital transformation of activities. The assessment was carried out on the basis of a statistical analysis of the spread of the main digital technologies in the Russian industry by areas of activity. As a result of the research, a number of trends and dependencies have been identified.

First, it is necessary to highlight a steady trend of continuous growth in the number of developed advanced production technologies during the 21st century. This ensures the necessary pace of digitalization.

Secondly, in the structure of the distribution of advanced production technologies by types of activities and management functions, activities for the production, processing and assembly of finished products, design and engineering, and the functions of communications and general management prevail. This indicates the active participation of industrial enterprises in the transition to Industry 4.0 and digitalization.

Thirdly, the presence of a stable ratio of the use of special software in the areas of activity has been established. During the entire analyzed period, the most successful in terms of digitalization are the areas of financial settlements in electronic form and the area of solving organizational, managerial and economic problems.

Fourth, the most indicative directions for the implementation of CRM, ERP, SCM - systems from the point of view of digital maturity showed a significant growth by more than 2 times by the end of the decade.

In general, it should be noted that the digital modernization of production implies the organization of the introduction of modern innovative technologies, the adaptation and development of new business models in the digital economy, which will entail a qualitative improvement in business processes, including the production of products and services.

**Acknowledgment.** The work is carried out based on the task on fulfilment of government contractual work in the field of scientific activities as a part of base portion of the state task of the Ministry of Education and Science of the Russian Federation to Perm National Research Polytechnic University (topic # *FSNM-2020–0026*).

# References

1. Arnold, C., Kiel, D., Voigt, K.-I.: How the industrial internet of things changes business models in different manufacturing industries. Int. J. Innov. Manag. **20**(8), art. no. 1640015 (2016). https://doi.org/10.1142/S1363919616400156
2. Gökalp, E., Martinez, V.: Digital transformation capability maturity model enabling the assessment of industrial manufacturers. Comput. Ind. **132**, art. no. 103522 (2021). https://doi.org/10.1016/j.compind.2021.103522
3. Akdil, K.Y., Ustundag, A., Cevikcan, E.: Maturity and readiness model for industry 4.0 strategy. In: Industry 4.0: Managing the Digital Transformation, pp. 61–94 (2018)
4. Mingaleva, Z., Danilina, H.: Significance of technological innovations for an increase of competitiveness of industrial companies. Life Sci. J. **11**(SPEC. ISSUE 8), 211–215 (2014)
5. Bikmetova, Z.M., Degtyareva, V.V., Makkaeva, R.S.-A.: Innovative development of the digital economy: a view of sustainability. In: Ragulina, J.V., Khachaturyan, A.A., Abdulkadyrov, A.S., Babaeva, Z.S. (eds.) Sustainable Development of Modern Digital Economy. RD, pp. 285–294. Springer, Cham (2021). https://doi.org/10.1007/978-3-030-70194-9_28
6. Bulavko, O.A., Tuktarova, L.R.: Information space concept of interaction between digital and innovative economy. In: Ashmarina, S.I., Mantulenko, V.V. (eds.) Current Achievements, Challenges and Digital Chances of Knowledge Based Economy. LNNS, vol. 133, pp. 11–17. Springer, Cham (2021). https://doi.org/10.1007/978-3-030-47458-4_2
7. Mingaleva, Z., Sevidova, E.: State regulation of the introduction of digital technologies in the oil and gas complex of Russia. J. Digit. Sci. **1**(1), 25–33 (2019). https://doi.org/10.33847/2686-8296.1.1_3
8. Pankov, A.A., Bratsun, D.A., Kameneva, A.L., Mingaleva, Zh.A., Oglezneva, S.A., Shipunov, G.S.: Development of theoretical and technological foundations and digital technologies for the design of functional composite materials, multifunctional nanocoatings and diagnostic information systems for monitoring highly loaded elements of aircraft structures. Research report (interim), Perm, 251 p. (2020)
9. Barbosa, G.F., Shiki, S.B., da Silva, I.B.: R&D roadmap for process robotization driven to the digital transformation of the industry 4.0. Concurr. Eng. Res. Appl. **28**(4), 290–304 (2020)
10. Barbosa, G.F., de Andrade Bezerra, W.: A knowledge tailor-made method oriented to robotization of process: a case study of aeronautical materials' drilling. J. Braz. Soc. Mech. Sci. Eng. **41**(4), 1–14 (2019). https://doi.org/10.1007/s40430-019-1679-x
11. On financial mechanisms for introducing the best available technologies in Russia. URL, http://www.mnr.gov.ru/docs/latonova.pdf. Accessed 16 Apr 2019

12. Griffy-Brown, C., Chun, M., Miller, H., Lazarikos, D.: How do we optimize risk in enterprise architecture when deploying emerging technologies? J. Digit. Sci. **3**(1), 3–13 (2021). https://doi.org/10.33847/2686-8296.3.1_1

13. Antipova, T.: Insights from some governments' budget functional expenditures for the fifteen years: 2005–2019. In: Antipova, T. (ed.) ICCS 2021. LNNS, vol. 315, pp. 63–73. Springer, Cham (2022). https://doi.org/10.1007/978-3-030-85799-8_6

14. Chen, L., Xie, X., Lu, Q., Pitt, M., Yang, J.: Gemini principles-based digital twin maturity model for asset management. Sustainability (Switzerland) **13**(15), 8224 (2021)

15. Aagaard, A., Presser, M., Collins, T., Skou, A.K., Jakobsen, E.M.: The role of digital maturity assessment in technology interventions with industrial internet playground. Electron. (Switzerland) **10**(10), 1134 (2021)

16. Asdecker, B., Felch, V.: Development of an industry 4.0 maturity model for the delivery process in supply chains. J. Modell. Manag. **13**(4), 840–883 (2018). https://doi.org/10.1108/JM2-03-2018-0042

17. Lu, Q., Xie, X., Parlikad, A.K.: Digital twin-enabled anomaly detection for built asset monitoring in operation and maintenance. Autom. Constr. art. no. 103277 (2020). https://doi.org/10.1016/j.autcon.2020.103277

18. Danilina, H., Mingaleva, Z.: Improving of innovation potential efficiency of industrial enterprises. Middle East J. Sci. Res. **13**, 191–194 (2013). https://doi.org/10.5829/idosi.mejsr.2013.13.sesh.1434

19. Moreno-García, C.F., Elyan, E., Jayne, C.: New trends on digitisation of complex engineering drawings. Neural Comput. Appl. **31**(6), 1695–1712 (2018). https://doi.org/10.1007/s00521-018-3583-1

20. Antipova, T.: Preface. In: Advances in Intelligent Systems and Computing, vol. 1352 (2021)

21. Haseeb, M., Hussain, H.I., Ślusarczyk, B., Jermsittiparsert, K.: Industry 4.0: a solution towards technology challenges of sustainable business performance. Soc. Sci. **8**(5), art. no. 154 (2019). https://doi.org/10.3390/socsci8050154

22. Mingaleva, Z., Mirskikh, I.: The problems of digital economy development in Russia. In: Antipova, T., Rocha, A. (eds.) DSIC18 2018. AISC, vol. 850, pp. 48–55. Springer, Cham (2019). https://doi.org/10.1007/978-3-030-02351-5_7

23. Zentner, H., Spremic, M., Zentner, R.: Measuring digital business models maturity for SMEs. In: IEEE Technology and Engineering Management Conference - Europe, TEMSCON-EUR 9488608 (2021)

24. Gileva, T.A., Galimova, M.P., Babkin, A.V., Gorshenina, M.E.: Strategic management of industrial enterprise digital maturity in a global economic space of the ecosystem economy. In: IOP Conference Series: Earth and Environmental Science, vol. 816, no. 1, p. 012022 (2021)

25. Brodny, J., Tutak, M.: Assessing the level of digital maturity of enterprises in the Central and Eastern European countries using the MCDM and Shannon s entropy methods. PLoS One. **16**, 253965 (2021)

26. Kapkaev, Y., Kadyrov, P.: Digital maturity of territories during digital transformation. In: E3S Web of Conferences, vol. 258, p. 01001 (2021)

27. Mittal, A., Slaughter, A., Bansal, V.: From bytes to barrels. The digital transformation in upstream oil and gas. A report by the Deloitte Center for Energy Solutions, 28 p. https://www2.deloitte.com/insights/us/en/industry/oil-and-gas/digital-transformation-upstream-oil-and-gas.html. Accessed 16 Apr 2021

28. Rosstat. Official website. https://rosstat.gov.ru/folder/14477. Accessed 08 Aug 2021

29. D'amico, G., Arbolino, R., Shi, L., Yigitcanlar, T., Ioppolo, G.: Digital technologies for urban metabolism efficiency: lessons from urban agenda partnership on circular economy. Sustain. (Switzerland) **13**(11), 6043 (2021)

30. Mingaleva, Z.: On digital development of russian urban transport infrastructure. In: Antipova, T., Rocha, A. (eds.) DSIC18 2018. AISC, vol. 850, pp. 29–35. Springer, Cham (2019). https://doi.org/10.1007/978-3-030-02351-5_4
31. Pernestål, A., Engholm, A., Bemler, M., Gidofalvi, G.: How will digitalization change road freight transport? Scenarios tested in Sweden. Sustain. (Switzerland). **13**(1), 304, 1–18 (2021)

# Digitalization and Modernization of the Industrial Production Management System Based on Lean-Green Approach

Zhanna Mingaleva[1]([✉]) [iD], Oksana Borisova[2] [iD], Denis Markov[1] [iD],
and Yuliya Grigorieva[1] [iD]

[1] Perm National Research Polytechnic University, Perm 614990,
Russian Federation
mingall@pstu.ru
[2] Gzhel State University, Moscow 140155, Russian Federation

**Abstract.** In connection with the transition to the information and digital economy, the widespread introduction of digital technologies, the combination of digital requirements with traditional approaches to management, as well as the concept of sustainable development, is of particular importance. This, in turn, influences the development of new management methods that meet the requirements of Industry 4.0. The article is devoted to the analysis of the areas of application of the modern lean-green approach in the framework of the digital transformation of the economy. Complex analysis and structural-logical synthesis of the main provisions of the concept of sustainable development, institutional theory, the concept of border work, business modeling methods were used as research methods. As a result of the study, it was revealed that the main obstacles to the wider use of digital technologies in modern Russian companies are in the framework of the regulatory and instrumental aspects. The most common obstacles to the current stage of digital transformation development are the complexity of coordinating various actions to carry out organizational change (normative aspect) and organizational complexity (instrumental aspect).

**Keyword:** Digitalization · Digital technologies

## 1 Introduction

The modern digital and innovative economy is based on the wide application of knowledge, the active use of the results of intellectual activity, the introduction of innovations, the use of digital technologies [1–6]. All this is associated with the operation of large amounts of data and information, which allow you to optimize and significantly speed up many production processes. However, "in the conditions of intensive development of technologies that allow us to move to a qualitatively new level of production, traditionally established approaches to managing business processes are losing their relevance and require additions" [7, p. 1].

Most often in this context, such conceptual approaches and production management practices are used as Lean Production [8], Quick Response Manufacturing - (QRM) [9], Kanban system [10], Kaizen [11], the theory of quality management (TQM) [12],

T. Antipova (Ed.): DSIC 2021, LNNS 381, pp. 48–56, 2022.
https://doi.org/10.1007/978-3-030-93677-8_5

Shewhart four-step cycle model (PDCA) [13], HADI cycle model [14] and others. Each of these concepts and management models has its own advantages and areas of application, however, the digital stage of development of the economy and society imposes special requirements on their application in modern conditions. One of these urgent requirements is the organization of a management system for enterprises and organizations, taking into account the massive distribution of digital technologies and its construction based on digital technologies.

In its broadest form, digitalization is considered the main way to achieve a new level and quality of production process management. The potential benefits of digitalization are clear and include increased productivity and operational safety, as well as cost savings [15]. However, there are currently no major solutions in the area of digitalization of business management. Moreover, as a rule, "digital" thinking and "digital" knowledge in practice is limited to obtaining information as a result of data analysis [16]. At the same time, digitalization is a much broader concept that includes not only digital (information and communication) technologies, but also a change in thinking and approaches to organizing business and production. Digitalization leads to serious structural changes in the ratio of sectors and industries of national economies, in its innovation and scientific systems [17, 18]. In this regard, the scientific literature actively discusses the social aspects of the implementation of big data technologies [19].

The role of quality management in corporate governance is growing. The researchers note that "in the conditions of improving production processes, the organization and implementation of a quality management system is one of the most important issues in the management system of individual business processes and the organization as a whole" [7, p. 1].

Another important area of transformation of management concepts and models is the integration of various classical management models with the concept of sustainable development. It is in this direction that many researchers see the greatest potential for adapting strategies and business models to the requirements of Industry 4.0. and digitalization of the economy [1].

## 2 Theoretical and Methodological Background

The study of modern scientific literature on the creation of new methods and models of management in the digital economy in accordance with the requirements of Industry 4.0. showed the presence of several of the most discussed and important directions in the development of management theory.

First, it is noted that most of the currently most widespread management models and approaches were originally created for the management of industrial production. "Historically, the manufacturing sector has acted as a vital source of innovation, making outsized contributions to research and the development of working practices and organisational models (e.g. lean manufacturing, Six Sigma, Total Quality Management, etc.) that have been later adopted and adapted across a wide range of other industries" [20, p. 25]. For example, management methods based on the four-stage Shewhart cycle (PDCA) have found wide application in many industries, and Chinese researchers have proven the feasibility of using the PDCA Cycle to manage the quality of construction projects [21]. H. Čierna, E.Sujová and M. Ťavodová showed the importance of Kaizen Philosophy for the development of Learner Business [22]. Expert risk management technologies based on QRM are applied in high-tech industrial enterprises [9].

The researchers note a similar trend in the case of lean manufacturing, "where the manufacturing sector, once more, has taken the research lead, followed by a slow transition into other industries such as construction (Banawi and Bilec, 2014; Sertye-silisik, 2014), aluminium electrolysis substations (Cluzel et al., 2010), agrifood (Folinas et al., 2014), ports (Esmemr et al., 2010), education (Ranky et al., 2012) and process (Vais et al., 2006)" [20, p. 25]. In modern research, a multi-level approach to achieving cyclicality in the context of lean manufacturing is being actively developed [23].

Therefore, the main problems and prospects for the development of management theory should be sought primarily in the manufacturing sector.

Modern researchers are studying how the Industrial Internet of Things is changing business models in various manufacturing industries [24], as well as developing new production management models based on knowledge management. So, to solve the problem of increasing the competitiveness of industrial enterprises in the digital era, various knowledge management models are increasingly being used, including the HADI-cycle [14, 25–27].

Secondly, in recent years, the issue of the influence of the environmental policy of enterprises and the environmental problems of the industry as a whole on the relationship between lean operations and the company's financial performance has been widely discussed [28]. These studies build on earlier work to address the challenge of adapting a lean management tool to the needs of industrial ecology [29].

However, a review of modern scientific literature in the field of solving topical organizational issues of implementing Lean Production tools at various enterprises showed that the issue of the relationship between Lean manufacturing and green manufacturing is still outside the scope of research. At the same time, it is green manufacturing in combination with modern management models that is the basis for the successful creation of a circular economy. The researchers note that "integrating lean of lean and green concepts holds promise for addressing the sustainability performance) [30].

Thus, the purpose of the study is to determine the areas of application of the modern lean-green approach in the digital transformation of the economy.

As a research method, a comprehensive analysis and structural-logical synthesis of the main provisions of the concept of sustainable development, institutional theory, the concept of border work, and methods of business modeling were used.

The analysis was carried out in three aspects:

- regulatory aspect
- instrumental aspect
- strategic aspect

As a result of the analysis, it is envisaged to determine the prospects for digital and innovative transformation of business models of industrial enterprises to achieve sustainable value creation, to create an innovative sustainable business model.

## 3   Research and Results

The innovation, digitalization and environmental sustainability now form the basis of behavior strategy and business modeling for both industrial enterprises and organizations in all other sectors and industries [31–33]. At the same time, the specified triad (innovation, digitalization, ecology) of the key foundations of the future development of enterprises and organizations should correspond to the traditional priorities of any business structure in making a profit and achieving operational efficiency. Integration of various approaches and development models involves identifying and clearly characterizing the main points of contact and intersection of various business strategies. For example, The Triple Layered Business Model Canvas (TLBMC), developed and proposed in 2016 by Joyce and Puckin, is a convenient tool for building sustainability-oriented and socially responsible business models [34].

The Triple Layered Business Model Canvas links three types of goals, values and performance: economic, environmental, and social. However, this model does not provide a clear answer to the question of the relationship between innovation, digital technologies and environmental sustainability. At the same time, it is these aspects that are extremely important, since the procedures for introducing digital technologies and simultaneously introducing innovations (including environmental ones) often face significant difficulties.

The study showed that the question of the need to integrate the concepts of lean and green production dates back to the beginning of the 21st century. During this period, the most significant and topical issues were the adaptation of the lean management tool to the needs of industrial ecology. Researchers working on these issues have concluded that "organizations that jointly implement lean and green achieve higher performance, particularly, environmental and operational" [20].

As recent studies have shown, a successful process of digital and innovative transformation is most often hampered by the complexity of the organizational structure of the company and the existing system of interconnections. Moreover, in many companies, especially large ones, corporate organizational structures have a number of disadvantages due to a set of relationships and connections within the firm [35].This situation is especially relevant for the conditions of digitalization, including due to the complexity of the corresponding software, as well as the complex process of its implementation in the organization [36].

A correct assessment of the complexity of the organizational structure of the company is especially important when the optimization of digital enterprise resource planning systems is carried out in order to obtain a higher return on the costs and revenues associated with innovation.

To harmonize the objectives of the various processes, it is advisable to apply the concept of border work [37]. This concept makes it possible to assess the borderline nature of innovative actions within a specific business model. Using this concept, it is possible to explore and evaluate various types of organizational changes in the boundaries between the internal goals of companies and their external stakeholders. The most important processes for assessment are the processes of obtaining information, negotiating, breaking and changing organizational boundaries. These processes affect mainly the instrumental and regulatory areas of management.

The main contradictions and obstacles to the wider use of digital technologies in modern Russian companies were identified as a result of the study. The most common for the current stage of development of digital transformation are the complexity of coordinating various actions to carry out organizational changes (normative aspect) and organizational complexity (instrumental aspect). A structured image of the main contradictions and obstacles in the development of digital technologies and their causes are shown in Fig. 1.

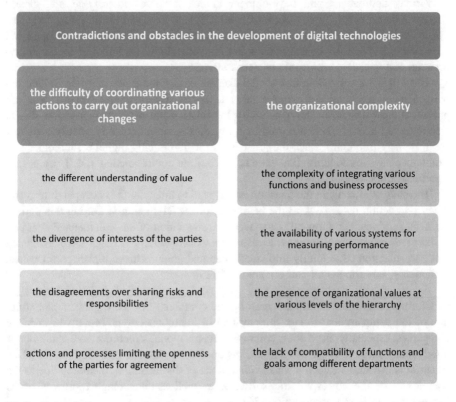

**Fig.1.** Contradictions and obstacles in the development of digital technologies. Source: compiled by the authors

As a result of the assessment of the main contradictions in the instrumental and regulatory spheres of management, it can be concluded that mutual changes in boundaries are necessary for organizing the participation of many stakeholders in order to improve organizations' understanding of value and to obtain the intended value. This is also confirmed by current research in this area.

Based on the analysis, the following key areas of activity can be identified, which can and should ensure the consistency of lean manufacturing and environmental friendliness approaches. Key areas of application of the lean-green approach are shown in Fig. 2.

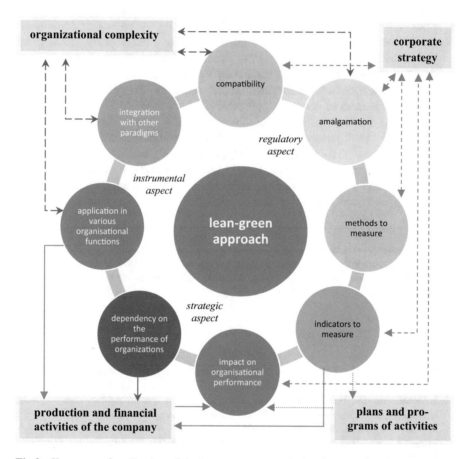

**Fig.2.** Key areas of application of the lean-green approach. Source: compiled by the authors

The diagram shown in Fig. 2 is a simplified version of the strategic roadmap for enterprise digitalization, reflecting the theoretical and methodological approaches to conducting a digitalization program based on a lean-green approach. Expanding and deepening the roadmap activities for particular operating segments will identify specific cutting-edge digital technologies to achieve definite business goals for companies.

Overall, the study confirmed that the lean-green approach can have a positive effect on various aspects of the productivity of lean-green and digital businesses. These findings are consistent with previous studies by Kitazawa and Sarkis, 2000 [38], Dües Dües, Tan and Lim, 2013 [39], Garza-Reyes, 2015 [20] etc.

# 4    Conclusions

The current stage of economic development is characterized by the expansion of areas of application of innovations in the field of technology and the growing possibilities for the interaction of various digital devices in the framework of production activities.

The study revealed that the main obstacles to the successful implementation of digital technologies in modern companies are the instrumental factor of the company's organizational structure and the normative factor of the complexity of coordinating various actions to carry out organizational changes. As a result of the study, it was found that the complexity of coordinating various actions to carry out organizational changes arises due to the different understanding of the subjects of value, diverging interests, sharing of risks and responsibilities, as well as existing processes and actions that limit the openness of actors for approval.

In order to eliminate a number of obstacles to the introduction of digital technologies into the management system of an industrial enterprise, a conceptual model of a roadmap for the digital transformation of an enterprise's business model during the transition to Industry 4.0 is proposed. Developing such a strategic roadmap will enable businesses to reap a range of benefits, such as assessing the digital maturity of each operating segment and identifying leading digital technologies to meet specific business goals.

**Acknowledgment.** The work is carried out based on the task on fulfilment of government contractual work in the field of scientific activities as a part of base portion of the state task of the Ministry of Education and Science of the Russian Federation to Perm National Research Polytechnic University (topic # *FSNM-2020-0026*).

# References

1. Berman, S.J.: Digital transformation: opportunities to create new business models. Strategy Leadersh. **40**(2), 16–24 (2012). https://doi.org/10.1108/10878571211209314
2. Lichtenthaler, U.: Building blocks of successful digital transformation: complementing technology and market issues. Int. J. Innov. Technol. Manag. **17**(3), art. no. 2050004 (2020). https://doi.org/10.1142/S0219877020500042
3. Mingaleva, Z., Mirskikh, I.: The problems of legal regulation of intellectual property rights in innovation activities in Russia (institutional approach). World Acad. Sci. Eng. Technol. **29**, 464–476 (2009)
4. Miura, K., Kobayashi, N., Miyake, T., Shirasaka, S., Masuda, Y.: A proposal of architecture framework and performance indicator derivation model for digitalization of quality management system. In: Chen, Y.-W., Tanaka, S., Howlett, R.J., Jain, L.C. (eds.) Innovation in Medicine and Healthcare. SIST, vol. 242, pp. 129–139. Springer, Singapore (2021). https://doi.org/10.1007/978-981-16-3013-2_11
5. Mingaleva, Z., Mirskikh, I.: The problems of digital economy development in Russia. In: Antipova, T., Rocha, A. (eds.) DSIC18 2018. AISC, vol. 850, pp. 48–55. Springer, Cham (2019). https://doi.org/10.1007/978-3-030-02351-5_7

6. Tamm, T., Seddon, P.B., Shanks, G., Reynolds, P.: How does enterprise architecture add value to organisations? Commun. Assoc. Inf. Syst. **28**(1), 141–168 (2011). https://doi.org/10.17705/1cais.02810

7. Antipov, D., Smagina, A., Klassen, N.: Information support of a quality management system in the context of digitalization of business processes. In: IOP Conference Series: Materials Science and Engineering, vol. 986, no. 1, art. no. 012025 (2020)

8. Aikhuele, D., Turan, F.: A conceptual model for the implementation of lean product development. Int. J. Serv. Sci. Manag. Eng. Technol. **9**(1), 1–9 (2018). https://doi.org/10.4018/IJSSMET.2018010101

9. Akatov, N., Mingaleva, Z., Klačková, I., Galieva, G., Shaidurova, N.: Expert technology for risk management in the implementation of QRM in a high-tech industrial enterprise. Manag. Syst. Prod. Eng. **27**(4), 250–254 (2019). https://doi.org/10.1515/mspe-2019-0039

10. Muris, L.J., Moacir, G.F.: Variations of the kanban system: literature review and classification. Int. J. Prod. Econ. **125**(1), 13–21 (2010). https://EconPapers.repec.org/RePEc:eee:proeco:v:125:y:2010:i:1:p:13-21

11. Markov, D.A., Markova, N.A.: Practices of Kaizen implementation at industrial enterprises. J. Ural State Univ. Econ. **19**(5), 130–140 (2018)

12. Marinkovic, V., Bekcic, S., Pejovic, G.B., Šibalija, T.V., Majstorović, V.D., Tasić, L.: An approach to TQM evaluation in pharma business. TQM J. **28**(5), 745–759 (2016). https://doi.org/10.1108/TQM-10-2015-0134

13. Gidey, E., Jilcha, K., Beshah, B., Kitaw, D.: The plan-do-check-act cycle of value addition. Ind. Eng. Manag. **3**(1) (2014). https://doi.org/10.4172/2169-0316.1000124

14. Mingaleva, Z., Deputatova, L., Akatov, N., Starkov, Y., Mitrofanova, E.: Application of HADI-cycle for providing sustainability of processes of knowledge and innovation. Entrep. Sustain. Issues **7**(2), 1628–1640 (2019). https://doi.org/10.9770/jesi.2019.7.2(58)

15. Lichtenthaler, U.: Profiting from digital transformation? Combining data management and artificial intelligence. Int. J. Serv. Sci. Manag. Eng. Technol. **12**(5), 68–79 (2021). https://doi.org/10.4018/IJSSMET.2021090105

16. Andriole, S.J.: Five myths about digital transformation. MIT Sloan Manag. Rev. **58**(3), 20–22 (2017)

17. Chen, S.-H., Ou, Y.-P.: Digital transformation and structural change in Taiwan's national innovation system. In: Industrial Development of Taiwan: Past Achievement and Future Challenges Beyond 2020, pp. 34–54 (2021)

18. Mingaleva, Z.A.: Structural modernization of economy and innovation development. World Appl. Sci. J. **20**(9), 1313–1316 (2012)

19. Balyakin, A., Taranenko, S., Nurbina, M., Titov, M.: Social aspects of big data technology implementation. J. Digit. Sci. **1**(1), 15–24 (2019). https://doi.org/10.33847/2686-8296.1.1_2

20. Garza-Reyes, J.A.: Lean and green – a systematic review of the state of the art literature. J. Clean. Prod. **102**, 18–29 (2015). https://doi.org/10.1016/j.jclepro.2015.04.064

21. Chen, Y., Li, H.: Research on engineering quality management based on PDCA cycle. In: IOP Conference Series: Materials Science and Engineering, vol. 490, no. 6, art. no. 062033 (2019). https://doi.org/10.1088/1757-899X/490/6/062033

22. Čierna, H., Sujová, E., Ťavodová, M.: Application of the Kaizen philosophy – a road to a learner business. In: International Scientific Days 2016. The Agri-Food Value Chain: Challenges for Natural Resources Management and Society, pp. 237–245 (2016)

23. Schmitt, T., Wolf, C., Lennerfors, T.T., Okwir, S.: Beyond "Leanear" production: a multi-level approach for achieving circularity in a lean manufacturing context. J. Clean. Prod. **318**, 128531 (2021)

24. Arnold, C., Kiel, D., Voigt, K.-I.: How the industrial internet of things changes business models in different manufacturing industries. Int. J. Innov. Manag. **20**(8), art. no. 1640015 (2016). https://doi.org/10.1142/S1363919616400156
25. Garstenauer, A., Blackburn, T., Olson, B.: A knowledge management based approach to quality management for large manufacturing organizations. EMJ – Eng. Manag. J. **26**(4), 47–58 (2014). https://doi.org/10.1080/10429247.2014.11432028
26. Mingaleva, Z., Deputatova, L., Starkov, Y.: Management of organizational knowledge as a basis for the competitiveness of enterprises in the digital economy. In: Antipova, T. (ed.) ICIS 2019. LNNS, vol. 78, pp. 203–212. Springer, Cham (2020). https://doi.org/10.1007/978-3-030-22493-6_18
27. Raudeliūnienė, J., Davidavičienė, V., Jakubavičius, A.: Knowledge management process model. Entrep. Sustain. Issues **5**(3), 542–554 (2018). https://doi.org/10.9770/jesi.2018.5.3 (10)
28. Manikas, A.S., Kroes, J.R., Foster, B.P.: Does the importance of environmental issues within an industry affect the relationship between lean operations and corporate financial performance? Sustain. Prod. Consum. **27**, 2112–2120 (2021)
29. Mason, R., Nieuwenhuis, P., Simons, D.: Lean and green supply chain mapping: adapting a lean management tool to the needs of industrial ecology. Progr. Ind. Ecol. Int. J. **5**(4), 302–324 (2008)
30. Bhattacharya, A., Nand, A., Castka, P.: Lean-green integration and its impact on sustainability performance: a critical review. J. Clean. Prod. **2361**. art. no 117697 (2019). https://doi.org/10.1016/j.jclepro.2019.117697
31. Gökalp, E., Martinez, V.: Digital transformation capability maturity model enabling the assessment of industrial manufacturers. Comput. Ind. **132**, art. no 103522 (2021). https://doi.org/10.1016/j.compind.2021.103522
32. Ustinova, Y.: The true and fair view concept: the palette of controversial points (of "worth banning" to "worth keeping"). J. Digit. Art Humanit. **2**(1), 39–47 (2019). https://doi.org/10.33847/2712-8148.2.1_4
33. Antipova, T.: Preface. Lecture Notes in Networks and Systems, vol. 186 (2021)
34. Joyce, A., Paquin, R.L.: The triple layered business model canvas: a tool to design more sustainable business models. J. Clean. Prod. **135**, 1474–1486 (2016)
35. Rodríguez, R., Molina-Castillo, F.J., Svensson, G.: The mediating role of organizational complexity between enterprise resource planning and business model innovation. Ind. Mark. Manag. **84**, 328–341 (2020)
36. Camargo, M., Dumas, M., González-Rojas, O.: Automated discovery of business process simulation models from event logs. Decis. Support Syst. **134**, 113–284 (2020)
37. Velter, M.G.E.: Sustainable business model innovation: The role of boundary work for multi-stakeholder alignment. J. Clean. Prod. **247**, 119497 (2020)
38. Kitazawa, S., Sarkis, J.: The relationship between ISO 14001 and continuous source reduction programs. Int. J. Oper. Prod. Manag. **20**(2), 225–248 (2020)
39. Dües, C.M., Tan, K.H., Lim, M.: Green as the new lean: how to use lean practices as a catalyst to greening your supply chain. J. Clean. Prod. **40**, 93–100 (2013)

# Digital Education

# Investigating the Use of Learning Analytics at South Africa's Higher Education Institutions

Palesa Maralitle Molokeng[(⊠)] and Jean-Paul Van Belle

University of Cape Town, Rondebosch, Cape Town 7700, South Africa
mlkpal002@myuct.ac.za, jean-paul.vanbelle@uct.ac.za

**Abstract.** Developments in educational technologies as well as the need by higher education institutions (HEI) to improve both the teaching and learning environment are contributing to the growth of learning analytics (LA). Research in LA has predominantly been exploratory in nature and a shift is needed for evaluative research that explores the impact and outcomes of LA. Moreover, there is limited research on the use of LA at HEI from a South African context with most research focusing on the Global North and Australia. The aim of this article is to respond to that research gap by presenting findings from an empirical study investigating the use of LA in SA. This article draws technology-organization-environment (TOE) framework and DeLone and McLean's IS success model to develop a theoretical integrated model. The outcome of the study seeks to help HEIs better recognize opportunities available to them through the use of LA and to guide future empirical research on exploratory studies of adoption of LA.

**Keywords:** Learning analytics · Learning analytics use · Higher education institutions · South Africa

## 1 Introduction

Online learning is growing in South Africa (SA) and this is evident in the number of higher education institutions (HEIs) offering online courses as well as an increase in private online learning providers [1]. Developments in technology are changing how learning is administered and the manner in which students are assessed [2]. Educators and researchers are continuously looking at ways to adopt new and robust data collection methods and platforms as institutions seek more data-driven decision making, and ways to improve learning systems [3].

Learning analytics (LA) has emerged as a powerful tool for improving teaching and learning and addressing a range of educational challenges such as; concerns over institutional retention, continuous improvement of the student learning experience through personalised learning and data-driven decision making [3]. LA is a growing multidisciplinary field with several definitions [4]. Some scholars define LA from the perspective of the student where generated data can be used as a tool to predict educational outcomes, while others define it as a means for educators to better understand student study behaviours in order to provide support and change student learning experiences [5]. This paper adopts the definition of LA as the "*measurement,*

T. Antipova (Ed.): DSIC 2021, LNNS 381, pp. 59–70, 2022.
https://doi.org/10.1007/978-3-030-93677-8_6

*collection, analysis and reporting of data about learners and their contexts, for purposes of understanding and optimising learning and the environments in which it occurs*" [6]. It is important to understand the extent to which LA is being used or could be used in online learning at HEIs in order to realise its potential benefits and/or effectiveness. The growing number of published work in LA in the last decade is indicative of the importance and growth of LA as a research field [7]. There is, however, limited research in the use of LA at HEI in South Africa with most research focusing on the Global North [8]. This study aims to address this gap by reporting on insights from educators, administrators and support staff on the level at which they are adopting LA to optimise the learning environment, both from a curriculum and student improvement perspective, and uncovering barriers that might exist in the use of LA.

The study aimed to meet this objective by addressing the following questions:

1. What type of student data can be and is currently being collected?
2. Is the data used for LA and if so, to what extent?
3. Is LA used to inform decision making?
4. What opportunities exist, if any, for more advanced LA?

Theoretical frameworks are starting to emerge in the field of LA [7] and for this reason, an integrated theoretical framework was applied for this study. The framework adopted the theory of the Technology-Organisation-Environment framework (TOE) and the updated DeLone and McLean IS success model.

This paper starts with a brief background of LA. Focus is drawn to the state of learning in South Africa to provide context of the phenomenon. The theories which focus on adoption and use of technologies are presented. An analysis discussion of the findings and are presented. The last section of the paper details the study contribution and recommendations for future empirical research in the field of LA within a SA context.

## 2   Literature Review

### 2.1   Learning Analytics Background

The changing educational landscape, the need for an educated workforce and increase in the cost of accessing higher education have led educators, policy makers and business people to explore and embrace emerging approaches to education [9]. Advancements in educational technology and the changes in administration of teaching and learning have prompted the occurrence of LA [10]. LA is a field that continues to develop as is evident in the growth of research in the field [11]. Most of the research focus has been on LA tools, data models and prototypes with less research focused on the behavioral elements of how LA complements education, and the improvements needed from both educators and students in order for the tools to better support their everyday learning experience [11]. The digital traces left by students can be analyzed and patterns identified on various elements that influence the students learning behavior, these elements can be both positive and negative [12].

The developing field of LA has, through the collection of large amounts of data, presented opportunities for educators and students alike to gain insights into various elements of teaching and learning [13]. LA helps tailor educational opportunities to the learner's needs, creates an environment of constant feedback between educators and students and improves the learning experience for students [2, 13]. LA methods and analysis results are observed to affect strategy and assist educators and policy makers in HEIs to make more informed decisions [10]. Using LA, educators are able to identify at risk students and apply the right intervention measures to enhance student's levels of achievement and retention [13]. Evidence based student learning has become increasingly important and educators are able to use the data to better understand students' engagement levels [2, 13].

Gaps have been identified in LA research. Ferguson, Brasher [11] argue that there is still limited empirical evidence on the effectiveness of LA at HEI. Furthermore, a large number of published work is significantly skewed towards the positive impact of LA while accounts of failure are not fully represented [11]. Research in the use of LA has also been exploratory with most studies focusing on the tools employed within a HEI as opposed to evaluating the level of use within a HEI [2].

## 2.2    Technology and Learning in South Africa

Higher education plays a key role in the sustainability and productivity of a society by providing skills that help improve the quality of life and help the society to grow [14]. The complex history of South Africa creates challenges for most South African HEIs as they seek to create an inclusive and equal educational society that meets the changing needs of learners and is on par with global HEIs [1]. SA's history has played a key role in the rate at which educational technologies have been adopted [1]. Many of the new generation of students are familiar and comfortable with technology and view it as part of their everyday lives, and expect institutions to incorporate technology in the learning environment [15].

The challenge in SA, however, is that due to existing inequalities some students are not exposed to technology and lack basic computer literacy skills when entering HEIs [16]. Online learning has been observed to be an option to mitigate these issues [1].. Online learning encourages availability of information and social learning communities which leads to the sharing of that information thus fostering creativity and independence of work amongst students [1]. Another challenge facing HEIs in SA in the use of online learning is inequality of access for students from low socioeconomic groups and continued disparity of both access and use [1]. In addition, there are concerns that students who lack digital capital are at risk of being marginalized [16].

## 2.3    Learning Analytics Opportunities and Challenges

New technologies have allowed educators and institutions access to data introducing opportunities to collaborate in designing single platforms that allow for the sharing of multiple data sets [12]. Research shows that although educators have access to vast amounts of student data, they do not always know how to use the data to make changes that will lead to improvements [11]. While LA presents opportunities for educators and students, education stakeholders need to be aware of the issues related to the use of LA in higher education [2]. Researchers have highlighted that the results and predictors from LA tools do not provide guidance on the course of action to be taken to improve learning and teaching [17]. There is no overwhelming evidence from the research of the efficiencies and improvements on the learning process resulting from the adoption of LA, instead, the institutions indicate an expectation of future benefits [11]. There is limited evidence of successful LA implementations [3, 11].

The nature of LA introduces concerns about the quality of the collected data and notes a risk that students may adjust their behaviors to trick the system, resulting in incorrect analysis and interpretations [18]. Ethical and privacy concerns have been highlighted by both educators and students in the use of LA [19]. Concerns have also been raised on the extensive scope institutions have in their data-collection activities, raising questions on the use and benefactors of the data [19]. Other existing challenges are in the design of LA tools to ensure they fulfil the intended objective without them being too rigid or restricting and imposing on the student [18].

## 3   Relevant Theories

In order to understand the use of LA at HEI and create the groundwork for research, this section provides an overview of the theories that the theoretical integrated model is based on.

### 3.1   Technology Organization Environment (TOE) Framework

The TOE framework has been used in different industries to understand analytics usage extent as well as explain adoption of innovations [20]. The three elements of the framework have also been shown to influence the way a firm identifies a need for a new technology and the resultant adoption of it [20]. Therefore the TOE framework is relevant when studying LA adoption and use [21].

### 3.2   DeLone and McLean's IS Success Model

The updated model presents six interconnected dimensions of information systems, namely; system quality, information quality, use, user satisfaction, service quality, individual impact and organizational impact [22]. The model has been cited by multiple studies and has been used in many IS studies as a measure of success and has been frequently adopted in evaluating success of e-learning systems [23]. The view is that these constructs each have an impact on the adoption of LA due to their independent contributions and serve as a guide in evaluating continual use of a technology within an organization.

### 3.3   Learning Analytics Integrated Conceptual Model

A conceptual model was developed to capture the salient factors in assessing LA adoption and use through integrating the TOE framework and the updated DeLone and

McLean IS success model. Using TOE's institutional and environmental context can potentially give insights into factors driving institutional adoption. The DeLone and McLean IS success model is a level lower and views LA use on an ongoing basis, such as at a course level. The DeLone and McLean model is concerned with continual use of a technology and provides a perspective of the benefits derived from the continual use of LA to support educational approaches and monitoring of student progress [24].

## 4  Research Methodology

This study adopted a positivist stance as it aimed to determine the use of LA in HEI without the researcher influencing the current state of use. The qualitative method of enquiry was adopted to answer the research questions. A deductive approach was followed in designing the study protocol guided by the conceptual model. While an inductive approach was followed in the initial data analysis, iterations of the analysis indicated a more deductive approach as the themes emerging were aligned to the conceptual model. This study adopted the purposeful sampling method and followed the case research strategy to gain insights into how participants are using LA. The cases that were selected were leading public HEI and commercial online learning providers in South Africa that were considered to have the most value-add from LA. These are top rated institutions as well as one institution with the largest student body in South Africa. Multiple cases were considered to establish whether the findings were common across the various cases, and to determine the extent of generalizability and reliability through inter-case comparison. Data was collected through semi-structured, voice recorded 60-min interviews and analysis of publicly accessible institutional documentation. The first few interviews were conducted at the institution's premises while subsequent interviews were conducted online using online conferencing tools. It was important to have a diverse representation of roles and having a balanced view across the roles helped identify differences and commonalities in views of the phenomenon. Table 1 illustrates the profiles of the interviewees who took part in the study.

**Table 1.** Study participants profiles

| Role | Description |
| --- | --- |
| Coordinator: Learning Technologies | Manages the portfolio of teams supporting teaching and learning processes including online learning environments and other software within the institution |
| Director/Head of Teaching and Learning | Reports to the executive and different stakeholders on the use of LA within the institution and serves as the LA practitioner within the institution |
| Head of Department (HOD) | Leads, manages and develops the department to ensure it achieves the highest possible standards of excellence in all its activities |

(*continued*)

**Table 1.** (*continued*)

| Role | Description |
|---|---|
| Course Convenors and Lecturers | Sets the aims and learning outcomes for the course and ensures delivery and assessment of the course |
| Institutional Researcher | Supports academics and management with information and statistics at a student, course and faculty level |
| Engineering Manager | Manages the various technology systems used within the institution to ensure optimum availability |
| Student Success Practitioner | Supports academic staff in their development and promotes teaching and learning best practices |

## 5  Findings

Thematic analysis was used to analyze the data and identify common themes and patterns. The analysis process followed a hermeneutic cycle for identifying codes that were categorized into the main themes of the study. This section presents a detailed data analysis by highlighting the main themes and subthemes from the study as illustrated in Table 2.

**Table 2.** Overview of the main and subthemes following data analysis

| Main theme | Subtheme |
|---|---|
| Technological factors | Information quality<br>System quality<br>Use of analytical tools |
| Organizational factors | Organizational culture<br>Educator capacity<br>Ethical use of student data |
| Environmental factors | External/regulatory reporting requirements<br>Vendor support/maturity of tools |
| Learning analytics use cases | Decision making<br>Holistic view of student cohort<br>Student support |

### 5.1  Technological Factors

The technological factors are the Information Technology (IT) related factors influencing the use or lack of use of LA within the institution. These were based on the components of the systems, physical architecture in place as well as the data and integration points within the institution. The institutions in the study collected a wide range of student data which included student registration data, student financial aid information, students accessing the learning management system (LMS) and students utilizing the library resources. Different stakeholders within the institution sought different meanings from the data based on their needs which influenced how they used the data.

The quality of the data collected determines the credibility of the insights derived from it. Educators at the various institutions found the LMS data to be more trustworthy than data captured on Microsoft Excel spreadsheets where information can be easily changed without requiring any authorization or view of the history of the changes. Data accuracy, completeness, consistency and ensuring that all student interactions with the institution are recorded were important elements of having a high level of data quality. Having accurate data captured and having data captured timeously were important in ensuring the usability of that data.

> *"In terms of collecting the data, one of the major challenges we had as a department that processes it, we don't capture any of the data ourselves .... And you'll find that they will make a change to the system that they don't log anywhere, and then suddenly your analyses are off."*
> *TR*

The interviewees viewed the quality of the technology systems as the reliability of the tools they use in the institution, the level of security of the tools, efficiency and availability of the tools to ensure that both educators and students can complete their tasks with minimal interruptions. Common across the various institutions was the type of data that was collected with interviewees all mentioning registration, financial, and LMS data to name a few. While all this information is common, there were differences in how it was stored and accessed, and the institution's ability to link the various data inputs to a specific student.

> *"There's no central warehouse, so the data are stored in different databases that often don't talk to one another"* *EO*

Interviewees across the various institutions found the dashboards on the LMS particularly helpful in giving a view of their student cohort. Educators were able to look at student engagement in the course, view student activity and link various activities to better understand a student's performance.

## 5.2 Organizational Factors

This theme explores non-technological aspects and relates factors such as the culture of the institution, influencers and or issues that may either drive the use of LA or hinder it. The level of senior-level support in creating a data-oriented organization where evidence drives decision making influenced the vigor, motivation and interest with which this was achieved. Strong leadership has played a pivotal role in the adoption of LA across the institutions where funding was provided and the needed time for experimenting, and learning is allocated. All the institutions had various initiatives as part of their strategy to be more evidence-driven in their decision making. One initiative that was common across the institutions was using data to identify courses that had high failure rates and keeping students from graduating.

> *"There is this high-impact modules project ...He heads how the improvement of these modules should happen ... he generates the data."* *NI*

Academic staff at institutions have, given their teaching and other responsibilities, little room and time for additional tasks. A barrier observed has been the lack of capacity by educators to delve deeper into their understanding of LA tools and a shortage of time to familiarize themselves with the complexities of the tools in order to get meaningful outputs and explore the various opportunities that the data presents.

> *"So, for lecturers to try and keep up with the expectations of the institution considering the success rates and they then must also do the research, they don't necessarily have time to sit and look at the numbers, and then say, if this is what is happening, how can I use this data to improve my offering." NI*

Ethics was an important aspect of LA with big responsibilities being put on institutions to ensure that student data is well protected, students are not prejudiced based on the outputs of the data and educators act responsibly in accordance with the insights derived from the data. All the institutions have very strict protocols in place in the use and access of student data.

## 5.3    Environmental Factors

Environmental factors relate to the external environment that the institutions operate in and the issues that may either drive or limit the adoption of LA. The cases in this study were all within SA and operate in the same physical environment. The regulatory reporting requirements are the same for HEIs and public institutions have to use the same systems to meet these requirements. Institutions in SA have a statutory reporting requirement to submit statistics about student enrolments and other statistics to the department of higher education. In South Africa, all institutions have to adhere to the Protection of Personal Information Act (POPIA) which aims primarily to safeguard the information collected by the institutions. Interviewees from the various institutions indicated that there are strict protocols in place to protect the data that the institution had, in turn protecting the privacy and confidentiality of students and educators.

> *"HEMIS is our higher education management information system ... the data that we provide there is part of statutory requirements ... they need to submit statistics about their student's enrolment, staff and space." DJ*

There were various tools used within the various institutions. While vendors are different, some tools served the same function and were used in similar ways while other tools were specific to an institution. All institutions have LMSs in place to run their courses and serves as a space for students to engage.

### Purpose of Collecting Data and Learning Analytics Use Cases

This theme examines why data is used and what opportunities exist with its use. To understand student behavior, descriptive data from the LMS and other data sources were used as input to get a broad perspective. The data collected is used to meet specific objectives and analyzed to answer specific questions. Most participants found value in the data when informing decision making related to the course and curriculum design and introducing changes on how a course was structured. As part of their goals to improve the design of their courses and to continuously improve teaching and learning and ensure students are meeting their learning goals, interviewees agreed that more use of data assisted them in meeting these goals.

*"The data we get, we go through it and then we use it to see how we can improve the quality of the course, but we also look at if the data we collect is actually telling us what we want to be able to improve upon the course." HI*

Early intervention and the correct type of intervention is said to help students achieve their learning objectives and improve retention rates. Institutions have devised interventions tailored to the situation that the student was in as opposed to applying a blanket solution in supporting students and helping them improve their performance. An early-alert mechanism, which uses algorithms to identify students who may be at risk early on, had been a helpful way of supporting students and giving them the help they need.

*"The university has developed an appropriate risk- prediction strategy that provides a framework for identifying at-risk and high-risk students within the institution based on current and historic data." InstB Annual Report*

The need for more specialized skills in the interpretation and mining of data has been highlighted as a major drawback as these skills are not necessarily available at the institution and the specialization of the skill also requires dedicated time and attention. While this is a missing skillset, the institutions do still mine and interpret the data with the limited skills available.

## 6  Discussion

The HEIs observed in this study all collect various types of student interactions however, the way the data was integrated varies across the institutions in accordance with their information systems architecture. The need of the HEIs in this study to become more technology focused has resulted in extensive use of their LMS as well as their student information systems (SIS), both of which were viewed as the core of any HEI's technological ecosystem [25]. The SIS and LMS were found to be key in the adoption of LA at the institutions due to extensive use by both students and academic staff. The findings suggest that LA has been used within institutions for making different types of decisions and to answer specific questions. In addition, data was found to be used widely by the different types of stakeholders to make non-routine decisions driven primarily by their needs. As corroborated by Ferguson, Brasher [11] there are many different stakeholders involved in LA with different objectives and different types of decision-making needs. The importance of pedagogy driving design of LA tools was found to be a key consideration in the extent to which the insights observed were credible.

The findings suggested that the assumptions made in the design of the institutions LA must be embedded in pedagogy to ensure that the LA system is effective. These findings are supported by Tsai, Rates [24] who posit that the extent to which LA deployment is informed by learning science impacts its effectiveness. The findings claim that LA was used by the different institutions to strategies to tackle the issue of students failing to complete their studies by flagging students at risk of failure or dropping out early and applying the necessary interventions. Dawson, Joksimovic [4] highlights the benefits that LA affords educators by identifying students that may be at risk early enough to intervene and demonstrate that the use of LA could accurately predict learner dropout and retention rates at early stages.

A shortage of skilled staff available to analyze and understand the meaning derived from the data at the institutions was found to be a concern at the institutions. This was found to be important in order to gain actionable insights from the student data. In their study, Avella, Kebritchi [2] claimed that the correct understanding of student data and having the right skills to interpret the data was important to assure educational benefits through the use of LA.

The findings suggest that having seamless integration and flow of data across various information systems within the organization and ensuring data is kept up to date influences the extent to which the data is used and trusted to yield reliable insights. Reyes [12] supports the findings in his claim that when data does not flow timeously and seamlessly, concerns have been raised on the accuracy of the insights observed and the completeness of the information.

The findings suggest that the degree to which the institutions collect data about students raises privacy and ethical concerns for both students and teachers and highlights the importance of adherence to legislation aimed at protecting personal information. Added to this is the need to adhere to the Protection of Personal Information Act (POPIA). Concerns regarding ethical and privacy issues in LA have been echoed by other studies with many researchers observing these to be key considerations in the successful adoption of an LA solution [25]. In addressing these challenges, Gasevic, Tsai [25] recommended having policies in place within the institution on the use of student data and LA as these would provide a way for institutions to have guidelines and frameworks that can be followed in ensuring that ethical and privacy issues are observed.

## 7   Summary

This study has given a comprehensive analysis on the factors promoting and/or hindering adoption of LA. These factors were identified and mapped up using a theoretical lens integrating the TOE framework and the updated DeLone and McLean IS success model. Many studies have been conducted in the field of LA and while it is still an area that is maturing, studies focusing on the state of LA from an SA perspective are scarce – most studies focus on Europe, USA and Australia [5, 24]. This study has given a high-level view of the emerging maturity of LA at four institutions in SA by establishing to what extent institutions are collecting student data, whether the data is being used for LA and highlighting the challenges and opportunities that exist with the use of LA at these institutions. The study has established that many different types of data are, in fact, being collected and different tools and systems exist in its collection, analysis and reporting. This study offers a perspective of where SA higher education institutions are in their adoption of LA.

The findings can help executive level stakeholders within HEI to better understand the challenges and barriers within institutions in their implementation and adoption of LA so they can design strategies, prioritize critical factors so that resources can be allocated accordingly. The findings can help non-academic staff to better understand the challenges faced by academics in their use of existing LA tools. This can encourage them to collaborate with academic staff when designing and implementing LA tools to ensure that they meet the needs of the users and that they are effective which will likely encourage adoption.

The findings have highlighted the need for LA policies within institutions and the importance of ensuring that the policy adheres to and supports regulations. The policies should also include ethical issues that are specific in the handling of student data as well as responsibilities in the use of the data.

Recommendations for future research are to explore in detail the effectiveness of LA use, the extent to which institutions use the insights and whether or not recommended interventions are implemented. In addition, future studies should investigate whether the interventions that are implemented lead to improved learner outcomes. The study did not focus on whether the actions that were taken were informed purely by the insights observed or if other factors were taken into consideration. These are areas that are unclear and require further research.

A detailed analysis of the tools used at institutions requires focus. Research in this area could evaluate how decisions on which LA tools to use are made, the approach taken when deciding on how said tool will be procured, whether built by the institution or bought off the shelf, customization of LA tools and the level of consultation, if any, in the choice and customization of tools. Future research could also incorporate the student's perspective to understand the benefits students derive from LA, understand their view on consent, ethical and privacy concerns. Ethical and privacy issues present opportunities for future research with a focus on establishing an institution's attitudes on ethical and privacy issues and how these impact stakeholders in various roles within the institution. Principles for ethics and privacy in LA could be an area of further research from the context of SA as this is an area that is being explored in research in Europe, USA and Australia.

# References

1. Ng'ambi, D., et al.: Technology enhanced teaching and learning in South African higher education–a rearview of a 20 year journey. Br. J. Edu. Technol. **47**(5), 843–858 (2016)
2. Avella, J.T., et al.: Learning analytics methods, benefits, and challenges in higher education: a systematic literature review. Online Learn. **20**(2), 13–29 (2016)
3. Bakharia, A., et al.: A conceptual framework linking learning design with learning analytics. In: Proceedings of the Sixth International Conference on Learning Analytics & Knowledge. ACM, Edinburgh (2016)
4. Dawson, S., et al.: Increasing the impact of learning analytics. In: Proceedings of the 9th International Conference on Learning Analytics & Knowledge, pp. 446–455. ACM, Tempe (2019)
5. Viberg, O., et al.: The current landscape of learning analytics in higher education. Comput. Hum. Behav. **89**, 98–110 (2018)
6. Siemens, G., Long, P.: Penetrating the fog: analytics in learning and education. EDUCAUSE Rev. **46**(5), 30 (2011)
7. Lester, J., et al.: Learning analytics in higher education. ASHE High. Educ. Rep. **43**(5), 1–145 (2017)

8. Prinsloo, P., Slade, S.: Ethics and learning analytics: charting the (un)charted. In: Lang, C., Siemens, G., Wise, A., Gaševič, D. (eds.) Handbook of Learning Analytics, pp. 49–57. SOLAR (2017)
9. Veletsianos, G.: Emergence and Innovation in Digital Learning: Foundations and Applications. Athabasca University Press, Edmonton (2016)
10. Zhang, J., et al.: Mapping the study of learning analytics in higher education. Behav. Inf. Technol. **37**(10–11), 1142–1155 (2018)
11. Ferguson, R., et al.: Research evidence on the use of learning analytics: implications for education policy. Joint Research Centre (2016)
12. Reyes, J.A.: The skinny on big data in education: learning analytics simplified. TechTrends **59**(2), 75–80 (2015). https://doi.org/10.1007/s11528-015-0842-1
13. Wong, B.T.M.: Learning analytics in higher education: an analysis of case studies. Asian Assoc. Open Univ. J. **12**(1), 21–40 (2017)
14. Kaliisa, R., Picard, M.: A systematic review on mobile learning in higher education: the African perspective. Turk. Online J. Educ. Technol.-TOJET **16**(1), 1–18 (2017)
15. Tshabalala, M., Ndeya-Ndereya, C., van der Merwe, T.: Implementing blended learning at a developing university: obstacles in the way. Electron. J. e-Learn. **12**(1), 101–110 (2014)
16. Brown, C., Czerniewicz, L., Noakes, T.: Online content creation: looking at students' social media practices through a connected learning lens. Learn. Media Technol. **41**(1), 140–159 (2016)
17. Gaševič, D., Dawson, S., Siemens, G.: Let's not forget: learning analytics are about learning. TechTrends **59**(1), 64–71 (2014). https://doi.org/10.1007/s11528-014-0822-x
18. Knight, S., Shum, S.B.: Theory and learning analytics. In: The Handbook of Learning Analytics, pp. 17–22 (2017)
19. Arnold, K.E., Sclater, N.:. Student perceptions of their privacy in leaning analytics applications. In: Proceedings of the Seventh International Learning Analytics & Knowledge Conference. ACM, Vancouver (2017)
20. Baker, J.: The technology–organization–environment framework. In: Dwivedi, Y., Wade, M., Schneberger, S. (eds.) Information Systems Theory. Integrated Series in Information Systems, vol. 28, pp. 231–245. Springer, New York (2012). https://doi.org/10.1007/978-1-4419-6108-2_12
21. Gangwar, H., Date, H., Ramaswamy, R.: Understanding determinants of cloud computing adoption using an integrated TAM-TOE model. J. Enterp. Inf. Manag. **28**(1), 107–130 (2015)
22. DeLone, W.H., McLean, E.R.: The DeLone and McLean model of information systems success: a ten-year update. J. Manag. Inf. Syst. **19**(4), 9–30 (2003)
23. Mohammadi, H.: Investigating users' perspectives on e-learning: an integration of TAM and IS success model. Comput. Hum. Behav. **45**, 359–374 (2015)
24. Tsai, Y.-S., et al.: Learning analytics in European higher education–trends and barriers. Comput. Educ. **155**, 103933 (2020)
25. Gasevic, D., et al.: How do we start? An approach to learning analytics adoption in higher education. Int. J. Inf. Learn. Technol. **36**(4), 342–353 (2019)

# Relationship Between Teaching Practice and Students' Understanding – Interpretation

Julia Belyasova[(⊠)] [iD]

The Louvain Catholique University, Ottignies-Louvain-la-Neuve, Belgium

**Abstract.** In this article, we seek to establish the relationship between teaching practice and students' comprehension-interpretation. We analyze the reliability of the meditative instrument developed for this purpose, namely «the Notebook of Reflective Pauses», the teaching practice and the repercussions that it could have on the quality of textual comprehension and interpretation of the readers. In order to achieve our goal, we built an evaluation grid that allowed us to estimate the level of comprehension-interpretation of the novel Since Your Death by the students-readers. We study the responses of Romanian students from Suceava high school, Belgian students from the Decroly general education school and Belgian students from the technical school of La Louvière.

**Keywords:** Youth literature · Comprehension · French as a foreign language · Teaching · Reading methodology

## 1 Introduction

Following several theorists in this field, we can affirm that reading is an activity that is difficult to assess because it does not manifest itself through direct clues, but only through those "that are provoked through the writings of work and exchanges around texts" [10]. We analyze the reliability of the meditative instrument developed for this purpose, namely "the Notebook of Reflective Pauses" [5], the teaching practice and the repercussions that it could have on the quality of textual comprehension and interpretation of the readers.

Our intention is to see how the first two facets of comprehension, namely precise explanation and meaningful interpretation, are worked out in the classroom.

In order to achieve our goal, we built an evaluation grid that allowed us to estimate the level of comprehension-interpretation of the novel Since Your Death [1] by the students-readers. We emphasize, "understanding" is not limited to a single operation. Some researchers, such as Séoud [9], distinguish several levels of understanding depending on whether it is a question of the literal, signified or evoked meanings of the work read. Grant Wiggins and McTighe [13], for their part, talk about six facets of understanding: explanation – accurate, interpretation – meaningful, application – effective, perspective – credible, empathy – sensitive, and self-knowledge-self-aware.

T. Antipova (Ed.): DSIC 2021, LNNS 381, pp. 71–82, 2022.
https://doi.org/10.1007/978-3-030-93677-8_7

Therefore, we will say that the student-reader achieves comprehension as a whole when he manages to interpret the text read, to apply his knowledge to it, to reveal the ability to appreciate different perspectives, to show his empathy, to get to know himself.

## 2  The Novel "Since Your Death" by Frank Andriat

The novel Since Your Death by Frank Andriat [1] is part of a group of socio-realistic stories. The Grasset-jeunesse edition and its Flashlight collection feature a multitude of authors (such as Marc Séassau, Olivier Ka, Martina Murphy and others) who, in their realistic novels, speak to adolescents about life: relationships between parents and children, first love, loneliness, discomfort, friendship, tolerance and many other subjects that interest them so much [4]. This literary genre offers, according to Françoise Lepage [7], an objective representation of reality rather than "an idealized image". This mirror-novel by Andriat [1] "got in tune with young people, describing them in their daily lives, painting them grappling with the family and social context of this turn of the century, sometimes confining to the study of cases or problem solving" [7].

It deals with a painful subject, that of the loss of a parent, which is particularly aimed at adolescents. The death of the other, that of the mother or the father, intervenes, according to Thaler and Jean-Bart [11], "in the initiatory journey of the hero". This ordeal is part of the adventure he is going through, it is an event that the character is forced to live with. Youth work that deals with this painful subject makes it possible to overcome the death of a loved one, to accept or reject it.

Ghislain, the main character of Andriat's novel [1], is also the narrator of this short text which tells of the difficult acceptance of his father's brutal death. He withdraws into himself, trying to erase this haunting separation from his thoughts. But the arrival of a first love helps her out of this deep psychological crisis. Little by little, Ghislain will manage to mourn and accept to live without his father and to keep him in his heart.

## 3  Evaluation Grid for the Overall Understanding of the Novel

Since we have imposed a series of questions on the students in the Notebook of Reflective Pauses [2], it is possible for us to discern the assumptions on which readers base their opinions, to see how they confirm, modify or refine their views.

How do you explain the hero's anger?

What changed in Ghislain's attitude after the conversation with Raphael's mother?

How do you understand Ghislain's changes?

What prompted the visit to the cemetery?

These questions relate to key ideas from Frank Andriat's Since Your Death [1] (Table 1).

**Table 1.** Evaluation grid of the comprehension-interpretation of the novel Since your death [1].

| Ques-tions | Responses received | | | | | |
|---|---|---|---|---|---|---|
| | Belgian students from the technical school of La Louvière | | Belgian students from the Decroly general education school | | Romanian students from Suceava high school | |
| How do you explain the hero's anger? | 1. "He thinks that other people don't know about the death of a loved one, so that gets on his nerves" | − | 1. "His father died when he was young" | ± | 1. "I believe that Ghislain's anger was due to the fact that he could no longer kiss his father and live happily with his family and that this situation was suddenly changed" | ± |
| | 2. "He doesn't want to talk about it with others because he thinks they wouldn't understand him" | − | 2. No response | − | 2. No response | − |
| | | | 3. "He was betrayed by life" | − | 3. "He is a boy who needs his father, so anger is normal" | + |
| | 4. "He was angry because of his father's death" | + | 4. No response | − | 4. 'Ghislain refused to contact the world around him because he didn't want the world to feel compassion for him" | ± |
| | | | 5. "I don't know how to explain it" | − | 5. "I think his anger was against life and its injustices; he didn't know why he must be so unhappy" | + |
| | | − | | ± | | ± |

*(continued)*

**Table 1.** (*continued*)

| Ques-tions | Responses received | | |
|---|---|---|---|
| | Belgian students from the technical school of La Louvière | Belgian students from the Decroly general education school | Romanian students from Suceava high school |
| What changed in Ghislain's attitude after the conversation with Raphael's mother? | 1. "It freed his mind. He no longer has the feeling of going around in circles in a dead end room" | 1. "He is no longer angry with his mother and accepts that we give the clothes" | 1. "He feels his mind is free and he doesn't have the desire to kill his father again in order to feel at ease" |
| | 2. "He finally understood that it wasn't his father's fault and that his mother was right about what she was doing"   − | 2. "He thinks more logically, in the interest of his own and that of his mother"   − | 2. "after the conversation with Raphaël's mother, he thought about it and he accepted his mother's decision to donate the clothes"   − |
| | | 3. "He is more understanding and open to his mother"   − | 3. "He thinks better of his words, he regrets his attitude"   ± |
| | 4. "He realizes that he must learn to live without his father, that he must not look for a culprit"   + | 4. "He begins to understand his mother"   + | 4. "He feels his mind is free and he doesn't have the desire to kill his father again in order to feel at ease"   ± |
| | | 5."He takes the time to think"   ± | 5. "He finally understands that his father will never leave him and that he will always remain in his head, in his heart"   + |

(*continued*)

**Table 1.** (*continued*)

| Ques-tions | Responses received | | | | | |
|---|---|---|---|---|---|---|
| | Belgian students from the technical school of La Louvière | | Belgian students from the Decroly general education school | | Romanian students from Suceava high school | |
| How do you understand Ghislain's changes? | 1. "Thanks to Amélie, he is happy" | | 1. "These changes are due to a change of stage: the second stage to re-stabilize: the return to reality" | ± | 1. "Since he went out with Amélie, the atmosphere has become lighter in his house because he changed the attitude thanks to his girlfriend" | ± |
| | 2. "He realizes he loves her and needs to move on" | + | 2. "Like an acceptance and something good" | − | 2. "Amélie made Ghislain forget that her father's death is also painful" | ± |
| | | | 3. "He accepts that his mother has the right to look at a photo of the man she loved" | − | 3. "With the help of the girl and the help of all the people who wanted her good" | ± |
| | 4. "The love he has for Amélie and the time allowed Ghislain to accept his father's death" | + | 4. "He understands that he can go on living without forgetting his father" | + | 4. "Amélie has managed to change Ghislain's attitude, the atmosphere has become lighter in her house" | ± |
| | | | 5. "He can leave himself to Amélie" | + | 5. "Ghislain says it took him four months to realize that his father's death is an indisputable reality" | + |

(*continued*)

**Table 1.** (*continued*)

| Ques-tions | Responses received | | | | | |
|---|---|---|---|---|---|---|
| | Belgian students from the technical school of La Louvière | | Belgian students from the Decroly general education school | | Romanian students from Suceava high school | |
| What prompted the visit to the cemetery? | 1. "He wanted to see the place where his father rests forever" | − | 1. "Because he accepted the death of his father" | ± | 1."The visit to the cemetery was prompted by Ghislain's statement:" you are dead, daddy, you are dead " | + |
| | | | 2. "All the above events and the acceptance of death" | + | 2. "The feeling of certainty, the fact that he finally accepted his father's death" | + |
| | 3. "He wanted to see where his father was" | − | 3. No response | − | 3. "The desire to meet his father and talk with his father" | − |
| | 4. "He wanted to see his father's resting place in order to realize he is dead" | + | 4. "He must accept the death of his father" | + | 4. "The visit to the cemetery was prompted by the admission of the death of his father" | + |
| | | | 5. "He accepted the death of his father" | + | 5. "The visit to the cemetery was prompted by Ghislain's statement: 'you are dead, daddy'" | + |

Because they are asked in terms of objectives, these questions allow us to achieve the set goal. By applying the approach of Mager [8], we specify to the students the two key instructions for the proposed task, namely those concerning the reading of the novel and the report in the Notebook of Reflective Pauses. We specify that the expected performance is to be able to read better, that is to say, to understand, interpret and appreciate the book at the same time. We have also described "what the pupil will do [concretely] when he demonstrates that he has achieved the objective, as well as the way in which [one can] verify that he has really achieved" [8]: "You write a document over the course of the reading, interrupting yourself several times to note elements that seem important, to express personal reactions (emotions, impressions, ideas), to establish hypotheses as to the next step". It is a question, according to Mager [8], of establishing the conditions under which the "global behavior" [8]. Of the readers must manifest itself and thus of defining the criteria of an "acceptable performance" [8].

Thus, the answers to the questions chosen above reflect four stages of grieving that the person must go through to achieve serenity after the loss of a loved one. In our opinion, this text is understood when the reader is able to grasp these four steps implied by the author. These questions give us the opportunity to estimate the overall comprehension of the text.

To facilitate the examination of the questionnaire, we will use a system of signs. In front of each answer, we will symbolize our evaluation by the signs "+", "−" or "±". The "+" means that the sentence contains, in our opinion, ideas that the author puts forward and that we perceive in his work. The "−" indicates that the answer is not what we expected or that it is not clear enough for us to be able to draw conclusions. Finally, "±" indicates the answers which present a sought-after idea but which are not sufficiently explicit to be able to be understood directly.

As for the data relating to the professors, we describe the characteristics of the declared teaching practices and assess their impact on the comprehension-interpretation of the text by the student-readers. We emphasize here that these are the teaching practices described by the professors themselves. It was impossible for us to observe what was actually practiced in class. Therefore, we take into account that there is probably a difference between declared activities and carried out activities.

## 4    The Case of Belgian Students from the Technical School of La Louvière and Their Teacher

Analysis of the responses shows that seven out of eleven students demonstrate a good understanding of the text. However, four of them are not always sure what they are saying and do not justify their ideas. The other four students give the impression that they do not understand the author's ideas at all. The answers of four students in this class testify to an interpretation that is correct or close to that which seems to us to be suggested by the writer.

Can you find explanations for these failures of understanding in the way the teaching was provided? The teacher in this group of students reads extensively from children's literature and gives a supporting role to classical texts. We believe that she has found a good balance between "heritage works" [12], carriers of general culture, and children's literature, which allows each student to find their way around. According to her, it is important that every reader can discover the character who has the same convictions and who shares the same ideas. Through her practice, she confirms these findings by saying that "at this age (13–14 years), they [these students] often identify with or deviate from the character and his behavior".

Frequent recourse to texts devoid of elements that are an obstacle for young readers (complexity of the text, abundance of literary and cultural implicit, difficulties in terms of language and style, in the sense of Delbrassine [6]) allows the technical school teacher to adopt a stable working framework, which can be summed up in four stages: oral questioning before working with the text, reading aimed at arousing pleasure, oral answers to questions and debate. We thus note that she favors the oral. It is for this reason that his students are not used to expressing their understanding of the text in writing. It seems to us that they don't always find the right words to spell out what they are thinking. All the students unanimously say that the text is easy to read and that they did not need a dictionary to guess the meaning of a few unfamiliar words. These testimonies make it possible to perceive the underlying teaching practice. The teacher teaches her students to look to the reference world and see the text as a whole to identify words they don't know [2].

## 5    The Case of Belgian Students from the Decroly General Education School and Their Teacher

The comparison between two groups of Belgian pupils is interesting to see how comprehension-interpretation is carried out in different contexts. In this case, rather than paying attention to the personal and emotional effects this novel has on its readers, we asked ourselves how well the four stages of grief described by the author were captured, and we compared the levels. understanding of both audiences.

This comparison is all the more interesting as we oppose two groups with the same first language - French - but who differ in their educational motivations and in their relationship to knowledge, strongly dependent on the cultural influences established by the educational institution [4].

Analysis of the answers they gave to the same questions as the first group shows that the majority of students (six out of eleven) demonstrate good or even very good understanding of the text. Two other students give the impression that they do not always grasp the author's ideas or that they cannot come up with the right words. Finally, three members of this group have not at all been able to give any meaningful meaning to this novel.

By comparing the two classes, we notice that twice as many students from the Decroly school expressed a good understanding of the text than those from the La Louvière school. Why ? We find explanations in the teaching practices of Professor de Decroly. Like her colleague from the technical school, she sent us the answers to the questionnaire which allow us to judge her working methods. First, she declares to devote a very important part (3/4 of the time) to literary analysis, and to organize the study of the text by emphasizing the work on the vocabulary. According to her, careful reading of the novel involves understanding the lexicon and particular syntactic phrases, and looking for clues. She always asks her pupils to justify their analysis by words or by precise sentences of the text approached. It draws their attention to morphological analysis, comparison with other types of words, semantic fields, reusing words in personal sentences, and finding the precise meaning of words in the text.

Then, the work on the text often ends with a synthesis on the blackboard or with a written account of impressions and reactions. Thus, the pupils are familiar with this kind of exercise which asks for the transfer of ideas, feelings on paper.

Finally, it should be noted that this teacher constantly offers students literary texts, press articles, contemporary texts and comics. Although works of children's literature do not fall under the recommended course for Grade 4 of secondary school, she made an exception by teaching the novel Since Your Death by Frank Andriat [1]. The students took this proposal seriously. Even if they did not like the novel very much, they did a decent job in mobilizing the concepts taught and they testified to their good comprehension-textual interpretation.

## 6  The Case of Romanian Students from Suceava High School and Their Teacher

As for the Romanian students, it seems difficult to us to say whether they really understood the text. Two members of this group were able to correctly explain the key moments of the narration and capture the author's intentions. A student did not catch the message sent by the text at all. In our opinion, its failure is caused by grammatical-lexical obstacles. This student found a lot of unfamiliar words and had to consult the dictionary, which greatly slowed down his reading. Therefore, he did not understand the text as a whole, did not like it very much, and found it difficult to read.

Analyzing the case of seven other students in this group, we are divided as to their textual understanding-interpretation because we cannot say that they did not understand the novel, but their understanding does not quite correspond to this. that we expected.

The comments of these students revolve around the feelings and feelings of the character, but they go no further in their interpretations, limiting themselves to the first ideas that come to them, to the first impressions they have had.

Again, we find the explanation for this fact in the teacher's teaching practice. According to his written statements, character identification, sentence analysis, and word meaning work are essential steps in the study of each text. This explains the participatory attitude of Romanian learners' reading behavior.

In addition, this teacher spends a lot of energy to learn the procedures which allow the pupils to identify the unknown words in the text by carrying out a syntactic, spelling, morphological analysis and by comparing segments of words, but it seems to us that his students have a lack when it comes to finding the hidden meaning of sentences and innuendo in the text. They have difficulty interpreting part of the story, a chapter, a story as a whole.

Finally, the teacher does not do a lot of written exercises based on the text read. In her home, impressions and reactions are usually reported orally. In this context, learners are less trained to find the right words to express what they are thinking. They are not encouraged to follow their reasoning to the end, whereas the development of their ideas could have led them to a better understanding and a more accurate interpretation of the novel.

We can now conclude on the relationships that we were able to establish between the teaching practices declared by the teachers and the comprehension-interpretation manifested by the three groups of students who read the novel Since your death by Frank Andriat [1]. We notice that Belgian technical school students as well as Romanian learners have difficulty expressing their ideas to testify to their good textual understanding. The cause seems to us to lie in the fact that the two teachers do not offer many written exercises that would stimulate them to develop interpretations with the right words for the story. On the other hand, the teacher of the general education school takes great interest in the exercises of expressing thoughts in writing. His students are therefore better prepared to put their perceptions on paper to interpret the text read.

The first two teachers (that of the technical school of La Louvière and that of the high school of Suceava) mentioned above believe that the participative attitude of the pupils and the identification with the characters prevail in the study of the text. As for our third teacher (that of the Decroly school), she devotes three quarters of the time of her lessons to training her students in literary analysis and textual interpretation. This explains why the majority of his pupils manifest a better comprehension-interpretation of the text than those of the two other groups.

## 7   Conclusion

We tested the mediation tools in five classrooms, with 62 students in total. The results obtained seem satisfactory to us. We were indeed able to validate our interfaces from the point of view of their use, to see that the observed observables were representative of the students' knowledge and that they were sufficiently rich and reliable to allow a diagnosis. We found correlations between the methods used by the teachers interviewed and the level of comprehension and the way of interpreting the novel read by the pupils.

We saw the teacher of the technical school of La Louvière. as an avid reader, who does everything possible to convey her passion to the students. They do not present as experts in analytical reading but they read, understand and interpret intuitively, being guided by their first impressions. By working with a class that is weak in reading, this teacher puts aside learning about literary analysis techniques. She focuses the teaching of reading on the development of literary taste and on the emotional reception of the novel.

From these results, it seems useful to us to recommend setting up the teaching of analytical reading through the Notebook of Reflective Pauses as a mediating instrument [3]. It seems to us possible that, presented as a learning tool (and not as a means of expressing feelings), it will prove to be very effective for readers weak in literary analysis.

On the other hand, following our analysis of the methodological practices declared by the teachers, we think that the professor from the Decroly school and her colleague from the technical school of La Louvière are teachers who put a lot of will in teaching the techniques of textual analysis. However, the results of their students are not the same. In the first case (Decroly school), the criticism of the text is more accentuated. It often focuses on classical works and is accompanied by written work. In the second case (La Louvière technical school), the textual, stylistic, referential analysis of children's novels, most often, is done orally and the students are not used to expressing their understanding of the text in writing.

Thus, by comparing these two classes, we have seen that the comprehension-interpretation of the novel by the students of the technical school of La Louvière seems more vague, less explicit. As for the students of the Decroly School, they did a good level of textual interpretation, although they did not like the proposed novel very much. In this case, we notice that this teacher spends a lot of time learning literary analysis, which is very effective in this context. It seems to us, moreover, that the shift towards the rational side of reading somewhat harms the emotional reception of the novel.

The Romanian teacher, in our opinion, focus her teaching of literary reading on the psychological development of the characters. Her goal is to teach students to express their critical attitude towards the novel, to explain the behavior of the characters, to justify their opinions.

Through the analysis of the work of her students, we noticed that lexico-grammatical obstacles are felt even among learners qualified as good readers. For this reason, this teacher aspires to teach her students the techniques that would allow them to overcome their language difficulties. In contrast, Romanian teacher tends to neglect the teaching of analytical reading of works. Therefore, students' interpretations generally focus on explaining the characters' behaviors and their feelings. The copies of these readers hardly contain any element of purely literary analysis.

In order to remedy this, we would like to recommend integrating the Notebook of Reflective Pauses as an instrument for learning analytical reading in the field. As we said above, it would appear to us to be effective in groups of weak readers. It might also be useful for foreign language readers. In both cases, the notebook of reflective pauses would help to overcome language obstacles, to apply literary analysis techniques and, at the same time, to express the emotions provoked by the novel.

The interest of our research is all the more important as it lies between two didactic fields: that of French as a first language and that of French as a foreign language. Our data analysis shows that low level first language readers and foreign language readers have similar comprehension-interpretation difficulties. Therefore, the use of the same methodological instrument (the Notebook of Reflective Pauses) could perhaps prove effective in the two different contexts.

# References

1. Andriat, F.: Since your death, Grasset - Jeunesse, Paris (2004)
2. Belyasova, J.: Comparative study of the reception of literary works of youth by audiences of French as a first language and French as a foreign or second language. Doctoral thesis in languages and letters. Catholic University of Louvain-la-Neuve (2014). https://dial.uclouvain.be/pr/boreal/object/boreal%3A141993/datastream/PDF_01/view
3. Belyasova, J., Martin, M.: Model of the communication process in the context of reading in French as a first language and French as a foreign language. JDS 2(1), 82–93. https://doi.org/10.33847/2686-8296.2.1_8
4. Julia, B., Raisa, T.: Particularities of language classes in a multi-cultural context. In: Antipova, T., Rocha, A. (eds.) DSIC18 2018. AISC, vol. 850, pp. 174–187. Springer, Cham (2019). https://doi.org/10.1007/978-3-030-02351-5_22
5. De Croix, S.: I read, I write: the practice of the journal of reading. Issues **50**, 121–129 (2001)

6. Delbrassine, D.: The novel for adolescents today: writing, themes and reception, Liège, Scéren-Crdp de l'Académie de Créteil and La Joie par les Livres, coll. Argos References (2006)
7. Lepage, F.: Youth literature 1970–2000, Montreal, Fides. In: Archives of Canadian Letters XI (2003)
8. Mager, R.F.: How to define educational objectives. Bordas, Paris (1977)
9. Séoud, A.: For a didactics of literature. Editions Didier, Paris (1997)
10. Tauveron, C.: What does it mean to evaluate literary reading? Case of pupils with reading difficulties. In: Daunay, B. (ed.) The Evaluation in Teaching of French: Resurgence of a Problematic, in Repères, no. 31, pp. 73–112 (2005)
11. Thaler, D.: Jean-Bart, A.: The stakes of the novel for adolescents. Historical novel, mirror novel, adventure novel, Paris, L'Harmattan (2002)
12. Therien, M. On the definition of literary literature and works to be offered to young people. In: Noel-Gaudreault, M. (ed.) Didactics of Literature. Assessment and Perspectives, Cap-Saint-Ignace (Quebec), Nuit Blanche, pp. 19–33 (1997)
13. Wiggins, G., McTighe, J.: Understanding by Design. ASCD (1998). https://goglobal.fiu.edu/_assets/docs/whatisbackwarddesign-wigginsmctighe.pdf

# Using Merrill's First Principles of Instruction to Reshape the Assessments Structure of a Fundamental First-Year Course

Pariksha Singh$^{(\boxtimes)}$ ⓘ, Kalisha Bheemraj ⓘ, Jayshree Harangee ⓘ, and Tania Prinsloo ⓘ

University of Pretoria, Pretoria 0002, South Africa
pariksha.singh@up.ac.za

**Abstract.** In this paper, a group a students taking a fundamental first-year course are scrutinized, where the method of assessment had to suddenly change from physically writing semester tests and examinations on campus, to applying Merrill's First Principles of Instruction to introduce continuous assessment in 2020. Educational environments are continuously evolving to increase the quality of teaching, learning and assessments with classes that have huge numbers. Academic Information Management (AIM 101), with 1000 students in 2020, made the transition to continuous assessment and the resulting data from the assessments clearly indicate that, not only do students perform better over the course of the semester and in the semester tests, they also attempt the assignments and tests more frequently, leading to an overall increase in distinctions obtained for the course. It was subsequently decided to keep the course online in future and to continue using continuous assessments. Future research will look at different learning styles of students taking fundamental courses and the role of lifelong learning in these courses.

**Keywords:** Higher education · Continuous assessment · Merrill's principles of instructions

## 1 Introduction

Higher Education in South Africa is working towards increasing the quality of education that is presented to all students. Educational environments are continuously evolving to increase the quality of teaching, learning and assessments with study groups that have a huge number of members [1]. Strategies and interventions are researched and tested regularly to determine the best fit to succeed with students that have many members [2]. At the University of Pretoria, a first year Academic Information Management (AIM 101) module comprises of approximately 1000 students. After the completion of AIM 101, the students should be able to find, evaluate, process, present and manage information resources for academic purposes using appropriate technology.

First year students at the university come from radically different backgrounds. Some students being privileged enough to own smartphones and technology devices like laptops and tablets, others that believe a computer mouse is an animal trapped inside a rectangular casing. This results in difficulty in teaching, learning and assessment of AIM 101 because students start with varying levels of computer literacy competencies. To bridge the gap the use of information and communication technology (ICT) is of great importance. We address this problem by having students complete a proficiency test in their first week. The results of this proficiency test determine if the student should complete a year or just a semester of a computer literacy course. Students' background and study choices are quite diverse and finding a way to bridge the gap between the different levels of students is becoming extremely difficult. Due to the global pandemic and traditional face-to-face classes, moving towards a hybrid online approach – "the new normal" [3] – and against this background a decision to review and reshape the current AIM 101 assessment model offering was reviewed.

## 2  Background

AIM 101 was conducted by means of face-to-face presentations in 2019 across two different campuses, in six laboratories that housed approximately 50 students each per laboratory session. Over 20 sessions of AIM 101 were conducted per week. Students booked a session to attend AIM 101 via the AIM department and completed all examinations and all other assessments during their booked session. AIM 101 is a fundamental course for all first-year students across nine faculties that carries six credits. AIM comprises two main sections, Navigating Information literacy as well as Computer Literacy. The skills that are taught in the subject remain valid throughout a student's academic career. The assessment model that was used for AIM 101 was the traditional 50% semester mark and 50% examination mark. In 2020 when the Covid-19 pandemic devastated all spheres of human activities, all teaching, learning and assessment throughout face-to-face institutes had to adapt expeditiously to an online hybrid approach [3], AIM 101 had to change its mode of teaching, learning and assessments so that the academic year could be successful. Academics, students, and support staff had to follow a revised teaching and assessment model for courses due to the online hybrid approach. AIM had to find the best, most student-centered way to run a course with 1000 AIM 101 first year students successfully.

Models were looked at where all teaching and learning were conducted online but final assessments was to be carried out at the university with the traditional proctoring for tests and. Unfortunately, South Africa went into a further level of lockdown and the schedules and methods that was going to be followed had to be discarded since students could not attend any classes in masses at the university. Other models with online examinations were proposed and we did not consider it as a feasible process for 50% of the students' mark being dependent on just one assessment, namely the examination. Factors that influenced this decision were that students did not have a device to work on, a part of the student population did not have a stable internet connection, load shedding and power interruptions were also significant factors [3]. We needed to find a way to deal with the flexibility of answering assessments within parameters of resources available to students. The university took care of devices and zero-rated the cost of data for systems used. At module level, we had to focus on teaching, learning and assessments to make sure students completed the academic year with an education which met all quality standards. Keeping this in mind the idea to move to continuous assessment was born.

## 3 Literature Review

### 3.1 Continuous Assessment

Continuous assessment is a type of educational evaluation used to quantify a student's understanding and comprehension throughout a course. One of the main outcomes of continuous assessments is feedback provided for the benefit of both the educator and student. This allows educators and students to identify gaps and refine skills within their learning process [4]. Additionally, feedback provides an experience-based opportunity for reflection. Reflection is a mental process that is considered to be deeper level of learning, as it integrates the student's ability to consider the product of work they have produced and critical thinking to form reasoned judgements regarding their strengths and weaknesses. This is the start of establishing one's self-knowledge to adapt one's learning style and determine one's course of action within the module [5].

However, as beneficial as feedback is, it is a strenuous task to manually provide individual feedback per student for each assignment or task. Educators of large groups of students prefer using an automatic tool that is able to instantly provide feedback and results to students which is correlated to an increased student performance [5]. In addition to grading assessments, a powerful and well-developed online grading tool may be able to offer training and practical activities to students. An example of a dominant tool used in the instruction of Microsoft (MS) Office is a product from Cengage Learning; Skills Assessment Manager (SAM) [6]. SAM plays an extremely important role in AIM 101 assessments.

### SAM

This product of Cengage Learning is designed to aid in teaching students how to use MS Office products. Educators are able to create tailor-made "Sections" to include training that align themes seamlessly to the desired module objectives and outcomes [7].

SAM integrates instructions within a MS simulation, providing real-world application-based examples to stimulate critical thinking and refine applied skills. Students can complete "Trainings" which tracks their progress and displays insights in their own gradebook. "Trainings" include graded practical training activities and sample projects. These graded practical training activities can be taken before sample project activities or assessments. These training activities consist of a three-step process; observe, practice and apply. In the first step, students observe a short example video of a specific instruction to be completed. Next, students are given the opportunity to complete the instruction with guided help with each step. Thereafter, students are expected to complete the instruction independently to reinforce the skill acquired. Once downloaded, sample projects can be completed offline on a Windows or Mac operating system. When the end product is submitted, SAM immediately provides a score and report, which details the errors or correctness completed for each instruction [6].

## 3.2    Related and Relevant Works

A study conducted in 2013 examined the impact of continuous assessments on final marks of a computer science module at a tertiary level, analyzing 1500 students across 25 different modules. It was determined that a strong correlation (0.57) exists between marks received throughout the semester for continuous assessments and final examination marks. In only three cases, the correlation coefficient was less than 0.4. In 24 out of 25 modules, student performed better in the continuous assessments than in the examination. The mean gain of final marks over examination marks were an overall average increase of 13%. This study demonstrated that instructors need to plan their instructional design to include continuous assessment for promoted learning [8].

In 2018, a study based on 245 medical students investigated if continuous assessments are able to identify at-risk students before the final examination, thus providing additional support where necessary. Poor performance in continuous assessments during the semester is statistically significant to the failure of examinations and ultimately the course. Instructors can use this knowledge early in the course, to identify their at-risk students to provide alternative or additional teaching support to assure learners receive appropriate training [9].

It is known that continuous assessments are incorporated into the instructional design to enhance information transfer and promote learning [10]. A study conducted in 2017 focused on the student's perception of using continuous assessment throughout the module at a tertiary level. The module consisted of continuous assessments throughout the semester with a final test at the end. A total of 177 students participated in the study by completing a questionnaire about their experience. Students responded with positive feedback. Stating that using continuous assessments:

- were beneficial to their learning style and studying methods;
- assisted them to identify their weak areas;
- improved their ability to answer questions in the final test;
- improved their overall academic success.

Instructors can use this information to confidently implement continuous assessments into their instructional design, with taking caution to the time interval and frequency of assessments given, preferably no more than one per week [10].

## 3.3    Merrill's First Principles of Instruction

Merrill defines a principle as a basic method in which the relationship that is always true, given the circumstances are appropriate, regardless of practice. Practice is defined as particular instructional activity. Merrill's First Principles of Instruction focuses on a problem-centered instruction and can be explained by using five principles related to provoking knowledge in a leaner. Merrill describes her principles in the most concise form [12]. Learning is promoted when:

1. learners are engaged in solving real-world *problems*;
2. existing knowledge is *activated* as a foundation for new knowledge;
3. new knowledge is *demonstrated* to the learner;
4. new knowledge is *applied* by the learner;

5. new knowledge is *integrated* into the learner's world.

These five principles can be graphically illustrated for clearer understanding, as seen in Fig. 1.

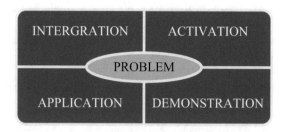

**Fig. 1.** Visual representation of Merrill's First Principles of Instruction [12].

Each principle is based on three defined corollaries.

*Principle 1: Problem-Centered:* learners are engaged in solving real-world problem [11, 13–16]

- Corollary – Show task: When a task or problem is clearly shown, learning is promoted;
- Corollary – Task level: Learners need to be engaged in the task or problem for learning to be promoted;
- Corollary – Problem progression: Learning is promoted when learners solve a sequence of problems related to one another.

*Principle 2: Activation:* existing knowledge is activated as a foundation for new knowledge [11, 13–16]

- Corollary – Previous experience: The foundation of new knowledge needs to be built of the learner's ability to recall, relate, describe or apply relevant knowledge from past experiences;
- Corollary – New experience: Learners need to be provided with relevant experience in order to create new knowledge;
- Corollary – Structure: Learning is promoted when learners recall a structure that organizes new knowledge.

*Principle 3: Demonstration: new knowledge is demonstrated to the learner* [11, 13–16]

- Corollary – Demonstration consistency: Optimal learning outcomes are achieved when the demonstration method used is consistent with the goal or objective;
- Corollary – Learner guidance: Learners need to be provided with a variety of material as to guide them to the appropriate outcomes for learning to be promoted;
- Corollary – Relevant media: Learners need to be exposed to a wide variety of unbiased informative media which will promote learning.

*Principle 4: Application: new knowledge is applied by the learner* [11, 13–16]

- Corollary – Practice consistency: Learning is promoted when the application of tasks is consistent with the objectives;
- Corollary – Diminishing coaching: For learning to take place, learners need to receive constructive feedback for their skills applied, including error detection and correction;
- Corollary – Varied problems: Learning is promoted when students actively practice how to solve a sequence of problems.

*Practice 5: Integration: new knowledge is integrated into the learner's world* [11, 13–16]

- Corollary – Watch me: Learners need to be given the opportunity to prepare for and present their new knowledge;
- Corollary – Reflection: The learning process is stimulated when learners are given the platform to reflect on, discuss and defend their new knowledge and skills;
- Corollary – Creation: Learning is promoted when learners can design, develop and discover new methods in which to apply their new knowledge and skills.

## 4   Methodology

### 4.1   Research Questions

The main research question in this paper is: *How can first-year students in a fundamental course be assessed using continuous assessment based on Merrill's First Principles of Instruction?* A subsequent question is: *How can this method of instruction lead to better student performance?*

### 4.2   The Concrete Model Pre-Covid-19

The AIM 101 module was serviced for first-year students at the University of Pretoria totaling approximately 1000 students who attended lectures in the IT labs until completion of the module. The module was designed so that the content was suited to students with a background in computer skills. There were approximately 12 weekly sessions of which three of these were allocated for semester tests and examinations. Students had to attend at least one two-hour contact session per week. The course content consisted of two components *Navigating Information Literacy* and *Navigating Office 2019*, which was covered in nine lectures (Fig. 2).

The semester mark involved seven compulsory assignments, each assignment was open for a duration of two weeks allowing three attempts each and only the highest mark was taken from the three attempts. Assignments were completed individually on either Blackboard or SAM, depending on the nature of the assessment students could complete the assignments before or on the last day of the due date.

The AIM 101 module also included two semester tests which contributed 30% each towards the semester mark. Semester test one was composed of two sections namely

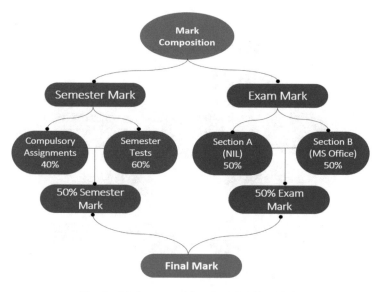

**Fig. 2.** Mark composition for AIM 101 2019.

*section A* which contained the theory aspect, and this involved an in-class based test. *Section B* was performed on a simulated system. Semester test two consisted of all the practical aspects that was thought throughout the semester. In this test students were provided with a real-world project from the SAM program. They were timed and submission was only allowed in class. The examination mark consisted of questions that were based on the theoretical concepts, *section A*, which carried 50% whilst the practical, *section B*, on the simulated system and this contributed to 50% of the examination mark. A minimum of 30% for the semester mark was required for the students to qualify for the examination. Only students with a valid excuse could take a make-up semester test or examination.

### 4.3 Assessments During Covid-19: Sink or Swim

From April 2020 the AIM 101 module was available online due to the Covid-19 pandemic. Because this undergraduate module shifted from contact sessions to being online students required an active learning strategy. Merrill's First Principle of Instruction [12] was used as a framework as it provided a logical, systematic method for implementing the requirements of the course content. This module conveyed a weighing of six credits, indicating that on average a student should spend approximately 60 notional hours to master the necessary skills including time for preparation for tests and assessments.

The mark contributions for the module contained a final mark represented in Fig. 5 below as there was no face-to-face contact sessions; therefore, monitoring the student's engagement with the module could be conducted through continuous assessments (Fig. 3).

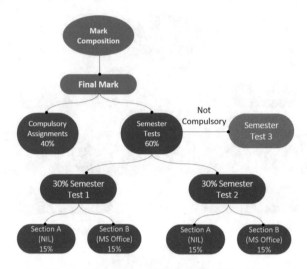

**Fig. 3.** Mark composition for AIM 101 2020.

As illustrated in Fig. 1 the first phase *activation* of Merrill's Principle of Instruction [12] was accomplished by allowing access to all course content including compulsory assignments all of which were made available with three attempts at the beginning of the semester and only closed at the end of the semester. This phase allowed students to gain prior knowledge of the content and it organized students thinking for what they were going to learn and experience in the course. Allowing access to the entire course content permitted students to identify their own strengths and weakness for the content. The weekly schedule provided them with an organized structure of what will be covered and required of them each week.

In the second phase of instruction [12], *demonstration*, of the theoretical content was presented through a Blackboard Collaborate platform, where the use of Power-Point lecture slides were used to present the content. Online sessions were presented every day and students had to attend at least one session per week. This platform allowed lessons to be recorded, so that students could revisit the content. In 2019 this was not possible due to limited university resources. SAM trainings also allowed *demonstration* of the practical component of the course content. Narrated PowerPoint videos were also created for both the theoretical and practical chapters of the course.

In the next phase of instruction students could *apply* what they learnt in the *demonstration* phase through compulsory assignments [12]. The compulsory assignments were designed in such a way that students had to apply their knowledge by answering content related questions.

*Integration* was the final phase of Merrill's First Principle of instruction [12] where both semester tests one and two were completed on a specific day for all students enrolled in the course. The semester test was opened for a duration of two days as this allowed all students to take the test without having to face any technical, connection or power related issues.

Each step found in Merrill's Principle of instruction allowed the AIM 101 module to fully engage with the student [12]. Blackboard collaborate sessions as well as discussion board. Students also had access to their lecturers contact details for further assistance with the course content. Table 1 summarizes how the AIM 101 module integrated Merrill's First Principle of Instruction [12] to the course structure.

**Table 1.** Merrill's framework incorporated in the AIM 101 module [12].

| Merrill's first principle of instruction | Approaches used in AIM 101 2020 |
|---|---|
| Problem- centered | Compulsory module that conveyed a weighing of six credits |
| Activation | All semester course material was made accessible to the students at the beginning of the semester |
| Demonstration | Used a Blackboard collaborate platform; Narrated PowerPoint lecture slides and SAM trainings |
| Application | Have students complete the compulsory assignments and extra exercises allowing a maximum of three attempts for each |
| Integration | Provided two semester tests to apply their knowledge and have an opportunity to reflect on their learning |

## 5   Findings and Discussion

A total of seven assignments were completed in both 2019 and 2020. The marks of these assignments can be directly compared. As shown in Fig. 4a below, students produced more proficient results in 2020. This indicates that the students achieved a higher level of understanding throughout the module. This can be largely accredited to the use of continuous assessments. Fig. 4b presents a summary of the total number of students that did not attempt any of the assessments. The results found in 2020 is significantly lower than in 2019, this shows that students prefer continuous assessments than a traditional examination-based setting. With the traditional examination settings used in 2019, students moved towards surface learning in which they memorized the

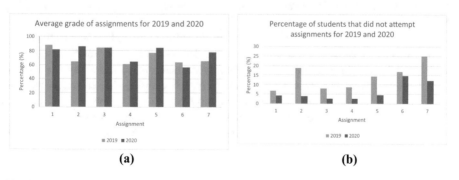

**Fig. 4.** (a) Average grade of assignments for 2019 and 2020. (b) Percentage of students that did not attempt assignments for 2019 and 2020

facts without understanding. The adoption of continuous assessments in 2020 led to an increase of student engagement with the module. This result also show that students had increased pressure to complete the continuous assessments as there was no examination mark to rely on.

In 2019 the continuous assessments were associated with deadlines and this showed that their engagement only peaked a few times through the year, which were usually the days before the assessment deadline.

A total of two semester tests, consisting of two sections each were assigned in both 2019 and 2020. As shown in Fig. 5a below, students produced more competent results in 2020. Figure 5b shows that in 2020 there was a significant increase with more students that completed the semester test two than in 2019. The detailed reports received from the continuous assessments provided feedback to the students so they could utilize this to improve on their problem areas. In this way students did also have reduced test anxiety because they were much more familiar of what was expected from them.

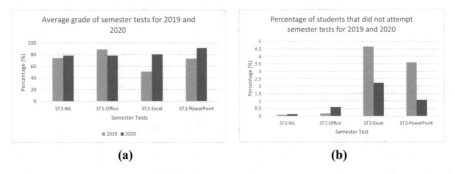

Fig. 5. (a) Average grade of semester tests for 2019 and 2020. (b) Percentage of students that did not attempt semester tests for 2019 and 2020.

All questions that were set for the semester tests in 2020 were scenario based; none of the questions used in the database was searchable on the internet. Therefore, the large number of questions loaded to the database allowed for randomization of the questions as this ensured that each student received a new set of questions when taking a test.

As seen in Fig. 6 below, the total number of distinctions in 2020 was significantly higher than in 2019 for the AIM 101 module. The continuous assessments engaged students with the task completely as they were carefully selected by the course instructors to match the learning outcomes for each chapter. Students also had the opportunity to maximize their number of attempts for each assignment. Semester test three was also made available for all students who wanted to improve their marks. The marks received from semester test three was used to replace the mark that was lower in the students' grade.

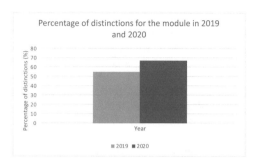

**Fig. 6.** Percentage of distinctions for the module in 2019 and 2020.

## 6  Conclusion and Future Research

It is concluded that the AIM 101 module was successfully delivered to students in 2020 after the Covid-19 lockdown restrictions made it impossible for them to return to campus for semester tests and examinations. The shift to continuous assessment, using Merrill's First Principles of Instruction led to an overall better performance of AIM 101 in 2020, with students performing better in 2020 over the seven different assignments, more students attempted the assignments, an overall higher grade was achieved in the semester tests and there was an increase in the number of distinctions compared to 2019. A decision has subsequently been made to keep fundamental modules in its current format of applying continuous assessment, even if students were to return to campus.

Future research will focus on how students can be taught and assessed online using various methods of instruction to accommodate different learning styles and the role of lifelong learning in the design of large online courses. Future research will also investigate how student engagement in an online module can improve. The creation of higher-order thinking assessments and tests can be explored, where students can learn skills that will help them succeed in the future. Research can also investigate what aspects contribute to a digital technology solution's success. In order to ensure rapid growth, research on different digital technology should be explored to ensure that each student has a fully personalised learning experience while also providing instructors and students with real-time evaluation and feedback.

## References

1. Government SA: Improving education, training and innovation. In: Education Do. South Africa (2014)
2. Pellegrino, J.W., Chudowsky, N., Glaser, R.: National research council. Committee on the Foundations of A. Knowing what students know: the science and design of educational assessment. National Academy Press, Washington (2001)
3. Prinsloo, T., Singh, P.: COVID-19: leapfrogging 8,000 students from face-to-face to online learning in three weeks. Commun. Assoc. Inf. Syst. **48**(1), 9 (2021)

4. Kalu, F., Dyjur, P.: Creating a culture of continuous assessment to improve student learning through curriculum review. New Direct. Teach. Learn. **155**, 47–54 (2018)
5. Ferreira, D.J., Da Silva, H.C., Melo, T.F.N., Ambrósio, A.P.: Investigation of continuous assessment of correctness in introductory programming. J. Educ. Technol. Soc. **20**(3), 182–194 (2017)
6. Cengage. Cengage Learning: SAM: Skills Assessment Manger United Kingdom2021. http://www.cengage.com/sam/. Accessed 15 Feb 2021
7. Machajewski, S.: Gamification strategies in a hybrid exemplary college course. Int. J. Educ. Technol. **4**, 1–16 (2017)
8. Pudaruth, S., Moloo, R.K., Chiniah, A., Sungkur, R., Nagowah, L., Kishnah, S.: The impact of continuous assessments on the final marks of computer science modules at the university of mauritius (2013)
9. Nagandla, K., Gupta, E.D., Motilal, T., Teng, C.L., Gangadaran, S.: Performance of medical students in final professional examination: can in-course continuous assessments predict students at risk? Natl. Med. J. India **31**(5), 293–295 (2018)
10. Assad, A.R., Asim, M.A., Heitham, M.M., Muntaser, M.A., Mohammed, A.M.: Students' perceptions toward continuous assessment in anatomy courses. J. Med. Sci. Health [Internet] **3**(2), 5–8 (2017)
11. Moore, M.G.: Theory of Transactional Distance. Routledge (1997). http://www.aged.tamu.edu/research/readings/Distance/1997MooreTransDistance.pdf. Accessed 15 Feb 2021
12. Merrill, M.D.: First principles of instruction. Educ. Technol. Res. Dev. **50**(3), 43–59 (2002)
13. Moore, M.G.: Handbook of Distance Education, 3rd edn. Routledge, New York (2013)
14. Simarmata, J., Djohar, A., Purba, J., Juanda, E.A.: Joint workshop of KP. In: 2nd International Conference on Mathematics STE, et al. Design of a Blended Learning Environment Based on Merrill's Principles. Journal of Physics, Conference Series, vol. 954, no. 1 (2018)
15. Lo, C.K., Lie, C.W., Hew, K.F.: Applying "first principles of instruction" as a design theory of the flipped classroom: findings from a collective study of four secondary school subjects. Comput. Educ. **118**, 150–165 (2018)
16. Tu, W., Snyder, M.M.: Developing conceptual understanding in a statistics course: Merrill's first principles and real data at work. Educ. Technol. Res. Dev. **65**(3), 579–595 (2017)

# From Learning Styles to Multiple Modalities in the Teaching of Computer and Information Literacy to First Year Students from Diverse Backgrounds

Pariksha Singh$^{(\boxtimes)}$ , Tania Prinsloo , and Machdel Matthee

University of Pretoria, Pretoria 0002, South Africa
pariksha.singh@up.ac.za

**Abstract.** The University of Pretoria presents an introductory computer and information literacy course to large groups of first-year students. This course was born from the pressure that all students should be computer literate and information literate to diminish the digital divide inequalities. The students come from very diverse socio-economic backgrounds with very different foundational knowledge. Current research shows that an effective way to teach such disparate groups is to match the diversity of students with a variety of instructional modalities. This research reports on the findings from a survey completed by 1 289 computer and information literacy students, to determine if specific students prefer specific learning styles or prefer distinct learning styles for certain curriculum sections. The results reveal that learning styles differ significantly across content areas in digital and information literacy education, such as theory and practicals. In addition, in general, students prefer a variety of different learning styles, which points towards the use of multiple modalities in teaching instead of focusing on individual students' learning styles.

**Keywords:** Information and computer literacy · Multiple modalities · Learning styles · Diverse backgrounds

## 1 Introduction

In a first-year computer and information literacy course at the University of Pretoria, approximately 9 000 students from diverse backgrounds are registered. Some students had Computer Application Technology (CAT) at school; however, this is a very small percentage of students, others come from areas without electricity, and they have never used a computer before. We wanted to make their learning experience as beneficial as possible; therefore, we considered different possible interventions in structuring the content. Learning styles showed promise because we targeted visual, aural, reading/writing, and kinesthetic learners.

We conducted a survey to determine if specific students prefer specific learning styles or prefer distinct learning styles for certain curriculum sections. We found that

T. Antipova (Ed.): DSIC 2021, LNNS 381, pp. 95–106, 2022.
https://doi.org/10.1007/978-3-030-93677-8_9

students, in general, prefer a variety of different learning styles, which points towards the use of multiple modalities in teaching instead of focusing on individual students' learning styles.

## 2  Background

South Africa is rich in its gold and diamond mining; unfortunately, it still faces many socio-economic problems such as high unemployment and poverty rates, social inequalities and degraded public services [1] To date, these issues are still prevalent in South Africa and affect the quality of life. While South Africa grapples with many inequalities, education should be a priority. Education will help improve human conditions and the 'country's economy [2, 3].

The University of Pretoria (the largest residential university in South Africa), introduced an introductory course – Academic Information Management (AIM) to a massive cohort of first-year students in 2003. The cohort of students in the AIM course has grown from 5363 in 2003 to 9000 in 2021. Students reside in nine different faculties across the university and come from very diverse backgrounds. Some students start university without ever seeing a computer and are expected to use a laptop in this Covid-19 era successfully. One purpose of AIM is to ensure that all first-year students can use a computer and be digitally literate. Twenty computer laboratories across three campuses consisting of over 1200 computer stations serviced the running of approximately 200 sessions of AIM every week when the university had face-to-face classes. All face-to-face lectures have now moved to the online Blackboard Collaborate environment where a stable internet connection and reliable smart device is the bare necessity [4]. However, many students do not live in places where internet connectivity is reliable and cannot afford a laptop to continue online education. Some students live in areas where there is no electricity or running water. The socio-economic factors play a critical role in 'students' success [1]. The university provided 2000 loan devices, campus accommodation for students most affected, and free data for Blackboard and third party systems in use to address these obstacles during the pandemic.

The content covered in AIM caters to novice learners to learn Windows, Computer Concepts, Word, Excel, PowerPoint, and Information Literacy concepts. These skills are considered essential for all students [5]. To cater to the large cohort of students, AIM uses Skills assessments systems and tools from a Learning Management System and interactive e-textbooks to aid in successfully running the modules.

## 3  Literature Review

### 3.1  Teaching Large Groups of Students with Diverse Backgrounds

In South Africa, higher educational institutions educate students to minimize the skills shortage and accomplish quality graduates through innovation and research [6]. In an attempt to help alleviate poverty and encourage access to higher education, quality is sometimes sacrificed [7, 8]. However, even achieving access has not been very successful as the learners, especially those who come from disadvantaged backgrounds, tend to drop out. The majority that remains in the system takes longer to complete their journey with unsatisfactory grades [9]. A critical problem affecting the quality of education is the underprepared learners entering universities without prior basic skills [9]. These students need the motivation to pursue higher education to be successful, evident in high dropout rates and longer completion times [6, 9]. The development of South Africa's economy lies primarily in higher education; large classes are a reality for higher educational institutions and threaten the quality of education. Limited funds and resources add to higher educational institutes' pressure to increase class sizes [10].

Large classes with very diverse students and very different language backgrounds lead to poor performances [11]. The challenge, therefore, lies in maintaining the academic standard while meeting each student at their individual skills levels. In such diverse student groups, researchers consider it good practice to match the diversity of students with a variety of instructional modalities [12].

### 3.2 Computer and Information Literacy Teaching

Computer Literacy is defined as "the knowledge and ability to use computers and related technology efficiently, with skill levels ranging from elementary use to computer programming and advanced problem solving"......Digital Literacy is a branch of computer literacy and is described as an "inter-related set of skills or competencies necessary for success in the digital age" [13].

A computer literacy course is essential to help bridge the digital divide at South African universities, for example, 82% of students from disadvantaged backgrounds in Limpopo, a rural area of South Africa, have access to a computer only when they enter university in their first year [14]. A study by Naidoo and Raju [15] identified that 66% of Extended Curriculum Program students at another university in South Africa did not have enough ICT skills for university and computer literacy training was essential for all students.

At the University of Pretoria, a computer and information literacy course had a traditional face to face method for teaching, learning and assessments prior to Covid-19. A hybrid method was introduced in March 2020 due to the Covid-19 pandemic [4]. The compulsory computer and information literacy course is used to bridge the digital divide of the South African students in their first year.

A study by Taylor, Goede, and Steyn [16] from a university in South Africa described the move from the traditional face to face approach to an e-learning environment approach. Support was available to students in a computer lab if it was required. All resources for computer literacy was readily available via the e-learning environment; however, students found this method of learning difficult.

### 3.3 Learning Styles

The process of accumulating information is characterized as learning; therefore, the main goal of teaching is to make the learning process easier [17]. Part of this procedure entails gaining an understanding of students learning behaviours. As a result, learning styles have become a popular topic in recent literature [18] with various styles being proposed to understand the dynamic process of learning [19]. A learning style is defined as a manner of thinking, processing information, and demonstrating learning that is preferred or ideal for an individual [19].

VARK is one of the most widely used learning model among the various learning style theories put forward, this model divides students into groups based on their sensory qualities [20]. In 1987, Fleming created this system to help students and others understand their specific learning preferences. The VARK model contained the following elements mentioned in Fig. 1 below.

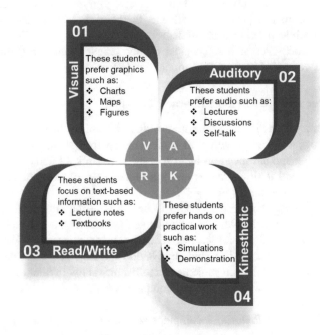

**Fig. 1.** The VARK model of learning styles (Khurshid, 2020).

VARK is an acronym for visual (V), aural/auditory (A), reading/writing (R), and kinesthetic (K), which are the modalities covered in Fleming's model of learning [21]. Visual students prefer graphics such as charts, maps, and diagrams. Students who are aural/auditory prefer lectures and discussions, while those who are read/write prefer text-based information, and those who are kinesthetic prefer hands-on practical work [20].

Fleming created a questionnaire for students to fill out to test the system and discover which group they would learn [21] Students could respond to each question with more than one option, resulting in a four-score profile, one for each modality [21]. The purpose of these questionnaires is to assess a student's sensory preferences for processing information [20].

### 3.4    The Use of Multiple Modalities in Teaching

According to Chaudry et al., [22] VARK is a learning system that encourages students to reflect on their learning styles and to develop a sense of awareness and motivation to learn. However, students rarely have a single learning style, and they may need more than one sensory modality to process information. Multimodal learners can be classified as bimodal, trimodal, or quad-modal, with various learning styles to choose from [20].

Multimodal learning occurs when all sensory modalities are engaged during learning. According to Bouchey et al., [23] this learning type enhances students' understanding and memory. Therefore, they can experience learning in several ways when combined, resulting in a diversified learning style. With the current improvement in learning environments, instructional materials can deliver in more than one sensory mode to cater to multimodal learning [23].

The study conducted by Chaudry et al., [22] revealed that dentistry students favored various learning modalities, with visual and kinesthetic preferences dominating in some groups. Even though the preclinical courses of this study believed that using a single or several modalities did not affect their learning performance, the clinical classes believed that using different learning modalities could increase performance. Therefore, multimodal learning techniques were favored in this study because they provided greater comprehension to the students [22]. Mirza and Khurshid [20] performed a comparative study on the impact of VARK learning model in higher education. The results of this study revealed that the VARK model only gives information on sensory modalities, not on the strengths and weaknesses of students or the use of facilitators. The literature emphasized that learners have inconsistent learning preferences. Recently researchers argue against individuals having specific learning styles. Instead they argue that use of multiple modalities in teaching as ideal for all learning brains [24, 25]. Therefore, if facilitators adopt multimodalities in their classes, all student needs can be addressed [20].

### 3.5    Research Questions

Do digital literacy students from diverse backgrounds have learning style preferences?

Do digital literacy students from diverse backgrounds prefer distinct learning styles for certain curriculum sections?

### 3.6    Research Methodology

We created an online survey for the students based on the VARK learning styles model. We made it available for students to then complete and submit the survey. The whole process was voluntary. Out of the 9 031 students, 1 289 students completed the survey, a response rate of 14,27%.

Data collected with the survey questionnaire was analyzed using exploratory analysis. Descriptive statistics refers to "statistically describing, aggregating and presenting the constructs of interest or associations between these constructs" [26] and was used to explain what was going on in the data. The descriptive data included the number of students choosing each specific option. We also conducted cross-tabulations, where we included a Chi-square test in order to evaluate if the groups have significant different results. We tested at a 5% level of significance. If the p-value is less than 0.05 then it indicates that there is enough evidence to show a significant difference. For this study, SPSS Version 27 was used.

# 4    Results

## 4.1    Main Findings

Considering the benefits and drawbacks of each of the four learning styles, it has been discovered that the majority of students use more than one learning style. Six questions were selected from the survey to investigate the students' learning styles. From the results 56% favored the Blackboard Collaborate sessions as a resource to master the AIM content, as indicated in Fig. 7 below. The Blackboard Collaborate sessions typically involve more than one modality. The other 44% preferred using either the textbook (26%), online PowerPoint videos (13%), Discussion board, and none of them referred to other sources to master the content (Fig. 2).

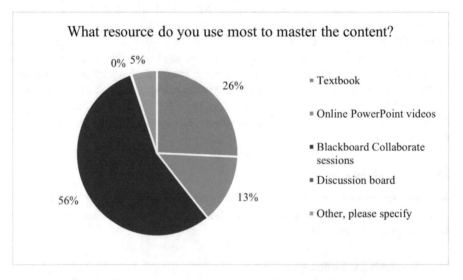

**Fig. 2.** The resources used most by students to master the content.

In the study, the majority of the students (N = 567/1289) preferred the Blackboard Collaborate sessions as they felt they learnt better when their lecturer explained the content, while some students (N = 546/1289) chose to use the textbook in conjunction with the Blackboard Collaborate sessions as they needed to see the lecture whilst making their notes. However, some students (157/1289) did prefer using only the textbook.

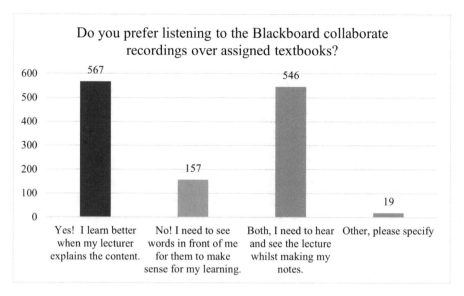

**Fig. 3.** Students indicated whether they prefer to listen to their recordings or study their textbooks.

When studying for the semester, 58% of students found themselves reading the textbook for the Navigating Information Literacy (NIL) content, 21% used the online PowerPoint videos, and 18% used the Blackboard collaborate sessions, and only 3% used other learning tools (Fig. 4).

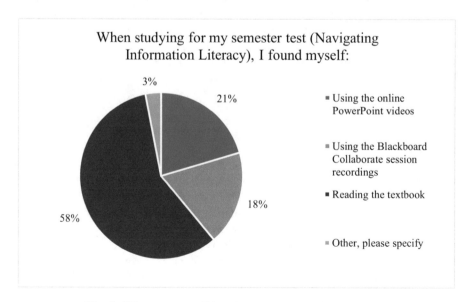

**Fig. 4.** What resources did students use to study for their tests?

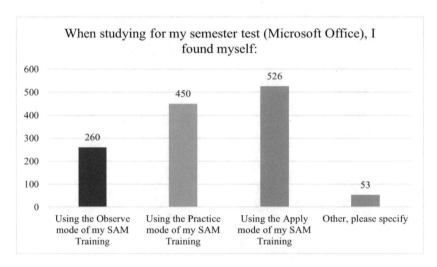

**Fig. 5.** The resources students used to master the content for the semester test.

For MS Office, the students (N = 526/1289) mostly learned through practical work, using the application mode of the SAM training. However, some students (N = 450/1289) also used the practice mode, leaving only 260 students who used the observe way for the SAM training.

For the practical component of NIL Chapter 3 searching, 675 students preferred the Blackboard Collaborate sessions and only 36 used other techniques to understand the practical searching content.

With reference to cross-tabulation, the majority of students (573) enjoyed using the Blackboard Collaborate sessions to learn the Microsoft Office content; using this platform, an interactive exercise was used to teach the content, students were asked to split their screen so that they could see, hear and practice the content. Blackboard Collaborate sessions were also the preferred platform for mastering the NIL content by 144 students.

## 4.2    Discussion

### Moving Towards Multiple Modalities
This comparative analysis reveals that learning styles differ significantly across content areas in digital and information literacy education, such as theory and practical. According to Mirza and Khurshid [19] students do not usually have a single learning style, and they may require more than one sensory modality for information processing. As a result, students can be classified as bi-modal, tri-modal, or quad-modal in several different ways. In this study, 56% of the students preferred the Blackboard Collaborate sessions to master the AIM content, as seen in Figs. 3, 6, and 7. The Blackboard Collaborate platform contained rich, high-quality video and audio that enabled a more engaging learning experience aligned with visual and audio learning modalities. The

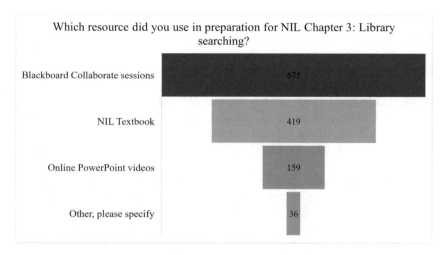

**Fig. 6.** The resources students used to study for a specific section of NIL.

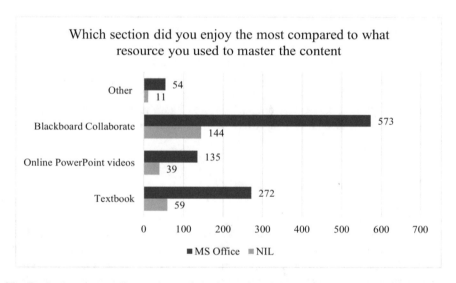

**Fig. 7.** Students responding to the sections they enjoyed the most compared to what resources they used to master the content.

medium had interactive tools to enforce student participation via real-time annotations and texts. This platform, therefore, covered the kinesthetic modality as well.

Students also preferred the Blackboard Collaborate sessions for studying the NIL content as lecturers presented the content verbally using a presentation tool. Therefore, students could not only hear but also see the lecturing content. The use of learning with multiple modalities was also used to prepare for the NIL Chapter 3: Library searching. Of the 1289 students who answered the survey, 675 learned using the Blackboard Collaborate sessions.

The second most favored learning resource used to master the content was the textbook, 26% of students preferred this learning tool as it contained animations to help explain the concepts, videos that showed scenarios to highlight real-world issues adapted for the visual learning style. The audio podcasts and media-rich content stimulated critical thinking to reinforce study skills and encourage critical thinking. The textbook mainly was adapted towards the read and write learning style as it contained activities, self-assessment exercises, and glossary pop-ups. The kinesthetic aspect of learning was addressed through the use of interactive infographics. Students (546) also preferred using the Blackboard Collaborate sessions recording in conjunction with the textbook.

The Online PowerPoint videos were mainly adapted towards the visual and audio learning styles, and 13% of the 1289 students that answered the survey preferred this as a learning tool to master the content. These videos engaged the students with interactive infographics. Narration and background music was used to explain the content to the students. These students were considered bimodal learners, and they required visual and auditory learning.

The Discussion board was not a preferred method used to master the content as it was only adapted to the read and write learning style. However, when it came to learning with a single learning style, students used kinesthetic learning for the Microsoft Office content (practical). The practice and apply mode of the SAM training was mostly preferred by students, as seen in Fig. 5 above. The observed mode in the SAM training was ideal for visual and auditory learning, and 226 students chose this type of learning. The practice mode was more aligned to auditory and read and write modalities. The apply mode was adapted towards kinesthetic learning, and from the results, 526 students used this mode when studying for the practical component of the semester test.

## 5   Conclusion and Future Research

For Microsoft Office, learning requires practice, and practice lends towards the kinesthetic learning style. From the study performed, it is noted that students learn from more than one modality but also prefer a particular modality when it comes to specific content. This pattern is also shown in other studies, where students need to perform practical elements learn better with the kinesthetic modality. To cater to all students learning needs, higher education should invest in creating policies and procedures for the inclusion of multiple-modular curriculums. This type of curriculum will lend to an inclusive classroom for all students.

## References

1. Bayat, A., Louw, W., Rena, R.: The impact of socio-economic factors on the performance of selected high school learners in the western Cape Province, South Africa. J. Hum. Ecol. **45** (3), 183 (2014)

2. Tapscott, C.: South Africa in the twenty-first century: governance challenges in the struggle for social equity and economic growth. Chin. Polit. Sci. Rev. **2**(1), 69–84 (2017). https://doi.org/10.1007/s41111-017-0055-1
3. Van Zyl, A.: The contours of inequality: the links between socio-economic status of students and other variables at the University of Johannesburg. J. Student Affairs Africa **4**(1) (2016). https://doi.org/10.14426/jsaa.v4i1.141
4. Prinsloo, T., Singh, P.: COVID-19: leapfrogging 8,000 students from face-to-face to online learning in three weeks. Commun. Assoc. Inf. Syst. **48**(1), 9 (2021)
5. Ranaweera, P.: Importance of information literacy skills for an information literate society. Mousaion **1**(39), 1 (2008)
6. Fisher, G., Scott, I.: Closing the skills and technology gap in South Africa background paper 3: the role of higher education in closing the skills gap in South Africa (2011)
7. Badat, S.: The challenges of transformation in higher education and training institutions in South Africa (2010)
8. Scott, C.L., Ivala, E.N.: Transformation of Higher Education Institutions in Post-apartheid South Africa. Routledge (2019)
9. Tewari, D.D., Ilesanmi, K.D.: Teaching and learning interaction in South Africa's higher education: some weak links. Cogent Soc. Sci. **6**(1), 1740519 (2020). https://doi.org/10.1080/23311886.2020.1740519
10. Hornsby, D.J., Osman, R., De Matos-Ala, J.: Large-class pedagogy: interdisciplinary perspectives for quality higher education: African sun media (2013)
11. Mulryan-Kyne, C.: Teaching large classes at college and university level: challenges and opportunities. Teach. High. Educ. **15**(2), 175–185 (2010)
12. Abbott, M.R., Shaw, P.: Multiple modalities for APA instruction: addressing diverse learning styles. Teach. Learn. Nurs. **13**(1), 63–65 (2018)
13. Tejedor, S., Cervi, L., Pérez-Escoda, A., Jumbo, F.: TDigital literacy and higher education during COVID-19 lockdown: Spain, Italy, and Ecuador. Publications **8**(4), 48 (2020). https://www.mdpi.com/2304-6775/8/4/48
14. Oyedemi, T., Mogano, S.: The digitally disadvantaged: access to digital communication technologies among first year students at a rural South African university. Afr. Educ. Rev. **15**(1), 175–191 (2018). https://doi.org/10.1080/18146627.2016.1264866
15. Naidoo, S., Raju, J.: Impact of the digital divide on information literacy training in a higher education context. South Afr. J. Libr. Inf. Sci. **78**(1), 34–44 (2012). https://doi.org/10.10520/EJC129280
16. Taylor, E., Goede, R., Steyn, T.: Reshaping computer literacy teaching in higher education. Interact. Technol. Smart Educ. **8**(1), 28–38 (2011). https://doi.org/10.1108/17415651111125496
17. İlçin, N., Tomruk, M., Yeşilyaprak, S.S., Karadibak, D., Savcı, S.: The relationship between learning styles and academic performance in TURKISH physiotherapy students. BMC Med. Educ. **18**(1), 1–8 (2018)
18. Cimermanová, I.: The effect of learning styles on academic achievement in different forms of teaching. Int. J. Instr. **11**(3), 219–232 (2018)
19. Pritchard, A.: Ways of Learning: Learning Theories for the Classroom. Routledge (2017)
20. Mirza, M.A., Khurshid, K.: Impact of VARK learning model at tertiary level education. Int. J. Educ. Pedag. Sci. **14**(5), 359–366 (2020)
21. Fleming, N., Baume, D.: Learning styles again: VARKing up the right tree! Educ. Dev. **7**(4), 4 (2006)
22. Chaudhry, N.A., Ashar, A., Ahmad, S.A.: Association of visual, aural, read/wite, and kinesthetic (VARK) learning styles and academic performances of dental students. PAFMJ **70**(Suppl-1), S58–S63 (2020)

23. Bouchey, B., Castek, J., Thygeson, J.: Multimodal learning. In: Innovative Learning Environments in STEM Higher Education: Opportunities, Challenges, and Looking Forward, pp. 35–54 (2021)
24. Newton, P.M., Salvi, A.: How common is belief in the learning styles neuromyth, and does it matter? A pragmatic systematic review. Frontiers **5**, 270 (2020)
25. An, D., Carr, M.: Learning styles theory fails to explain learning and achievement: recommendations for alternative approaches. Pers. Individual Differ. **116**, 410–416 (2017)
26. Woehr, D.J., Loignon, A.C., Schmidt, P.B., Loughry, M.L., Ohland, M.W.: Justifying aggregation with consensus-based constructs: a review and examination of cutoff values for common aggregation indices. Organ. Res. Methods **18**(4), 704–737 (2015)

# Teaching, Assessing and Promoting Lifelong Learning – A Case of First Year University Students, in a Developing Country During a Pandemic

Komla Pillay$^{(\boxtimes)}$ ⬤ and Pariksha Singh ⬤

University of Pretoria, Pretoria, South Africa
{komla.pillay,pariksha.singh}@up.ac.za

**Abstract.** The Covid-19 pandemic resulted in a major technological disruption for Universities. This paper addresses the reflections of the staff regarding teaching, assessing and lifelong learning for the first year "Pastel for Accounting" module during the COVID-19 pandemic. The Covid-19 pandemic forced universities to move from face to face teaching and assessing to a purely online mode of delivery. This paper uses the Reflective Practice Cycle to provide insight into the teaching and learning experience for the Pastel module, from the perspective of the teaching team. The insight provided included the 2020 academic year when the initial disruption took place. The insight then continued on to the 2021 academic year where changes were made for the module to be fully online. The paper further discusses the ways in which lifelong learning was encouraged in the renewed online module. The outcomes of this research can assist other modules at other universities with possible ideas of how to conduct a fully online service module. Future research should investigate the results of the expected changes for 2022.

**Keywords:** Covid-19 · Reflective practice cycle · Lifelong learning

## 1 Introduction and Background

COVID-19 had a tremendous impact on higher education institutions [1]. This paper addresses the reflections of staff regarding teaching, assessing and lifelong learning for the first year "Pastel for Accounting" (Pastel) module during the COVID-19 pandemic. The Covid-19 pandemic forced universities to move from face-to-face teaching and assessing to a purely online mode of delivery. Universities worldwide had to look for and adopt suitable online technologies to ensure all module goals were achieved. South African universities had to do this with many more additional challenges present in a developing country. Students had to evacuate university residences and many had to return home to poor, rural communities where basic services are a challenge. Many students had to share their small personal spaces with large extended families [2]. Students had to find ways in which to continue their online education and the University had to facilitate the online learning process taking into consideration the diversely resourced student population.

T. Antipova (Ed.): DSIC 2021, LNNS 381, pp. 107–117, 2022.
https://doi.org/10.1007/978-3-030-93677-8_10

"Pastel for Accounting" Students (INF183) is a compulsory first-year module with a student cohort of more than 2000. The purpose of the module is to introduce students to an accounting software tool to assist in managing a business. Students are not introduced to new accounting terminology or processing, but rather for the students to practice what they have already learned in their accounting module, using the Pastel accounting software. The 2020 academic year had a student cohort of 2211, and the 2021 academic year has a student cohort of 2247. In 2020, students had face to face classes and physical demonstrations for the first term of the year with the rest of the year being online. Students had sufficient training to use the university Learning Management System (LMS) and to use the Pastel accounting software. In 2021, all teaching, practical session and assessments were online. This student cohort did not see the campus, some did not have any computer skills and they had to install the software on their own.

## 2    Research Approach

This paper qualitatively reflects on the teaching, assessing and promotion of lifelong learning during the COVID-19 pandemic. This paper used the Reflective Practice Cycle framework to analyze the Pastel module during the Covid-19 pandemic. This paper also used the six ways to build lifelong learners, as outlined by Wabisabi Learning, to explain how the renewed Pastel module was presented to ensure the incorporation of lifelong learning. The research questions addressed in this paper are:

1. What are the reflections of the teaching team prior to and during Covid-19?
2. How did the Pastel module aim to create a lifelong learner during online teaching and assessment during Covid-19?

## 3    Literature Review

### 3.1    Reflective Practice Cycle

The researchers aimed for critical reflection in order to engage in reflective practice. The chosen reflective model was the Reflective Practice Cycle. This reflective cycle starts with simple reflection and progresses on to reflective practice. At the beginning, the focus is mainly on objective, descriptive, single-loop learning, thereafter shifting to more subjective, analytical, double-loop learning practice. The focus is on practitioners reflecting on "assumptions, values and paradigms making sure that they are not bound to the existing way of framing problems" [3]. The analytical stage progresses onto critical reflection which is more interactive and multi-facetted. The reflective practice cycle was adapted from the work of Kolb (1984) and Gibbs (1988). Kolb laid the foundation for the general view of experiential learning; while Gibbs built on this with reflective practice. The framework's main focus is on reflective practice, on a continuum from simple to critical reflection. It is important to clarify the level, extent, form and context, of the reflective practice, particularly in terms of achieving the planned results. There needs to be a clear purpose to such reflective practice. Figure 1 outlines the process followed for reflective practice with step 1 being "Reflect on and describe the initial learning situation or knowledge"; step 2 being "Analyse & evaluate the learning situation or knowledge"; step 3 being "Reflect on & discuss the current learning situation or knowledge"; and step 4 being "Identify the future learning situation or knowledge". The detail and questions to be addressed for each step is outlined in Fig. 1. Step 4 does not mean that the process has ended, there can be a continuation back to step 1.

Reflective practice is "learning through and from experience towards gaining new insights of self and practice" [4]. Reflection is a core part of teaching and learning. The purpose is to make one conscious of one's professional knowledge by "challenging assumptions of everyday practice and critically evaluating practitioners' own responses to practice situations" [4].

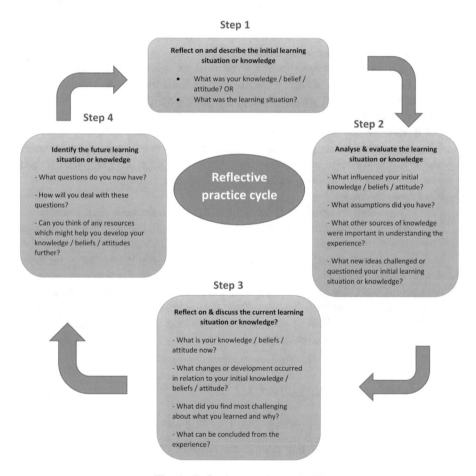

**Fig. 1.** Reflective practice cycle [3]

### 3.2   Six Ways to Create a Lifelong Learner

The researchers used Wabisabi's six ways to create a lifelong learner to redesign the Pastel module.

1. **Encourage Learning Ownership**
   Students are essentially responsible for their own learning. Once students finish University, students need to learn on their own. Students must be given the opportunity to learn on their own, this is the only time when, what they learn will stick with them. Educators are encouraged to drive independence, greater self-esteem and pride in one's own success as rewards for encouraging learning ownership.
2. **Turn Mistakes into Opportunities**
   Learning from ones mistakes is an important lifelong learning skill to master. Mistakes enable a better understanding of the problem, enabling improved ways of thinking and problem solving. Learners are inherently tough and fragile at the same time, they need to be shown that mistakes are opportunities.
3. **Stash a few Go-To Learning Tools**
   Learners have different tools that they use in order to aid their learning. This can range from constant repetition to more creative tools like creating rhymes. More modern tools like blogs, podcasts and online discussions can also foster a hunger for knowledge acquisition.
4. **Let them take the teaching reins**
   It has been scientifically proven that learning retention is improved when imparting one's own knowledge to someone else. Create a system to foster peer to peer learning, promoting learning ownership and knowledge sharing.
5. **Find Time to Play**
   Play is important to learning. If learning is fun and enjoyable, then learners will have positive associations with acquiring knowledge and creating new knowledge. There is much personal growth to be achieved if learners positively associate learning with an adventure of sorts, instead of a compulsory task forced on them.
6. **Set Learning Goals**
   All learning activities have an end goal in mind. This end goal must be clear to the learner. Furthermore, the value of reaching the end goal must have meaning and be useful to the learner. Setting goals is a vital lifelong learning skill that motivates learners to acquire new knowledge [5].

## 4   Analysis and Findings Part 1

### 4.1   Our Created Rainbow - Reimaging Teaching and Learning for Pastel

This section analyses the teaching prior to and during COVID-19 using the Reflective Practice Cycle.

*Step 1 – Reflect on and describe the initial learning situation or knowledge*
The lecturing team held the belief that classes needed to be physical and structured with student oriented activities. They further emphasized the time students spent on a task in a controlled disciplinary focused environment.

The Learning situation in 2020 was that students had face to face classes and physical demonstrations for the first term of the year. Resources required for the module were provided – the physical textbook together with the required software needed to do the practical sessions. A strict timetable was followed where students attended their chosen timeslot routinely every week. Students had access to the University's computer labs or alternatively could have the software installed on their personal devices. IT support was provided on campus for the labs and the personal devices.

Student queries were typically handled during class time with an option to email the lecturing team or physically consult during prescheduled consultation hours.

*2 – Analyse & evaluate the learning situation or knowledge*
The COVID-19 pandemic was the organizational disruptor that altered the 2020 learning situation for Pastel. The University had no other option but to offer all modules in an online environment. Due the high number of students registered for this module, even when the University began allowing selected students onto campus, Pastel was not afforded that opportunity due to the student numbers being in excess of 2000. The pandemic allowed the teaching team to assess and rework their initial belief on how the course needs to be structured. The course was structured in a physical environment as it has always worked in that manner. It was also assumed that due to the University being in a developing country, it would not be possible to migrate to online due to the anticipated infrastructure challenges.

*3 – Reflect on & discuss the current learning situation or knowledge?*
After the 2020 lockdown, many students did not have devices and left textbooks at the university residence not realizing they will not return to campus. The module was not designed for online, interim remedies and problem patching continued for the rest of 2020 in order to get the students through the module. Assistance was provided to students for the remainder of the 2020 academic year, in the form of loaned laptops, a scanned electronic textbook and sourced generic practice software videos from Pastel. In 2021, all teaching and learning moved to an online only environment. In 2021, students received devices with the NAFSAS bursaries (Government funded loan and bursary scheme). These devices were able to be acquired prior to the start of the academic year. All resources, interactive textbooks and software was available on the University's LMS. The schedule changed to meet the flexible nature of online learning. The look and feel of the LMS was also updated to feel more welcoming and provide more information. Navigation bars were more descriptive, icons were added to better convey information in an easy to understand manner. The number of Blackboard Collaborate sessions (live lecturing and tutoring sessions conducted on the LMS) were increased to build more flexibility into the module. Students were allowed to attend any number of collaborate sessions, whereas in the physical teaching environment, students could only attend one lecturing session that they signed up for. Recordings were available for students that missed classes.

All teaching resources were also only available online via email, classes and online discussion boards. Primary communication moved to the discussion board communication tool instead of individual emails, this allowed students to receive a faster response and there were FAQs that students could refer to.

*4 – Identify the future learning situation or knowledge*
The future learning situation involves moving to a cloud version of the accounting software Pastel – this has been resolved to be the case for 2022. The online technical support proved to be more difficult with students using a varying list of devices, the Apple Macbooks were not compatible with the 2021 version of software released for students to use. The cloud software is designed for all different platforms and operating systems. The teaching schedule will also change to better accommodate online teaching, by using "spacing", teaching small chunks of information at a time, repeated for a student centered approach.

Lecturing staff are now more open minded to varying teaching avenues. New approaches are bound to experience challenges with initial roll out – time and effort can resolve these challenges ultimately resulting in a more efficient teaching and learning solution.

## 4.2    The Testing Challenge – Assessing the Pastel Module

This section analyses the assessment prior to and during COVID-19 using the Reflective Practice Cycle.

*Step 1 – Reflect on and describe the initial learning situation or knowledge*
In 2020, term 1, practical and theoretical assessments were done during class time. Assistant lecturers and IT Lab personnel were physically available to administer and assist to resolve any technical issues during the test. Some aspects of the test were automatically set to be marked by the system and other questions were marked electronically by the lecturers. The lecturing staff believed this to be an effective assessing method as the possibility of cheating was eliminated. This was a time consuming way of assessing, as the tests needed to be conducted during an entire week as this way of testing required a lot of physical, technical and human resources. The original question bank mostly tested the surface learning of the student.

*2 – Analyse & evaluate the learning situation or knowledge*
The initial way of testing was a tried and proven method. However, due to the COVID-19 pandemic limiting access to campus for students, alternate ways of testing had to be investigated. The original database of questions was easily accessible to students in an online environment. When tests were done physically on campus, this did not pose an issue as students did not have access to the Internet while doing examinations. The database of questions were re-engineered for online assessing, the renewed questions required students to demonstrate a higher order level of thinking. The renewed questions were original scenario-based which did not allow students to easily find answers on the Internet. The students needed to apply their mind to the given scenario. Different scenarios were used and questions were randomized.. It was thought that this will prevent students from attempting to cheat.

*3 – Reflect on & discuss the current learning situation or knowledge?*
The scenario-based testing expanded the thinking and expectations of the students. While the scenario-based testing did assist in alleviating cheating in the online assessments, it did not completely eradicate it. The University takes any form of cheating very serious, to the point of expelling students from the University. But proving cheating allegations or suspicions become a very challenging task in an online environment.

*4 – Identify the future learning situation or knowledge*
The questions that the teaching team have with regard to assessment is if the students are all doing their own assessments for the entire duration of the assessment. The teaching staff has therefore started the process of investigating the use of "remote proctoring technologies" to ensure that all exams are free of cheating. Remote proctoring technologies enable live video monitoring, artificial intelligence proctoring and browser lockdown. Live video monitoring allows student actions to be continuously recorded during the exam to check authenticity. Artificial intelligence proctoring allows students to be photographically identified and verified. Browser lockdown enables the lockdown to browsers like IE and Chrome to prevent students from printing, copying, going to other websites, or accessing other applications during the exam session [6]. The university has already started trialing this software in some modules. The teaching team feel that this proctoring software together with the cloud based assessing will ensure the rigor of the exams as if they were taking place physically on campus. Investigation is continuing with the use of both the cloud based accounting package solution and the proctoring software working together.

Section 4 answered the first research question: What are the reflections of the teaching team prior to and during Covid-19?

## 5   Analysis and Findings Part 2

### 5.1   Lifelong Learning – The Way Forward

This section discusses how the online Pastel module addresses the six ways to create a Lifelong Learner as explained in the Literature Review. The dynamics of creating a lifelong learner differ in a physical and online environment. In keeping with attaining, Sustainable Development Goal 4, promoting lifelong learning, the teaching team ensured the adherence to creating lifelong learners during the reflection process.

1. **Encourage learning ownership**
   The teaching team encourage learning ownership by offering choices, encouraging voices and leadership [7]. The university's LMS also supports learning ownership. It is user friendly and easy to use [8]. Blackboard is especially useful in terms of accessibility of unit materials [9].

   *Build ownership by offering choices.*
   Students can attend any scheduled lecture, any number of times. All lecture sessions are recorded and are available for download immediately after class. They can also access the recorded lectures. They get to determine when they have achieved

understanding. They can also work at their own pace as the lecturing and assessment schedules for the year, are released on the first day of the academic calendar. Students are also afforded flexibility with learning, as they can download module content and class recordings at any time. The time available for the completion of short assessments is one week, they can fit this in with their own personal and university schedules.

The Discussion Board is available 24 h a day, 7 days a week, students can post messages at any time and will be answered within a 24 h turnaround time. Students will also have access to questions already answered on the Discussion Board which very often already answers the question that the student had. One on one consulting is also available where students need to book a scheduled slot. Students are able to use the always available resources, like videos, for revision or enforced learning.

*Build ownership by encouraging voice.*
The teaching allow students to make the choice of attending online sessions, listening to the recording or using the videos of each lesson. Pre-recorded videos are available at the start of the academic year. Advanced students can use them to work ahead, novice students can use them for reinforcement of content. Other students can use it for revision. Students are strongly encouraged in class and online announcements to have their voices heard. Various avenues exist for them to achieve this, use of chat or mic function in the live classes, use of 24/7 online Discussion Board or by emailing the necessary support and academic staff.

*Build ownership by encouraging leadership.*
Students are the future leaders. Most of the present generation embrace leadership opportunities when asked to present in class or share their work on their virtual screens. The teaching team also uses the LMS feature of breakout rooms to encourage small groups of peer learning and assessment.

2. **Turn Mistakes into opportunities**
Students are afforded the opportunity to join working groups that experienced problems with the same sections. These opportunities are facilitated by online revision sessions that focus on specific problems at a time. Peer to peer teaching also assists in this regard. In the physical environment, stronger students would be allocated to mentor students that have experienced hurdles. In the online environment, this process is still being fine-tuned. The plan is to group students that experience the same problem into the same online breakout session, then get a student who is academically strong in that section to facilitate the session. There are some online challenges with this, the availability of the affected students versus that of the proposed facilitator.

Blackboard's (LMS) adaptive release function also allows students with below average performance to complete additional assignments – progression through the content is only permitted once a certain mark has been achieved for previous assignments. This function does not only have to be used to support at-risk students but can be incorporated as the normal course of module content release. The LMS provides good peer support and peer coaching which contributes to enhancing peer learning. Furthermore, it is used as a tool to facilitate student-centered learning and it promotes lifelong learning and active engagement [10].

3. **Stash a Few Go-To Learning Tools**

Blackboard allows more modern tools like blogs, podcasts and online discussions to foster the students' hunger for knowledge acquisition. Student blogs are encouraged especially for the resolution of the technical difficulties experienced with the software installations and features. The online Discussion Board is addressed daily, quickly clarifying any misunderstandings that students may have. Students also independently find YouTube videos that address their issues, they then share these resources with other students who find themselves in similar situations.

Whatsapp groups have also proved to be an efficient learning tool. It is easily accessible and almost all students have the mobile application. Disclaimers are provided by staff that these resources are used at their own risk. Despite the disclaimer in place to protect the university from any parent or student legal action, the sharing of these resources are not in any way discouraged.

4. **Let them take the teaching Reins**

Students are constantly encouraged to take the teaching reins – explaining the easy to grasp concepts during the online sessions. This builds their confidence to more easily share their thinking. Students are also divided into small groups using the LMS's breakout room function where peer to peer learning takes place. Each group is assigned an exercise to which they must use the literature to solve. At the end of the breakout sessions, students return to main teaching room, where each group will share and justify to the class, their solution.

Much planning also goes into the flipped classroom approach exercises where students get to explain smaller chunks of teaching material, with the lecturer steering the course of events explaining the complex tie in content.

Students also value the connections made with other students which is facilitated by the LMS [11].

5. **Find Time to Play**

Kahoot is a game based learning platform that allows student engagement [12]. The teaching team makes use of Kahoot to interact with students and to also get an idea if concepts taught during class was actually understood. No marks are allocated to playing this game, it is just a fun exercise with no expectations. Lifelong learning is enhanced or promoted if the learning outcomes of two or three modules are simultaneously assessed. Work is in progress with the AIM first year module that teaches digital literacy to combine some or part of the assessment. An idea on how this can be achieved, develop a game to be played solving a particular accounting dilemma, at the same time, the students navigation and use of the game can determine if students are in fact digital literate, which is the outcome of the AIM module.

6. **Set Learning Goals**

This module built into the schedule released at the start of the academic year, the learning goals expected for each teaching week. Apart from the learning goals being specific per week, administrative goals are also outlined. An example of a snippet of the schedule is depicted in Fig. 2.

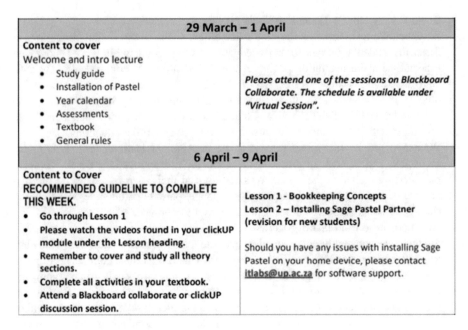

**Fig. 2.** Example of learning goals from the schedule [13].

This Sect. 5 answered the second research question: How did the Pastel module aim to create a lifelong learner during online teaching and assessment during Covid-19?

## 6    Conclusion

The Covid-19 pandemic resulted in a major technological disruption for Universities. This paper uses the Reflective Practice Cycle to provide insight into the teaching and learning experience for the Sage Pastel module, from the perspective of the teaching team. The insight provided included the 2020 academic year when the initial disruption took place and when the year commenced with a physical environment in mind. The insight then continued on to the 2021 academic year where changes were made for the module to be fully online. The paper further discusses the ways in which lifelong learning was encouraged in the renewed online module. The outcomes of this research can assist other modules at other universities with possible ideas of how to conduct a fully online service module. Future research should investigate the results of the expected changes for 2022, step 4 of the Reflective Practice Cycle. Future research can also extend this kind of reflective research to other institutions in developing countries to determine if the teaching teams' experience was similar. It would also be interesting to see how different the reflections of the teaching team would be from developed countries.

# References

1. Naciri, A., Baba, M., Achbani, A., Kharbach, A.: Mobile learning in higher education: unavoidable alternative during COVID-19. Aquademia **4**(1), ep20016 (2020). https://doi.org/10.29333/aquademia/8227
2. Francis, D., Valodia, I., Webster, E.: Politics, policy, and inequality in South Africa under COVID-19. Agrar. South: J. Polit. Econ. **9**(3), 342–355 (2020). https://doi.org/10.1177/2277976020970036
3. Wight, I., Kellett, J., Johannes, P.: Practice reflection learning: work experience in planner education. Plann. Pract. Res. (2016). https://doi.org/10.1080/02697459.2016.1222109
4. Finlay, L.: Reflecting on 'reflective practice'. Practice-based Professional Learning Paper 52, The Open University (2008)
5. Wabisabilearning: 6 ways to create a Lifelong Learner (2016). https://wabisabilearning.com/blogs/critical-thinking/6-lifelong-learning-skills. Accessed 8 Aug 2021
6. Melimu: Our Remote Proctoring Technologies Ensure Cheat Free Exams (2021). https://www.melimu.com/onlineexamsa/?gclid=Cj0KCQjw6s2IBhCnARIsAP8RfAhfR_cb6DDQoRDvisfYXBMph_x51fMxMWh76swARVCNR_yzxsH_53caAh6EEALw_wcB. Accessed 10 Aug 2021
7. Blackburn, B.: 3 ways to encourage ownership of learning. Middleweb (2016). https://www.middleweb.com/32620/3-ways-to-encourage-ownership-of-learning/
8. Lin, S., Persada, S., Nadlifatin, R.: A study of student behavior in accepting the blackboard learning system: a technology acceptance model (TAM) approach. In: Proceedings of the 2014 IEEE 18th International Conference on Computer Supported Cooperative Work in Design (CSCWD), pp. 457–462 (2014). https://doi.org/10.1109/CSCWD.2014.6846888
9. Heirdsfield, A., Walker, S., Tambyah, M., Beutel, D.: Blackboard as an online learning environment: what do teacher education students and staff think?. Aust. J. Teach. Educ. **36**(7) (2011). https://doi.org/10.14221/ajte.2011v36n7.4
10. Bradford, P., Porciello, M., Balkon, N., Backus, D.: The blackboard learning system: the be all and end all in educational instruction? J. Educ. Technol. Syst. **35**, 301–314 (2007)
11. Liaw, S.: Investigating students' perceived satisfaction, behavioral intention, and effectiveness of e-learning: a case study of the blackboard system. Comput. Educ. **51**, 864–873 (2008). https://doi.org/10.1016/j.compedu.2007.09.005
12. Unknown: What is Kahoot? (2018). https://kahoot.com/what-is-kahoot/. Accessed 5 Aug 2021
13. University of Pretoria: INF 183 Module Guide (2021)

# BeAwareStopHPV: Design and Usability of a Consumer-Focused Educational Web-Based System for Women Living in Saudi Arabia

Hind Bitar[1]([⊠]) and Sarah Alismail[2]

[1] King Abdulaziz University, Jeddah, Saudi Arabia
hbitar@kau.edu.sa
[2] Claremont Graduate University, Claremont, CA 91711, USA

**Abstract.** Concerning the cancer burden in Saudi Arabia, studies have pre-dicted a 6 to 10-fold increase in 2020 and 2030, respectively, compared to 2004. The most common genotypes of human papillomavirus (HPV) are 16 and 18, responsible for almost 72.4% of all cervical cancer cases in Saudi Arabia; inhibiting this infection will be fruitful in preventing the onslaught of the cancer. Although considerable effort has been devoted to exploring the knowledge, perception, and awareness of women in Saudi Arabia regarding HPV, its vac-cine, and cervical cancer, the majority of research was descriptive in nature. Consequently, there is a lack of intervention-based studies targeted at promoting knowledge and awareness of HPV and cervical cancer screening among women in Saudi Arabia. In this study, we developed and assessed the usability of the BeAwareStopHPV website to assist in increasing the levels of knowledge regarding HPV and cervical cancer among women in Saudi Arabia. Our findings yielded high usability. Before this website can be disseminated and used largely by women, it should be tested in an experimental study.

**Keywords:** HPV · Cervical cancer · Usability · Women's knowledge and awareness · Health education website

## 1 Introduction

Concerning the cancer burden in Saudi Arabia, studies have predicted a 6 to 10-fold increase in 2020 and 2030, respectively, compared to 2004 [1]. Given the country's current and estimated future burden, it is vital to prioritize cancer control. It must be specifically noted that the incidence of cervical cancer has increased to 31.4% within 10 years [1]. According to [2], cervical cancer is ranked as the 9th most common cancer among women in Saudi Arabia between the ages of 15–44. The predicted increase in cervical cancer rates in Saudi Arabia will cause resource consumption to increase, which could, in turn, increase the economic burden associated with cancer in Saudi Arabia [1].

T. Antipova (Ed.): DSIC 2021, LNNS 381, pp. 118–127, 2022.
https://doi.org/10.1007/978-3-030-93677-8_11

The most common genotypes of HPV are 16 and 18, responsible for almost 72.4% of all cervical cancer cases in Saudi Arabia [2]; inhibiting this infection will be fruitful in preventing the onslaught of the cancer. One such way is by taking HPV vaccinations [3], as they will protect Saudi people against more than two-thirds of the cervical cancer cases [4]. Another way to decrease the number, especially in developed countries, is by performing the Pap smear/Pap test, a cervical cancer screening program which assists in decreasing the mortality and incidence rates around 70–80% as reported by [5]. Pap smear screening, recommended globally for all women who are sexually active, can detect the cytological abnormalities of the cervix—that is, pre-invasive and invasive disease lesions—at an early stage, when these lesions can be managed, and progression halted [5]. Even though all these facts and benefits regarding HPV infections, cervical cancer, HPV vaccinations, and Pap smear screening are available and well-known in Saudi Arabia, there is a lack of knowledge and awareness about these diseases and their preventive methods [3, 6].

Although considerable effort has been devoted to exploring the knowledge, perception, and awareness of women in Saudi Arabia regarding HPV, its vaccine, and cervical cancer [6–10], the majority of research tends to be descriptive in nature. Moreover, most people globally, including Saudi people, use the internet to seek health information. The internet is a resource and tool that can be used to promote and seek heath information [11]. However, [11] reported that from 144 investigated web pages, providing health and medical information about epilepsy, only a few were deemed good enough in quality terms. Another study assessed the quality and readability of Arabic web-based interventions that provide COVID-19 related health information; this concluded that most of these easily accessible Arabic web-based interventions did not meet the quality standards [12]. To the best of our knowledge, there is a lack of intervention-based studies targeted at addressing barriers and promoting knowledge and awareness of HPV, its vaccine, and cervical cancer screening among women in Saudi Arabia. More specifically, there is a lack of consumer-focused, web-based interventions developed or evaluated in the country. This situation calls for more attention to be directed toward Arabic intervention-based studies that aim to educate and promote women's awareness of HPV itself, its vaccine, and cervical cancer screening tests. Thus, the main aim of this research is to develop and evaluate an Arabic, consumer-focused educational web-based system to educate women from Saudi Arabia about HPV and cervical cancer.

This research study sought to answer the following question: How useable is a newly developed Arabic consumer-focused educational website in terms of effectiveness, efficiency, and satisfaction to users? This research study fills this gap by investigating the usability of a newly developed consumer-focused web-based intervention, named BeAwareStopHPV. Following exploratory interviews with women and physicians, the BeAwareStopHPV website was developed and tested for its usability through iterative refinement based on feedback obtained from public women and physicians.

This research is significant because it can improve Saudi women's public health by increasing the levels of knowledge and awareness regarding the stigmatized illness HPV and the global health crisis regarding cervical cancer. This study presents its potential implications for both research and practice. It addresses the research gap in developing a consumer-focused technological intervention for cervical cancer control targeting Arabic speakers among Saudi women. This study adds to the growing body of knowledge on women's health education, with a specific focus on IT intervention development and prevention.

## 2 The Role of Technology

The occurrence of cervical cancer has increased in Saudi Arabia and globally. This calls for action to decrease the number of cases. One possible action that IS/IT researchers may take is to educate women about HPV and cervical cancer by using technological solutions. Designing and developing an IS/IT intervention that addresses Saudi women's need to educate themselves regarding HPV and cervical cancer is an emerging study that requires additional research. Although a growing body of evidence has called for the engagement of consumers/end users in the development of such an intervention, such an engagement has rarely been implemented in intervention-based studies [13]. We sought to ensure the inclusion of evidence-based content in an engaging, easy to use format by including potential users in the design, development, and evaluation processes.

Many studies have proven the benefits of digital technology in the healthcare sector, especially in educating people about sexually transmitted diseases (STDs), such as HPV. As stated by [14], there are several benefits of using digital health technology. For example, using technology can be a cost-effective means to enhance HPV knowledge, increase vaccination, and remove barriers. According to [15], there is still an issue with the HPV vaccine uptake level; it is still low in the Netherlands. Thus, the authors developed a tailored and interactive web-based intervention to improve the HPV vaccination acceptability and informed decision-making (IDM) that specifically targeted mothers. The authors were mainly concentrating on the evaluation of the intervention's working mechanisms, using a randomized controlled trial (RCT) experimental design. The results supported the idea that mothers who complete the intervention's tasks have a positive effect on their daughters' vaccine uptake [15]. However, to our knowledge, the authors did not build or develop this interactive web-based intervention using a consumer-focused approach. They mainly focused on providing the participants with tailored feedback regarding HPV vaccination with the aid of two virtual assistants.

[16] conducted a 3-arm prospective RCT research study to evaluate Outsmart HPV, which is a tailored web-based HPV vaccination intervention designed for young gay, bisexual, and other men who have sex with men (YGBMSM) since there is high incidence rate of anal cancer among this population. Some aspects from the protection motivation theory, information-motivation-behavioral skills model, and the minority stress model were used in Outsmart HPV framework to develop the web-based. Outsmart HPV provided the users with some information regarding HPV prevalence, HPV vaccines, HPV schedule, and HPV-related diseases among YGBMSM, supported with some content about the logistics of getting the vaccine. Outsmart HPV also supported users with text messages including an interactive vaccination reminder. The target population is YGBMSM aged 18 to 25 years who live in the United States and

have not received the HPV vaccine. As reported by [16], this study is considered as an ongoing study. The authors' future aim is to fill the gap by promoting the HPV vaccine using Outsmart HPV.

## 3    Research Methods

This study aimed at developing and investigating the usability of an Arabic consumer-focused website to educate women about HPV and cervical cancer. To do so, this study went through four phases (see Fig. 1). The framework for the design process was informed by the expanded user-task-context matrix incorporating eHealth literacy [17], focusing on two domains: the access to technology that best suits individual needs and the feel that using technology is beneficial. The BeAwareStopHPV prototype was first evaluated qualitatively, focusing on its design and content by medical experts in the areas of gynecology, family medicine, and internal medicine. The usability of the developed website was then investigated quantitatively in accordance with ISO 9241-11 standard in terms of effectiveness, efficiency, and satisfaction by inviting real users to use it during interview session. After each phase, the website was refined based on the participants' feedback. All participants signed the consent form before data collection and the study protocol was approved by the Institutional Review Board (IRB) at Claremont Graduate University. Participation in this study was voluntary and no incentives were provided for participation.

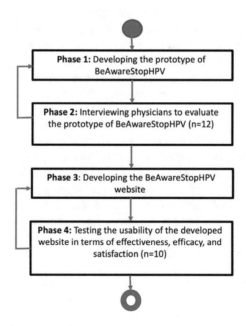

**Fig. 1.** BeAwareStopHPV building and evaluation phases

## 3.1  Study Procedures

The sampling techniques used to recruit the participants were convenience and snowball. The convenience sampling technique was used since this aids in choosing the participants who are willing to contribute to the purpose of the study. Besides, participants' easy accessibility, so that they are available at a given time [18]. The snowball sampling technique was also used since it aids researchers to increase their sample size and accesses to new participants through referrals [19].

This study is part of an ongoing research project aimed at developing and evaluating a consumer-focused educational website targeted at women for HPV and cervical cancer awareness. We conducted a qualitative study to understand the women's needs about the content and design of the website [20], which informed the development of the prototype (phase 1). In phase 2, a group of 12 physicians are involved in the prototype testing through Zoom, a virtual platform, to evaluate the prototype's content and design. All the interviews were audio-recorded, transcribed, and reviewed for accuracy. After each interview, the research team discussed the physicians' feedback, and adjustments were made to the prototype, accordingly. In phase 3, the website was built considering the users' needs. During this phase, permissions were obtained from sources to use illustrative images on the website when needed.

Phase 4 encompassed evaluating the usability of the BeAwareStopHPV website. The usability assessment encompassed two aspects: test scenario and completion of an SUS for the developed website. Specifically, the usability of the website was tested in accordance with ISO 9241-11 standard [21]. It is a quantitative method that tests three main attributes, which are effectiveness, efficiency, and satisfaction. This standard is suitable for applying to new technologies, such as the BeAwareStopHPV website. These three attributes were measured by recruiting 10 women from the public and asking them to perform three different tasks: task 1 (What is HPV?), task 2 (How could you prevent cervical cancer?), and task 3 (How can you prevent contracting HPV?). The effectiveness and efficiency were measured by task completion success of hypothetical scenarios on the website. The task success portion was crucial to understanding the users' paths to complete tasks, errors, time spent, and when and where any points of confusion or frustration were encountered. The test scenario started with an introduction to the test session process (briefing), followed by the test session itself, and concluded with a debriefing (short discussion). The test sessions were conducted virtually via Zoom platform in Arabic and were video recorded. Participants were encouraged to verbalized their thoughts while they are completing the tasks. To measure effectiveness, the number of errors that occurred during each task were counted, as well as determining the completion status [21]. The completion of each task was coded as follows: (1) task failed, (2) task completed successfully with help, and (3) task completed successfully without help. To measure the efficiency, the average time for each task was calculated to determine the level of effort that a user needed to achieve the goal. The satisfaction was measured using the SUS instrument, an established method of evaluating overall usability by allowing users to report their responses and feelings [19]. The SUS included 10 items, each on a five-point Likert scale from 1 (strongly disagree) to 5 (strongly agree). The SUS scores range from 0 to 100, with scores above 70 considered acceptable. The higher the score, the better usability the system has [21].

## 4  Results

In the prototype evaluation, 12 physicians were interviewed. They highlighted some issues and recommendations for improving the website's content and design. For example, several recommended changing the background color from light gray to white while others highly recommended adding visuals next to the text to simplify the information for the user, e.g., an image showing the Pap smear screening procedure. The physicians' characteristics are shown in Table 1.

**Table 1.** Physicians' characteristics

| Gender, no. (%) | |
|---|---|
| Male | 5 (41.6) |
| Female | 7 (58.3) |
| Specialty, no. (%) | |
| Ob/Gyn | 7 (58.3) |
| Family medicine | 4 (33.3) |
| Internal medicine | 1 (8.3) |

For usability testing, a total of 10 participants were consecutively recruited. All ten participants completed tasks 1 and 3 successfully without help from the test moderator, while nine completed task 2 successfully without help, and one participant failed to complete the task because she couldn't find the answer using the search bar on the website's home page. Tasks 1 and 3 were performed with zero errors detected, while in task 2, two errors were detected from two participants. Task completion, average time taken to complete the tasks, and error rate are presented in Fig. 2. Table 2 shows the participants characteristics during phases 4 and 5.

**Table 2.** Participants' characteristics for usability testing

| Characteristics | |
|---|---|
| Age range | 27–44 |
| Gender, no. (%) | Female 10 (100) |
| Nationality no. (%) | Saudi 10 (100) |

As shown in Fig. 2, participants could fulfill task 1, "What is HPV?" and task 3, "How can you prevent contracting HPV?" within the estimated time limit (average 3.7 and 8.3 s, respectively); however, task 2, "How could you prevent cervical cancer?" took longer than expected with an average of 15.8 s. The reason could be that task 2

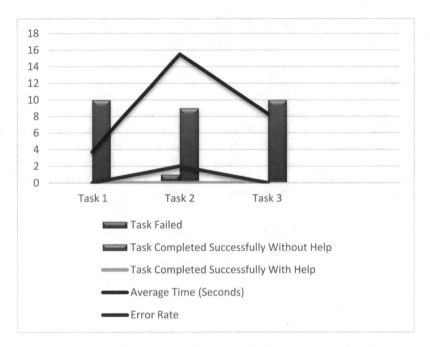

**Fig. 2.** The effectiveness and efficiency for the BeAwareStopHPV website

included an indirect method of how to answer it: the participants have to click on "what is cervical cancer?" box to find the answer. Looking at task 1 and task 3, the answer could be directly found from the boxes' label on the home page. The average SUS score for the ten participants was 97 (in the acceptable range), which indicated a good satisfaction level across the participants. Figure 3 provides examples of the website's pages after refinement is made in accordance with the participants' feedback.

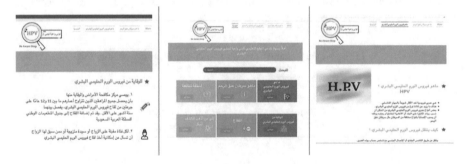

**Fig. 3.** Some web pages of the BeAwareStopHPV website

# 5 Discussion and Conclusion

To the best of our knowledge, BeAwareStopHPV is the first website that has been developed based on the preferences of women in Saudi Arabia to increase their awareness and knowledge about HPV and cervical cancer. Even though several research studies have applied different interventions to increase knowledge [22], appointment attendance [23], or HPV vaccinations [24], the issue of lack of knowledge, and cervical cancer cases still evolving, which highlight the need to develop and evaluate an educational website based on users' preferences that would assist in solving the problem under study.

The design of BeAwareStopHPV website was strengthen by the user-based and expert-based feedback gained to identify women's design preferences to inform the development of and to assess the usability of the website from their views and per- spectives. We conducted several rounds of interviews to ensure the BeAwareStopHPV website could be adopted and refined to incorporate the information needs and design preferences of target users.

The usability of the BeAwareStopHPV website was evaluated quantitatively and qualitatively by interviewing experts, and test sessions with potential real users. Our findings yielded positive assessments. This website is a promising solution that women in Saudi Arabia can use to increase their awareness and knowledge about HPV infection and cervical cancer. This website can also be used by the Saudi Arabian Ministry of Health, other health authorities, and policy makers as an educational tool to promote HPV knowledge. The limitations within this study were due to the convenient and snowball sampling; indeed, findings should be interpreted with caution in terms of generalization. Another limitation was the small sample size. However, three to five participants considered sufficient for usability studies [25]. Before this website can be disseminated and used largely by women, as a future direction, we will test the effectiveness of the BeAwareStopHPV website by conducting a randomized control trial.

**Acknowledgment.** "The authors extend their appreciation to the Deputyship for Research & Innovation, Ministry of Education in Saudi Arabia for funding this research work through the project number IFPRC-071-612-2020" and King Abdulaziz University, DSR, Jeddah, Saudi Arabia.

Also, the authors would like to thank Wed Abogazah, Noura Alsanbi, Rahaf Al-batati, and Layan Abdulate for their help and support during the website's development phase.

# References

1. Alattas, M.T.: Cancer control priorities and challenges in Saudi Arabia: a preliminary projection of cancer burden. Lung **6**, 1–70 (2019)
2. HPV Information Center Saudi Arabia Human Papillomavirus and Related Cancers, Facts Sheet 2018 (2019). https://hpvcentre.net/statistics/reports/SAU_FS.pdf. Accessed 16 Apr 2021

3. Gari, A., et al.: The awareness of the HPV's association with cervical cancer and the HPV vaccine among Saudi females. Life Sci. J. **9**, 2538–2546 (2012)
4. Alsbeih, G.: HPV infection in cervical and other cancers in Saudi Arabia: implication for prevention and vaccination. Front. Oncol. **4**, 65 (2014)
5. Al Khudairi, H., Abu-Zaid, A., Alomar, O., Salem, H.: Public awareness and knowledge of pap smear as a screening test for cervical cancer among Saudi population in Riyadh city. Cureus **9**(1), 1–8 (2017)
6. Jradi, H., Bawazir, A.: Knowledge, attitudes, and practices among Saudi women regarding cervical cancer, human papillomavirus (HPV) and corresponding vaccine. Vaccine **37**(3), 530–537 (2019)
7. Hussain, A.N., et al.: Attitudes and perceptions towards HPV vaccination among young women in Saudi Arabia. J. Fam. Community Med. **23**(3), 145 (2016)
8. Sait, K.H.: Knowledge, attitudes, and practices regarding cervical cancer screening among physicians in the Western Region of Saudi Arabia. Saudi Med. J. **32**(11), 1155–1160 (2011)
9. Malibari, S.S.: Knowledge about cervical cancer among women in Saudi Arabia. Egypt. J. Hosp. Med. **70**(10), 1823–1825 (2018)
10. Almehmadi, M.M., Salih, M.M., Al-Hazmi, A.S.: Awareness of human papillomavirus infection complications, cervical cancer, and vaccine among the Saudi population. A cross-sectional survey. Saudi Med. J. **40**(6), 555–559 (2019)
11. Alkhateeb, J.M., Alhadidi, M.S.: Information about epilepsy on the internet: an exploratory study of Arabic websites. Epilepsy Behav. **78**, 288–290 (2018)
12. Halboub, E., Al-Ak'hali, M.S., Al-Mekhlafi, H.M., Alhajj, M.N.: Quality and readability of web-based Arabic health information on COVID-19: an infodemiological study. BMC Public Health **21**(1), 1–7 (2021)
13. Hodges, P.W., Setchell, J., Nielsen, M.: An internet-based consumer resource for people with low back pain (MyBackPain): development and evaluation. JMIR Rehabil. Assistive Technol. **7**(1), e16101 (2020)
14. Johnson-Mallard, V., Darville, G., Mercado, R., Anderson-Lewis, C., MacInnes, J.: How health care providers can use digital health technologies to inform human papillomavirus (HPV) decision making and promote the HPV vaccine uptake among adolescents and young adults. BioResearch Open Access **8**(1), 84–93 (2019)
15. Pot, M., Paulussen, T.G., Ruiter, R.A., Mollema, L., Hofstra, M., Van Keulen, H.M.: Dose-response relationship of a web-based tailored intervention promoting HPV vaccination: process evaluation. J. Med. Internet Res. **22**, e14822 (2020)
16. Reiter, P.L., et al.: A web-based human papillomavirus vaccination intervention for young gay, bisexual, and other men who have sex with men: protocol for a randomized controlled trial. JMIR Res. Protoc. **9**(2), e16294 (2020)
17. Kayser, L., Kushniruk, A., Osborne, R.H., Norgaard, O., Turner, P.: Enhancing the effectiveness of consumer-focused health information technology systems through eHealth literacy: a framework for understanding users' needs. JMIR Hum. Factors **2**(1), e3696 (2015)
18. Etikan, I., Musa, S.A., Alkassim, R.S.: Comparison of convenience sampling and purposive sampling. Am. J. Theor. Appl. Stat. **5**(1), 1–4 (2016)
19. Waters, J.: Snowball sampling: a cautionary tale involving a study of older drug users. Int. J. Soc. Res. Methodol. **18**(4), 367–380 (2015)
20. Bitar, H., Alismail, S.: The Perceived Information Needs of Women Searching for HPV and Cervical Cancer eHealth Educational Technologies: An Exploration of Enablers and Inhibitors. Unpublished manuscript (under review)

21. Georgsson, M., Staggers, N.: Quantifying usability: an evaluation of a diabetes mHealth system on effectiveness, efficiency, and satisfaction metrics with associated user characteristics. J. Am. Med. Inform. Assoc. **23**(1), 5–11 (2016)
22. Bitar, H., Ryan, T., Alismail, S.: The Effect of Implementing an SMS Messaging System to Overcome the Lack of Transmission of HPV Facts in Saudi Arabia (2020)
23. Linde, D.S., Andersen, M.S., Mwaiselage, J.D., Manongi, R., Kjaer, S.K., Rasch, V.: Text messages to increase attendance to follow-up cervical cancer screening appointments among HPV-positive Tanzanian women (Connected2Care): study protocol for a randomised controlled trial. Trials **18**(1), 1–10 (2017)
24. Francis, D.B., Cates, J.R., Wagner, K.P.G., Zola, T., Fitter, J.E., Coyne-Beasley, T.: Communication technologies to improve HPV vaccination initiation and completion: a systematic review. Patient Educ. Couns. **100**(7), 1280–1286 (2017)
25. Nielsen, J., Landauer, T.K.: A mathematical model of the finding of usability problems. In: Proceedings of the INTERACT 1993 and CHI 1993 Conference on Human Factors in Computing Systems, pp. 206–213 (1993)

# Digital Engineering

# Intelligent Scheduling in MES with the Fuzzy Information and Unclear Preferences

Artem Vozhakov[(⊠)] [iD]

Perm National Research Polytechnic University, Perm 614990, Russia

**Abstract.** The problem of manufacture operational planning inside the work-shop is considered, taking into account the available operational data on the state of production and the accumulated knowledge about the distribution of resources using the knowledge base. A mathematical formulation of the operational planning problem is proposed, which makes it possible to form optimal plans taking into account the main constraints on the available resources and the fuzzy nature of the optimality criteria. A description of the composition of attributes and normative and reference information required for operational planning is given. The problem of obtaining the values of attributes necessary for planning based on analogs found by the method of "k-nearest neighbors" is considered. To use the previous experience, the specialization-function of performers and equipment is introduced, formulated as an adaptive search task based on the knowledge base. An example of the practical implementation of the operational management system at the workshop level using a modified ERP is considered. Provides a general description of the interface and the mechanics of working with the system.

**Keywords:** Small-scale production · Manufacture execution system · Operational planning · Knowledge base · Classification · Adaptive search · Mathematical modeling · Heuristic algorithms · Enterprise resource planning

## 1 Introduction

The third industrial revolution created a reality characterized by volatility, uncertainty, complexity and ambiguity, the so-called VUCA. In all areas of human life, the changes that have taken place have caused many problems, for the solution of which, as a rule, now there is not even a theoretical base, especially the practical tools of solution [1, 2]. Manufacturing companies specializing in complex high-tech products, consisting of thousands of parts, have been implementing production planning systems (MRP II, APS) [3–5], manufacture execution systems (MES) and lean manufacturing (TPS) [6–8] to ensure a controlled and efficient production process. Despite this, such industries have always been characterized by a high level of costs (losses) [9], a long production cycle, low efficiency [2, 9]. In order to be competitive in the modern world, such companies must offer customers a constantly updated range of products, production according to customer requirements in minimal batches, tight production times, frequent changes in the customer order portfolio, and frequent changes of suppliers. All this requires building a truly flexible production that can quickly adapt to a changing

T. Antipova (Ed.): DSIC 2021, LNNS 381, pp. 131–143, 2022.
https://doi.org/10.1007/978-3-030-93677-8_12

environment. This requires new production management tools, devoid of the constraints dictated by the systems of the previous generation. Such a system should allow ensuring the most flexible and at the same timing effective work of production in conditions of constant changes in the order book, a changeable situation in production, and insufficient information. The article proposes an approach to developing an intelligent production planning system within a shop, based on the study of spinning production experience, identifying analogies, and forming a knowledge base.

As will be shown bellow, the use of combined planning mechanisms (manual, automatic) and filling in attributes based on "similar operations" enables to reduce the requirements for the quality and completeness of regulatory information by an order of magnitude. This allows you to shorten the long period of data preparation [6, 7] to almost zero, using the system here and now. The employee responsible for the formation of shift tasks can always adjust the automatically generated tasks, which in the future will also be taken into account during the subsequent operation of the system. Therefore, the implementation of such a system [8, 9] can be performed much faster than the classical MES [6, 7] and with fewer costs for preparing the initial data.

## 2   Concept

Operational production management is a complex task, for the solution of which it is necessary to analyze gigantic volumes of information and thousands of parameters. The first manufacturing execution systems (MES – manufacturing execution system) [6, 7] appeared in the late 1980s for automating and optimizing operational planning, capacity utilization management and operational accounting of production progress. This is a specialized application software designed to solve problems of synchronization, coordination, analysis and optimization of product output within the framework of any production. Main functions of MES [6, 7]:

1) RAS (Resource Allocation and Status) – monitoring the status and allocation of resources in real time. Resource management: technological equipment, materials, personnel.
2) ODS (Operations/Detail Scheduling) – operational detailed planning. Provides the ordering of production orders based on the sequence, attributes, characteristics and regulations associated with the specificity of products and production technology. The goal is to set up a production schedule with minimal equipment changeovers and parallel operation of production facilities to reduce time to finished product and downtime.
3) DPU (Dispatching Production Units) – production dispatching. Manages the flow of items in the form of jobs, orders, series, batches, and work orders. Dispatch information is presented in the order in which the work must be performed and changes in real time as events occur at the shop floor. This makes it possible to change the given schedule at the level of workshops. Includes functions of elimination of rejects and waste processing, along with the ability to control labor costs at each point in the process with data buffering.

The MES-system receives the scope of work, which is either represented by MRP II [4, 8] at the scheduling stage or issued by the APS-system and builds more accurate schedules for the equipment, and monitors their implementation online. In this sense, the goal of the MES-system is not only to fulfill the specified volume with the specified deadlines for the fulfillment of certain orders, but also to fulfill it as best as possible in terms of the economic performance of the shop [10, 11].

For the effective operation of the planning algorithms [12–14], a high level of quality of reference information (SRI) is required, a list of attributes that must be specified, and an increased operational load on the shop staff associated with the need to keep an online record of the performance of all technological operations with many the sameness of the input parameters [6]. These conditions are the main reason for failures in the implementation of MES systems [6, 7]. This article will present an alternative approach to building a planning system, which is based on the experience of employees, identifying analogies, reducing the quality requirements for reference data, intellectualizing the manual planning process and using the knowledge base to form optimal schedules of operations.

Let's consider the problem of production management at the level of a workshop using the example of a small-scale machine-building enterprise. At such enterprises, as a rule, there is a planning system (usually ERP) [4]. The task of the in-shop management is to ensure the performance of technological operations on all batches of semi-finished products (parts, assemblies, etc.) within the time frame established by the planning system [5]. This, in turn, leads to the need to form operational plans for performing technological operations with reference to employees and equipment and disciplined execution of planned technological operations.

For each batch of semi-finished products, the following attributes must be defined [6, 7]:

– semi-finished product produced (type, product of destination, and other properties);
– the order to which the batch belongs;
– batch size (quantity);
– planned time of completion of processing in the shop;
– priority of the party.

For each batch of semi-finished products entering the shop, a list of technological operations is predetermined. The complete list of possible attributes of technological operations is extremely extensive, but hereinafter we will proceed from the minimum composition of attributes:

– batch of semi-finished products to which the operation belongs;
– section, where the operation must be performed in accordance with the technology;
– type of operation;
– equipment on which the operation can be performed (including analogues);
– total time of the operation;
– preparatory and final time;
– equipment changeover time (depends on the parts of the predecessors);
– used tools and equipment (availability will depend on the possibility of performing the operation);

- real qualifications and specialization of the worker;
- technical condition of a piece of equipment;
- minimum batch/frequency of processing;
- other parameters affecting the process.

As a rule, enterprises do not have a complete and up-to-date database of all technological operations that are carried out in production. The reasons for this are as follows:

- data is generated for a long time, in various software packages and on paper, collecting such data can be quite laborious;
- data is constantly changing, equipment, tools are changing, new processing technologies are being developed, personnel do not have time to update the data;
- data may come from external sources and be incomplete;
- errors in working with data and the human factor;
- lack of a reliable way to verify data.

To form a complete database for all technological operations requires laborious preparatory work, the completeness and quality of which directly depends on the success of the entire project [7]. The preparatory stage can drag on for years, and if the products of the enterprise are constantly changing [9–11], this process can become an eternal attempt to catch up with something that is constantly moving forward with greater speed [9].

Extremely important for the operation of the production management system is not only regulatory information, but also data on the actual performance of operations (statistical data). It is extremely important to establish a high-quality collection of actual data on the operation of production in manual and automatic modes [5]:

- actual time of the beginning and end of operations;
- performer;
- piece of equipment;
- the actual number of operations performed;
- recorded deviations and defects;
- deviations in the complexity of implementation;
- other parameters.

A workshop is usually divided into production sections, which are production units that combine a number of tasks, grouped according to a certain criterion, carrying out part of the overall production process for the manufacture of products or maintenance of the production process.

The sections are created according to two principles:

1) Technological. The site consists of the same type of equipment (a group of lathes, a group of milling, drilling machines). Workers on the site perform a certain type of operation. The production of certain types of products is not assigned to workplaces. This type of site is typical for small-scale and single types of production organization.
2) Subject-closed. At such a site, various types of equipment are used, which are located in the course of the technological process. Workplaces specialize in the

manufacture of a certain type of product (parts). The site employs workers of various specialties. A variety of this type of site are production lines. This type of section is typical for large-scale and mass production, its work is more efficient than a section created according to the technological principle.

Consider the attributes that are accumulated in the system and constitute the initial knowledge base for analysis:

1) Semi-finished product (part, assembly).
2) Technological operation.
3) Type of operation.
4) The duration of the operation.
5) The number of operations to be performed.
6) Contractor.
7) A piece of equipment.
8) The actual number of operations performed.
9) The actual duration of the execution.
10) Actual number of operations with deviations.

The provided analytics is enough to reveal the following patterns in the course of statistical analysis [12, 13]:

1) Technological operations similar in properties (used to determine missing attributes).
2) Specialization of the contractor and equipment (may differ from the originally declared). Determined by the frequency of assignment of a performer per piece of equipment.
3) Specialization of the contractor for semi-finished products.
4) Specialization of equipment for semi-finished products.
5) Specialization of the contractor by types of operations.
6) Specialization of equipment by type of operations.
7) Average percentage of semi-finished products with deviations in the context of performers and equipment.
8) The average percentage of exceeding the complexity of the implementation.
9) Average percentage of final marriage.

When constructing analytics, it should be borne in mind that the production environment is constantly changing, as well as the level of training of employees and the state of equipment. Thus, when constructing analytics, results that are more recent should be taken into account with greater weight than old ones [14, 15]. A linear relationship of weight loss with a zero point at 12 months is proposed [15].

Based on the above conditions, the information system should perform the following intellectual functions based on the generated knowledge base [16]:

1) To identify technological operations analogs that are close in properties to each other and, according to statistics, are planned in a similar way [17];
2) Filling in the missing information for operations of analogs, based on the verified values of operations of analogs (the received data have a preliminary status and are

used only for automatic planning with confirmation by the user of the planning results);
3)  Automatically generate daily tasks for the area;
4)  Automatically focus on operations with an increased risk of noncompliance and rejection.


## 3   Mathematical Model Description

It is believed that at the beginning of planning, a complete list of batches of semi-finished products to be processed in the workshop $P_i$, *where* $i \in \overline{1,P}$ is the number of batches in production. The batch sizes of $PC_i$ are known [17, 18].

For convenience, we represent the time scale in the form of a work shift number $d$, where the current shift has a number d = 0. The next shifts have numbers $d \in 1,2,3...$ and the previous shifts have negative numbers $d \in -1,-2,-3....$.

The planned completion date for the processing of a batch of semi-finished products in the shop is set in the form of a shift number $d$ and is indicated for all batches of semi-finished products $u_i$.

For each batch of semi-finished products, a list of operations $o_{ij}$ is defined. The list is formed in such a way that at the time of drawing up the schedule, all fully completed technological operations are excluded from it, thus, the first operation of the batch ($j = 1$) is in fact the first unfulfilled operation in the batch. At the same time, a situation is possible in which a batch of semi-finished products in production is divided into several. For such cases, for each operation, the number of semi-finished products in the batch is indicated for which this operation must be performed $oc_{ij}$. In this case, technological operations must be performed strictly in order. Thus, the number of subsequent operations is always greater than or equal to the number of predecessor operations:

$$oc_{ij} \leq oc_{ij+1} \leq PC_i \tag{1}$$

For each operation, the total complexity of the operation is determined for the total number of semi-finished products available for execution $tr_{ij}$ and the section for the operation $st_{ij}$ is indicated.

Known is the list of workers and/or teams of workers for assigning shift tasks $W_k, k \in \overline{1,W}$ – the total number of performers in the shift. Also known is the list of available equipment $M_c, c \in \overline{1,M}$ – the number of equipment units. For each operation, the technology determines the requirements for the profession and the category of the worker, the type of equipment used. For each operation, we introduce subsets of employees $W_k^{ij}$ and equipment $M_c^{ij}$ that meet the requirements of the technological process for this operation. The duration of the $Dl$ change is known. The maximum output of each performer is determined $Dl_k$.

The classic approach to planning in the MES is to distribute the list of tasks among the performers, while observing clear constraints [18–21]. This approach does not take into account the volatility and fuzzy nature of information [22].

Let us introduce a function of the priority of processing a batch of semi-finished products, which takes on lower values with an increase in the priority of processing of semi-finished products. If there is not enough time left for processing to process on time, the function takes negative values:

$$U(i) = Dl \times u_i - \sum_j tr_{ij} - \rho \times \sum_j sign(o_{ij}), \text{where } \rho \text{ is the sojourn time} \quad (2)$$

The sojourn time parameter characterizes the average time of interoperative sojourn and transportation of semi-finished products from one place of execution to another. This parameter is a conditional value and is selected expertly.

The solution to the standard planning problem in the MES is a detailed plan for the execution of operations until the full completion of the production of all current orders [18–21]. Instead, we only consider the next steps to be taken, thus limiting the planning horizon to one or two days in advance. The solution to the problem of planning the shift-daily task will be filling the array of data tuples $T_s$ of the following type:

$$T_s = (d_s, o_s, ot_s, w_s, m_s, tr_s), \quad (3)$$

where $o_s \in o_{ij}$ is the assigned operation,
$ot_s \leq oc_{ij}$ – quantity to be processed within the task,
$w_s \in W_k^{ij}$ – the assigned executor of the operation,
$m_s \in M_c^{ij}$ – the assigned equipment of the operation,
$tr_s = tr_{ij} \times \frac{ot_s}{oc_{ij}}$ – the labor intensity of the issued works.

Note that a number of natural restrictions are imposed on the shift task:
The number of issued transactions should not exceed the number of transactions in the plan:

$$\sum_{o_s=o_{ij}} ot_s \leq oc_{ij} \quad (4)$$

The complexity of the work assigned to the performer should not exceed the maximum permissible value:

$$\sum_{w_s=W_k} tr_s \leq Dl_k \quad (5)$$

The complexity of the work assigned to the equipment should not exceed the maximum permissible value:

$$\sum_{m_s=M_c} tr_s \leq Dl \quad (6)$$

The order of operations should not be violated:

$$\sum_{o_s=o_{ij}} ot_s + PC_i - o_{ij} \leq \sum_{o_s=o_{ij+1}} ot_s + PC_i - o_{ij+1} \quad (7)$$

Statistics on the execution of shift jobs is accumulated in the actual arrays of data tuples $F_a$ of the following form

$$F_a = (d_a, o_a, ot_a, w_a, m_a, tr_a, of_a, ob_a, trf_a) \tag{8}$$

where the following are added to the previously described variables:

$of_a \leq oc_{ij}$ – the number of actually processed semi-finished products.

$ob_a \leq oc_{ij}$ – the number of operations with deviations.

$trf_a \leq oc_{ij}$ – actually, achieved labor intensity.

Based on the statistics, a function of specialization of operations for equipment and performers can be built, which evaluates data for the period of accounting statistics $D$. It is considered that the value of specialization is higher if in the near future there were successfully performed operations of the specified type by the specified performer on the specified equipment. The value of the function decreases if there are statistics on the execution of operations with deviations:

$$Sp\left(o_{ij}, W_k, M_c\right) = \sum_{\substack{ot_a = of_a \\ tr_a = trf_a \\ d_a \in \overline{-D,0}}}^{o_{ij}, W_k, M_c} sign(ot_a) \times \left(1 + \frac{d_a}{D}\right)$$

$$- \sum_{\substack{ot_a \neq of_a \\ tr_a \neq trf_a \\ d_a \in \overline{-D,0}}}^{o_{ij}, W_k, M_c} sign(ot_a) \times \left(1 + \frac{d_a}{D}\right) \tag{9}$$

Consider particular criteria for the optimality of a shift task:

The urgency of the assigned work should be as high as possible:

$$J_1 = \sum U(T_s) \rightarrow \min \tag{10}$$

The risk of non-execution of operations due to non-execution of previous operations should be minimal, i.e. first, the first operations should be planned for operation, and only then the subsequent ones. In this case, it is assumed that the risk of non-execution of operations increases linearly depending on the number of previous operations in the shift task:

$$J_2 = \sum j(T_s) \rightarrow \min \tag{11}$$

The level of specialization of the assigned operations for the performer and equipment should be maximum (the criterion is reduced to minimization):

$$J_3 = -\sum Sp(T_s) \to \min \tag{12}$$

The problem can be solved by choosing one of the optimization criteria as the "main criterion" based on an assessment of the current state of production. Alternatively, you can introduce a generalized criterion for the optimality of a shift task [15, 22, 23].

The posed problem of forming optimal shift tasks (1)–(12) is a nonlinear optimization problem with a specialization function based on the knowledge base. These circumstances make it impossible to solve it by analytical methods and require the development of new algorithms based on intelligent technologies [15, 23].

## 4  Algorithm for Solving

The problem of forming optimal shift tasks is a nonlinear problem with a fuzzy optimization criterion, which makes it impossible to solve it by analytical methods. Consequently, it is necessary to develop an empirical algorithm for solving the problem that is sufficiently effective so that its software implementation would make it possible to find close to optimal solutions to the problem in a reasonable time [24].

Consider one of the possible general algorithms for solving the problem. This algorithm consists of the following sequential steps:

1) Choosing a strategy for forming a shift task based on an analysis of the general list of operations to be performed.
2) Formation of a sorted set of operations in accordance with the selected strategy.
3) Sequential planning of operations from one or more sets in proportion to the weighting coefficients of the membership function.

At the stage of choosing a strategy for forming a shift task, the following values are calculated:

– $Fw$ – general fund of working time of performers
– $Fm$ – equipment working time fund
– $Fwm$ – the minimum of the values of $Fw$ and $Fm$
– $Ftr$ – is the total labor intensity of all operations in all batches of semi-finished products
– Ftr1 – is the total labor intensity of performing only the first operations in batches of semi-finished products.

Based on the calculated values, a strategy for forming shift tasks is determined (brief rules for choosing a strategy are given in the description of strategies):

1) "Optimal distribution" – this strategy is used if the total labor intensity of all available operations is below the fund of working time of performers and equipment ($Ftr < Fwm$). When using this strategy, to solve the problem, it is enough to sort the list of operations according to the urgency of execution, and sequentially find for each operation the optimal specialization in terms of performer and equipment from the available resources and assign the operation.

2) "Distribution by equipment" – this strategy is used if the total labor intensity of all available operations is lower than the fund of working time of performers ($Ftr <$ $Fw$), but higher than the fund of equipment operation ($Ftr > Fm$). When using this strategy, it is important to optimally distribute operations across the available equipment, minimizing the risks of downtime and disruption of the plan. To solve the problem, operations are also sorted by urgency, and the optimal equipment for performing the operation is sequentially calculated, the equipment is fixed and the optimal performer on this equipment is calculated, after which the operation is assigned.

3) "Distribution by performer" – this strategy is used if the total labor intensity of all available operations is lower than the fund of equipment working time ($Ftr < Fw$), but higher than the fund of work of performers ($Ftr > Fm$). When using this strategy, it is important to optimally distribute operations among performers, minimizing the risks of downtime and disruption of the plan. To solve the problem, the operations are also sorted according to the urgency of execution, and the optimal executor for performing the operation is sequentially calculated, the executor is fixed and the optimal equipment for the selected executor is calculated, after which the operation is assigned.

4) "First or urgent" – this strategy is used if the total labor intensity of all available operations is higher than the fund of working time of equipment and performers ($Ftr > Fw$, $Ftr > Fm$), but the complexity of performing the first operations is lower than the fund of working time of equipment or performers ($Ftr1 < Fw$ or $Ftr1 < Fm$). In practice, this strategy is quite common. To solve the problem, it is necessary to form two lists of operations – the first operations, and the most urgent operations with the total labor intensity $Fwm$. Further, operations are planned that were included in both lists. Then, at each iteration, the values of the generalized optimality criterion are compared for the first operations from the two lists – the one for which the criterion value is preferable is assigned. Operations that are not available for assignment are skipped. A shift task is formed until the equipment or performers are fully loaded, if required, the list of operations can be expanded with the next most urgent operations.

5) "High level of work in progress" – this strategy is used if the total labor intensity of the first operations is higher than the fund of working time of equipment or performers ($Ftr1 > Fmw$). In practice, this is the most common situation that speaks of the insufficient quality of production management. When using this strategy, to solve the problem, it is necessary to sort the list of operations according to the urgency of execution, and sequentially assign operations until the equipment or performers are fully loaded.

## 5    Implementation

It was decided to implement the system as a separate module on the 1C technological platform capable of being embedded in 1C: ERP [1]. To implement the task, the functional of the standard solution was modified in such a way that while maintaining the existing functionality, the system allows generating shift tasks automatically in accordance with the algorithm for solving the problem of optimal production control at the operational control level using the knowledge base [1].

The system allows you to plan a schedule for the execution of operations (formation of tasks for workers with an indication of the equipment) by sections of the shop, monitor the execution of tasks and take control actions. In real production, as a rule, they operate with short-term tasks, a maximum of 2 days in advance, and according to the results of the work of each day, the tasks are specified because the work has been completed.

The principle of the system is as follows:

1) Before the formation of shift tasks in the system, a list of employees and equipment available for work in the current shift is formed in a semi-automatic mode. The system can itself propose a list of employees, based on the division of employees by areas and taking into account information from the enterprise access control and management system. The list of equipment should also be generated automatically, with the exception of equipment under repair or out of service for another reason. The equipment can be rigidly tied to the employee (the choice of the employee automatically selects the equipment, or vice versa, the choice of the equipment automatically selects the employee).

2) Next, a list of technological operations available for planning is formed.

3) An attempt is made to select unfilled attributes of operations based on matching attributes of "similar operations" (the procedure for matching analogs is based on the method of k-nearest neighbors, automatically filled data are highlighted in gray).

4) The work interface is located in one window, in one part of which there is a list of employees and equipment, and in the other - a list of operations available for scheduling, quick filters.

5) The assignment of operations to the performer and equipment is carried out both manually (by quick commands and drag and drop), and partially or completely automatically. Automatic formation of the plan is carried out based on the solution of the optimization problem (1)–(12). The values of the specialization function are used both when assigning work to a contractor and when accepting work - the system automatically highlights operations with a low level of specialization in color in order to focus attention on operations that potentially carry execution risks with deviations from the specified standards.

6) In the process of performing operations, all available information is collected. Before starting the operation and upon completion, the worker reads the barcode of the operation using a barcode reader installed at each workstation. Most of the other information comes from SCADA.

7) The results of the system's operation are accumulated and analyzed in order to gradually transfer the system to semi-automatic and automatic modes of operation.

Separately, it should be noted the mechanism of selection of unfilled attributes of operations based on matching attributes of "similar operations". The analog selection procedure is based on the standard "k-nearest neighbors" method using the Euclidean metric most common for this problem [13]. It was hypothesized that the value of an unfilled attribute is filled in if the value of the reference type for all analogs is the same, or the mean value of a continuous variable, if the spread of the values of the attribute of analogs does not exceed 5%. In the process of fine-tuning the algorithm, the details were additionally divided according to their importance using weighting factors. In practice, such a simple approach made it possible to speed up the formation of the regulatory framework by an order of magnitude and significantly speed up the reconciliation of data.

The introduction of automatic generation algorithms was carried out gradually, as a sufficient array of statistical data was accumulated. In the enterprise that generated the case study, this period lasted six months. At the time the system began functioning, the number of manual corrections and cancellations of operations was $\sim 20\%$, which is a fairly large percentage of planning inaccuracy. In addition, up to 20% of all assigned operations based on the results of work per shift were marked as uncompleted, i.e. the percentage of tasks completed was $\sim 80\%$. Based on the results of three months of operation, it was possible to reduce the number of adjustments to $\sim 10\%$ and increase the percentage of tasks completed to $\sim 92\%$.

In addition, all shift tasks, generated manually at the initial period of operation, were analyzed according to the optimality criterion. The analysis revealed that only 5% of manually generated shift tasks are preferable to the automatically generated shift tasks with the lowest values.

## 6   Conclusion

Using specialized algorithms to generate Facebook news feed and recommended You-Tube videos has long proven its worth. Now is the time [1] to apply similar approaches in managing complex production in order to increase the flexibility and efficiency of production processes [15], which will allow companies to focus on updating the product range and individual customer needs. The use of combined planning mechanisms (manual, automatic) and filling in attributes based on "similar operations" allows to reduce the requirements for the quality and completeness of regulatory information by an order of magnitude. Companies will be able to launch new parts and products into production faster and at lower costs, ensure production in minimal batches in a short time [1], without significantly increasing planning mistakes. At the same time, a simple and effective scheduling algorithm allows you to perform operational rescheduling as often as necessary, which minimizes the costs associated with frequent changes in the customer's order portfolio. The operational production management system described in the article with an intelligent production planning system inside the shop has proven its effectiveness in real production conditions. At the same time, the system has extremely low requirements for the quality and composition of regulatory and reference information, which will allow the launch of such a system in the shortest possible time. The employee responsible for the formation of shift tasks can always adjust the automatically generated tasks, which in the future will also be taken into account during the subsequent operation of the system. The proposed system is able to significantly reduce labor costs for the formation of shift tasks, improve the quality of planning, increase the speed of response to changes in production and reduce the number of deviations that arise with specialization of performers and equipment [16].

# References

1. Mangey, R.: Advances in Mathematics for Industry 4.0, 1st edn. Academic Press (2020)
2. Vozhakov, A.: The practice of creating intelligent manufacture management systems based on a ERP. In: Antipova, T. (ed.) ICADS 2021. AISC, vol. 1352, pp. 327–339. Springer, Cham (2021). https://doi.org/10.1007/978-3-030-71782-7_29
3. Browne, J., Harhen, J., Shivnan, J.: Production Management Systems: An Integrated Perspective, 2nd edn, 284 p. Addison-Wesley Publishing Company (1996)
4. Sumner, M.: Enterprise Resource Planning. Pearson, 1st edn, 208 p. (2004). ISBN 0131403435, 978-0131403437
5. Gaither, N., Frazier, G.: Production and Operations Management, 8th edn, 874 p. Southwestern College Publishing, Cincinnati (1999)
6. McClellan, M.: Applying Manufacturing Execution Systems, 1st edn, 208 p. CRC Press (1997)
7. Meyer, H., Fuchs, F., Thiel, K.: Manufacturing Execution Systems: Optimal Design, Planning, and Deployment, 274 p. McGraw Hill, New York (2009)
8. Altemir, D.: Lean MRP: Establishing a Manufacturing Pull System for Shop Floor Execution Using ERP or APS, 109 p. Independently Published (2018)
9. Suri, R.: It's About Time: The Competitive Advantage of Quick Response Manufacturing, 228 p. CRC Press (2010)
10. Ohno, T.: Toyota Production System: Beyond Large-Scale Production, 176 p. CRC Press (1988)
11. Takeda, H.: The Synchronized Production System, 263 p. Kogan Page Publishers (2006)
12. Deisenroth, M.: Mathematics for Machine Learning, 398 p. Cambridge University Press (2020)
13. Phillips, J.: Mathematical Foundations for Data Analysis, 303 p. Springer, Cham (2021)
14. Kroese, D., Botev, Z., Taimre, T., Vaisman, R.: Data Science and Machine Learning: Mathematical and Statistical Methods, 532 p. Chapman and Hall/CRC (2019)
15. Gitman, M., Stolbov, V., Gilyasov, R.: Management of Socio-technical Systems, Taking into Account Fuzzy Preferences, 272 p. LENAND (2011)
16. Vandeput, N.: Data Science for Supply Chain Forecast, 237 p. (2018)
17. Womack, J., Jones, D.: Lean Thinking, 1st edn, 350 p. Taylor & Francis (1996)
18. Frost: World Manufacturing Execution Systems (MES) Market. Frost & Sullivan: Mountain View (2010)
19. Liu, Q., Dong, M., Chen, F., Lv, W., Ye, C.: Single-machine-based joint optimization of predictive maintenance planning and production scheduling. Robot. Comput. Integr. Manuf. **55**, 173–182 (2019)
20. Sivasundari, M., Suryaprakasa, K., Raju, R.: Production, capacity and workforce planning: a mathematical model approach. Appl. Math. Inf. Sci. **13**(3), 369–382 (2019)
21. Kim, S., Lee, Y.: Synchronized production planning and scheduling in semiconductor fabrication. Comput. Ind. Eng. **96**, 72–85 (2016)
22. Hu, Z., Hu, G.: A multi-stage stochastic programming for lot-sizing and scheduling under demand uncertainty. Comput. Ind. Eng. **119**, 157–166 (2018)
23. Mahdieh, M., Clark, A., Bijari, M.: A novel flexible model for lot sizing and scheduling with non-triangular, period overlapping and carryover setups in different machine configurations. Flex. Serv. Manuf. J. **30**, 884–923 (2017)
24. Wolosewicz, C., Dauzère, P., Aggoune, R.: Lagrangian heuristic for an integrated lot-sizing and fixed scheduling problem. Eur. J. Oper. Res. **244**(1), 3–12 (2015)

# Particular Problems of Gas Diffusion in Polyethylene

Vladimir D. Oniskiv[1]($\boxtimes$) (iD), A. Yu. Yakovlev[1] (iD), L. M. Oniskiv[1] (iD),
and I. M. Gitman[2] (iD)

[1] Perm State National Research Polytechnic University, Perm, Russia
[2] University of Sheffield, Sheffield, UK

**Abstract.** The problem of oxygen and ethylene diffusion in the framework of classical Fick's law are considered. It is assumed that the material is placed in a limited sealed container. This leads to the need to take in account the parameters of the gas in the free volume. The problem has a non-stationary and nonlinear character. The temperature of the medium is assumed to be constant. Numerical results describing the process of oxygen evacuation from polyethylene and the process of ethylene diffusion into the material are obtained. The problem of ethylene diffusion is solved for the case of the formation of molecular bonds during gamma-ray irradiation due to absorbed ethylene. The finite element method was used to solve the problems. The main characteristics of the noted processes are determined. The results are of practical value for determining the irradiated time of polyethylene, which provides the required level of crosslinking of macromolecules.

**Keywords:** Polyethylene · Gas diffusion · Gas limit concentration · Fick's second law · Boundary conditions · Ethylene absorption · Oxygen degasification degassing time

## 1 Introduction

Crosslinked polyethylene (PEX) is one of the products of the radiation synthesis of carbon-based nanostructures. Products made from PEX are widely used in different areas (aerospace and car industries, fiber-optic link, oil and gas industries, etc.). Particularly, PEX is used in manufacturing heat-shrink tubes and corrosion protection tapes. Also, it is successfully used as a material for high-tension and low-tension electricity cable insulation, pipes for water supply and heating, prosthetic devices and implants, catheters and endoscopes, etc. The life span of these things amount 20–25 years. The main PEX advantages are improved physical and mechanical properties, increased melting temperature, chemical resilience and insulation resistance, a shape memory effect [1, 2]. Besides the irradiation crosslinking method, peroxide and silane-induced crosslinking methods are used for making PEX [2]. When the peroxide crosslinking method is used the polymer chain bonds formation happens by means of organic peroxides. Silane-induced crosslinking method is assuming to use organic silanes with a small amount (0.1–0.2%) of peroxide. We assume that irradiation polyethylene crosslinking method has great potential. It should be added that PEX

T. Antipova (Ed.): DSIC 2021, LNNS 381, pp. 144–153, 2022.
https://doi.org/10.1007/978-3-030-93677-8_13

products made by means of irradiation do not produce any radiation hazard, because accelerated electrons with the energy less than 10 MeV and gamma rays do not initiate nuclear reactions.

The main advantage of the irradiation crosslinking method is the chemical purity of PEX products, which makes it possible to use them in such fields as pipe production for drinking water transportation [3], or making prosthetic devices. In addition, the irradiation crosslinking method has some more advantages in comparison with other methods [2]:

1. less power consumption;
2. less floor space;
3. possibility to control the crosslinking degree by means of adjusting the total radiation dose.

However, the main disadvantage of the crosslinking method is long-time duration of the irradiation process, i.e. the time that is needed to reach a required total dose might equal tens or hundreds of hours. Therefore, technological enhancement of the crosslinking method is required.

It is well-known fact that the polyethylene has semi-crystalline structure. At the same time the required to formation new polymer chain bonds radical mobility is significantly higher in the amorphous regions than in the crystalline ones. It is also should be noted that the gas diffusion occurs most actively in the amorphous regions. Thus, polyethylene macromolecule radicals and gas molecule radicals have the highest probability of polymer chain bonds formation. The experimental data [4, 5] shows that the polyethylene crosslinking in the atmosphere of hydrocarbon gases might significantly increase the speed of crosslinking process. It is also possible to accelerate crosslinking by placing the samples in hydrocarbon gases before the crosslinking process until they reach limit gas concentration.

Thus, the purpose of this study is to construct a model of oxygen diffusion in polyethylene. Assessment of the feasibility of the double vacuuming process. In addition, the process of ethylene diffusion in the material can be combined with the process of gamma-ray irradiation. The theoretical model of the process is important for assessing the feasibility of the proposed innovation.

## 2    Nonstationary Gas Diffusion in Polyethylene Problem

During the irradiation process, polyethylene tube rolls are located in the rotating airtight containers. In this case, the rotating axis is parallel to the line along which gamma radiation sources are positioned [6]. Gamma irradiation induces chemical reactions between polyethylene radicals and oxygen, which was absorbed by polyethylene. The polymer oxidation yields to significant physical and chemical properties degradation (polymer chain scission, physical and mechanical properties degradation, insulation resistance reduction, etc.). It is necessary to minimize the oxygen concentration in the polyethylene samples to reduce the oxygen molecules negative influence in the irradiation crosslinking process. That can be reached by means of material vacuumization (degasification) process in the containers. Polyethylene in-gassing occurs on the next stage of the technological process.

We assume that material takes the region $\Omega \subset R^3$ and diffusion processes can be described by Fick's second law [7]. We suppose that tubes material is homogeneous and isotropous. Temperature variations during the diffusion process are insignificant therefore these variations should be ignored. Then the gas diffusion problem can be described by this equation:

$$\frac{\partial C(x,t)}{\partial t} = D\nabla \cdot (\nabla C(x,t)), x \in \Omega \tag{1}$$

where $C(x,t)$ is the gas concentration in the polyethylene tube roll, $D$ is the diffusion coefficient, which is assumed to be constant, $t$ is the time, $x$ is the position vector of any point from the $\Omega$ region. It is obvious that $D$ is different for any gas-material pair. We assume that interphase processes in the gas-polyethylene border goes faster than mass transport diffusion processes in material. In this situation, the gas concentration at the subsurface region quickly reaches the equilibrium value. Hence it is possible to use the first-type boundary conditions and the gas concentration on the surface layers is equal to the limit concentration of this gas in polyethylene in given conditions (temperature, pressure):

$$C(x,t) = KP(t), x \in \partial\Omega \tag{2}$$

where $K$ is the solubility coefficient, $P$ is the penetrant partial pressure. Initial conditions should be added to Eqs. (1), (2):

$$C(x,0) = C_0, x \in \Omega \tag{3}$$

Furthermore, for the region of the container that does not contain polyethylene (let it be marked as $\Omega_s$) these equations have a place:

$$P(t) = f(C(x,t)), P(0) = P_0 x \in \Omega_s \tag{4}$$

The function $f(.)$ is determined by the gas law. In this work, the classic one is used: $P = \rho RT$ where $R$ is the universal gas constant, $T$ is the absolute temperature, $\rho$ is the density. Some solving results can be found in work [8].

The limit gas concentration in the material and diffusion coefficient could be determined both by using experimental data [9, 10] and by using mathematical modelling methods such as molecular dynamics [11]. In addition, it should be noted, that the length of polyethylene tubes is much bigger than the tubes diameter, so it is possible to ignore end effects because of that fact. Also, the tubes thickness is much smaller than diameter, so three-dimensional geometry could be simplified to a one-dimensional problem.

## 3   The Modelling Results of Gas Diffusion in Polyethylene

In this work, the time, required for polyethylene tube roll to absorb ethylene ($C_2H_4$) until it reaches the limit concentration in the pressure range from 1 to 3 atmosphere at 300K temperature. The time, required for the preliminary deoxygenation of poly-ethylene film tube into the free space, is also calculated. Some coefficient values that are required for calculations are shown in Table 1.

We assume that the polyethylene film roll is homogeneous, 1.5 mm thick with fraction crystallinity of 0.69 and the density of amorphous region equal to 0.86 g/cm$^3$ [7]. The results of the modelling of the preliminary deoxygenation process are shown in Fig. 1.

**Table 1.** Parameter values that were used in calculations [7, 9, 10]

| Gas | Diffusion coefficient, cm$^2$/s | Solubility coefficient, mol/cm$^3$ * Pa |
| --- | --- | --- |
| Oxygen ($O_2$) | $4.7 * 10^{-7}$ | $2.389 * 10^{-10}$ |
| Ethylene ($C_2H_4$) | $3.3 * 10^{-8}$ | $8.3 * 10^{-11}$ |

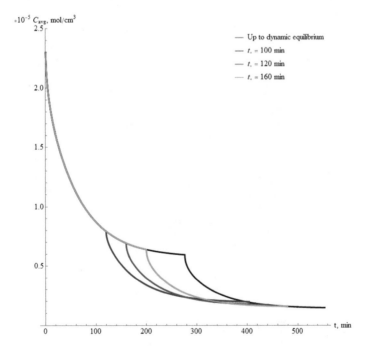

**Fig. 1.** Average oxygen concentration in the polyethylene changing with the time during the deoxygenation process (with the intermediate oxygen removal at different time)

To achieve the minimal of oxygen concentration it makes sense to do vacu-umization process in several stages. It involves dynamic equilibrium reaching between the degasification and gas penetration processes in the diffusion process. With all this going on, changing the time, when intermediate oxygen removal from the free space occurs, can significantly affects the total degasification process passing time. Degasi-fication process goes the most intensively when there is huge concentration gradient on the border, so intermediate oxygen removal time moving towards the beginning of the degasification process significantly increases total process dynamics. The results shows that it is possible to achieve average oxygen concentration up to $0.15 * 10^{-5}$ mol/cm$^3$ with the intermediate oxygen removal occurs at the dynamic equilibrium ($t \approx 280$ min) and the total time of the deoxygenation about 550 min. The intermediate oxygen removal time changing to $t = 100$ min allows us to reach average oxygen concentration up to $0.22 * 10^{-5}$ mol/cm$^3$ when the degasification process ends at $t = 385$ min. It can be noted that the resulting difference in the time of the processes is more significant than the difference in the achieved average concentrations of oxygen molecules in polyethylene. The gas concentration index was averaged by the volume of the material.

The results of the ethylene absorption process are shown in Fig. 2. It can be seen that the partial gas pressure affects the diffusion speed. The material gas saturation is becoming apparent with the time. It is obvious that penetrant average concentration values are appreciably different at each condition (pressure).

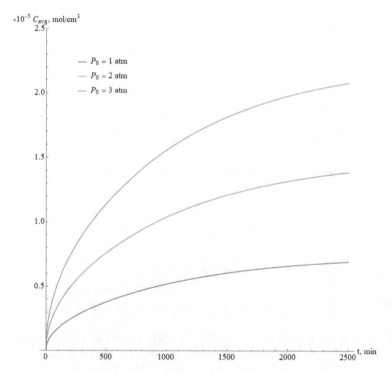

**Fig. 2.** Average ethylene concentration in polyethylene changing with the time during the ethylene absorption process at different ethylene pressure

The results of the ethylene degasification after the material saturation until it reaches $0.69 * 10^{-5}$ mol/cm$^3$ ethylene concentration at 1 atm. pressure, until it reaches $1,39 * 10^{-5}$ mol/cm$^3$ at 2 atm. pressure and until it reaches $2,08 * 10^{-5}$ mol/cm$^3$ at 3 atm pressure are shown in Fig. 3. The process goes sufficiently dynamic. In all cases average ethylene concentration significantly decreases in 16 h after the process begins. It should be noted that ethylene removal from the free space does not happen in these calculations.

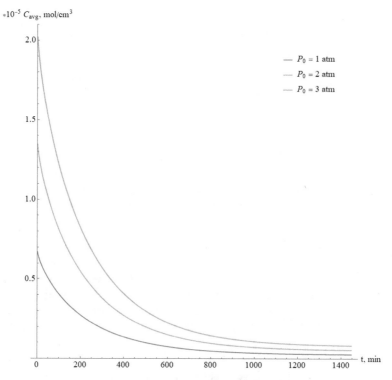

**Fig. 3.** Average ethylene concentration in polyethylene changing with the time during the ethylene degasification process with the initial conditions that were reached at different pressures

The ethylene partial pressure in free space changes with the time are shown in Fig. 4. It can be seen that the pressure asymptotically approaches the value of 0.27 atm. at the initial concentration equals $2.08 * 10^{-5}$ mol/cm$^3$, the value of 0.18 atm. at the initial concentration equals $1.39 * 10^{-5}$ mol/cm$^3$ and the value 0.09 atm. at the initial concentration equals $0.69 * 10^{-5}$ mol/cm$^3$. The diffusion process duration and the gas concentration during the degasification process estimate is important for irradiation crosslinking parameters calculation.

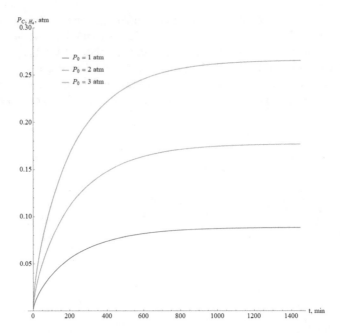

**Fig. 4.** Ethylene pressure in free space changing with the time during the ethylene degasification process with the initial conditions that were reached at different pressures

## 4   Nonlinear Gas Diffusion in Polyethylene Problem

In the process of irradiation of polyethylene and ethylene dissolved in it, radicals are formed. These radicals are the main material for creating intermolecular bonds. As a result, the part of the dissolved gas leaves the diffusion process and takes a part in crosslinking process. It is possible to consider it by adding the nonlinear member to the Eq. (1):

$$\frac{\partial C(x,t)}{\partial t} = D\nabla \cdot (\nabla C(x,t)) - kC(x,t), x \in \Omega \tag{5}$$

where $k$ describes interaction between the gas and environment and $k$ units are $sec^{-1}$ [12]. As follows from Eq. (5), the gas concentration decreasing by means of polymer chain bonds formation is irreversible. The crosslinking process goes most actively in the amorphous regions because polymer chains activity is maximum [13]. The $k$ coefficient is definitely affected by the irradiation intensity, which is also affects the radical formation intensity. It should be added that irradiation crosslinking decreases the diffusion coefficient because polymer chain bonds block gas spreading [14], but in this work, the diffusion coefficient is assumed to be constant. Boundary conditions (2)–(4) without any changes are used to solve (5) equation.

## 5  Nonlinear Problem Modelling

The time, required for polyethylene tube roll to absorb ethylene ($C_2H_4$) until it reaches the limit concentration with different $k$ values at 1 atmosphere pressure and 300 K temperature. Coefficient values that are required for calculations are shown in Table 1. The modeling results is obtained by solving Eq. (5) with (2)–(4) boundary conditions by using finite element method.

The results of the simulation are shown in Fig. 5. The average concertation that is corresponding with gas saturation decreases with $k$ value increasing. This concentration decreases up to three times when the $k$ value reaches the $10^{-4}$ order value.

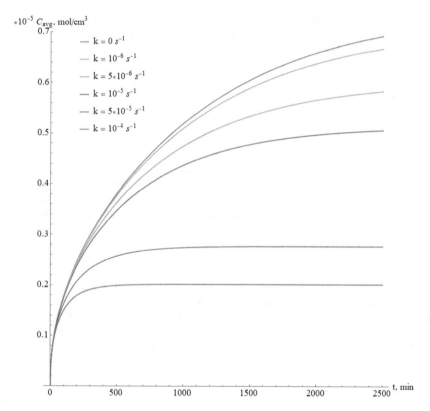

**Fig. 5.** The change in the average concentration of ethylene during diffusion in polyethylene for various values of $k$

# 6   Conclusion

The results made it possible to obtain polyethylene deoxygenation practical recommendations, including the start time of repeated degassing and the resalting oxygen concentration. In addition, understanding the gas diffusion processes is very important to do the crosslinking processes assessment during material gamma irradiation. The presence of radicals of the penetrant gas molecules in the required concentration significantly stimulates the formation of the structure. The direction of further research is related to use of the presented models to assess the parameters of the irradiation efficiency in the environment of various hydrocarbons. It is also necessary to take into account the influence of the macromolecule crosslinking process on the diffusion coefficient. Of course, this will have an impact on the final results.

**Acknowledgment.** The research was carried out with the financial support of the Government of the Perm Region in the framework project: "Models, methods and digital technologies for creating functional composite and polymer materials by processing them with concentrated gamma-ray streams in various gas environments" (grant № C- 26/581).

# References

1. Kniazev, V.K., Sidorov, N.A.: Irradiated polyethylene in engineering. Chemistry, Moskov, vol. 374 (1974)
2. Skroznikov, S.V.: Structural and mechanical properties formation regularities in crosslinked polyolefins for cable engineering. Ph.D. thesis. Moscov, vol. 149 (2015)
3. Holder, S.L., Hedenqvist, M.S., Nilsson, F.: Understanding and modelling the diffusion process of low molecular weight substances in polyethylene pipes. Water Res. **157**, 301–309 (2019)
4. Shyichuk, A., Tokaryk, G.: Simulation-assisted evaluation of acetylene effect on macromolecular crosslinking rate under polyethylene irradiation. Macromol. Theory Simul. **12**(8), 599–603 (2003)
5. Golubenko, I.S., Prokop'ev, O.V., Dalinkevich, A.A.: Method of radiation cross-linking polyolefin products. Russian Federation Patent no. 2004104113/04 (20016)
6. Oniskiv, V.D., Stolbov, V., Khatiamov, R.K.: On one control problem of the process of gamma irradiation of the polyethylene. Appl. Math. Control Sci. **3**, 119–130 (2019)
7. Mehrer, H.: Diffusion in Solids, p. 654. Springer, Heidelberg (2007). https://doi.org/10.1007/978-3-540-71488-0
8. Oniskiv, V.D., Yakovlev, A.: Some results of solving the non-stationary problem of gas diffusion in polyethylene. Appl. Math. Control Sci. **2**, 41–51 (2021)
9. Meshkova, I.N., Ushakova, T.M., Gul'tseva, N.M.: Determination of ethylene and propylene solubility constants in polyethylene and polypropylene and their application for calculation of kinetic parameters of gas-phase and suspension homo-polymerization and copolymerization of olefins. High-Mol. Compd. Ser. A **46**(12), 1996–2003 (2004)
10. Gerasimov, G.N., Abkin, A.D., Khomikovskii, P.M.: Mechanism of the heterogeneous polymerization of ethylene under the action of ionizing radiation. High-Mol. Compd. **5**(4), 479–485 (1963)

11. Börjesson, A., Erdtmana, E., Ahlström, P., Berlin, M., Andersson, T., Bolton, K.: Molecular modelling of oxygen and water permeation in polyethylene. Polymer **54**(12), 2988–2998 (2013)
12. Bekman, I.N.: The Diffusion Mathematic: School-Book, vol. 400. OntoPrint, Moskov (2016)
13. Kozlov, G.V., Zaikov, G.E.: Gas diffusion in semi-crystalline polyethylene and in it's flux. High-Mol. Compd. Ser. B **45**(7), 752–758 (2003)
14. Kozlov, G.V., Burya, A.I., Zaikov, G.E.: The structure influence on gas diffusion in polyethylene. Rep. Natl. Acad. Sci. Ukraine **2**, 138–142 (2007)

# Information System for Controlling the Process of Gamma Irradiation of Composite Materials

Daniil Oniskiv⬛, Valerii Stolbov$^{(\boxtimes)}$⬛, and Vladislav Dolgirev⬛

Perm State National Research Polytechnic University, Perm, Russia

**Abstract.** The process of gamma irradiation of polymer materials in order to improve their operational properties is considered. The necessity of automating the process on the based on the development of mathematical and software for controlling the distribution of radiation sources inside the chamber is justified. Localization of sources should ensure a minimum amount of uneven exposure. The problem of optimizing the process of gamma-ray irradiation of polyethylene products due to the distribution of radiation sources in an industrial installation is considered. A heuristic algorithm is proposed to determine the optimal distribution of sources, which ensures minimal power losses inside the chamber where the material is irradiated. The algorithm is implemented as an information decision support system for controlling the irradiation process, taking into account the processed material and a set of radiation sources with a given activity. Functional and structural models of the information system are presented. Some optimization results and demonstration examples of information system operation that have practical application are presented. Using the results allows you to significantly reduce the exposure time of the polymer material.

**Keywords:** Process of gamma irradiation · Polymer materials · Distribution of radiation sources · Heuristic algorithm · Information system · Discrete optimization

## 1 Introduction

The process of gamma irradiation, which is widely used in various branches of mechanical engineering to improve the operational properties of various polymer and composite materials is considered [1–3]. For example, high-quality materials obtained on the basis irradiated polyethylene, are widely used in electro and radio engineering. Irradiation of polyethylene with gamma quanta leads to the appearance of more complex bonds in it. Such changes in polyethylene are observed when it is irradiated to a dose of $5 \cdot 10^6$ Rad. By gamma irradiation dose above $5 \cdot 10^6$ Rad polyethylene goes into a non-melting state and does not melt even at temperature of 300 °C. At a higher temperature, polyethylene passes into rubberlike state. At a radiation dose of $2 \cdot 10^9$ Rad polyethylene completely passes into an amorphous state. The crystalline regions are preserved only when irradiated with small doses. By changing the conditions of the technological process of irradiation, the structure of initial polymer, it is possible to

T. Antipova (Ed.): DSIC 2021, LNNS 381, pp. 154–162, 2022.
https://doi.org/10.1007/978-3-030-93677-8_14

obtain optimal variants of practically new materials that have valuable properties. Irradiated polyethylene is used for insulation of mounting wires, coaxial compounds cables in high-frequency devices [4].

Products with the shape memory effect are widely used in the technique and various technologies of modern production [5]. Such industries as the electrical, cable and construction are among the largest consumers of this type of product. In addition to chemical technologies and procedures for irradiation with directed fluxes of accelerated electrons, gamma irradiation is used to create molecular crosslinked polyethylene [6–9]. This technology is used to create additional molecular bonds using gamma irradiation from source cobalt 60 - $CO^{60}$ and has a number of specific advantages related to the quality of resulting product. Moreover, there are certain design features and technological limitations of the irradiation process. The radiation dose required for the crosslinking effect should have a well-defined value, which should be achieved as evenly as possible over the entire length of the irradiated material located in a special chamber. In order to guarantee the elimination of defects in a real production (lack of insoluble gel fraction in the material), the exposure time is calculated according to the minimum value of the dose rate, which depends on the location of ionizing radiation sources [10]. Optimal placement of point-based sources is a complex problem that requires the use of modern mathematical methods and information technologies. This makes it necessary to develop an information system (IS) to support decision-making in the management of the irradiation process, depending on the irradiated material, as well as data on the activity of existing sources and their possible location.

## 2   Conceptual Model of the Information System

The need to create this IS exists at the enterprises of the nuclear industry in connection with the desire to automate the technological process of irradiation of materials with gamma quanta. The irradiation of polymers occurs in special chambers, where the sources are located. Their placement of sources in the chambers often depends on the experience of the technologist and does not have an optimal character. This leads to the partial loss of the sources' useful power. However, the irradiation process is quite complex and requires not trivial calculation. This is due to the fact that according to technological requirements, each point of the irradiated material must receive a certain radiation dose, which is 10 Mrad, but the placement takes place without preliminary calculations and the material receives different radiation doses at different points. For example, achieving the required dose of 10 Mrad at the material boundary leads to excessive irradiation in the middle. This circumstance is a consequence of the isotropic nature of gamma radiation. With a uniform distribution of sources along the length of the chamber, the material placed at the extreme points will not be sufficiently irradiated.

These arguments were the main ones in decision-making on the IS creation. The IS should manage this process in a way that the material in the chamber receives a uniform dose of radiation. This will reduce the total time of exposure, decrease production costs and lead to increased profits.

In other words, the main task of the developed information system is the automation functions, that were previously performed by people. At the same time, automation requires that these functions would be performed more accurately, faster and cheaper, thereby making production more efficient.

The users of this information system will be the engineers of the enterprise who manage the processes of irradiation of the material, creating the optimal mode.

The IS under development should perform the following main functions:

1. Receiving, processing and storing source information.
2. Optimal distribution of gamma radiation sources within one camera and between cameras.
3. Calculation of the expose time in each chamber, depending on the material and the specified radiation dose.
4. Visualization of results and preparation of reports (Fig. 1).

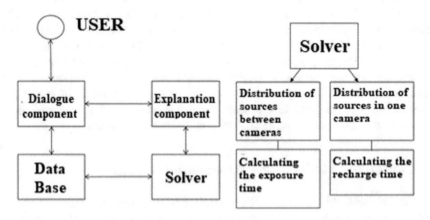

**Fig.1.** The IS architecture

Consider the tools that allow you to implement the described information system. Let's start with a description of the database. The database is needed in this system in order to store information about cameras and their sources. The database consists of two entities: camera and source, linked to each other by a foreign key «idKamera» (Fig. 2).

Foreign key (Foreignkey) provides an unambiguous logical relationship between entities. In addition, each entity (table) has a unique attribute (column) that identifies each record as much as possible. This attribute is called the primary key (Primarykey). In essence "Sources"this key is an attribute "idSources", but in essence "Kamera» – «idKamera».

Each DB entity contains specific information about itself. For example, the entity "Sources"has fields such as the source number (idSources), camera number (idKamera), upload date (Purchase_date), source activity (Activity), number of sources (Count). Entity "Kamera"contains only information about the camera number for distributing radiation sources to cameras.

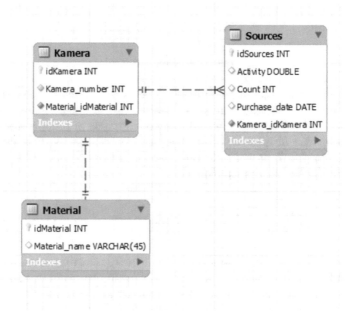

**Fig. 2.** Database scheme

DBMS Oracle Corporation – MySQL was chosen for this project. The choice is made in favor of this DBMS because this information system does not have a large database, so there is no need to use "large" products such as MS SQL Server, PostgreSQL and others. The prototype of the information system will be used, presumably, on one computer. However, if you need to transfer the database to another device, this will not be difficult, because you can dump the database in just a few steps. Also, the main advantage of this DBMS is a simple interface, which makes it easier for users to learn and work with this system.

The programming language selected for this system is C#. The choice is based on the fact that this programming language is supported by the products Microsoft. Besides, C# distributed for free with the development environment VisualStudio, which is annually updated, making life easier for programmers who maintain the IS.

The main module of this information system is - solver that allows, depending on the irradiated material and the required radiation dose, to find the optimal distribution of radiation sources and exposure time by solving the corresponding discrete optimization problem.

## 3   Formulation and Algorithm for Solving a Discrete Problem of Optimizing the Distribution of Radiation Sources

Let there be $N$ sources of gamma radiation and $M$ placement points ($N \gg M$). Denote by $A_i$ – activity of the $i$ - th radiation source. At the moment $t = 0$ it is known for each $i$ and is marked as $A_{0i}$. The sources are placed in special tubes, each contains no more than five of them. The tubes are located along a certain straight-line x at some distance from the irradiated material. The following optimization problem is considered: to find such an optimal distribution of ionizing radiation sources over the tubes during the irradiation period $T$, at which the objective function that characterizes the power loss during irradiation reaches its minimum:

$$W = \int_0^T \int_0^l \left( \sum_{i=1}^N \Gamma_{CH} * A_{0(i)} * \frac{2^{\left(\frac{-t}{T_n}\right)}}{r_i^2(x)} - P_{min}(t) \right) dxdt \rightarrow min,$$

where $\Gamma_{CH}^\infty$ – ionization gamma constant, $T$ – exposure time, $l$ – length of the source location line, $T_n$ – the half-life, which for Co$^{60}$ is 5.2 years old, $r_i$ - distance from $i$-th source to the point with the coordinate $x$ on the surface of the material, $P_{min}$ – minimum power value for length, $i$ – summation index for all sources from 1 to $N$.

Note that power losses occur due to the uneven distribution of power along the length due to the peculiarities of radiation and different activity of sources. Obviously, the power value will be higher in the middle of the camera, and lower at the edges. Depending on the distribution of sources at each time point $t$ a certain distribution of the irradiation power over the length of the chamber is realized, for which the minimum value can be found:

$$P_{min}(t) = \min_{0 \le x \le l} (P(x,t)) \; \forall t \in [0, T].$$

This value is the base value for calculating the exposure time of the irradiated material, which determines the required radiation dose for the entire material. Due to the uneven distribution of radiation sources, some of the material receives an excessive dose, which can be conditionally called "lost" and needs to be minimized.

A more detailed statement of this discrete optimization problem is given in the works [10]. It should be noted that this task is related to NP-complex optimization problems [11]. Intelligent algorithms are usually used to solve such problems [12, 13]. Therefore, two algorithms for solving it were proposed. The first one is based on the proposed heuristics and allows us to significantly reduce the set of acceptable source distribution options based on the characteristics of the irradiation process [14], and the second one is based on the Monte Carlo method [15]. As calculations have shown, the first algorithm requires longer calculations, but gives more accurate results that provide an approximation to the results obtained by full search on test cases. Both algorithms are included in the solver and they can be used at the user's choice.

# 4   IS Testing Results

Initially, we will consider the case of a typical distribution of radiation sources inside the camera. Consider a camera that contains 34 tubes. It is required to distribute 48 sources of ionizing radiation of two types. There are 24 sources of each type. The first type of sources has activity on the purchase date of 4.54 * $10^{13}$ Bq. The second type is 7.23 * $10^{13}$ Bq. Accordingly, we have a resource limit: the camera tubes must contain exactly 48 sources with 24 elements of each type. The number of sources in the tube is 3.

All conditions for calculating capacities and losses are set, so you can plot the so-called typical distribution (which is usually used in the company). A typical distribution was chosen that would ensure a uniform arrangement of elements in the camera (see Table 1). The following designations are used here: 0 - no source, 1 - type 1 source, 2-type 2 source.

**Table 1.** Typical distribution of sources in the camera

```
2 2 2 2 2 2 2 2 2 2 2 2 1 1 1 1 1 1 1 1 1 1 2 2 2 2 2 2 2 2 2 2 2 2
0 0 0 0 0 0 0 0 0 0 0 0 1 1 1 1 1 1 1 1 1 1 0 0 0 0 0 0 0 0 0 0 0 0
0 0 0 0 0 0 0 0 0 0 0 0 0 0 1 1 1 1 0 0 0 0 0 0 0 0 0 0 0 0 0 0 0 0
```

Figure 3 shows a screenshot obtained by the information system results visualization module. It shows the power distribution along the length of the camera, taking into account the time when sources are located in the camera and the decrease in activity (with a discrete time step of 1 year).

The following power loss results were obtained:

- Instantaneous power loss at t = 0 (in red): 0.0918 Gr/s.
- Instantaneous power loss at t = 1 (green): 0.0851 Gr/s.
- Instantaneous power loss at t = 2 (brown): 0.0706 G/s.
- Instantaneous power loss at t = 3 (blue): 0.0618 Gr/s.
- Instantaneous power loss at t = 4 (in black): 0.0542 Gr/s.

Based on this study, we can conclude that the prototype of the information system is adequate. This graph of the distribution of power losses qualitatively coincides with the data that were provided by the enterprise as an example. Specialists irradiated the polymer material in the chamber for a certain time, during which the effect of irradiation on the material was investigated, and a similar power loss graph was obtained. Therefore, we can conclude that the system is working correctly.

Now consider the case optimal distribution of the same radiation sources. It was found using a heuristic algorithm for solving the original discrete optimization problem. It should be noted that the algorithm based on the Monte Carlo method gave a similar result, which confirms the correctness of the approximate solution of the original optimization problem (Table 2).

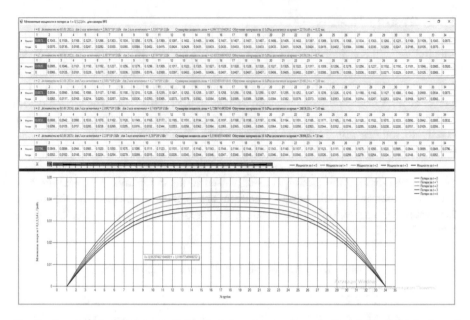

**Fig. 3.** Screenshot of ionizing radiation (graphs of power losses with a typical distribution of radiation sources)

**Table 2.** Optimal distribution of sources

```
2 2 2 1 2 2 2 1 1 1 1 0 2 2 1 1 2 2 1 1 2 2 0 1 1 1 1 2 2 2 1 2 2 2
1 2 2 1 1 0 0 0 0 0 0 0 0 0 0 0 0 0 0 0 0 0 0 0 0 0 0 0 0 1 1 2 2 1
2 1 0 1 0 0 0 0 0 0 0 0 0 0 0 0 0 0 0 0 0 0 0 0 0 0 0 0 0 1 0 1 2
```

The resulting source distribution is characterized by the following power loss graph, shown in Fig. 4.

With this distribution, the following power loss results were obtained:

- Instantaneous power loss at t = 0 (in red): 0.024 Gr/s.
- Instantaneous power loss at t = 1 (green): 0.0215 Gr/s.
- Instantaneous power loss at t = 2 (brown): 0.0188 G/s.
- Instantaneous power loss at t = 3 (blue): 0.0165 G/s.
- Instantaneous power loss at t = 4 (in black): 0.0145 G/s.

From the above results, it can be seen that with an optimal distribution of power sources, power losses are reduced by about 4 times compared to their typical distribution used in the enterprise. This confirms the possibility of using the developed IS to improve the efficiency of the irradiation process.

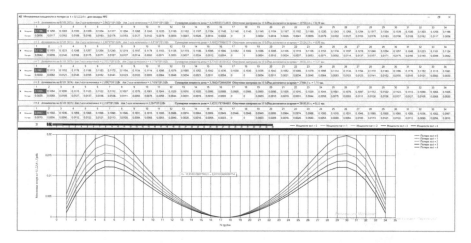

**Fig. 4.** Screenshot of ionizing radiation (power loss graphs for optimal distribution of radiation sources)

## 5   Conclusion

The main goal of the work was the task of developing an information system that would allow controlling the gamma irradiation of various materials. In this case, the field of equal radiation doses should be as uniform as possible along the axis of the container. The search for the optimal placement of radiation sources was solved by creating some heuristics. The results that were obtained are quite close to the calculations using the Monte Carlo method. The conclusion is made about the suitability of the information system for the design of a real technological process. The system certainly has limitations associated with a specific irradiation technology. The expansion of the capabilities of the created information system can be associated with the calculation of the optimal placement of radiation sources simultaneously for several irradiation chambers.

**Acknowledgments.**   The research was carried out with the financial support of the Government of the Perm Region in the framework project: "Models, methods and digital tecynologies for creating functional composite and polymer materials by processing them with concentrated gamma-ray streams in various gas enviroments" (grant № C-26/581).

## References

1. Kablov, B.N.: Modern materials - the basis of innovative modernization of Russia. Metals Eurasia **3**, 10–15 (2012)
2. Gumenyuk, N.S., et al.: Application of composite materials in shipbuilding. Mod. High-Tech Technol. **8**, 116–117 (2013)
3. Weng, D., et al.: The influence of Buckminsterfullerenes and their derivaties on Polymer properties. Eur. Polymer J. **35**, 867–878 (1999)

4. Zhilkina, N.V., Larin, Yu.T., Vorob'ev, V.M.: Investigation of the influence of gamma radiation on the physical and mechanical characteristics of polymer materials for protecting optical cable shells. Cables Wires **3**, 11–15 (2004)
5. Knyazev, V.K., Sidorov, N.A.: Irradiated polyethylene in engineering. Chemistry, Moskov (1974)
6. Shaimukhametova, I.F., Shigabieva, Y., et al.: Effect of gamma irradiation on the gel-forming ability of cross-linked acrylic polymer. Chem. High-Ones Technol. **54**(3), 122–125 (2020)
7. Aguilar, M.R., Roman, J.S.: Smart Polymers and their Applications, 1st edn, vol. 584. Woodhead Publishing (2014)
8. Chmielewski, A.G., Al-Sheikhly, M., Berejka, A.J., Cleland, M.R., Antoniak, M.: Recent developments in the application of electron accelerators for polymer processing. Radiat. Phys. Chem. **94**(1), 147–150 (2014)
9. Berejka, A.J., Cleland, M.R., Walo, M.: The evolution of and challenges for industrial radiation processing. Radiat. Phys. Chem. **94**(1), 141–146 (2014)
10. Oniskiv, V.D., Stolbov, V., Khatiamov, R.K.: On one control problem of the process of gamma irradiation of the polyethylene. Appl. Math. Control Sci. **3**, 119–130 (2019)
11. Court, B., Figen, J.: Combinatorial System Optimization. Theory and Algorithms. Publishing House: ICNMO, vol. 720 (2015)
12. Gladkov, L.A., Kureychik, V.V., Kureychik, V.M.: Genetic Algorithms (Electronic Resource). Fizmatlit, Moskov, vol. 365 (2010)
13. Shtovba, S.D.: Ant algorithms. Exponenta pro. Math. Appl. **4**, 70–75 (2003)
14. Oniskiv, V., Stolbov, V., Oniskiv, L.: Optimization of the processing of functional materials using gamma irradiation. In: Antipova, T., Rocha, Á. (eds.) DSIC 2019. AISC, vol. 1114, pp. 489–498. Springer, Cham (2020). https://doi.org/10.1007/978-3-030-37737-3_42
15. Stolbov, V.Y., Dolgirev, V.D., Oniskiv, D.V.: Optimization of the process of gamma irradiation of polyethylene based on the monte Carlo method. In: Antipova, T. (ed.) ICCS 2021. LNNS, vol. 315, pp. 471–479. Springer, Cham (2022). https://doi.org/10.1007/978-3-030-85799-8_41

# Digital Environmental Sciences

# Financial Support Measures for Organic Agricultural Products Producers as an Element of Sustainable Development of Ecological Production

Irina Korostelkina[✉] [iD], Natalia Varaksa[iD], Victoria Gordina[iD],
Oksana Fokina[iD], and Mikhail Korostelkin[iD]

Orel State University, Naugorskoe Highway, 40,
302020 Orel, Russian Federation

**Abstract.** Today, the principles of sustainable development of society and the economy are actively supported in the world through the greening of processes and relations. One of the promising ecological directions of economic development is organic production, the role of which is to preserve and improve ecosystems and organisms. The coronavirus pandemic has once again shown the importance of forming a healthy immune system of the population around the world, which actualizes the importance of the development of the organic sector. In general, the role of organic farming, as a basis, both in the production and consumption of food in Russia remains still insignificant. Increasing the importance of organic eco-culture is possible only through state regulatory measures, including in the form of financial support. The purpose of this article is to consider the dynamics of the organic products market, identify threats and opportunities for the development of organic production in Russia with the application of the best international practices and propose financial support measures for the growth of the organic sector. The methodological tools of the research are general and private scientific methods, tools of graphical interpretation and comparative analysis. The article provides a theoretical overview of the research of the problem under consideration, analyzes the state of the organic products market in the world, gives an overview of existing methods and measures to support organic producers in a number of foreign countries, on the basis of which a system of financial support for this segment in Russia is proposed.

**Keywords:** Agriculture · Organic products · Financial support · Development of the organic sector · State support measures · Crowdfunding

## 1 Introduction

In modern conditions of environmental deterioration, people are increasingly thinking about the benefits of natural agricultural products grown without synthetic mineral fertilizers, growth regulators and artificial food additives, that is, the so-called «organic products». The recent events of 2020 related to the coronavirus pandemic have once

T. Antipova (Ed.): DSIC 2021, LNNS 381, pp. 165–176, 2022.
https://doi.org/10.1007/978-3-030-93677-8_15

again emphasized the importance of the natural immunity of the nation, and the immunity of each person depends, first of all, on the products that he uses. The global trend of recent years is an increase in the share of production and consumption of organic products, as well as an increase in the area for organic farming compared to traditional agriculture. According to the report of the Research Institute of Organic Agriculture FiBL and IFOAM - Organics International, presented at the world exhibition BIOFACH 2021, the area of organic farmland in the world increased by 11 million hectares and amounted to 72.3 million hectares [1, 2]. The report is based on a study of data on 187 countries with organic activity. The global organic food market grew more than fivefold from 2000 to 2019 and amounted to 106.4 billion euro [2]. In Russia, the market of organic products is just beginning to develop, a regulatory framework has been created, regions are actively involved in the process of organic farming, which has its own characteristics.

Organic farming is based on the use of natural means and substances to combat various pests and weeds, as well as to fertilize the soil. Organic production requires a conversion period, usually three years, during which there is an adaptation of agricultural technologies, the use of new forms of biological fertilizers, high-quality seeds, special preparations for plant protection, bio-feeds and special preparations for animals. This period is associated with additional costs for the agricultural producer and is accompanied by a decrease in production volumes. In this regard, for the normal functioning and further development, increasing the pace of production, an agricultural producer needs state support and, above all, its financial component. Therefore, the study of financial support measures for producers of organic agricultural products is important and necessary at the present stage of global development and ensuring food security of the state. The study of scientific works on this topic showed that at present there are many studies devoted to the development of organic agriculture, however, insufficient attention has been paid to improving financial support for agricultural producers of organic products.

An analysis of scientific works on this topic showed that at present there are many studies devoted to the development of organic agriculture, however, insufficient attention has been paid to improving financial support for agricultural producers of organic products. In this regard, the purpose of the study is to study the world market for organic products and identify threats and opportunities for the development of organic production in Russia, applying the best world practices, as well as developing and justifying financial support measures for the growth of the domestic organic sector.

## 2   Materials and Methods

The methodological basis of the research was a set of general scientific approaches and methods, in particular, when writing an article, justifying the relevance of the topic, obtaining results and formulating conclusions, methods of scientific search, analysis, comparison, systematization and classification were used. The provisions of the study are also reasoned using private scientific methods. In the course of the work, a systematic approach was used to identify the key areas of financial support for agricultural producers of organic products. A theoretical review of the research of the problem under consideration was carried out using the tools of graphical interpretation and comparative analysis. The study of the state and development of organic production in Russia was conducted using SWOT analysis. Measures of financial support for organic producers are justified through the use of expert methods, as well as content analysis.

## 3 Theoretical Review

Considering the theoretical basis of the research, we can not that for many years domestic and foreign scientists have raised questions about the potential of organic agriculture and its importance for solving problems of environmental protection and ensuring food security. The success of the development of organic production depends on many factors that generally determine its competitiveness. The issues of the sustainability of agricultural producers and the competitiveness of organic agriculture were studied by such scientists as J. Henning, L. Baker, P. Thomassin [3], C. Benbrook, S. Kegley, B. Baker [4], W. David Crowder, John P. Reganold [5], A.S. Shcherbakova (Ponomareva) [6], etc. As a result, it is established that the development of this promising segment requires a well-thought-out state policy, including infrastructure, legal, information, financial and other support tools.

Each country has its own peculiarities of the development of the organic products market and measures of state support, which are reflected in the research of such scientists as O. Minkova, A. Kalinichenko, O. Gorb [7], S. Roljevic Nikolic, P. Vukovic, B. Grujic [8], etc. In their works, scientists note that the main obstacle to the development of organic production is the weak development of environmental infrastructure and insignificant support from various services. Effective incentive mechanisms include subsidies during the conversion period, advisory and organizational support, a clear legislative framework, an effective control and certification system [7]. Only systematic state support is crucial for accelerating the growth of organic agriculture [8]. W. Luczka, S. Kalinowski, N. Shmygol they believe that the state support policy opens up great opportunities for using the potential of organic farming and contributes to sustainable development, thanks to which it will have both environmental and socio-economic benefits [9].

Interesting from a practical and theoretical point of view are the studies of scientists devoted to the application of the world experience in the development of organic production to the conditions of functioning of domestic agricultural producers. Thus, T. M. Polushkina, Yu. Akimova, S.A. Kochetkova [10] distinguish the formation of a culture of organic consumption based on the development of the concept of organic production as a key element of state regulation. V.F. Pivovarov, A.F. Razin, M.I. Ivanova, R.A. Meshcheryakova, O.A. Razin, T.N. Surikhina, N.N. Lebedeva [11], studying the system of legal regulation of organic production, they focus on the need to adapt the domestic legal system to international requirements in order to recognize it by international partners and reduce technical barriers to export, and they see the solution to the problems of organic production development in effective financial support. Despite the available research of scientists, today it is necessary to further analyze the trends in the development of the organic products market and the mechanisms of support from the state in order to increase the share of this segment and increase its attractiveness. In addition, the contribution of existing support measures at the national level for the organic sector as an element of sustainable development of ecological production is not sufficiently evaluated, which requires the development of new effective tools aimed at stimulating the interest of all participants in this process.

## 4   Results

The world market of organic crops is developing rapidly, there is an increase in the number of producers of organic products, the agricultural area used and the share of consumption of these products by the population is growing. The leading positions in terms of consumption of organic products are occupied by the United States – more than 44 billion euro or 42.03% of total world consumption. Germany and France are allocated from the EU countries – more than 11 billion euro. The fourth place belongs to China with a consumption volume of 8.5 billion euro, which is 7.99% of world consumption [2]. Sales of organic products in the EU have increased by 128% over 10 years, which is explained by the high level of consumer confidence, who are willing to spend 84 euros a year on organic products. In accordance with the EU's long-term plan until 2030, an indicator of an increase in agricultural land for organic production by 25% is set [12]. Denmark and Switzerland account for the largest consumption of organic products per capita (see Fig. 1).

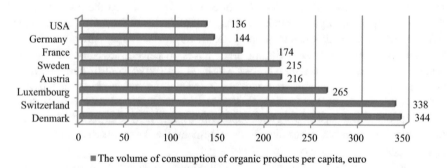

■ The volume of consumption of organic products per capita, euro

**Fig. 1.** The volume of consumption of organic products per capita by countries of the world (compiled by the authors on the basis of data [2])

The number of organic producers is increasing due to the profitability of this market. In 2019, India is the leader in this indicator – more than 1.3 million producers, Uganda-more than 210 thousand, Ethiopia – more than 203 thousand producers [2]. At the same time, the peculiarity of the organic products market is the fact that the countries that are the main consumers of organic products cannot produce it them-selves. For example, there are enough lands in Europe where classical agricultural technologies work, but it is not possible to transfer them to organic agriculture in a short time. The use of land in Africa and Latin America is complicated by the lack of innovative technologies for the development of this area. Therefore, the importance of the development of organic agriculture in Russia and the CIS countries and the interest of foreign investment in this segment of agriculture is increasing. In Russia, organic production began to develop more than 10 years ago. Currently, about 120 manufacturers carry out their activities in this segment. According to the data of the unified state register of organic producers, 72 organic producers were included in 2021. Of course, such a number of organizations is not enough for a significant increase in market share, while there is an increased demand for products of this type. To change this situation in order to approach the indicators of the EU countries, Russia needs to certify about 300 enterprises annually, which will allow it to take positions among the world market leaders in the next ten years [13]. Thus, this type of agriculture is developing very slowly in Russia, although a number of positive trends can be noted in recent years.

To classify the manufacturer's products as organic, its certification is required. In the countries of the European Union, there is a single organic standard, the symbol of which is the marking «Green Leaf», in the USA – USDA Organic, in Italy – ICEA [14]. The Federal Law «On Organic Products and on Amendments to Certain Legislative Acts of the Russian Federation» dated 03.08.2018 No. 280-FZ», which regulates the production procedure and defines labeling standards, has entered into force in Russia [15]. At the moment, there are 4 national standards in this area in Russia, the unified state register of organic producers, 8 certification bodies conduct it according to interstate GOST 33980–2016, 17 certification bodies conduct it according to international standards of the EU, the USA and Japan [16].

According to experts, Russia's potential in organic production is now realized by 1%, which requires active government measures to regulate and support this segment, especially during the conversion period, which is three years and is accompanied by an increase in costs and a decrease in the productivity of farms engaged in growing organic products.

The growth of organic production in Russia is also limited by the fact that the largest share of investments in organic production is private investment. The state currently allocates insufficient funds for the development of this area of agriculture. The cultivation of organic products is mainly carried out by peasant farms (KFH) and small agricultural organizations, which are more flexible in terms of changing production processes compared to holding companies. However, the transition to organic production (conversion) requires high costs for certification, the development of environmentally friendly technologies, the purchase of seeds and organic fertilizers, which is often beyond the power of small farms and it is difficult for them to develop without financial support from the state.

Currently, the Russian market of organic agricultural products has its own characteristics. Organic products in Russia have a high marginality, which ensures their high price, and only after the successful completion of the conversion period by the manufacturer. European organic producers are ready to work on a low margin, ensuring high competitiveness of their products on the market in the absence of bureaucratic barriers at all levels. Domestic producers of organic products can be competitive in international markets due to the low cost of labor resources. Russia has a high export potential, but now the country's share in the world production of organic products is only 0.2% [13]. The most popular Russian organic products in the EU countries are cereals, oilseeds, wild berries, mushrooms, as well as legumes. Promising crops for the development of organic farming are flax, sunflower, rapeseed and corn.

Studies have shown that the main motive for the consumption of organic products in Russia is health care. However, less than 1% of the population noted the constant consumption of organic products (during the survey) [13]. It is important to understand that organic matter is not a product of mass demand, which imposes its own specifics and forms factors that restrain the formation of demand: 1) Most consumers have difficulty identifying organic products; 2) High price of organic products. The difference in prices of 30–40% between organic and non-organic products is too large. A difference of 15% due to an increase in market share would solve the problem; 3) Low level of public awareness about organic products and their advantages; 4) Demand is formed to a greater extent in large cities with a high population density and a higher level of income of citizens than in other regions. Based on the analysis, it is possible to identify the strengths and weaknesses of the state, opportunities and threats of the development of organic production in Russia (see Fig. 2).

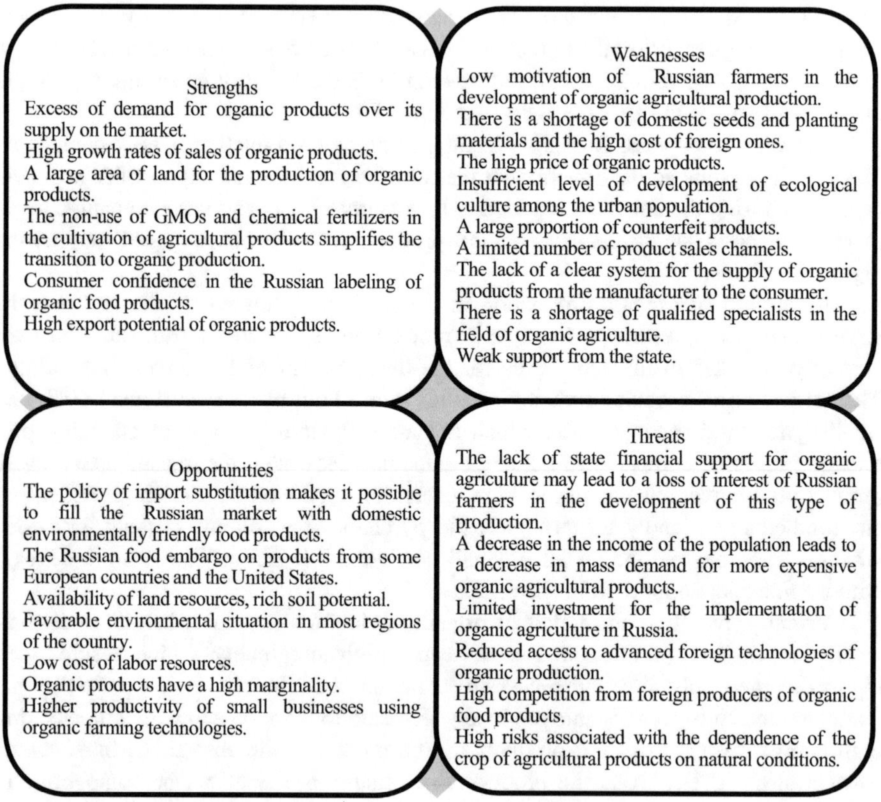

**Strengths**
Excess of demand for organic products over its supply on the market.
High growth rates of sales of organic products.
A large area of land for the production of organic products.
The non-use of GMOs and chemical fertilizers in the cultivation of agricultural products simplifies the transition to organic production.
Consumer confidence in the Russian labeling of organic food products.
High export potential of organic products.

**Weaknesses**
Low motivation of Russian farmers in the development of organic agricultural production.
There is a shortage of domestic seeds and planting materials and the high cost of foreign ones.
The high price of organic products.
Insufficient level of development of ecological culture among the urban population.
A large proportion of counterfeit products.
A limited number of product sales channels.
The lack of a clear system for the supply of organic products from the manufacturer to the consumer.
There is a shortage of qualified specialists in the field of organic agriculture.
Weak support from the state.

**Opportunities**
The policy of import substitution makes it possible to fill the Russian market with domestic environmentally friendly food products.
The Russian food embargo on products from some European countries and the United States.
Availability of land resources, rich soil potential.
Favorable environmental situation in most regions of the country.
Low cost of labor resources.
Organic products have a high marginality.
Higher productivity of small businesses using organic farming technologies.

**Threats**
The lack of state financial support for organic agriculture may lead to a loss of interest of Russian farmers in the development of this type of production.
A decrease in the income of the population leads to a decrease in mass demand for more expensive organic agricultural products.
Limited investment for the implementation of organic agriculture in Russia.
Reduced access to advanced foreign technologies of organic production.
High competition from foreign producers of organic food products.
High risks associated with the dependence of the crop of agricultural products on natural conditions.

**Fig. 2.** SWOT analysis of the state and development of organic production in Russia (compiled by the authors)

The analysis showed that along with the strengths and opportunities for the development of organic production in Russia, there are disadvantages and threats. Among them: low motivation of Russian farmers in the development of organic agricultural production; high price of products, insufficient level of development of ecological culture among the urban population; shortage of qualified specialists in the field of organic agriculture; limited investment for organic agriculture, and others. Moreover, it can be noted that eliminating most of the threats and leveling the weaknesses of the development of organic production in Russia requires comprehensive state financial support. In foreign countries, direct and indirect methods of financial support for organic production are used. Direct methods of financial support mainly involve large cash payments during the transitional stage of organic agriculture to carry out environmental activities, as well as to meet the requirements of organic standards. Indirect support involves compensation for part of the costs of certification of organic production, scientific research, information support for participants in the market for organic products, promotion of this area through various sources, etc. As studies of foreign experience of state support for organic producers have shown, significant funds are allocated in the USA and EU countries for the development of the sector and state subsidies are provided to farmers [17]. In foreign countries, direct and indirect methods of financial support for organic production are used. In the United States, there is direct support in the form of subsidies for the costs associated with certification and support for farmers during the conversion period, compensation for the costs of marketing organic products, providing tax discounts, etc. In addition, support for organic producers is provided within the framework of the Environmental Quality Incentive Program( EQIP), the annual maximum amount of financial assistance is $ 20 thousand, and in the long term (for 6 years) - $ 80 thousand. There is also financial support for the organic products market at the level of individual states, including on the basis of co-financing with federal programs. Indirect financial support in the United States is associated with the provision of compensation for part of the costs of certification of production, an information campaign to promote organic products, etc. In the EU countries, financial support for the organic products market is carried out in three main directions: with in the framework of the Rural Development Program (RDP), whose activities are carried out in accordance with the second fundamental element of the CAP (Pillar II); direct support for the production and development of agri-food markets in accordance with the first fundamental element of the CAP (Pillar I); within the framework of national and regional agricultural support programs.

So, in Germany, in the first year of the conversion period, the manufacturer receives the maximum amount of subsidies-about 500 euros per hectare, in the second year – about 300 euros, in the third year-200 euros. An important area of support in the organic products market in the EU countries is the support of investments in organic products, certification services, professional development of employees, consulting and information support, the development of education in the profile, public procurement of organic products, marketing, scientific research. Some EU countries (Austria, Belgium, Cyprus, Estonia, France, Greece, Italy, Malta, the Netherlands, Poland, Portugal, Slovenia, most regions of Spain) compensate for the costs of certification in accordance with the RDP. In Denmark, organic certification takes place on a free basis [18]. Thus, state support for the production of organic products in developed countries occurs in several directions at once: 1) science (financing of research in the field of organic production, educational programs on organic agriculture), 2) production (direct per-hectare payments during the transition to organic agriculture, subsidies to cover the costs of certification, preferential lending), 3) realization (support for certification of exporters, development of e-commerce platforms, support for the export of organic products), 4) information (financing of an information campaign to promote organic products, consulting support, information support for product manufacturers).

In Russia, measures of financial support for organic producers are just beginning to develop. A common support measure is concessional lending, and free certification of organic production by Roskachestvo is provided for small and medium-sized busi-nesses from April 2020 to the end of 2021 [19]. Russian manufacturers can receive compensation for the costs of certification of export-oriented products. The regions are actively involved in the development of organic production in the country and take such support measures at their level as reimbursement of costs for the purchase of fertilizers, plant protection products from pests and biological products for animals, preferential taxation, subsidies for various purposes.

In Russia, federal support measures have not yet been formed on a systematic basis for producers of organic agricultural products. The existing general support measures that are now available to ordinary agricultural producers are insufficient, since they do not take into account the specifics of organic production. For example, banks are reluctant to issue preferential loans to organic producers due to high risks. We need special credit programs with state guarantees. In our opinion, for the active develop-ment of organic production in Russia, a state program for the development of organic agriculture is necessary, including a set of federal measures of financial support for organic producers. In particular, state financial support is needed not only for producers-exporters of organic products (now they have a preferential position), but also for all farms operating in this area, since organic production is associated with high costs and high risks.

Organic production involves a long transition period and requires the adaptation of agricultural technologies, the use of high-quality seeds, new forms of biological fer-tilizers, special preparations for plant protection and for animals. During this transition period (the conversion period), subsidies for certification are necessary for all organic producers, and not only for exporters, since the costs of these procedures amount to approximately 300 thousand rubles per year. Also, given that ordinary agricultural enterprises receive subsidies for the purchase of mineral fertilizers, organic producers have the right to count on subsidies for the purchase of biological products and biofertilizers. It is advisable to adopt the experience of foreign countries and introduce a mechanism of per-hectare support during the conversion period, when the enterprise moves from classical to organic production, and the amount of subsidies should be greater than classical support, since costs are higher in this area.

Indirect support mechanisms will be important for organic producers in the form of providing tax benefits during the conversion period, obtaining preferential loans for the purchase of agricultural machinery on special conditions, with a deferred repayment of the principal debt after the conversion period.For the accelerated development of organic production, it is necessary to allocate subsidies for the training of specialists in the field of certification, for the development of training programs for organic agriculture in universities, the creation of advanced training programs on the basis of agricultural universities and institutions of additional professional education. State financial support is also necessary for the implementation of targeted research programs in the field of organic farming. These are the creation of special varieties of agricultural crops, the development of biological products, biofertilizers, special agricultural technologies on the basis of existing certified organic farms, agricultural universities and research institutes.

Also, in our opinion, state financial support for the sale of organic products is necessary, and funds should be allocated not only to promote Russian organic products to international markets, but also to bring products to the Russian consumer. And this involves conducting a broad information campaign to explain to consumers the essence of organic products, its differences from ordinary farm products, the benefits and benefits for the health of citizens. It is also an interaction with distributors, retail chains, processors of organic products. The purpose of such interaction is to establish an acceptable price for organic products, at which Russian citizens will be able to buy it. Preferences for competitive purchases of organic agricultural products by educational and medical institutions are also very important.

An important factor in the active development of the organic direction in agriculture is the information support for the production of organic products. Therefore, funds should be allocated for information support of organic producers: standards, certification conditions, labeling of organic products, measures to support producers. An equally important role is played by consulting support for organic producers on the organization of activities, the use of agricultural technologies, advanced domestic and foreign experience. Consulting support can be organized in the form of training seminars, forums with the involvement of leading practitioners of organic farming. At the same time, it seems appropriate to create a state digital platform «Organic Agriculture» by analogy with the «My Business» platform for representatives of small and medium-sized businesses in order to more effectively provide organic producers with the necessary information. On this platform, using regional resources, it is possible to accumulate information for organic producers in various areas: doing business, regulatory framework, standards, certification, technologies, consultations, best practices, training, support measures, etc. Thus, having studied the foreign experience of state financial support for organic producers, support measures existing in Russia, as well as taking into account the necessary conditions for the development of Russian organic agriculture, we can propose the following scheme of state financial support (see Fig. 3). In connection with the presented scheme of state financial support, we note that subsidies, tax incentives, preferential lending and other measures of state financial support could stimulate the growth of the number of Russian organic production enterprises and contribute to the development of the sector as a whole. However, in addition to state support for producers of organic agricultural products, such a method of obtaining financial support as crowdfunding is popular. Crowdfunding is a way of popular financing, when people interested in the idea of a project or business invest in the latter on a voluntary basis. Fundraising within the framework of the crowdfunding mechanism takes place on electronic platforms where author's projects that require funding are already placed.

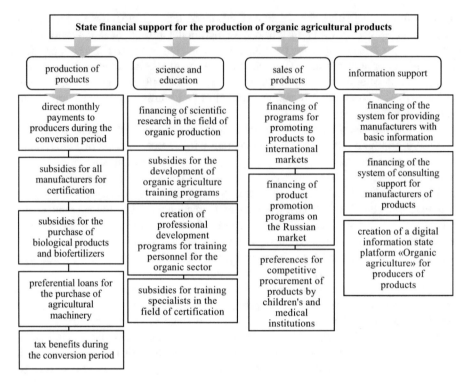

**Fig. 3.** Measures of state financial support for the production of organic agricultural products in Russia (compiled by the authors)

Organic producers also use crowdfunding, but so far these are isolated examples. Thus, in Russia, the financing of a farm engaged in organic farming using the crowdfunding mechanism is carried out not only through organized electronic platforms, but also social networks. This is how, for example, the family livestock farm «Planet of Cows» works in the Omsk region (Russia) [20]. Its feature is organic animal husbandry without violence against animals. Raising funds through crowdfunding platforms is possible both at the stage of creating an organic production, and at the stage of maturity of this business. A distinctive feature of the Russian market in this segment is that small farms resort to such a source of financing and they can often count on insignificant amounts of collected financial resources. While, for example, in the United States, large holding companies are counting on crowdfunding financing – at the end of 2020, the greenhouse organic company Edible Garden® announced the launch of an investment campaign through the Republic crowdfunding platform, hoping to raise more than $ 1 million to activate its business.

# 5  Conclusion

Thus, the conducted research has shown the importance and significance of organic production as a priority direction of greening the economy and ensuring food security. The high costs and risks of organic production require an effective state policy of regulation and support for this segment. The expediency of expanding measures of state financial support for organic agricultural production, the use of public-private partnership in financing large investment projects in this area, as well as the use of such a modern method of financial support as crowdfunding is justified.

Studies of foreign experience of state financial support for organic agricultural producers in the United States and the EU and comparison with the experience of Russia have revealed the following distinctive features: in the US and the EU, direct and indirect state financial support for organic producers is carried out; in the US, the amount of financial support is generally lower than in the EU, with most of the allocated funds being indirect support, among which payments for research and reimbursement of certification costs prevail; in the United States there is direct support for organic producers at the federal level, as well as support at the state level, including on the basis of co-financing with federal programs; EU countries have a higher level of financial support for organic producers than the USA. EU subsidies are allocated from funds supporting the implementation of the Common Agricultural Policy (CAP) of the EU countries, as well as from regional and national sources; Russian producers of organic products can use measures of state financial support, which are already applied in the agricultural sector (for example, preferential lending), separate support measures are provided specifically for producers of organic products (for example, compensation for the costs of certification of export-oriented products), but the federal there is no program of direct and indirect support for organic agricultural producers yet; at present, only some regions of Russia envisage their own measures of financial support for organic producers.

The analysis made it possible to substantiate the advisability of expanding measures of state financial support for Russian organic agricultural production, the use of public-private partnerships in financing large investment projects in this area, as well as the use of such a modern method of financial support as crowdfunding.

The creation of a federal target program for the development of organic agriculture in Russia, including a set of measures of state financial support, will contribute to the active growth of organic production and the country's economy as a whole, as well as the solution of the most important tasks of sustainable development of rural areas: the introduction of unused (low-value) lands into circulation, the creation of additional jobs in rural areas (due to high labor costs in this type of agricultural production), providing the population with high quality natural products.

# References

1. Official website of the Research Institute of Organic Agriculture FiBL. https://www.fibl.org/. Accessed 21 July 2021
2. FiBL & IFOAM: The World of Organic Agriculture. https://www.organic-world.net/yearbook/yearbook-2021.html. Accessed 22 July 2021
3. Henning, J., Baker, L., Thomassin, P.: Economics issues in organic agriculture. Can. J. Agric. Econ. **39**, 877–889 (2008)
4. Benbrook, C., Kegley, S., Baker, B.: Organic farming lessens reliance on pesticides and promotes public health by lowering dietary risks. Agronomy **11**, 1266 (2021)
5. Crowder, D.W., Reganold, J.P.: Financial competitiveness of organic agriculture on a global scale. PNAS **112**, 7611–7616 (2015)
6. Shcherbakova (Ponomareva), A.S.: Organic agriculture in Russia. Siberian J. Life Sci. Agric. **4**, 151–173 (2017)
7. Minkova, O., Kalinichenko, A., Gorb, O.: Research of organic agricultural production support in Poland. Technol. Audit Prod. Reserves **2**(5), 50–54 (2017)
8. Roljevic, N.S., Vukovic, P., Grujic, B.: Measures to support the development of organic farming in the EU and Serbia. Econ. Agric. **1**, 323–336 (2017)
9. Luczka, W., Kalinowski, S., Shmygol, N.: Organic farming support policy in a sustainable development context: a polish case study. Energies **14**, 4208 (2021)
10. Polushkina, T.M., Akimova, Y., Kochetkova, S.A.: State regulation of organic agriculture development. J. Environ. Manag. Tourism **3**, 429–439 (2016)
11. Pivovarov, V.F., et al.: Regulatory support for the organic market (in the world, EAEU countries, Russia). Vegetable Crops Russia **1**, 5–19 (2021)
12. Action plan for organic production in the EU. https://ec.europa.eu/info/food-farming-fisheries/farming/organic-farming/organic-action-plan_en. Accessed 26 July 2021
13. The organic market in Russia in 2020. https://rosorganic.ru/files/Analiz%20organic%20RF%202020%20%D0%B3.pdf. Accessed 26 July 2021
14. Official website of the International Movement for Organic Farming IFOAM. https://www.ifoam.bio. Accessed 28 July 2021
15. Federal Law №280-FZ of 03.08.2018 «On Organic Products and on Amendments to Certain Legislative Acts of the Russian Federation». http://www.consultant.ru/document/cons_doc_LAW_304017/. Accessed 29 July 2021
16. Russia's potential in organic agriculture has been realized only by 1%. https://soz.bio/potencial-rossii-v-organicheskom-sel/. Accessed 26 July 2021
17. Financial Resources for Farmers and Ranchers. https://www.usda.gov/topics/organic/financial-resources-farmers-and-ranchers. Accessed 28 July 2021
18. Organic agriculture: innovative technologies, experience, prospects: a scientific analytical review. https://soz.bio/wp-content/uploads/2019/11/rosinformagrotekh-verstka-organic-cx-2019.pdf. Accessed 25 July 2021
19. Roskachestvo has developed measures to support small and medium-sized enterprises producing organic products. https://roskachestvo.gov.ru/news/roskachestvo-razrabotalo-mery-podderzhki-malykh-i-srednikh-predpriyatiy-proizvoditeley-organicheskoy/. Accessed 25 July 2021
20. ECO farm «Planet of cows». https://planetakorov.ru/. Accessed 28 July 2021

# Operational Management Practices in Onshore Fields Oil and Gas Producers to Fulfill Structural Integrity Requirements of the Installations and of Safety of Operations

Raymundo Jorge de Sousa Mançú⬤, Luís Borges Gouveia(✉)⬤,
and Silvério dos Santos Brunhoso Cordeiro⬤

Universidade Fernando Pessoa (UFP), 4249-004 Porto, Portugal
qualidade@ufp.edu.pt

**Abstract.** The objective of this work was to carry out a comparative analysis of operational management practices in onshore oil and natural gas producing fields to meet the requirements of the SGI: ISO 9001:2015, ISO 14001:2015, ISO 45001:2018 and technical regulations of operational safety management systems (RTSGSO) and structural integrity of producers well and injector well installations (RTSGI and RTSGIP), defined by the National Petroleum Agency (ANP) in Brazil. The methodology was an exploratory and descriptive research, with a quali-quantitative, bibliographic, documentary approach, a multiple case study method in 5 (five) producing fields "A"; "B"; "C"; "D" and "E". In the results of the comparative analysis of category A-KTM on knowledge transfer and management for own employees and contractors in the producing fields, a total of 175 responses was obtained, with 76 (43%) "*not conform*" responses, with relevant amounts and percentage values of deviations in all fields. The results of category B-PIW on the integrity of the installations and safety of operations of wells, with a total of 310 responses, 146 (47%) of which are "*not conform*" responses, with producer fields "B", "C" and "D" being the most critical, with an average equal to 35 (55%) "*not conform*" responses. It is concluded that all the producing fields "A"; "B"; "C"; "D" and "E" presented similar values of "not conform" variables, being systemic in category A-KTM, as well as in category B-PIW, with higher mean values of "*not conform*" variables observed in producer fields "B"; "C" and "D", both with potential risks of incidents, accidents, near misses, non-conformities in internal and external audits and with probability of infraction notice/multi-million fines by the regulatory and supervisory agency ANP in Brazil.

**Keywords:** Operational management practices · Structural integrity of installations · Safety of operations

## 1 Introduction

In the exploration and production (E&P) process, oil and natural gas producing wells and injectors for produced water, natural gas and steam are operating in locations spread over large geographical areas, exposed to weather, vandalism, theft, producing fluids (oil, gas and water) with medium and high pressures, temperatures, oil, gas and water flows, presence of contaminants in the emulsion, such as: $H_2S$ (hydrogen sulfide), $CO_2$ (carbon dioxide), water with high salinity, sand and others, which contribute to the process corrosion in equipment and pipes, in addition to the need to perform various activities and critical tasks that can characterize loss of primary containment (leaks to the environment), incidents, accidents and/or near accidents, with risks to the health of workers and generation of waste (by-products), in the onshore and offshore concessions/fields producers oil and natural gas, being necessary to adopt operational management practices for the training and comp ethics of own and contracted employees, safety of operations and structural integrity of the installations, which minimize risks, as decisive requirements in the preservation of life and a basic element in the search for environmental excellence.

Therefore, the concessions/producing fields need to carry out studies and risk analysis of the processes, equipment, activities and tasks of oil exploration and production, comply with the legal requirements defined in the technical regulations of the regulatory and inspection bodies, such as the National Petroleum Agency (ANP), which emerged as mandatory recently, and also meet the voluntary requirements defined in the ISO standards of the quality management systems ISO 9001:2015 (QMS), environment ISO 14001:2015 (EMS), safety and health at work ISO 45001:2018 (SHMS), through the incorporation in the global management of the organization, application of the best safety of operations practices and integrity of oil production installations and equipment recognized by the oil industry, to minimize possible deviations or failures of systems, installations, production processes, products and services, and thus guarantee the structural integrity of the installations and trust it equipment in mature fields producing oil and natural gas in Brazil [1, 2].

Therefore, the objective of this work was to carry out a comparative analysis of operational management practices in onshore oil and natural gas producing fields to meet the requirements of the SGI: ISO 9001:2015, ISO 14001:2015, ISO 45001:2018 and technical regulations of operational safety management systems (RTSGSO) and structural integrity of productors well and injector well installations (RTSGI and RTSGIP), defined by the National Petroleum Agency (ANP) in Brazil.

The paper is organized in five sections plus references, including this introduction. Section 2 provides a brief literature review where main concepts are presented. Section 3 describes the followed methodology for the research. In Sect. 4 contains the results presentations and its discussion, being the core for this paper. Finally, in Sect. 5, the final remarks are made.

## 2    Literature Review

The oil exploration process (E&P) is characterized by the activities of research, prospecting, drilling and production of oil and natural gas through onshore (onshore) and offshore (wells), with the large volumes of production coming from offshore oil and natural gas producing wells and with the lowest extraction costs [2, 3].

In the oil industry, the area of exploration and production of oil and natural gas encompasses the activities of exploration, production development and production of oil and natural gas, in terrestrial (onshore) and maritime (offshore) production fields, through the process of elevation of oil and gas, collection, separation, treatment and transfer of oil to the refinery, movement and processing of gas for a distribution company [4], and these present safety risks, impact on the environment and occupational health [2].

In the northeast of Brazil, onshore (onshore) and offshore (offshore) oil producing fields have a low share in national oil production, with more than thirty years in operation, characterized as mature fields, structured by thousands of producing wells (Fig. 1) and injection wells (Fig. 2) and ducts/pipes for the flow of production (Fig. 3), with old installations, processes and equipment and with several critical operational activities and tasks, which can characterize the risk of incidents, accidents, fluid leaks (oil, water, gas and steam), with impacts on the health of workers, the environment and the organization's image [2, 4, 5].

**Fig. 1.** Production well (So urce: Mançú, Gouveia and Cordeiro, 2020).

**Fig. 2.** Injection well (Sou rce: Mançú, Gouveia and Cordeiro, 2020).

**Fig. 3.** Ducts/Piping (Source: Ma nçú, Gouveia and Cordeiro, 2020).

The installations and equipment of the producing wells and injectors that produce and move fluids (oil, water and gas) with the presence of contaminants and abrasives drained through pipes and ducts (pipelines, gas pipelines and mains), from the oil reservoir to the storage tanks located in the oil collection stations, with their various activities and critical tasks, require formal documented standardization for training, qualification and qualification of own and contracted employees and control of operational processes, systematization of daily routines, inspections and maintenance in installations and equipment, to ensure the integrity of the installations, safety of operations, environmental and health preservation of workers, quality of services and products, as defined in the technical regulations RTSGSO, RTSGI and RTSGIP of ANP and in the ISO standards of management systems integrated (MSI): ISO 9001:2015; ISO 14001:2015 and ISO 45001:2018 [2].

In the technical regulations of safety of operations management systems (RTSGSO), structural integrity management of installations (RTSGI) and well integrity management (RTSGIP) aim to establish mandatory requirements for implementation in the global management of exploration and production organizations (E&P) and application in concessions/producing fields aiming at safety of operations and structural integrity of installations and offshore and onshore (onshore) production Installations for oil and natural gas, protection of human life, the environment, integrity the assets of the Union, third parties and the Contract Operator [2, 6–9].

The technical regulations RTSGSO and RTSGIP (Fig. 4) of the ANP are structured with 17 (seventeen) Management Practices (MP), which cover the various organizational disciplines and the RTSGI with 3 (three) chapters, with chapter 1 defined the general provisions, in chapter 2 present the requirements for the organization and safety of operations and in chapter 3 requirements to guarantee the structural integrity of the installations, with common and specific requirements among these 3 (three) mandatory regulations, to guarantee safety of operations and the integrity of the installations of producer and injector wells and other Installations, processes, equipment of the oil production concessions/fields [2, 7, 9].

| ANP Resolution No. 43/2016 |
|---|
| **Technical Regulation of the Onshore Well Integrity Management System (RTSGIP) for the Production of Oil and Natural Gas** |
| **Chapter 2: 17 (seventeen) Management Practices (MP)** |
| MP n°  1: Culture of Safety, Commitment and Managerial Responsibility |
| MP n°  2: Involvement of the Workforce |
| MP n°  3: Skills Management |
| MP n°  4: Human Factors |
| MP n°  5: Selection, Control and Management of Contracted Companies |
| MP n°  6: Monitoring and Continuous Improvement of Performance |
| MP n°  7: Audits |
| MP n°  8: Information and Documentation Management |
| MP n°  9: Incidents |
| MP n° 10: Well Life Cycle Stages |
| MP n° 11: Critical Elements of Well Integrity |
| MP n° 12: Risk Analysis |
| MP n° 13: Well Integrity |
| MP n° 14: Well Control Emergency Planning and Management |
| MP n° 15: Procedures |
| MP n° 16: Management and Changes |
| MP n° 17: Environmental Preservation |

**Fig. 4.** RTSGIP Management Practices of mandatory application in E&P (Source: Adapted from ANP 2016, Gouveia; Mançú and Cordeiro 2020).

The common and specific requirements of the ANP technical regulations can be integrated into a correlation matrix, and later structured with the other management systems of the oil producing fields, to fulfill mandatory legal and statutory requirements and voluntary requirements of good industry practices. petroleum, with a focus on the integrity of installations, safety of operations and health at work, and also with the objective of facilitating the systematic implementation of these requirements in all phases of the life cycle of oil exploration and production (E&P) processes and natural gas, from the project to the decommissioning [2].

## 3   Methodology

The methodology applied as to the objectives is classified as exploratory and descriptive research, with a qualitative and quantitative approach, with the strategies being bibliographic and documentary research, a multi-case study method in five oil and natural gas producing fields, located in northeastern Brazil, classified as: field "A"; "B"; "C"; "D" and "E". Exploratory research aims to develop, explain and modify concepts and ideas for the formulation of later approaches and descriptive research seeks to accurately describe the facts and phenomena of certain realities [10, 11]. The case study, on the other hand, is an empirical research that allows the investigation of organizational and administrative processes, focusing on the case and in a holistic perspective, with application in different fields and the researcher has no control over events and variables [11–14].

As for the data collection instruments, 10 (ten) checklists (CLs) were applied in loco on the technical documentation of the installation, the management and transfer of knowledge for the training of the own operation technicians and contracted operators, and on safety operational and integrity in total of 50 (fifty) oil and natural gas producing wells, and produced water injector wells (PIW), from the five producing fields "A"; "B"; "C"; "D" and "E", with 10 wells of each producing field, with a total of 97 closed questions structured in 2 (two) CLs, with the following "C - conform" answer options; "NC - not conform" and "NA - not applicable", structured in 2 (two) categories, being: category A-KTM to evaluate the management and transfer of knowledge to own and contracted employees in the oil producing fields with 35 questions closed; and category B-PIW to verify the integrity of the Installations and equipment and safety of operations of oil producing wells and fluid injectors, with 62 closed questions, applied by 5 (five) contracted operators who work in the monitoring and monitoring of wells in the fields producers and under the guidance of the researcher. For data processing, SPSS version 22 software was applied to quantify and total on-the-spot checks in the producing fields and the percentage values "conform"; "not conform" and "not applicable".

## 4 Results and Discussion of Data Collected from Checklists (CLs) Applied by Own and Contracted Employees

To better distribute the 97 questions that characterize quality, environment, operational safety, integrity and health at work (QEOSI&HW) in the processes of lifting and draining oil, composed of oil producing wells, fluid injection wells, pipelines/pipelines, satellites, multipaths and accessories, defined in requirements of ISO standards of management systems integrated (MSI) and of technical regulations (RTSGI and RTSGIP) of the ANP, checklists (CLs) under investigation, they were structured in 2 (two) categories and variables: A-KTM and B-PIW, as follows:

- In the A-KTM category, 35 questions were structured with the aim of verifying on the operational fronts the management and transfer of knowledge in the producing fields; and
- In the B-PIW category, 62 questions were organized to investigate the integrity of the installations and the safety of operations of oil producing wells, fluid injection wells, pipes/ducts, satellites, multipaths and accessories.

The CLs were applied in loco on the operational fronts by 5 (five) contracted operators performing activities and operational tasks in the producing fields, with experiences varying from 1 to 5 years in the company, training in a medium technical level, performing daily routines of inspection, follow-up and monitoring of producer and injector wells, under the guidance of the researcher.

## 4.1 Results of Category A-KTM Checklists in Loco on Management and Knowledge Transfer in Oil Producing Fields "A"; "B"; "C"; "D" and "E"

In the studies with the application of in loco checklists in the operational fronts of oil and natural gas producing fields "A", "B", "C", "D" and "E" we observed an overall total of 175 responses in the category A-KMT on knowledge management and transfer and an overall average of 35 responses per producer field, where we counted the overall total of 76 (43%) "not conform" responses, 54 (31%) "conform" and 45 (26) responses %) "not applicable" responses. By producer field, we verified an average of 15 (44%) "not conform" responses, 11 (30%) "conform" responses and 9 "not applicable" responses. In this context, we highlight as variables of greatest concern and relevance the high percentage values of the "not conform" responses, as they were systemic in all the researched producing fields, shown in Table 1.

**Table 1.** Results of the variables of the checklists of category A-KTM on the management and transfer of knowledge of the producing fields "A", "B", "C", "D" and "E"

| VARIABLE | RESULTS OF IN LOCO CHECKLISTS ON OPERATIONAL FRONTS ON KNOWLEDGE MANAGEMENT AND TRANSFER (KMT) | FIELD A | FIELD B | FIELD C | FIELD D | FIELD E | TOTAL |
|---|---|---|---|---|---|---|---|
| KMT 01 | Did people change management carried out before the Voluntary Dismissal Program (VDP)? | C | C | NC | NC | NC | 5 |
| KMT 02 | People change management serves for management and knowledge transfer? | NC | NC | NC | NC | NC | 5 |
| KMT 03 | Accomplished the setting of employees after admission? | C | C | C | C | C | 5 |
| KMT 04 | Training of employees carried out in the standards of the Safety Manuals? | C | C | C | C | C | 5 |
| KMT 05 | Training of employees carried out in the standards of 15 Safety, Environment and Health (SEH) guidelines? | C | C | C | C | C | 5 |
| KMT 06 | Did they train employees in the standards of the Emergency Response Plan (ERP), Local Emergency Plan (LEP) and Fixed Fire Fighting System (FFFS)? | C | C | C | C | C | 5 |
| KMT 07 | Did employee training be carried out in the Operation Manual (MO) standard? | C | C | C | C | C | 5 |
| KMT 08 | Have employees been trained in the Execution (EP) procedures? | C | C | C | C | C | 5 |
| KMT 09 | Are Execution Procedures (EP) with an updated review? | C | C | C | C | C | 5 |
| KMT 10 | Are guideline and implementation standards available for consultation at the facility? | NC | NC | NC | NC | NC | 5 |
| KMT 11 | Do employees have a key and password to access the standards through the System? | NC | NC | NC | NC | NC | 5 |
| KMT 12 | Are the contractor's standards updated based on the contractor's standards? | NC | NC | NC | NC | NC | 5 |
| KMT 13 | Are the contractor's standard operating procedures (SOP) printed? | NC | NC | NC | NC | NC | 5 |
| KMT 14 | Standards (SOP) of the contracted company identified the aspects and impacts of Safety, Environment and Health (SEH)? | NC | NC | NC | NC | NC | 5 |
| KMT 15 | Employee has the 10 Golden Rules badge? | NC | NC | NC | NC | NC | 5 |
| KMT 16 | Does the Bulletin and Checklist have traceability with the standard number? | NC | NC | NC | NC | NC | 5 |
| KMT 17 | Training conducted in Operational Safety Documentation - OSD (DLU, Flowcharts; Plants; LCE; APR and HAZOP)? | NC | NC | NC | NC | NC | 5 |
| KMT 18 | Land Unit Description (DLU) available for consultation? | NC | NC | NC | NC | NC | 5 |
| KMT 19 | List of Critical Elements (LCE) available for consultation? | NC | NC | NC | NC | NC | 5 |
| KMT 20 | Are Process and Engineering Flowcharts available for consultation? | NA | NA | NA | NA | NA | 5 |
| KMT 21 | Is the Area Electrical Classification Plant available for consultation? | NC | NC | NC | NC | NC | 5 |
| KMT 22 | Is the floor plan of the "As Biult" installation available for consultation? | NA | NA | NA | NA | NA | 5 |
| KMT 23 | Are APR and HAZOP Risk Studies available for consultation? | NC | NC | NC | C | C | 5 |
| KMT 24 | Is the Safety Record Book (SRB) available and without erasures? | NA | NA | NA | NA | NA | 5 |
| KMT 25 | Is the Fire Department Inspection Certificate (FDIC) valid and available? | NA | NA | NA | NA | NA | 5 |
| KMT 26 | Is the Operating Permit valid and available for consultation? | NA | NA | NA | NA | NA | 5 |
| KMT 27 | Is the Risk Map updated and does internal commission of accident prevention (ICAP) management period app | NA | NA | NA | NA | NA | 5 |
| KMT 28 | Is the Environmental Operating License (EOL) valid and available for consultation? | NA | NA | NA | NA | NA | 5 |
| KMT 29 | Is the Inadvertent Blocking Device (IBD) Checklist from the previous month available for consultation? | NA | NA | NA | NA | NA | 5 |
| VARIABLE | RESULTS OF IN LOCO CHECKLISTS ON OPERATIONAL FRONTS ON KNOWLEDGE MANAGEMENT AND TRANSFER (KMT) | FIELD A | FIELD B | FIELD C | FIELD D | FIELD E | TOTAL |
| KMT 30 | Is system inhibition control or critical equipment available for consultation? | NA | NA | NA | NA | NA | 5 |
| KMT 31 | Do employees issue the Work Permit (PT) on the intranet system? | C | C | C | C | C | 5 |
| KMT 32 | Is the Service Ticket registered on the intrenet System? | NC | NC | NC | NC | NC | 5 |
| KMT 33 | Is the Service Ticket registered / formalized in an occurrence book? | C | C | C | C | C | 5 |
| KMT 34 | Does Service Pass include critical items defined in the contractor's standard? | C | C | C | C | C | 5 |
| KMT 35 | Do employees have access to the production control and well test system? | NC | NC | NC | NC | NC | 5 |
| | TOTAL RESPONSES BY FIELD AND TOTAL OVERALL | 35 | 35 | 35 | 35 | 35 | 175 |
| | TOTAL ANSWERS "C - CONFORM" BY OIL PRODUCING FIELD | 11 (31%) | 11 (31%) | 10 (29%) | 11 (31%) | 11 (31%) | 54 (31%) |
| | TOTAL ANSWERS "NC - NOT CONFORM" BY OIL PRODUCING FIELD | 15 (43%) | 15 (43%) | 16 (46%) | 15 (43%) | 15 (43%) | 76 (43%) |
| | TOTAL ANSWERS "NA - NOT APPLICABLE" BY OIL PRODUCING FIELD | 9 (26%) | 9 (26%) | 9 (26%) | 9 (26%) | 9 (26%) | 45 (26%) |
| | LEGEND: "C" - CONFORM; "NC" - NOT CONFORM and "NA" - NOT APPLICABLE. | | | | | | |

RESULTS OF IN-LOCO CHECKS CARRIED OUT IN NORTHEAST OIL AND NATURAL GAS PRODUCING FIELDS FIELDS A, B, C, D AND E - WELL TEAM — CATEGORY A - WELL TEAM

Source: Research Data

In the results of the responses in category A-KTM, the following "*not conform*" variables were observed for the transfer of knowledge: lack and/or failure in managing change of people at admission and/or at the beginning of a service provision contract. operation to supply pensions, voluntary dismissals and/or staff turnover, with actions pending/delayed; lack of training; unavailability of technical documentation of production processes, producing wells and injector wells; unavailability of execution standards for consultation; unavailability of printed operating procedures for immediate consultation; and lack of access to the computerized standards system. The not conformities identified can compromise the maintenance of quality, integrity of the installations, preservation of the environment, safety of operations and health at work.

The variables of category A-KTM with "*not conform*" responses referring to the management and knowledge transfer requirements for own and contracted employees verified in a systemic way from 5 (five) studied producing fields were:

- KMT 01 Did people change management carried out before the Voluntary Dismissal Program (VDP)?
- KMT 02 Personnel change management serves for knowledge management and transfer?
- KMT 10 Are guideline and execution standards available for consultation at the facility?
- KMT 11 Employees have a key and password to access the standards through the System?
- KMT 12 Contractor's standards are updated based on the contractor's standards?
- KMT 13 Standard Operating Procedures (SOP) of the contractor are printed?
- KMT 14 Standards (SOP) of the contracted company identified the aspects and impacts of Safety, Environment and Health (SEH)?
- KMT 15 Employee has the 10 Golden Rules badge?
- KMT 16 Bulletin and Checklist have traceability with the standard number?
- KMT 17 Training conducted in Operational Safety Documentation - OSD (DLU, Flowcharts; Plants; LCE; APR and HAZOP)?
- KMT 18 Description of Land Unit (DLU) is available for consultation?
- KMT 19 List of Critical Elements (LCE) available for consultation?
- KMT 21 Electrical Area Classification Plant is available for consultation?
- KMT 23 Risk Studies APR and HAZOP are available for consultation?
- KMT 32 Service Pass registered on the intranet system?
- KMT 35 Employees have access to the production control and well test system?

## 4.2 Results of Category B-PIW Checklists in Loco on Installation Integrity and Safety of Operations in Oil Producing Fields "A", "B", "C", "D" and "E"

In the results of the on-site checklists with questions of category B-PIW on the integrity of the facilities and operational safety of the producing and injector wells, oil and gas pipelines, water mains, satellites, multiway and accessories of oil producing fields "A"; "B"; "C"; "D" and "E" we verified an overall total of 310 responses and an overall average of 62 responses per producer field, being identified in the overall total 146

(47%) "not conform" responses, 120 (39%) "conform" responses and 44 (14%) "not applicable" responses. By producing field, we found in producing fields "B", "C" and "D" an average of 34 (55%) responses "not conform", producing field "A" with 25 (40%) and "E" with 19 (31%) "not conform" responses, average of 24 (39%) "conform" responses and average of 9 (14%) "not applicable" responses. However, we observed important variables of operational safety and integrity of facilities with high percentage values of "not conform" responses, in a systemic manner and with greater concern in the "B", "C" and "D" producing fields, and with lower values percentages, but also relevant in producer fields "A" and "E", defined in Table 2.

In the B-PIW category, they presented high amounts and percentage values of "*not conform*" responses by field "A", "B", "C", "D" and "E", with values of greater prominence in the fields "B", "C" and "D", and with lower but representative values. in the "A" and "E" producing fields, both considered of concern for the structural integrity of the Facilities and for the safety of operations of oil producing wells, fluid injector wells (water, gas and steam), oil and gas pipelines, satellites and multi-way outlets for oil production and fluid injection, as these are requirements legal and were systemic in the studied producing fields, and also these Installations and processes produce and drain aggressive fluids, move gas and high loads, with exposure to bad weather, with great geographical extension, locations close to federal, state and municipal highways, rural communities, urban centers, schools and areas sensitive to environmental impacts.

The variables of category B-PIW with two or more "not conform" responses to the requirements related to the structural integrity of the installation and safety of operations identified in the production wells and injection wells of the studied production fields were:

- PIW 01 Do the drain valves (Annular and Flow Tee) of well have CAP/PLUG?
- PIW 02 Does the well have an "Out of Operation" or "Under Maintenance" sign?
- PIW 03 Do the rackets have an adhesive/label with technical manufacturing data?
- PIW 04 Do the rackets have a "Racquet" warning sign?
- PIW 05 Flange bolts have leftover thread after the nuts?
- PIW 06 Does the production/injection line have tagging (TAG)/well number?
- PIW 07 Does the production/injection line have signs with a fluid type range?
- PIW 08 Does the production/injection line have a flow direction arrow sign?
- PIW 09 Are well slopes/slopes free from erosion?
- PIW 10 Do well slopes/slopes have erosion protection vegetation?
- PIW 11 Is the well location free from high vegetation?
- PIW 12 Is the forefoot (box) with a low fluid level?
- PIW 13 Is the production line pressure transmitter and indicator (PTI) operational?
- PIW 14 Manometer installed in a well is calibrated and operational?
- PIW 15 Satellite and/or multivia has a cap/plug on the drain valves?
- PIW 16 Are the well automation equipment operational?
- PIW 20 Does the well have a protective grid or cage around the installation?
- PIW 22 Does the well location have a complete fence?
- PIW 23 Does the well lease have a complete gate?

**Table 2.** Results of checklist variables related to the integrity of the installation and safety of operations of wells in producing fields "A", "B", "C", "D" and "E"

| VARIABLE | CATEGORY B - WELL TEAM<br>RESULTS OF IN-LOCO CHECKLISTS ON OPERATIONAL FRONTS OIL-PRODUCING WELLS AND FLUID INJECTOR WELLS (PIW) | FIELD A | FIELD B | FIELD C | FIELD D | FIELD E | TOTAL |
|---|---|---|---|---|---|---|---|
| | RESULTS OF IN-LOCO VERIFICATIONS PERFORMED IN PRODUCING FIELDS OF OIL AND NATURAL GAS FROM THE NORTHEAST - FIELDS A, B, C, D AND E - WELL TEAM | | | | | | |
| PIW 01 | Do the drain valves (Annular and Flow Tee) of well have CAP / PLUG? | NC | NC | NC | NC | NC | 5 |
| PIW 02 | Does the well have an "Out of Operation" or "Under Maintenance" sign? | NC | NC | NC | NC | NC | 5 |
| PIW 03 | Do the rackets have an adhesive / label with technical manufacturing data? | NC | NC | NC | NC | NC | 5 |
| PIW 04 | Do rackets have a "Racquet" warning sign? | NC | NC | NC | NC | NC | 5 |
| PIW 05 | Flange bolts have leftover thread after the nuts? | NC | NC | NC | NC | NC | 5 |
| PIW 06 | Does the production / injection line have tagging (TAG) / well number? | NC | NC | NC | NC | NC | 5 |
| PIW 07 | Does the production / injection line have signs with a fluid type range? | NC | NC | NC | NC | NC | 5 |
| PIW 08 | Does the production / injection line have a flow direction arrow sign? | NC | NC | NC | NC | NC | 5 |
| PIW 09 | Are well slopes / slopes free from erosion? | NC | NC | NC | NC | NC | 5 |
| PIW 10 | Do well slopes / slopes have erosion protection vegetation? | NC | NC | NC | NC | NC | 5 |
| PIW 11 | Is the well location free of high vegetation? | C | NC | NC | NC | NC | 5 |
| PIW 12 | Is the forefoot (box) with a low fluid level? | C | NC | NC | NC | NC | 5 |
| PIW 13 | Is the production line pressure transmitter and indicator (PTI) operational? | C | NC | NC | NC | NC | 5 |
| PIW 14 | Is the manometer installed in a well calibrated and operational? | C | NC | NC | NC | NC | 5 |
| PIW 15 | Does satellite and / or multivia have a cap / plug on the drain valves? | NC | NC | NA | NC | NC | 5 |
| PIW 16 | Are the well automation equipment operational? | NC | NC | C | NC | NC | 5 |
| PIW 17 | Is the well lease exempt from civil construction waste? | C | NC | NC | C | NC | 5 |
| PIW 18 | Does the fortress have retaining walls / dykes? | C | C | NC | C | NC | 5 |
| PIW 19 | Is satellite and / or multivia free of high vegetation? | C | C | NA | NC | C | 5 |
| PIW 20 | Does the well have a protective grid or cage around the installation? | NC | NC | NC | NC | NA | 5 |
| PIW 21 | Does satellite and / or multivia have an operational thermometer? | NA | NA | NA | NC | NA | 5 |
| PIW 22 | Does the well location have a complete fence? | NC | NC | NC | NC | C | 5 |
| PIW 23 | Does the well location have an intact gate? | NC | NC | NC | NC | C | 5 |
| PIW 24 | Is the well installation free from corrosion? | NC | NC | NC | NC | C | 5 |
| PIW 25 | Installation flanges and well valves have all screws? | NC | NC | NC | NC | C | 5 |
| PIW 26 | Is the well location free of ferrous scrap? | C | NC | NC | NC | C | 5 |
| PIW 27 | Is the well lease exempt from open trenches? | C | NC | NC | NC | C | 5 |
| PIW 28 | Are the drainage channels intact? | NC | NA | NC | NC | C | 5 |
| PIW 29 | Are the drainage channels clean and free of vegetation and / or sand? | NC | NA | NC | NC | C | 5 |
| PIW 30 | Is the well installation free of leaks? | C | C | NC | NC | C | 5 |
| PIW 31 | Does satellite and / or multivia have a fence, a closed gate with a padlock? | NC | NC | NA | NC | C | 5 |
| PIW 32 | Does the satellite and / or multivia have a calibrated pressure gauge? | NC | C | NA | NC | C | 5 |
| PIW 33 | Does the satellite and/or multiway have a nameplate / tagging (TAG)? | C | C | NA | NC | C | 5 |
| PIW 34 | Is satellite and / or multivia integrated without corrosion? | C | C | NA | NC | C | 5 |
| PIW 35 | Does a shut-off valve and / or a well drain with a handwheel have a chain with a padlock? | NC | NC | C | NC | C | 5 |
| PIW 36 | Is the well base free of oily residue? | C | NC | C | NC | C | 5 |
| PIW 37 | Is the well lease accessible? | C | C | C | NC | C | 5 |
| PIW 38 | Is the well automated? | C | C | C | NC | C | 5 |
| PIW 39 | Does the well have an identification plate with tagging (TAG) / well number? | C | NC | NC | C | C | 5 |
| PIW 40 | Is the well identification plate intact, with readable data? | C | NC | NC | C | C | 5 |
| PIW 41 | Does the well identification plate have the company's identification? | C | NC | NC | C | C | 5 |
| PIW 42 | Does the well nameplate have risk signs? | C | NC | NC | C | C | 5 |
| PIW 43 | Does the well identification plate have Individual Protection Equipment (IPE) signage? | C | NC | NC | C | C | 5 |
| PIW 44 | Does the foreground have a protective grid on the floor around the wellhead? | C | C | NC | C | C | 5 |
| PIW 45 | Does a well stopped out of operation have rackets? | NC | NC | NA | C | C | 5 |
| PIW 46 | Is satellite and / or multivia free of leaks? | C | NC | NA | C | C | 5 |
| PIW 47 | Is satellite and / or multivia free of oily residue? | C | NC | NA | C | C | 5 |
| PIW 48 | Does satellite and / or multivia have an operational pressure gauge? | NC | C | NA | C | C | 5 |
| PIW 49 | Are the sealing joints of the wells (Mechanical Pumping and Progressive Cavity Pumping) intact, with no leaks? | NC | C | C | C | C | 5 |
| PIW 50 | Does a well that is stopped in maintenance have rackets? | NA | NA | NA | NA | NA | 5 |
| PIW 51 | Is the well thermometer in the steam influence zone calibrated? | NA | NA | NA | NA | NA | 5 |
| PIW 52 | Is the well thermometer in the steam influence zone operational? | NA | NA | NA | NA | NA | 5 |
| PIW 53 | Does satellite and / or multivia have a calibrated thermometer? | NA | NA | NA | NA | NA | 5 |
| PIW 54 | Does satellite and / or multivia have a fence and an integral gate? | C | C | NA | NA | C | 5 |
| PIW 55 | Does satellite and / or multivia have tagging (TAG) / No. on the production line, test and pig? | C | C | NA | C | C | 5 |
| PIW 56 | Does satellite and / or multivia have any screws on the valve flanges? | C | C | NA | C | C | 5 |
| PIW 57 | Do satellite and / or multivia screws have excess thread threads on the nuts? | C | C | NA | C | C | 5 |
| PIW 58 | Is satellite and / or multivia free of construction waste? | C | C | NA | C | C | 5 |
| PIW 59 | Is satellite and / or multivia free of ferrous scrap? | C | C | NA | C | C | 5 |
| PIW 60 | Is the access lane to the well passable for light vehicles? | C | C | C | C | C | 5 |
| PIW 61 | Is the access road to the well passable for the Onshore Production Probe (OPP)? | C | C | C | C | C | 5 |
| PIW 62 | Is the well location free of oily residue? | C | C | C | C | C | 5 |
| | TOTAL RESPONSES BY FIELD AND TOTAL OVERALL | 62 | 62 | 62 | 62 | 62 | 310 |
| | TOTAL ANSWERS "C - CONFORM" BY OIL PRODUCING FIELD | 32(52%) | 20(32%) | 9(14%) | 22(36%) | 37(60%) | 120 (39%) |
| | TOTAL ANSWERS "NC - NOT CONFORM" BY OIL PRODUCING FIELD | 25(40%) | 35(56%) | 32(52%) | 35(56%) | 19(31%) | 146 (47%) |
| | TOTAL ANSWERS "NA - NOT APPLICABLE" BY OIL PRODUCING FIELD | 5(8%) | 7(11%) | 21(34%) | 5(8%) | 6(9%) | 44 (14%) |
| | LEGEND: "C" - CONFORM; "NC" - NOT CONFORM and "NA" - NOT APPLICABLE. | | | | | | |

Source: Research Data

- PIW 24 Is the well installation free from corrosion?
- PIW 25 Installation flanges and well valves have all the screws?
- PIW 26 Is the well lease free of ferrous scrap?
- PIW 27 Is the well lease exempt from open trenches?
- PIW 28 Are the drainage channels intact?
- PIW 29 Are the drainage channels clean and free of vegetation and/or sand?
- PIW 30 Is the well installation free of leaks?
- PIW 31 Satellite and/or multivia has a fence, a closed gate with a padlock?
- PIW 35 Shutoff and/or well drain valve with handwheel has chain with padlock?
- PIW 36 Is the well base free of oily residue?
- PIW 39 Does the well have an identification plate with tagging (TAG)/No. of the well?
- PIW 40 Is the well identification plate intact, with legible data?
- PIW 41 Does the well identification plate have the company's identification?
- PIW 42 Does the well nameplate have risk signs?
- PIW 43 Does the well identification plate have Individual Protection Equipment (IPE) signage?
- PIW 45 Well stopped out of operation does it have rackets?

### 4.3 Comparative Analysis of the Results of the Checklists (CLs) of Categories A-KTM and B-PIW of the Producing Fields "A", "B", "C", "D" and "E"

For the analysis of the results of the 97 questions defined in the checklists by categories A-KTM and B-PIW, composed of 35 questions in the category A-KTM and 62 questions in the category B-PIW the quantities of responses were defined and the percentage values were calculated, referring to the five oil and natural gas producing fields "A", "B", "C", "D" and "E", as follows:

- In the A-KTM category, which evaluates the management and transfer of knowledge to the well operation teams, there were responses in the producing fields "A", "B", "D" and "E" 11 (31%) "*conform*" responses, and in the producing field "C" 10 (28%) "*conform*" responses and overall total of 54 (30%) "*conform*" responses. For "*not conform*" responses, 15 (43%) responses were found in producing fields "A", "B", "D" and "E", in the producing field "C" 16 (46%) responses, and an overall total of 76 (43%) "*not conform*" responses. In relation to the "not applicable" responses, 9 (26%) "*not applicable*" responses were verified in all the studied producing fields and a grand total of 45 (26%) "*not applicable*" responses. The amount and percentage values of "*not conform*" responses identified in the producing fields "A", "B", "C", "D" and "E" in knowledge management and transfer to employees were systemic and relevant, due to the risks involved in exploration and production activities and tasks of oil and natural gas, demonstrated in Fig. 5.

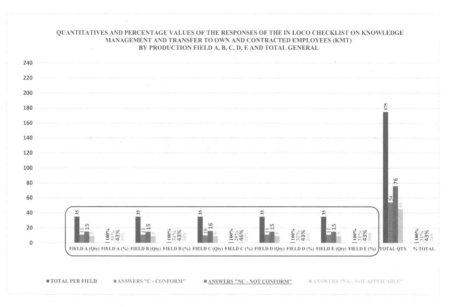

**Fig. 5.** Quantitatives and percentage values of the responses in the local checklist on the management and transfer of knowledge (KMT) to the own and contracted employees of the producing fields "A", "B", "C", "D" and "E". (Source: Research Data).

- In the B-PIW category of operational integrity and safety of the producing well and an injector well, in the producing field "A" 32 (52%) responses were identified as "*conform*", in the producing field "B" 22 (32%) responses "*conform*", in the producing field "C" 9 (14%) "*conform*", in the producing field "D" 22 (35%) "*conform*", in the producing field "E" 37 (60%) "*conform*" and overall total of 120 (39%) answers "*conform*". For "*not conform*" responses were observed in producing field "A" 25 (40%), in the producing field "B" and in the producing field "D" 35 (57%) "*not conform*" responses, in the producing field "C" 32 (52%) "*not conform*" and in the producing field "E" equal to 19 (31%) "*not conform*" and overall total of 146 (47%) "*not conform*" responses. Regarding the answers "*not applicable*" were identified in the producing fields "A" and "D" 5 (8%) "*not applicable*", in the producing field "B" 7 (11%) answers "*not applicable*", in the producing field "C" 21 (34%) answers "*not applicable*", in the producing field "E" 6 (9%) answers "*not applicable*" and overall total of 44 (14%) answers "*not applicable*". The high amounts and percentage values of "*not conform*" responses identified in the producing fields "B", "C" and "D" in the integrity of the installations and in safety of operations were considered of concern, due to the volumes produced, flows, pressures, temperatures and contaminants present in the fluids (petroleum, gas and produced water), which can characterize risk of incidents (near accident and/or accident) in the activities and tasks of exploration and production of oil and natural gas, observed in Fig. 6.

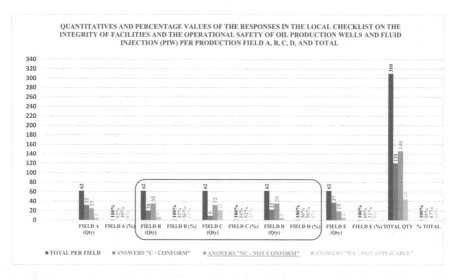

**Fig. 6.** Quantitatives and percentage values of the responses in the local checklist on the integrity of installations and the safety of operations of oil production wells and fluid injection (PIW) per production field "A", "B", "C", "D", "E" and Total. (Source: Research Data).

In total by category of producing fields "A", "B", "C", "D" and "E", high quantitative and percentage values were observed in category A-KTM equal to 76 (43%) "*not conform*" responses and in category B-PIW 146 (47%) "*not conform*" responses, and overall total of categories A-KTM and B-PIW equal to 222 (46%) "*not conform*" responses, shown in Table 3. In the comparative analysis between the two categories studied (A-KTM and B-PIW) high quantitative and percentage values of "*not conform*" responses were observed, and it is necessary to apply operational management practices in the producing fields, to correct deviations and develop training programs. continuous improvement for quality assurance, preservation of the environment, safety of operations, integrity of installations and health at work (QEOSI&HW).

**Table 3.** Comparative analysis of the categories A-KTM and B-PIW of the checklists applied in the fields of producer "A", "B", "C", "D" and "E", with answer options: "*conform*"; "*not conform*" and "*not applicable*"

| Categories: A-KTM and B-PIW | | Oil Productions Fields | | | | | | | | | | Total by Category | |
|---|---|---|---|---|---|---|---|---|---|---|---|---|---|
| | | A | | B | | C | | D | | E | | | |
| | | QTD | % | QTD | % | QTD | % | QTD | % | QTD | % | QTD | % |
| A-KTM - Knowledge Transfer and Management | *conform* | 11 | 31% | 11 | 31% | 10 | 28% | 11 | 31% | 11 | 31% | 54 | 31% |
| | *not conform* | 15 | 43% | 15 | 43% | 16 | 46% | 15 | 43% | 15 | 43% | 76 | 43% |
| | *not applicable* | 9 | 26% | 9 | 26% | 9 | 26% | 9 | 26% | 9 | 26% | 45 | 26% |
| Total Category A-KTM | | 35 | | 35 | | 35 | | 35 | | 35 | | 175 | |
| B-PIW - Integrity of installations and Safety of operations of Production and Injector Wells | *conform* | 32 | 52% | 20 | 32% | 9 | 14% | 22 | 35% | 37 | 60% | 120 | 39% |
| | *not conform* | 25 | 40% | 35 | 57% | 32 | 52% | 35 | 57% | 19 | 31% | 146 | 47% |
| | *not applicable* | 5 | 8% | 7 | 11% | 21 | 34% | 5 | 8% | 6 | 9% | 44 | 14% |
| Total Category B-PIW | | 62 | | 62 | | 62 | | 62 | | 62 | | 310 | |
| Total by Field and Overall | | 97 | | 97 | | 97 | | 97 | | 97 | | 485 | |
| Total "*conform*" - A-KTM and B-PIW | | 43 | 45% | 31 | 32% | 19 | 20% | 33 | 34% | 48 | 50% | 174 | 36% |
| **Total "*not conform*" - A-KTM and B-PIW** | | 40 | 41% | 50 | 52% | 48 | 49% | 50 | 52% | 34 | 35% | 222 | 46% |
| Total "*not applicable*" - A-KTM and B-PIW | | 14 | 14% | 16 | 16% | 30 | 31% | 14 | 14% | 15 | 15% | 89 | 18% |

(Source: Research Data).

In the results of the responses to the on-the-spot checklists applied in the five producer fields, the percentage values of the "not conform" responses in the A-KTM category - knowledge management and transfer were observed to be most critical, as they are systemic in all fields studied producers "A", "B", "C", "D" and "E"; and in category B-PIW - integrity of installations and safety of operations of producing and injector wells, higher percentage values of "*not conform*" responses were identified in producers fields "B"; "C" and "D", thus characterizing potential risks of "*not conform*" in internal and/or external audits, for incidents (accidents or near misses), impacts on the environment and health at work.

## 5 Final Considerations

In the results of the responses of Category A-KTM variables on knowledge management and transfer (KTM) to own and outsourced employees, this being the basis for quality assurance, operational safety, facilities integrity, environmental preservation and occupational health were observed quantitative and relevant percentage values of items related to "*not conform*" legal requirements in all five "A" producer fields; "B"; "C"; "D" and "E", characterized as systemic and of potential risk to quality, safety, environment and occupational health (QHSE), incidents (accidents and near misses) on the operational fronts.

The results of the responses of Category B-PIW variables on the integrity of the facilities and operational safety in producing wells and injector wells (PIW) were verified in the "B" producing fields; "C" and "D" higher quantitative and relevant percentage values of items related to "*not conform*" legal requirements; and in producer fields "A" and "E" with lower quantities and percentage values of "*not conform*" items, but systemic, with locations, installations and operational pipelines in remote uninhabited areas, close to sensitive areas (river, stream, lake, mangroves) draining oil, gas and water produced with high salinity, exposed to vandalism and theft, risk of loss of primary and secondary containment with possibilities for leaks, of great impact on the environment, as well as risks to operational safety, integrity of facilities and equipment, people, heritage and the image of the producing fields.

Variables with "*not conform*" legal requirements items identified in Categories A-KTM and Category B-PIW required the opening of corrective work (OS) or maintenance (OM) orders, requested by the operation teams of the producing fields, with registration and control in SAP/R3-ERP, after applying the checklists in loco in the operational fronts. However, in the event of non-compliance with these "nonconforming" items, they may characterize "a notice of infraction/multi-million-dollar fines to producing fields", by the auditors of the regulatory and supervisory body, the National Petroleum, Natural Gas and Biofuels Agency (ANP) in Brazil.

It is recommended to automate all processes and equipment, apply specialist information and communication technology (ICT) systems, online checklists with digital data collectors, for periodic and systemic on-site application in installations and wells in onshore fields and maritime (offshore) oil and natural gas producers, with the structuring of formal procedures, including good operational management practices, action programs to achieve targets and correct deviations, with performance indicators, forecast control system versus performed, to assess and develop continuous improvements.

For Knowledge Management and Transfer (KTM) to prepare a formal execution procedure with description of activities and tasks and annual matrix of lectures on risk perception of production processes, technical courses and non-technical courses, with photos/images and videos of real cases of short duration of operations, in addition to normative courses on legal requirements, for the qualification of own employees and outsourced service providers, through the execution of training service contracts, with specialist companies and instructors with theoretical and practical knowledge, technical and management training and experience (=>10 years) in activities applicable to Oil Exploration and Production (E&P).

Develop courses on technical documentation of facilities and processes for producing oil and natural gas, such as engineering and process flowcharts, results of process risk studies - APR and HAZOP, area classification plant, "As Built" plant, systems, equipment and critical operating procedures, with internal instructors, with preparation, monitoring, monitoring and control of the management matrix of knowledge transfer programs, with definition of internal instructors, workloads and also lesson plan, exercises for fixing knowledge and attendance lists with date, name, registration and signature of participants for employees and contractors. To assess the effectiveness of training by the supervisor responsible for the installation and processes, an on-site checklist must be drawn up per employee, with application throughout the year and monitoring by performance indicators, to ensure competence proficiency, continuous improvement quality, environmental preservation, safety and health at work (QHSE).

As for Operational Safety and Integrity of Producing Wells and Injecting Wells (PIW) installations, prepare a formal documented execution procedure with definition of activities and tasks; develop internal audits and periodic planned inspections, with a multidisciplinary team, semiannual and annual application according to the criticality of operational processes and equipment, using Information and Communication Technology (ICT), forms/digital checklists with data collection through PDA (Personal Digital Assistant) or cell phone with android or other system, specialized software/application for data management, storage and control, with communication and data transfer capability to the computerized Enterprise Resource Planning system - SAP/R3-ERP and treatment of deviations and/or not conform. As well as carrying out contracts for the provision of corrective and preventive maintenance services with the acquisition and application by the supplier of components of existing technologies, such as: plugs/caps, screws with nuts, valves of various diameters, painting material, signaling, identification/ tagging, equipment protection grids and others, in order to ensure the operational safety and integrity of facilities, processes and operational equipment, continuous improvement in producer wells, injection wells, other facilities, equipment, pipelines and with follow-up and monitoring by performance indicators.

# References

1. Mançú, J.S.: Proposed application of FMEA in the installation of the Submerged Centrifugal Pumping System (BCS): The cases of companies providing services to the oil industry, pp. 24–33 (2013). http://www.senaicimatec.com.br/wp-content/uploads/2017/03/dissertacao_jeanderson_de_souza_mancu.pdf. Accessed 23 Nov 2020. (in Portuguese)
2. de Sousa Mançú, R.J., Gouveia, L.B., dos Santos Brunhoso Cordeiro, S.: Comparative analysis of installation integrity and safety of operations in wells, in the perception of operators of onshore oil producing fields. In: XL National Meeting of Production Engineering - "Contributions of Production Engineering to the Management of Sustainable Energy Operations", ENEGEP 2020, Foz do Iguaçu, Paraná, Brazil, 20–23 October, pp. 1–5 (2020). https://doi.org/10.14488/enegep2020_tn_sto_345_1774_40776. (in Portuguese)
3. Oliveira, J.B. (org.): Study of the oil and natural gas production chain in Espírito Santo. Espírito Santo: SEBRAE, p. 174 (2007). (in Portuguese)
4. Mançú, R.P.: Performance of Inspections in Oil Producing Wells: Comparison between Manual and Digital Data Collection. Savior. Dissertation. Graduate Program in Management and Industrial Technology - Faculty of Technology Senai Cimatec - Bahia, pp. 24–27 (2018). http://repositoriosenaiba.fieb.org.br/handle/fieb/873. Accessed 30 Nov 2020. (in Portuguese)
5. Thomas, J.E., et al.: (organizer). Fundamentals of Petroleum Engineering, 2nd edn, p. 271. Interscience, Petrobras, Rio de Janeiro (2004). (in Portuguese)
6. National Agency for Petroleum, Natural Gas and Biofuels - ANP. ANP Resolution 43, dated 12.6.2007 - Official Diary of The Union - DOU 12.7.2007 - Corrected DOU 10.12.2007 and DOU 12.12.2007. Annex - Technical Regulation of the Safety of operations Management System (RTSGSO), p. 24 (2007). (in Portuguese)
7. National Agency for Petroleum, Natural Gas and Biofuels - ANP. ANP Resolution No. 2, of 14.1.2010 Official Diary of the Union - DOU 18.1.2010. Annex - Technical Regulation of the Structural Integrity Management System for Onshore Oil and Natural Gas Production Installations (RTSGI), p. 46 (2010). (in Portuguese)

8. National Agency for Petroleum, Natural Gas and Biofuels - ANP. ANP Resolution 46, dated 11.16.2016 - Official Diary of The Union - DOU 3.11.2016 - Ratified DOU 7.11.2016. Annex - Technical Regulation of the Well Integrity Management System (RTSGIP), p. 19 (2016). (in Portuguese)
9. Gouveia, L.B., de Sousa Mançú, R.J., dos Santos Brunhoso Cordeiro, S.: Proposed integration of the technical regulations of systems of management of safety of operations and structural integrity of installations, defined by the ANP of Brazil. Int. J. Adv. Eng. Res. Sci. (IJAERS) 6(7), 197–209 (2019). https://doi.org/10.22161/ijaers.6724. ISSN 2349-6495 (P) | 2456-1908 (O)
10. Gil, A.C.: Methods and techniques of social research/Antonio Carlos Gil. - 6. ed. 7. reimpr. São Paulo: Atlas, p. 200 (2016). ISBN 978-85-224-5142-5. (in Portuguese)
11. Lacerda, M.R.: Research methodologies for nursing and health: from theory to practice/Organizers: Maria Ribeiro Lacerda, Regina Gema Santini Costerano - Porto Alegre: Moriá, p. 511 (2015). ISBN 978-85-99238-17-2. (in Portuguese)
12. Yin, R.K.: Case study: planning and methods; translation: Cristhian Matheus Herrera. - 5. ed. Porto Alegre: Bookman, p. 290 (2015). ISBN 978-85-8260-231-7. (in Portuguese)
13. de Martins, G.: Scientific research methodology for applied social sciences/Gilberto de Andrade Martins, Carlos Renato Theóphilo. 3. ed. São Paulo: Atlas, p. 247 (2016). ISBN 978-85-97-00811-1. (in Portuguese)
14. Gil, A.C.: How to design research projects/Antonio Carlos Gil. - 6. ed. São Paulo: Atlas, p. 173 (2017). ISBN 978-85-97-01261-3. (in Portuguese)

# Information and Analytical Technologies for Emergency Decision Making

Gennady Ross$^{(\boxtimes)}$ (iD) and Valery Konyavsky (iD)

Plekhanov Russian University of Economics, Moscow, Russia

**Abstract.** A methodology for the modeling of various versions of emergency situations (ES), which provides an economic analysis of overcoming their consequences, is considered. The models based on new information technologies, which allow to perform the economic evaluation of various emergency development scenarios, are at the core of such analysis. Optimization models are stated in the form of discrete programming problems, for the solution of which a modified method of dynamic programming, implemented in "Combinatorics" module of "Decision" tool system, is proposed. These models form the basis of intellectual component of information and analytical support for decision-making for the prevention and elimination of ES. The practical implementation was performed on the basis of real data with a comprehensive analysis of scenarios for the development of various measures to prevent fires in the Shatura Region.

**Keywords:** Optimization models · Dynamic programming · Discrete programming · Emergency scenario modeling · Economic analysis of scenarios · "Decision" tool system

## 1 Introduction

The issues of ensuring safety and protection of critical facilities, population and environment from natural and man-made threats and emergency situations (ES) as well as terrorist acts are becoming a priority security area, and the tasks of their prevention require a comprehensive solution. The efficient provision of safety of infrastructure facilities and population requires timely and full information about the exposure of territories of a particular region to hazardous natural and man-made processes and risks as well as the nature and tendencies of their development for the system of monitoring and forecasting of ES.

Within the framework of the automated control system (ACS) for emergency situations, the information and analytical system (IAS) ensures the collection and processing of information, which is required for the implementation of control functions, on the basis of modern information technologies (IT) [1]. The control is aimed at the provision of stable condition of potentially hazardous facilities (PHF) in terms of occurrence of ES. For this person, various scenarios of the occurrence and development of specific ES are developed, the most probable of them are determined, the efficient personnel emergency plans as well as the manpower and resources required for ES

T. Antipova (Ed.): DSIC 2021, LNNS 381, pp. 193–205, 2022.
https://doi.org/10.1007/978-3-030-93677-8_17

localization are developed for these situations. The comparison of the manpower and costs, being at the disposal of a facility, with the resources it requires allow to determine the shortage of resources, possible external sources of obtaining the ones as well as the plans for mobilizing internal and external resources, etc. As a result, a bank of situational plans for localizing ES of various types being possible for specific types and territorial entities of various levels, shall be created. The analysis of such banks allows to single out the invariant necessary actions, the description of which forms a set of expert modules (of artificial intelligence) in the decision support system (DSS) when developing the situational plans. These solutions are aimed at increasing the efficiency of measures for ES prevention, localization and response [2, 3].

The purpose of this study is to develop a comprehensive methodology for optimizing the financing of various activities for the prevention and elimination of consequences in the emergency area, as well as on the technology of its practical use with the help of the dialog tool system "Decision" (module "Combinatorics"). The complexity of solving the optimization problem is associated with the need to take into account the specifics of interrelated activities and the large dimension of the problem. This methodology is based on a modified dynamic programming algorithm, which makes it possible to obtain suboptimal solutions in real time if the dimensionality of the problem is large enough (about 200 measures). The decision-making procedure is exemplified by the optimization of funding for firefighting activities in Shatura District of Russian Federation. Current study based on the results of research by leading research teams from the RAS, MoD, Ministry of Emergency Situations, Ministry of Communications and Ministry of Economic Development [4]. In this aspect, more and extended fires, in particular in regions with high rates of deforestation and other types of land use change, will have disastrous impacts on the environment and climate, and also on human populations and their culture. We need a long-term commitment with conservation and sustainable development, independent of specific government mandates and as a true State policy, where fire management is a central issue. This needs to go beyond protected areas – where advances have been made in recent years – but should also include private lands and funds [5–8].

## 2 The Tasks of Optimization of Measures for Emergency Prevention and Response in ES Area

The tasks for the prevention and elimination of emergency situations are diversified in their content, in particular, these tasks include [4]:

- material and technical support for ES prevention and response;
- financial support of federal, regional, sectoral and local programs for ES prevention and response;
- the prevention and elimination of ES affecting the CIS countries as well as the near and far abroad countries;

- the provision of services (including the paid ones) at the request of other ministries and departments for solving various tasks, for example, for solving migration issues, namely, the accommodation of refugees and forced migrants as well as "the recalled ones" (i.e. military personnel, the members of their families and service personnel changing the place of deployment and residence);
- prompt decision-making upon ES response.

The whole set of possible actions aimed at the prevention of ES and overcoming their consequences is presented in the form of a package of measures [9]. If possible, it is advisable to have a complete list of possible measures of different content and scope, so that one could choose a set adjusted to almost any specific situation from this list. Initial information is presented in the form of an excessive list of possible measures, the list of logical connections between the measures, the data on the availability of resources that can be attracted for the implementation of the measures [3, 10].

Upon that, there are two possible options for setting the task: to make up a set of measures ensuring the maximum of preventable damage with given resource constraints, or to provide a predetermined level of preventable damage with a minimum amount of spendable resources.

All indices characterizing the measures shall be additive. Among these indices of the measures, there may be indicators that characterize:

- the costs of material resources, labor resources by profession, and financial resources;
- direct and indirect material damage prevented by this measure. The direct damage shall be understood as the assessment of tangible assets destroyed and (or) damaged as a result of the direct impact of ES, in monetary terms, as well as the measures taken for saving people and tangible assets.

The indirect ES damage shall be understood as the assessment of costs of ES response in monetary terms;

- each measure can be presented in several alternatives differing in the amount of costs and damage to be prevented (in the simplest case, the minimum and maximum value of costs and damage is indicated);
- the financial costs of a measurement can be allocated between the financing sources (from the budgets of federal entities, ministries and departments, etc.);
- the amount of prevented damage can be distributed by the following lines: housing stock, social and cultural facilities, public utilities, transport, communications, agriculture, forestry, and industry.

The Instruction on Accounting and Reporting for the Use of Financial Resources for the Emergency Response Measures contains the necessary measures and financing sources for the laying and cleaning of gravity water conduits up to fire-hazardous forests, peat companies and other facilities (see Table 1).

**Table. 1.** Form 1: Emergency prevention and response measures

| № | The name of the resources required for the implementation of this measure, financing sources, and preventable damage | min Nat. unit | min (c.u.) | max Nat. unit | max (c.u.) |
|---|---|---|---|---|---|
| | TYPE OF ES: Fire in the Shatura Region. | No. 1. WATER CONDUIT | | | |
| | Measure summary: the laying and cleaning of gravity water conduits 2 m wide, 1 m deep and 50 - 200 km long up to fire-hazardous forests, peat companies and other facilities | | | | |
| | The scope of work: min 100 thous. cub. m.; max 400 thous. cub. m. | | | | |
| 1 | Material and technical resources | | 1000 | | 4000 |
| 2 | Labor input | | 115 | | 460 |
| 3 | Total: Costs of material, labor and financial resources in particular by financing sources: | | 1115 | | 4460 |
| 4 | From the budgets of federal entities | | 152 | | 610 |
| 5 | At the expense of ministries and departments | | 763 | | 3050 |
| 6 | At the expense of enterprises, institutions, organizations, associations (joint-stock companies, concerns, corporations, etc.) | | 140 | | 560 |
| 7 | From the Government Reserve Fund | | 60 | | 240 |
| 8 | At the expense of insurance organizations | | | | |
| 9 | Preventable damage by: | | | | |
| 10 | Housing stock | | 600 | | 2400 |
| 11 | Social and cultural facilities | | 600 | | 2400 |
| 12 | Public utilities | | 600 | | 2400 |
| 13 | Transport | | 3000 | | 12000 |
| 14 | Communications | | 600 | | 2400 |
| 15 | Agriculture | | 7500 | | 30000 |
| 16 | Forestry | | 7500 | | 30000 |
| 17 | Water economy | | 600 | | 2400 |
| 18 | Industry | | 9000 | | 36000 |
| 19 | Total damage in case of failure to implement the measure | | 30000 | | 120000 |

Source: Authors' experience.

This form is applicable for all types of emergency situations. The form is filled in for the illustration of one of the measures. When specifying the initial data, it shall be borne in mind that the types of technical facilities and resources (by type) as well as the labor costs (by profession) can be specified at the discretion of specialists with the specifics of the measure being taken into account, however, their nomenclature and itemization shall be the same for all measures aimed at ES of this type. Form 1 is filled with conditional initial data for one fire prevention measure in the Shatura Region.

Solving the problem of forming the optimal version of the project represents the assignment problem having quite clear economic meaning, however, it is a considerably complicated discrete programming problem from a mathematical point of view [11]. When solving the one, it is important to take into account that some measures involve a certain sequence of implementation, or are alternative. All this does not fit into the formulation of the classical assignment problem. Discrete optimization methods are quite diverse and are divided into 3 main classes: prune methods, branch-and-bound method, combinatorial methods [12, 13]. Universal methods are not effective enough, which prompts the development of special optimization algorithms taking into account the specifics of the problems under consideration. Therefore, the paper [3] presents a modified dynamic programming method for solving Boolean problems, which is implemented in "Combinatorics" module of "Decision" tool system.

## 3 Logical Conditions, Resource Constraints and the Aims of Economic Decisions

*Methodology for the Building of Logical Connections Network.* Let us assume that there are several proposals for building separate objects and their subsequent use. The proposals come both from the experts of the organization and from outside. The proposals are made independently and inconsistently, and some of them contradict or exclude the others. There are several well-developed ones among the proposals while the rest exist only in the form of an idea and the expected effect of the ones raises doubts, which are discussed on an on-going basis. The rejection or acceptance of a proposal depends on many reasons, including economic efficiency, informal connections between the participants, etc. Within this framework it is necessary to agree on the possibility of their arrangement taking into account the management style, and the principles of division of responsibility between contractors and customers.

In terms of meaning, logical conditions can be diverse. Complex logical conditions can be represented using only 2 types of logical connections, for example, negation and conjunction (i.e. "and"), or negation and disjunction (i.e. "or") [5].

Let us consider the presentation of any proposal by a sentence P as a set of 3 types of logical connections: or (exclusive disjunction); if … then … (implication; at least one of … must be present). We use the following symbols to show the logical connections: $\cup, \rightarrow, \{\}$. $P_1 \cup P_2$ notation means that $P_1$ and $P_2$ sentence contradict each other (mutually exclude each other). $P_3 \rightarrow P_4$ notation means that $P_4$ sentence is the consequence of $P_3$ sentence. $\{P_5, P_6\}$ notation means that any of $P_5$ or $P_6$ sentences possesses a certain property, for example, being true. The logical connections, which are presented by means of the following symbols: $\cup, \rightarrow, \{\}$, are called logical formulas. Independent propositional variable $P_7$ is a logical formula as well.

Each sentence $P_i$ can be true (T) or false (F). Complex sentences obtained by using logical connections can be true or false as well. This is determined by means of so-called truth tables, which are combined in the form of Table 2:

**Table 2.** Truth table

| $P_1 \cup P_2$ | $P_1$ | $P_2$ | $P_3 \rightarrow P_4$ | $P_3$ | $P_4$ | $\{P_5, P_6\}$ | $P_5$ | $P_6$ |
|---|---|---|---|---|---|---|---|---|
| F |  | T | T | T | T | T | T | T |
| T |  | T | F | T | T | F | T | T | F |
| T |  | F | T | F | F | T | T | F | T |
| T |  | F | F | T | F | F | F | F | F |

Source: Authors' compilation

We can observe from Table 2 that complex sentence $P_1 \cup P_2$ is false only if both sentences $P_1$ and $P_2$ are true, i.e. they exclude each other. $P_3 \rightarrow P_4$ is false if $P_4$ sentence is true upon $P_3$ being false. The implication allows for $P_4$ sentence to be true only if $P_3$ sentence is true. Finally, $\{P_5, P_6\}$ connection is false in case when both sentences $P_5$ and $P_6$, being included in it, are false, i.e. any sentence must be necessarily established. An example of a logical network is shown in Fig. 1.

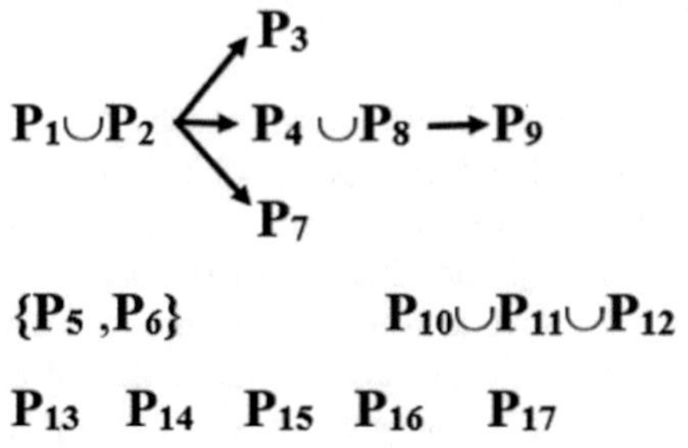

$$P_1 \cup P_2 \longleftrightarrow \begin{matrix} P_3 \\ P_4 \cup P_8 \longrightarrow P_9 \\ P_7 \end{matrix}$$

$$\{P_5, P_6\} \qquad P_{10} \cup P_{11} \cup P_{12}$$

$$P_{13} \quad P_{14} \quad P_{15} \quad P_{16} \quad P_{17}$$

**Fig. 1.** Logical network of connections. Source: Authors' design.

In this logical network $P_{13}$, $P_{14}$, $P_{15}$, $P_{16}$, $P_{17}$ sentences are logically independent of other sentences, in other words, they do not form complex sentences with other sentences. All other sentences form 3 connected subnetworks (3 complex sentences). The first subnetwork includes $P_1$, $P_2$, $P_3$, $P_4$, $P_7$, $P_8$, $P_9$ sentences. The second subnetwork includes $P_5$ and $P_6$ sentences. The third subnetwork includes $P_{10}$, $P_{11}$, $P_{12}$ sentences.

In addition to logical conditions, decision-making is also conditioned by resource constraints and economic and business aims. Each certain decision $P_i$, i = 1, 2, 3, ..., n requires certain costs Zi and is aimed at obtaining a certain result $R_i$. In general, the problem is therefore to find the optimal complex decision consisting of such a set $\Omega$ of sentences $P_i$, i $\in \Omega$ (where $\in$ is the symbol standing for "belongs to"), which firstly do not contradict the logical conditions, secondly do not require total costs excessing the allowable (limited) value, and thirdly provide the maximum cumulative result.

*Mathematical Formulation of the Problem.* Let us denote the Boolean variable by $\alpha$, namely the variable which can take on only one of two values: 0 or 1, and let us denote the number of the sentence by i. Let us assign Boolean variable $\alpha_i$ to each sentence $P_i$ as follows: $\alpha_i = 1$ when $P_i$ is true, and $\alpha_i = 0$ when $P_i$ is false. Using Boolean variables, the rules for establishing the truth of complex sentences can be presented in the form of algebraic relations. In particular, complex sentence $P_1 \cup P_2$ is equisignificant to inequality:

$$\alpha_1 + \alpha_2 < 1 \tag{1}$$

Truely, one can see from the inequality that $\alpha_1$ and $\alpha_2$ cannot be equal to 1 simultaneously. Other combinations of values are allowed. This exactly corresponds to the condition that $P_1$ and $P_2$ are not true simultaneously.

Complex sentence $P_3 \rightarrow P_4$ is equisignificant to inequality:

$$\alpha_3 > \alpha_4 \tag{2}$$

The possibility of $\alpha_4$ being equal to 1 only under the condition that $\alpha_3 = 1$ corresponds to the condition that $P_4$ can be true only under the condition that $P_3$ is true. Complex sentence $\{P_5, P_6\}$ is equisignificant to inequality:

$$\alpha_5 + \alpha_6 > 1 \tag{3}$$

According to condition (3), $\alpha_5 + \alpha_6$ cannot be equal to 0 simultaneously. This expresses the requirement that $P_5$ and $P_6$ are not false simultaneously.

The resource constraint can be expressed by using the Boolean variables:

$$\sum_{i=1}^{n} Z_i * \alpha_i < V \tag{4}$$

where V is total volume of limited resource, for example, total cost volume.

The aim can be expressed in the same way:

$$\sum_{i=1}^{n} R_i * \alpha_i \rightarrow \max(\min) \tag{5}$$

In total (1)–(5) conditions represent the formal statement of the problem formulated above.

В рассмотренной выше задачи финансирования мероприятий, связанных с ликвидацией пожаров в Шатурском районе Московской области, в качестве критерия оптимизации был использован параметр минимизации ущерба, а ограничениями являлись необходимость обеспечения логических связей между мероприятиями и суммарного объем финансовых затрат, выделенных для тушения пожаров.

## 4  Result of Experimental Research for Damage Prevention in Emergency Situations

Table 3 provides a list of main standard measures for the prevention and suppression of forest fires.

**Table 3.** Standard measures for the prevention of ES

| No. | Description | Name |
|---|---|---|
| 1 | The laying and cleaning of gravity water conduits | Water conduit |
| 2 | The construction and repair of offtake regulators | Offtake |
| 3 | The construction of artificial reservoirs and water storages with the required water reserve | Water reservoirs |
| 4 | The construction of improved dirt roads | Roads |
| 5 | The construction of accessing and passing sites | Sites |
| 6 | The organization of control over the technical condition of all types of vehicles | Transport |
| 7 | The arrangement of communications | Communications |
| 8 | The maintenance of electric power transmission lines passing through fire hazardous areas | Electric power transmission lines |
| 9 | The arrangement of patrol service for the purpose of detecting fire sources | Patrol |
| 10 | The construction of fire break barriers with 2 flame hight wide using bulldozers or tractors with trailed implements | Barriers |
| 11 | The creation of reserve mobile detachment equipped with fire engines, motor pumps and offroad vehicle | Detachments |
| 12 | The creation and maintenance of the required stock of fuels and lubricants | Fuels and lubricants |
| 13 | The preparation of public buildings, private houses and institutions of the area for the reception of the evacuated population | Accomodation |
| 14 | The construction and equipment of lookout towers | Lookout towers |
| 15 | The arrangement of control over the technical condition of railway vehicles and adjacent areas | Railway control |
| 16 | The arrangement of fire-fighting public awareness campaigns via mass media | Public awareness campaigns |
| 17 | The arrangement of helicopter patrols | Helicopter |
| 18 | The laying and cleaning of firebelts | Firebelts |
| 19 | The cutting, cleaning and removal of the forest stand | Forest stand |
| 20 | The planting of hardwood belt | Hardwood belts |
| 21 | The construction of fire ditches down to the firebreak layer in order to protect valuable forest areas and production facilities | Fire ditches |
| 22 | The creation of firefighting squads, detachments and stations, equipped with the required technical devices and equipment, in emergency situations | Firefighting squads |
| 23 | Clean burn (counterfiring) | Clean burn |
| 24 | The evacuation of people from settlements and institutions | Evacuation |
| 25 | The creation of medical aid posts for the provision of medical aid | Medical aid post |

(*continued*)

**Table 3.** (*continued*)

| No. | Description | Name |
|-----|-------------|------|
| 26 | The inspection of wood peat deposits with the installation and replacement of barriers, the placement of prohibitory signs | Signs |
| 27 | The protection of farm animals | Animals |
| 28 | Agricultural crop protection | Plants |
| 29 | The arrangement of food supply, water supply, medical care and other services for evacuated population | Life sustenance |
| 30 | Preparing hospital beds to receive the injured | Hospital beds |
| 31 | Increased stuffing of ES and civil defense services | ES staff |
| 32 | Increased stuffing of fire guard garrison | Staff of fire guard garrison |
| 33 | Increased equipment of fire guard garrison | Equipment |

Source: Authors' elaboration

Let us consider the building of a network of logical connections between standard versions using the example of measures for the prevention and suppression of forest fires. Table 1 provides a list of main standard measures for ES prevention and response[1], the data of which are an example illustrating the very principle of the arrangement of standard measures for ES prevention and response.

When developing logical connections between the measures, they are linked to a specific situation. When solving a specific practical problem, for example, when developing an emergency plan for a specific area for the upcoming season, it is necessary to concretize each measure both in one or several versions. This concretization shall result in the completion of Form 1 (Table 1) for each possible version for each measure.

Let us consider an example of the arrangement of measures in the Shatura Region. Here, mass forest and peat fires can cover an area of 10,000 hectares and induce the combustion of 50 peat storage piles, each of which has a volume of 52.5 thousand cubic meters. In case of mass forest and peat fires, it is necessary to evacuate 11 thousand people from 14 settlements. At the same time, all resources (the financial, material and labor ones) are insufficient.

Based on this example, it can be seen that the following is required for the optimal investment of resources upon ES response of any kind (not only in relation to forest fires):

1. To concretize each measure in several alternative versions being different in the amount of resource costs and the achieved effect. For example, "WATER CON-DUIT" measure can be performed in the range from 50 km to 200 km. It is clear

---

[1] On additional measures for increasing fire safety level and preventing emergencies of intermunicipal and regional (the Moscow Oblast) nature, natural disasters and elimination of their consequences in the territory of the Moscow Oblast: The Resolution of the Moscow Oblast Government No. 268/10 d/d April 4, 2006.

that the laying of a water conduit 50 km long is not enough, but at the same time, the long length of the water conduit will require withdrawing from the implementation of some other measures. These versions are alternative. Form 1 (Table 1) provides for a minimum and maximum value, however, if it is necessary, 3 or more versions can be presented. Different versions of the same measure are assigned different numbers and a logical connection is established: "Exclude each other,...,". For example, "WATER CONDUIT" measure with its minimum length can be assigned No. 1, and the same measure with its maximum length shall be assigned additional measure No. 34 in Table 3, and a logical connection shall be established.

2. It shall be taken into account that some measures are connected by the sequence of implementation. For example, the construction of offtakes makes sense only if water conduits are laid. This is expressed by the logical connection "If... then...": $(1) \rightarrow (2)$. Upon that, the scope of measures shall be taken into account. Thus, for example, if "WATER CONDUIT" measure with its maximum scope is assigned No. 35, then logical connection $(34) \rightarrow (35)$ shall be provided for along with logical connection $(1) \rightarrow (2)$.

3. Some measures shall be performed without fail at least in their minimal scope, which is expressed by the following logical connection: "There shall be at least one...". In particular, this refers to the stock of fuels and lubricants. If the measure expressing the maximum stock of fuels and lubricants is assigned number 36, then the following logical connection shall be provided for: {12,36}.

4. The measures that are logically independent of the rest ones are possible.

Taking into account the said peculiarities, a logical network reflecting the content of measures and local specifics is built [3].

**Table 4.** ES prevention measures (c.u.)

| Expense designation | Fuels and lubricants (c.u.) | Federal budget expenditures (c.u.) | Local budget expenditures (c.u.) | Preventable damage (c.u.) |
|---|---|---|---|---|
| Water conduit, min | 12 | 150 | 350 | 1700 |
| Water conduit, max | 17 | 250 | 570 | 3900 |
| Offtake | 55 | 350 | 670 | 7500 |
| Water reservoir | 35 | 270 | 730 | 5900 |
| Roads | 45 | 450 | 350 | 6700 |
| Sites | 9 | 30 | 150 | 790 |
| Transport | 11 | 55 | 95 | 670 |

Source: Data from Authors' experience

*Problem Solving Technology Using Combinatorics Module.* Let us turn to the example considered above. Table 4 lists possible measures in a certain fire hazardous area.

Upon that, "Water conduit, min" and "Water conduit, max" are alternative versions for the construction of a water conduit. It makes sense to build "sites" only upon the availability of appropriate roads. The total cost of fuels and lubricants shall not exceed 180 c.u., the limit of financing from the federal budget is 1,500 c.u., the limit of financing from the local budget is 2,500 c.u.

In order to make up a set of measures ensuring maximum preventable damage, let us load Combinatorics module and enter the data from Table 2 on "Data" sheet. We select such indices as "Fuel and lubricants", "Federal budget" and "Local budget" as limited resources and set an appropriate limit for them, and "Preventable damage" index is selected as the target one (see Table 4).

**Table 5.** The network of logical connections between the measures

| 1 | Name of measures | Logical connections between the measures | Rules for the presentation of logical measures |
|---|---|---|---|
| 2 | Water conduit, min | [2, 3] | Excluding measures |
| 3 | Water conduit, max | (6) → (7) | Measure (6) "roads" can be performed only after the implementation of measure (7) "sites" |
| 4 | Offtake | | |
| 5 | Water reservoir | | |
| 6 | Roads | | |
| 7 | Sites | | |
| 8 | Transport | | |

We introduce logical connections between the measures (see Table 5) according to the following rules: independent measures P,S, ...; excluding measures [A,B, ...]; conditioning measures (X,Y, ...) → (Q,U, ...); at least one of ... {K,L, ...}, and solve maximum problem (see Table 6).

**Table 6.** The sample of optimal set of measures

| Fuel and lubricants | Federal budget | Local budget | Preventable damage | The optimal set of measures | | | | |
|---|---|---|---|---|---|---|---|---|
| 161.00 | 1 350.00 | 2 470.00 | 24 790.00 | Conduit | Offtake | Water reservoir | Roads | Sites |
| Constraint 180,00 | Constraint 1 500,00 | Constraint 2 500.00 | | | | | | |

We can observe from Table 6 that it is possible to achieve the scope of prevented damage of 24 790 c.u. taking into account all the constraints. The optimal set of measures includes: *a water conduit, an offtake, a water reservoir, roads and sites.*

It is possible to investigate the impact of various constraints. We will cancel the constraints on financing from the federal budget and from the local budget and perform an optimization calculation for maximum. The modeling result showed that it is possible to achieve the scope of prevented damage of 25 460 c.u. The optimal set of measures will include: *a water conduit, max; an offtake, a water reservoir, roads, sites, and transport.* Upon that, the required financing from the federal budget fits into the existing constraint. It is only necessary to increase the financing from the local budget a little bit (from 2 500 to 2 565 USD).

The model considered above also allows to determine the optimal versions for a set of measures that minimize the resource consumption. This calculation is of interest if there are particularly hard resource constraints, for example, for fuels and lubricants.

# 5  Conclusion

To sum up, the presented methodology is universal and allows in an interactive mode, depending on the arisen ES, to carry out a choice of this or that criterion of decision-making and possible restrictions. In the present work the emphasis is made on practical use, considered above methodology, revealing of the list of various parameters and ways of financing of actions on the prevention and suppression of forest fires (in particular the suppression of fires in Shatura area). The procedure of formation of logical connections between measures is shown, which gives reliable guarantee from acceptance of not rational decision.

The analysis of the proposed methodological approach to choosing the optimal version of financing for the set of ES response measures allows us to conclude that:

- it is universal and applicable for solving any economic, organizational, engineering and technical, scientific research problems, which are converted to (1)–(5) formulation. In other words, "Combinatorics" module is applicable in all situations, which require making up a set of logically interconnected elements that does not contradict the logical conditions, resource constraints and ensures the maximum (or minimum) of targeted index;
- the procedure for forming a logical network of interconnection of various measures and the modified method of dynamic programming, implemented in "Combinatorics" module of the Decision tool system, are substantiated;
- the efficiency is confirmed experimentally via the example of measures for the prevention and suppression of fires in the Shatura Region. The comparison of the obtained model results with actual data has confirmed the adequacy of the models.

However, the proposed technique has some limitations. The scope of the information management system in this study is limited to an after-the-fact approach. In reality, a fire disaster can be controlled using a preventive maintenance approach. In the future, this system should be extended to serve as an integral disaster management system that is capable of acquiring and analyzing various aspects of both internal and external environments. This would be accomplished by not only obtaining real-time dynamic information through sensing technology integration but also integrating the system with other simulation programs to analyze the diffusion paths of smoke similar to [14].

# References

1. Artamonov, A.S., Plotnikov, V.A., Simonova, M.A.: The improvement of emergency response management system. Risk Manag. Prob. Technosphere **4**(24) (2012)
2. Kulba, V.V., Shulz, V.L., Chernov, I.V.: The methods and mechanisms of planning and management in emergency situations. Trends Manag. J. **2**, 134–155
3. Lichtenstein, V.E., Ross, G.V.: Information Technologies in Business. Practicum: The Application of Decision System in Solving Applied Economic Problems: Textbook, Finance and Statistics, 560 p. (2009)
4. Security of Russia. Legal, Socioeconomic and Scientific and Technical Aspects. System Research of Emergencies. International Humanitarian Public Fund "Znaniye" Publishing House, Moscow, 864 p. (2015)
5. Pivello, V.R., et al.: Understanding Brazil's catastrophic fires: Causes, consequences and policy needed to prevent future tragedies. Perspect. Ecol. Conserv. **19**(3), 233–255 (2021). https://doi.org/10.1016/j.pecon.2021.06.005
6. Mingaleva, Z., Deputatova, L., Starkov, Y.: Values and norms in the modern organization as the basis for innovative development. Int. J. Appl. Bus. Econ. Res. **14**(9), 5799–5808 (2016)
7. Mingaleva, Z., Balkova, K.: Problems of innovative economy: forming of «innovative society» and innovative receptivity. World Acad. Sci. Eng. Technol. **59**, 838–843 (2009)
8. Mingaleva, Z., Shironina, E., Buzmakov, D.: Implementation of digitization and blockchain methods in the oil and gas sector. In: Antipova, T. (ed.) ICIS 2020. LNNS, vol. 136, pp. 144–153. Springer, Cham (2021). https://doi.org/10.1007/978-3-030-49264-9_13
9. Yamalov, I.U.: Information support for decision-making in emergency response based on management scenarios modeling. Inf. Technol. **6**, 51–58 (2005)
10. Lichtenstein, V.E., Ross, G.V.: Information Technologies in Business. Practicum: The Application of Decision System in Micro- and Macroeconomics: Textbook, Finance and Statistics, 560 p. (2008)
11. Shaptala, V.G., et al.: Emergency Modeling Fundamentals: Textbook under General Editorship of V.G. Shaptala. Belgorod: Belarusian State Technological University Publishing House, 166 p. (2010)
12. Ross, G., Konyavsky, V.: Models and methods of identification of threats related to the uncontrollability of capital flows. In: Antipova, T. (ed.) ICIS 2019. LNNS, vol. 78, pp. 242–250. Springer, Cham (2020). https://doi.org/10.1007/978-3-030-22493-6_22
13. Ross, G., Konyavsky, V.: Financial costs optimization for maintaining critical information infrastructures. In: Antipova, T. (ed.) ICIS 2020. LNNS, vol. 136, pp. 188–200. Springer, Cham (2021). https://doi.org/10.1007/978-3-030-49264-9_17
14. Jung, S., Cha, H.S., Jiang, S.: Developing a building fire information management system based on 3D object visualization. Appl. Sci. **10**, 772 (2020). https://doi.org/10.3390/app10030772

# Ecological Motives to Motivate OTT Video Streaming Subscribers to Reduce Video Quality

Frank G. Goethals[1,2(✉)]

[1] IESEG School of Management, 3 rue de la Digue, 59000 Lille, France
f.goethals@ieseg.fr
[2] LEM-CNRS 9221, 3 rue de la Digue, 59000 Lille, France

**Abstract.** This paper shows that OTT video streaming platform subscribers can be motivated to decrease video quality on the basis of ecological motives. People's intent to reduce video quality on an OTT video streaming platform is higher if they perceive we are approaching the limit of what the earth can support; if they believe that reducing video quality on the platform helps protecting against global warming; and if they don't perceive an external party can control global warming. People's intent to reduce video quality on an OTT video streaming platform is lower if they perceive that reducing video quality would significantly degrade their experience. The data shows subscribers often overlook the ecological footprint of using OTT platforms. The paper suggests that educating people on the state of the planet, the ecological footprint of Internet usage and the options and responsibility they have as an individual, could lead to less data traffic and thus a smaller ecological footprint and cost savings for the OTT companies.

**Keywords:** OTT video quality · New Ecological Paradigm · Ecological footprint

## 1 Introduction

Over-the-top (OTT) video streaming usage is increasing. OTT Platforms, such as Netflix, Hulu, Amazon Prime Video and Disney+ have to invest in server capacity to deal with that rising demand [1]. This is not only costly for those companies, which invest heavily in technical solutions to lower the amount of data that needs to be transferred [1], but it is also leading to a bigger environmental footprint [2]. Indeed, while the environmental impact of increased Internet use may be overlooked, the use of online services such as videostreaming platforms, online meeting platforms and online music streaming services has a significant environmental cost [2].

Lowering the amount of data that is being streamed, would be good for the OTT companies' costs and for the environment. One way to achieve that, is by lowering the video quality. For instance, Netflix subscribers' data usage is about 7 GB per hour in Ultra HD (or 4K), 0.7 GB per hour for standard video quality, and 0.3 GB per hour for basic video quality. The difference is non-negligible, given that Internet usage – across the globe – has an estimated carbon footprint ranging from 28 to 63 g $CO_2$ equivalent

T. Antipova (Ed.): DSIC 2021, LNNS 381, pp. 206–216, 2022.
https://doi.org/10.1007/978-3-030-93677-8_18

per gigabyte (GB). The related water and land footprints range respectively from 0.1 to 35 L/GB and 0.7 to 20 cm$^2$/GB [2]. If 70 million subscribers would reduce the video quality of their streaming services, this would reduce $CO_2$ emissions with 3.5 million tons per month, which is equivalent to eliminating 1.7 million tons of coal, or approximately 6% of the total monthly coal consumption in the USA [2].

Given the reasons above, it is interesting to know what can motivate subscribers to reduce video quality. The study at hand investigates antecedents of users' intent to reduce video quality on OTT platforms and concentrates on the environmental part of the story; something that seems to have been overlooked in earlier studies. Most studies to date concentrate on delivering higher quality of experience and technical solutions such as advanced video compression [1] rather than on changing subscribers' behavior. Indeed, the common perception is that subscribers always want more [1] and it is not mainstream to suggest subscribers could be motivated to lower the video quality.

Section 2 justifies the research model. Next, the research method is presented and Sect. 4 shows the results of the data analysis. The results are discussed in Sect. 5. Limitations of the study are presented in Sect. 6.

## 2   Research Model

A recent study on the overlooked environmental footprint of increasing internet use suggested that "increased Internet use has some hidden environmental impacts that must be uncovered" [1, p. 1] and that small actions such as reducing the quality of streaming services and turning off video during a virtual meeting can significantly reduce the environmental footprints of Internet use. The majority of people are not aware of the environmental footprint associated with video streaming. Our study investigates whether people are motivated to reduce video quality on an OTT video platform if they know that reducing video quality on such platforms leads to lower $CO_2$ emissions.

First, it is important to recognize that people look differently at $CO_2$ emissions. Some people see high $CO_2$ emissions as a problem, while others don't. The New Ecological Paradigm (NEP) focusses on people's beliefs about the existence of limits to growth, humanity's ability to upset nature's balance, and humanity's right to rule over the rest of nature [3]. Environmental believes can influence behavior. For instance, a positive correlation was found between people's NEP scores and their willingness to pay for renewable energy expansion [4]. The NEP takes into account several aspects of ecological worldviews. Given that the study at hand concentrates on $CO_2$ emissions, this paper concentrates on the 'limit to growth' facet of the NEP and hypothesizes:

Hypothesis 1: A person's intent to reduce video quality on an OTT video streaming platform is higher if that person perceives there is a limit to the amount of activity that the earth can support.

When people face threats, in casu threats related to global warming, people seek to control their fear or the danger [5]. The danger control process aims to mitigate the threat through the adoption of adaptive coping behavior. According to Protection Motivation Theory (PMT), a person's motivation to show adaptive coping behavior (in this case, against global warming), depends on the perceived response efficacy.

Response Efficacy is defined as "the belief that the adaptive [coping] response will work, that taking the protective action will be effective in protecting the self or others" [6, p. 411]. That is, if a person believes that reducing video quality on an OTT video streaming platform will work against global warming, that person is more likely to take that action. In line with PMT, the following hypothesis is put forward:

Hypothesis 2: A person's intent to reduce video quality on an OTT video streaming platform is higher if that person believes taking that action will help protecting against global warming.

The PMT recognizes that a person's intent to show protective behavior can be decreased because of costs associated with that behavior. Response Costs have been defined as "beliefs about how costly performing the recommended response will be to the individual" [7, p. 109]. Response Costs can be any costs associated with the protective behavior, and are thus not limited to financial costs for instance [6]. In frame of our study, reducing video quality comes with costs in the sense that subscribers may perceive a significant degrade in their video experience [8]. The following hypothesis is thus based on PMT:

Hypothesis 3: A person's intent to reduce video quality on an OTT video streaming platform is lower if that person perceives higher responses costs associated with a video quality reduction.

According to Attribution Theory, people's behavior depends on the expected outcome of showing that behavior [9]. These expected outcomes are a function of people's attributional styles. Attributional styles show in the way people explain successes and failures. Amongst others, the attributional style is determined by the perceived controllability of the cause [10]. This perceived controllability concerns two *separate* dimensions [11]: is the cause controllable or uncontrollable by the person (Personal Control) and is the cause controllable or uncontrollable by other people (External Control). While the issue of personal control can be related to the Response Efficacy mentioned above, the following hypothesis recognizes people may believe the cause of global warming can be controlled by someone else (external control). The latter could turn these people passive. Therefore, the following hypothesis is put forward:

Hypothesis 4: A person's intent to reduce video quality on an OTT video streaming platform is lower if that person perceives other people can control the causes of global warming.

Figure 1 shows the research model. Gender, age and data cap on the Internet plan were included as control variables.

## 3   Research Method

A questionnaire was posted on Amazon Mechanical Turk, open for Netflix subscribers living in the USA. Respondents received a financial fee for participation, in line with what is common on that platform. 100 responses were gathered. A question was included to test whether participants had carefully read the vignette. On the basis of this question 14 responses were discarded, so that 86 responses remained after data cleaning. The following vignette was presented to the respondents:

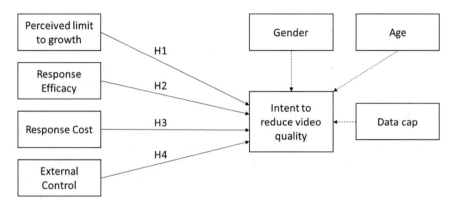

**Fig. 1.** Research model

"Netflix subscribers can adjust the video quality on Netflix. Data usage is

- 7 GB per hour in Ultra HD (or 4K),
- 0.7 GB per hour for standard video quality, and
- 0.3 GB per hour for basic video quality.

The higher the video quality that Netflix servers need to send, the more computer servers are needed to deal with that demand and thus the more electricity is consumed. Netflix currently uses more than 100 000 computer servers to stream videos.

A recent study by the University of Maryland, Purdue University, MIT and others suggests that if you watch 4 h of Netflix video streaming at the highest quality (Ultra HD) a day for 30 days; this has a monthly carbon footprint of about 117 lb $CO_2$. However, the study argues that by lowering the video quality from HD to standard, the monthly footprint would drop with 111.5 lb to 5.5 lb $CO_2$, saving the emissions of driving a car from Baltimore to Philadelphia (about 100 miles).

If 70 million streaming subscribers were to lower the video quality of their streaming services, there would be a monthly reduction in 7716 million pounds of $CO_2$. This is the equivalent of eliminating 3747 million pounds of coal per month, or approximately 6% of the total monthly coal consumption in the USA.

Studies argue that global warming will lead to more extreme weather events such as hurricanes, food and water shortages and increased disease spreading, conflict and war."

After reading the vignette, respondents were asked questions to measure the different constructs in the model. Measures for the Intent to reduce video quality (three items such as "I intend to reduce video quality on Netflix in the next month."), Response Efficacy (three items such as "Reducing video quality on Netflix works for protection against global warming.") and Response Cost (three items such as "Reducing video quality on Netflix decreases the convenience afforded by Netflix.") were based on the measures used in the PMT study of [12]. Measures were adapted to the Netflix context.

The External Control was measured in line with the Revised Causal Dimension Scale (CDSII) of [11]. This implies respondents were first asked to write an answer to the following question: "In general, what are the main reason(s) for global heating (i.e., climate change) according to you?". The, respondents then were asked "Think about the reason or reasons you have written above. The items below concern your impressions or opinions of this cause or causes. Select one number for each of the following questions: Is the cause(s) something… [Over which others have control … over which others have no control], [Under the power of other people … not under the power of other people], [Other people can regulate … other people cannot regulate].

Three items were used to measure the perceived Limits to Growth, based on the related dimension of the NEP scale using a 5-point Likert scale (from Strongly Disagree to Strongly Agree) in line with [3]: "We are approaching the limit of the number of people the earth can support", (reverse coded:) "The earth has plenty of natural resources if we just learn how to develop them" and "The earth is like a spaceship with very limited room and resources".

Gender and age were included as control variables. An additional variable was included to control for the data cap in the respondent's data plan. A 6-point scale was used for this, ranging from "There is no cap; I have unlimited Internet data" to "There is a cap that very frequently poses problems".

## 4 Data Analysis

The survey also included a few questions that gave us some relevant insights apart from testing the research model. Those questions are presented first. Then, the assessment of the research model is presented.

When asked whether they had ever considered that streaming high quality video on Netflix has a bigger ecological footprint than streaming lower quality video; 73 respondents answered *no* while 13 answered *yes*. Furthermore, respondents were asked twice whether they agreed on the statement "Streaming high quality video on Netflix has a bigger ecological footprint than streaming lower quality video.": once before reading the vignette and once after reading it. Before reading the vignette, the average response was 4.27, while it was 2.27 after reading the vignette (a value of 1 meaning "strongly agree" and a value of 7 meaning "strongly disagree" and a value of 4 "neither agree nor disagree"); which is a statistically significant difference ($p = 0.000$). This confirms the idea that the ecological footprint of using OTT platforms is often overlooked by subscribers, and confirms the respondents had read the vignette.

It is interesting to note that 30 of the 86 respondents said they were (moderately to strongly) inclined to reduce the video quality on Netflix in the next month (as measured with the dependent variable). Hence, the often-made assumption that subscribers are always only interested in getting "more", does not seem to hold in all cases anno 2021.

In what follows, the research model presented in Fig. 1 is assessed. Given the limited sample size, SmartPLS (v3.3.3) was used to analyze the data. In what follows, first the results of the assessment of the measurement model are presented and then those concerning the assessment of the structural model.

## 4.1    Assessment of the Measurement Model

The average variance extracted (AVE) is well above 0.5 for all constructs (lowest AVE of 0.686 for Response Cost) and the Composite Reliability is well above 0.7 (lowest score of 0.866 for Response Cost). Cronbach's alpha was above 0.7 (lowest alpha 0.776 for Response Cost). Discriminant validity is established in line with the Fornell-Larcker criterion (see Table 1). The cross-loadings (see Table 2) also show that discriminant validity is established. Also, the Heterotrait-Monotrait Ratio was in all cases well below 0.9, confirming discriminant validity is established. The highest Variance Inflation Factor value (VIF) is 1.330, leading to the conclusion there is no problem with multicollinearity.

**Table 1.** Inter-correlations of latent variables (square root of AVE on diagonal).

|  | External Control | Data cap | Gender | Intent | Limits to Growth | Response Cost | Response Efficacy | Age |
|---|---|---|---|---|---|---|---|---|
| External Control | 0.864 | | | | | | | |
| Data cap | 0.367 | 1 | | | | | | |
| Gender | 0.057 | 0.037 | 1 | | | | | |
| Intention | 0.144 | 0.109 | 0.2 | 0.98 | | | | |
| Limits to Growth | −0.298 | −0.142 | 0.096 | 0.302 | 0.859 | | | |
| Response Cost | 0.124 | 0.143 | −0.028 | −0.383 | −0.25 | 0.828 | | |
| Response Efficacy | −0.191 | −0.12 | 0.077 | 0.444 | 0.363 | −0.407 | 0.967 | |
| Age | 0.096 | −0.009 | 0.065 | 0.179 | 0.133 | −0.058 | 0.098 | 1 |

## 4.2    Assessment of the Structural Model

Table 3 shows the results of testing the structural model. The $R^2$ of the model is 0.38. All four hypotheses are confirmed.

As hypothesized (H1), a person's intent to reduce video quality on an OTT video streaming platform is higher if that person perceives we are approaching the limit of what the earth can support ($p < 0.05$).

As expected on the basis of PMT, H2 was confirmed: the higher the perceived response efficacy, the higher the intent to lower the video quality. That is: the test suggests that a person's intent to reduce video quality on an OTT video streaming platform is higher if that person believes that reducing video quality on Netflix works for protection against global warming ($p < 0.01$).

H3, suggesting a negative relation between response costs and the intent to reduce video quality was also confirmed ($p < 0.01$). The more a person perceives that reducing video quality on Netflix would significantly degrade his/her Netflix experience, the less likely that person is to lower the video quality.

**Table 2.** Cross-loadings.

|  | External Control | Data cap | Gender | Intent | Limits to growth | Resp. Cost | Resp. Eff. | Age |
|---|---|---|---|---|---|---|---|---|
| External Control1 | **0.971** | 0.305 | 0.082 | 0.17 | −0.244 | 0.102 | −0.139 | 0.091 |
| External Control2 | **0.642** | 0.296 | −0.061 | 0.034 | −0.401 | 0.19 | −0.345 | 0.089 |
| External Control3 | **0.941** | 0.409 | 0.043 | 0.105 | −0.286 | 0.111 | −0.182 | 0.086 |
| Data cap | 0.367 | **1** | 0.037 | 0.109 | −0.142 | 0.143 | −0.12 | −0.009 |
| Gender | 0.057 | 0.037 | **1** | 0.2 | 0.096 | −0.028 | 0.077 | 0.065 |
| Intention1 | 0.15 | 0.13 | 0.196 | **0.974** | 0.291 | −0.384 | 0.436 | 0.178 |
| Intention2 | 0.13 | 0.098 | 0.198 | **0.981** | 0.271 | −0.371 | 0.423 | 0.15 |
| Intention3 | 0.142 | 0.092 | 0.193 | **0.984** | 0.325 | −0.369 | 0.446 | 0.195 |
| LimitsToGr1 | −0.285 | −0.152 | 0.092 | 0.11 | **0.732** | −0.277 | 0.207 | −0.021 |
| LimitsToGr2 | −0.208 | −0.162 | 0.067 | 0.299 | **0.919** | −0.191 | 0.395 | 0.185 |
| LimitsToGr3 | −0.315 | −0.078 | 0.102 | 0.295 | **0.912** | −0.232 | 0.291 | 0.103 |
| Response Cost1 | 0.23 | 0.227 | 0.016 | −0.22 | −0.338 | **0.731** | −0.254 | 0.12 |
| Response Cost2 | 0.136 | 0.09 | −0.026 | −0.24 | −0.208 | **0.811** | −0.308 | −0.139 |
| Response Cost3 | 0.024 | 0.085 | −0.044 | −0.426 | −0.148 | **0.93** | −0.413 | −0.087 |
| Response Efficacy1 | −0.177 | −0.098 | 0.039 | 0.394 | 0.303 | −0.341 | **0.965** | 0.082 |
| Response Efficacy2 | −0.201 | −0.14 | 0.078 | 0.383 | 0.312 | −0.41 | **0.969** | 0.102 |
| Response Efficacy3 | −0.179 | −0.112 | 0.1 | 0.494 | 0.42 | −0.422 | **0.967** | 0.099 |
| Age | 0.096 | −0.009 | 0.065 | 0.179 | 0.133 | −0.058 | 0.098 | **1** |

**Table 3.** Assessment of the structural model (*p < 0.05; **p < 0.01).

|  | Path coefficient |
|---|---|
| Response Efficacy → Intention | 0.318** |
| Response Cost → Intention | −0.244** |
| External Control → Intention | 0.233* |
| Limits To Growth → Intention | 0.19* |
| Gender → Intention | 0.127 |
| Data cap → Intention | 0.12 |
| Age → Intention | 0.078 |

Finally, in line with H4, the data confirms that a person's intent to reduce video quality on an OTT video streaming platform is lower if that person perceives other people can control the causes of global warming. (Please note that External Control was reverse coded, explaining the positive sign.) The idea that an external party can control global warming seems to lower people's intent to take action themselves ($p < 0.05$).

The control variables gender and age had no significant relation with the dependent variable. The same goes for the data cap. The latter is not surprising, given the fact that respondents could already take their data cap into account before taking the survey. (Dropping this control variable from the model had no impact on the results.)

## 5  Discussion

OTT video platform subscribers do not always want "more" quality; some can be motivated to reduce the video quality (apart from their data cap in their Internet plan) on the basis of ecological motives. The results above give a number of indications of how OTT platform subscribers can be motivated to transfer less data. Such a reduction would lower the OTT platform's costs, but also decrease the environmental footprint. While the vignette in the paper at hand concentrated on the drop in $CO_2$ emissions, there are also other benefits in terms of other environmental costs, such as a drop in water consumption and a smaller land footprint [2]. A drop in data transfers is thus an important objective, be it for financial reasons or for ecological reasons.

The data shows that most respondents had never considered that streaming high quality video on Netflix has a bigger ecological footprint than streaming lower quality video. It seems important to educate subscribers on this. Governments might for example suggest a warning message if videos are streamed at high quality and even ask for a consent, somewhat like the European GDPR obliges websites to inform users about the use of cookies and to ask for their consent.

Given that respondents who perceive a high response efficacy are more likely to be more economical with data transfers, it is important to signal to subscribers that their actions make a difference. If 70 million subscribers would reduce the video quality of their streaming services, this would eliminate approximately 6% of the total monthly coal consumption in the USA [2]. Getting such significant results is only possible if individuals can be convinced to change their actions. A qualitative question in our survey asked why subscribers had (not) changed the video quality in the past. The most typical answers to this question are "I did not know I could" and "I never thought about it". As earlier studies in the field of Information Systems suggested [13], default settings are important. Besides informing OTT subscribers that data transfers come with ecological costs, informing and reminding the subscribers that they can consciously choose the video quality is important as well.

Related to this, the study shows that respondents who perceive there are limits to the amount of activity the earth can support are more likely to reduce the video quality. Educating and convincing citizens with scientific facts about the state of the earth can thus have an impact on their behavior. The study at hand confirms that the more citizens subscribe to the New Ecological Paradigm, the more likely they are to show environmental friendly behavior and – in frame of this study – to reduce data transfers.

This data shows that if subscribers believe that *others* have control over the causes of global warming, the subscribers are less likely to reduce the video quality. Instead, they seem to rely on others to resolve the issues related to global warming and they are less likely to take action themselves. In line with the suggestion above to inform people of the ecological impact of their actions, it is thus relevant to emphasize that subscribers are in control themselves and are responsible themselves for lowering the data transfers and that they should not assume others will resolve the issue.

Finally, the data confirms the idea that people are less likely to reduce video quality if they perceive doing so would decrease their experience on the platform. While it may be hard to convince people to watch videos at a lower quality than what their screen supports; it is important to look forward and it may be interesting for governments to try to steer future screen sales. That is, they could try to limit the sales of higher resolution screens (which otherwise could lead to further increases in data traffic). For instance, governments may consider taxes to include environmental costs in prices of future 8K or even 16K televisions. Similarly, governments could consider taxes on data transfers so that the prices for data transfers include the environmental costs; which are typically hidden today.

## 6   Limitations

The paper at hand used data gathered in a single country: the USA. It would be interesting to gather data in different countries so as to make a comparison. For instance, the New Ecological Paradigm score tends to fluctuate across countries (14). It has been shown that the public attitude towards green investments is more positive if ecological sensitivity rises [14]. In a similar vein, maybe OTT platforms could consider setting a lower default video quality in regions with high NEP scores; leaving subscribers the option to increase video quality if desired.

The current study only investigates one specific type of internet usage. The use of other online services is also on the rise. Similar to the Netflix-context, music streaming services such as Spotify typically offer multiple streaming settings (e.g. ranging from 24 Kbit/s to 320 Kbit/s). Lowering the audio quality may be perceived as degrading the experience less than decreasing video quality does. Moreover, people tend to listen repeatedly to the same songs. Repeatedly streaming the same song has a bigger carbon footprint than downloading it once and storing it locally. Further research is needed to investigate whether users could be motivated to download music/stream them at lower quality for environmental reasons.

In a work context, Zoom usage increased from 10 million daily meeting participants end 2019 to 300 million by May 2020 [15]. Zoom uses many thousands of servers to enable this. Before the covid19 crisis, Zoom largely relied on its own data center. When Covid19 kicked in, Zoom's own data center could not handle the traffic anymore, so they started using 5000 to 6000 servers from Amazon and servers from the Oracle cloud [16]. All these servers consume a significant amount of electricity; something which is typically not considered by Zoom users. Turning off the video in

Zoom meetings would lower the carbon footprint of online meetings significantly [2]. The user's decision process may, however, be much more complex in such settings. Further research is needed to determine whether environmental concerns may also have an impact on the way people use Zoom.

## 7   Conclusions

While this is hardly recognized, the data shows that people can be motivated to decrease video quality on OTT video streaming platforms, apart from the data cap on their Internet plan. More specifically, ecological motives can motivate people to do so. People's intent to reduce video quality on an OTT video streaming platform is higher if they perceive we are approaching the limit of what the earth can support; if they believe that reducing video quality on the OTT platform works for protection against global warming; and if they believe there is no external party which can control global warming. People's intent to reduce video quality on an OTT video streaming platform is lower if they perceive that reducing video quality would significantly degrade their experience. The data shows the ecological footprint of using OTT platforms is often overlooked by subscribers. The paper suggests that educating people on the state of the planet, the ecological footprint of internet usage and the options and responsibility they have as an individual, could lead to less data traffic and thus a smaller ecological footprint and cost savings for the OTT companies.

## References

1. Harmonic: Delivering Exceptional Quality of Experience for OTT. Streaming Media Magazine July/August 2017, pp. 37–40 (2017)
2. Obringer, R., Rachunok, B., Maia-Silva, D., Arbabzadeh, M., Nateghi, R., Madani, K.: The overlooked environmental footprint of increasing internet use. Resour. Conserv. Recycl. **167**, 1–4 (2021)
3. Dunlap, R.E., Van Liere, K.D., Mertig, A.G., Jones, R.E.: Measuring endorsement of the New Ecological Paradigm: a revised NEP scale. J. Soc. Issues **56**(3), 425–442 (2000)
4. Ntanos, S., Kyriakopoulos, G., Skordoulis, M., Chalikias, M., Arabatzis, G.: An application of the New Environmental Paradigm (NEP) scale in a Greek context. Energies **12**(2), 239 (2019)
5. Leventhal, H.: Findings and theory in the study of fear communications. In: Berkowitz, L. (eds.) Advances in Experimental Social Psychology, vol. 5, pp. 119–186. Academic Press, New York (1970)
6. Floyd, D.L., Prentice-Dun, S., Rogers, R.: A meta-analysis of research on protection motivation theory. J. Appl. Soc. Psychol. **30**(2), 407–429 (2000)
7. Milne, S., Sheeran, P., Orbell, S.: Prediction and intervention in health-related behaviour: a meta-analytic review of protection motivation theory. J. Appl. Soc. Psychol. **30**(1), 106–143 (2000)
8. Karim, S., He, H., Laghari, A.A., Madiha, H.: Quality of Service (QoS): measurements of video streaming. Int. J. Comput. Sci. Issues **16**(6), 1–9 (2019)

9. Martinko, M.J., Zmud, R.W., Henry, J.W.: An attributional explanation of individual resistance to the introduction of information technologies in the workplace. Behav. Inf. Technol. **15**(5), 313–330 (1996)
10. Weiner, B.: An attributional theory of achievement motivation and emotion. Psychol. Rev. **92**, 548–573 (1985)
11. McAuley, E., Duncan, T.E., Russell, D.W.: Measuring causal attributions: the revised causal dimension scale (CDSII). Pers. Soc. Psychol. Bull. **18**(5), 566–573 (1992)
12. Boss, S.R., Galletta, D.F., Lowry, P.B., Moody, G.D., Polak, P.: What do systems users have to fear? Using fear appeals to engender threats and fear that motivate protective security behaviors. MIS Q. **39**(4), 837–864 (2015)
13. Goethals, F.: Exploring the choice for default systems. Int. J. Technol. Hum. Interact. **13**(1), 21–38 (2017)
14. Ntanos, S., Arabatzis, G., Tsiantikoudis, S.: Investigation of the relationship between ecological sensitivity and renewable energy investment acceptance by using the NEP Scale. In: Proceedings of the CEUR Workshop Proceedings, Greece, vol. 2030, pp. 561–570 (2017)
15. Inside Zoom's Infrastructure: Scaling Up Massively With Colo and Cloud. https://datacenterfrontier.com/inside-zooms-infrastructure-scaling-up-massively-with-colo-and-cloud/. Accessed 25 Aug 2021
16. Ask Eric Anything. https://www.youtube.com/watch?v=tlC-sEdqY48&feature=youtu.be&t=1236. Accessed 25 Aug 2021

# Digital Finance, Business and Banking

# Brand Valuation of the Russian Bank: Interbrand Model

Darya Rozhkova[1]([✉]) [ID], Nadezhda Rozhkova[2] [ID],
Daniela Gonzalez Serna[3] [ID], and Uliana Blinova[1] [ID]

[1] Financial University under the Government of the Russian Federation,
125993 Moscow, Russia
[2] State University of Management, 109542 Moscow, Russia
[3] Trade Finance, Singapore 189350, Singapore

**Abstract.** At the moment, it is definitely believed that a brand of a product or service is one of the prerequisites for a company's successful development, as well as for long-term operation in the market. Assessment of brand value is still one of the most relevant and controversial research subjects in marketing and finance. The choice of a particular model for brand assessment is based on goals and objectives (e.g. mergers and acquisitions, licensing), as well as the completeness of sources and inputs, and of course, the choice depends on reliability of results. The main problem with valuation is that brands are unique, so it is difficult to determine their market value. The basis for the investigation which was launched was the dramatically different results obtained by two independent agencies, Interbrand and Brand Finance, for PJSC "Sberbank" (a Russian bank) in 2013. In this paper, by analyzing the Interbrand brand equity valuation methodology, we have calculated the brand value for Sberbank in 2019–2020. The analysis is carried out by comparison of selected brand values (Amazon, Google, Apple, Microsoft, Sberbank) provided by Interbrand and Brand Finance. The findings of this study demonstrate the necessity for addressing different brand valuation models, taking into consideration country, industry specificity and overall demand as well as overall brand valuation goal. Results demonstrate an existence of drastic deviation of brand values and an existence of different dynamic.

**Keywords:** Brand valuation · The Interbrand model · The Brand Finance model · Russian banks

## 1 Introduction

Nowadays it is widely recognized that brands play an important role in creating and sustaining the financial performance of companies. Against the backdrop of intense competition and virtually unlimited business opportunities, strong brands help companies differentiate in their target market and convey to the consumer a need to buy the product or service due to brand loyalty (Pessemier 1959; Jacoby and Chestnut 1978; Reichheld 1996; Chaudhuri and Holbrook 2001). In an environment where functional differences between products are minimized, brands create a basis for distinguishing

© The Author(s), under exclusive license to Springer Nature Switzerland AG 2022
T. Antipova (Ed.): DSIC 2021, LNNS 381, pp. 219–230, 2022.
https://doi.org/10.1007/978-3-030-93677-8_19

themselves from similar offerings. The ability of brands to endow a product, service or company with emotional value beyond functional is a source of value creation through individual human behavior (Voskanyan et al. 2021; Nikitina 2021). Over the past 30 years, there has been a shift in the sources of value creation from tangible assets (buildings, land, equipment) to intangible assets (skilled personnel, patents, brands) (Daum 2007; Porter and Kramer 2011).

Over the last few years, an increasing number of companies and experts have been trying to find an adequate brand assessment model (Aaker 2009). From an economic point of view, a brand can be identified as an intangible asset that allows companies to earn excess profits by instilling the consumer's desire to overpay, sometimes contrary to common sense. Emphasis is on the privileges of the brand (Wood 2000).

At the same time, this concept is often confused with brand equity. Brand equity represents perceptions of a customer about a company (a consumer-based focus), this measurement is more abstract. It indicates how successful a brand is and what a brand is worth to a customer, whereas brand value (a financial-based approach) is a financial representation of brand's worth (Raggio and Leone 2007). However, for the purpose of this investigation we consider these differences irrelevant when brand measurement mixed methods (and mixed concepts accordingly) are used (Tarakçi et al. 2021).

Many methods to valuate brand have been developed (Simon and Sullivan 1993; Knowles 2003; Aaker 2009; Reyneke et al. 2014). However, due to the subjective and intangible nature of brands, no method provides users with sufficient confidence in the accuracy of the rating. Not all methods are universal, and the use of different approaches to evaluate the same brand leads to significant disparities in results.

In this article we aim to address this relevant issue, taking in mind that there are 2 main agencies that provide an annual report and brand ratings – Brand Finance and Interbrand. We underline that Brand Finance provides a report that includes the most valuable and strongest banking brands (500 brands), whereas Interbrand shows a report that includes the world's most valuable brands (100 brands).

PJSC "Sberbank" (hereafter Sberbank) ranked first for several years in a row, according to the annual Brand Finance Russia 50 ranking. Additionally, the bank is considered to be a leader in the economic sector. That makes it suitable and interesting to evaluate the brand value of Sberbank. From one side, there is a practical need to apply several methods for brand valuation due to the inconsistency of results and possibility to take into consideration different factors. Thus, the brand value of Sberbank, calculated by Brand Finance in 2013 is equal to $14,160 mln. For the same year a report "Best Russian brands 2013" by Interbrand indicated the brand value of Sberbank to be equal to $3,267 mln. The deviation is $10,893 mln. (more than 4 times).

From the other side, it should be noted that Russian brands have never been included in the world ranking of Interbrand, despite their high value. Theoretically, many Russian brands could enter the global ranking as, financially speaking, their value is quite high. However, there are some requirements that still cannot be met by brands in Russia. First of all, it is implied in the name of the ranking that the brands that are evaluated must be global. The agency ranks companies only if a third of revenue comes from operations outside of the brand's home region. The brand must also be present on at least three continents (Asia, Europe, and North America). Not a single Russian domestic owner of a large brand has yet to meet these requirements.

Following the research path and a practical need, the article addresses the issue of Sberbank brand valuation using the Interbrand model. The purpose of the research is to assess brand using Interbrand methodology, check whether it applicable or not (taking into consideration Russian experience and particularities).

The paper proceeds as follows. Section 2 reviews theoretical base and evolution of Brand Finance methodology. Section 3 provides methodology of research and includes short analysis of Sberbank's brand development. In Sect. 4 we have estimated the value of Sberbank's brand using the Interbrand model for 2019–2020. Last part provides with results evaluation, summarize main problems.

## 2 Theoretical Foundations

The formulary (mixed) approach is widely used by consulting companies and other researchers from around the world in order to evaluate the brand value. The mixed approach combines methods from various assessment approaches (monetary aspect, consumer-based aspect), forming a unique combination and leveling the disadvantages of each previous approach (Kim et al. 2003; Firat and Badem 2008). The mixed approach is quite common in the modern market and we will further consider the main models that are included in it.

Brand Finance is a UK based consulting company that developed the model in 1999. The model is based on the allocation of a brand premium. First, it is necessary to identify the factors that are drivers for the company and determine the influence of the company's brand on each of the drivers in order to determine the aggregated role of the brand in the company's value. Further, it is necessary to find the discount rate (Ra), it is calculated using the Capital Asset Pricing Model (CAPM – capital asset pricing model) (formula 1):

$$Ra = rrf + (rm - rrf)\,\beta \tag{1}$$

where:

- rrf – risk-free rate;
- rm – expected return of the market;
- $\beta$ – measure of a stock's risk (volatility of returns).

Ultimately, this is the added value from the brand, discounted over time. The model correlates only with the past data of the company, and it is important to understand the future of the company and its attractiveness in the market. At the same time, this method allows for the combining of marketing analyses and the company's financial methods.

Interbrand is a consulting firm specializing in areas such as brand strategy, brand analytics, brand valuation, etc. Interbrand identifies brand revenue and capitalizes on it, making adjustments accordingly. The firm develops brand assessment based on financial analysis, brand role and brand strength, a model developed in 1998 by J. Murphy (Murphy 1998).

The model includes several steps:

1) Calculate the cash flow generated by intangible assets;
2) Find the share of the brand in the cash flow from intangible assets;
3) Find a brand multiplier.

Interbrand attempts to determine the discount rate using a brand index that is based on factors with different weights. From 2010 on the official web-site of Interbrand, brand strength factors were reviewed and they are based on external and internal dimensions (for example direction, participation, trust etc.). In Table 1, we will present the evolution of factors that are used and their weights (Lomax and Raman 2007; Interbrand 2021).

**Table 1.** Brand strength factors (evolution of methodology).

| Before 2010 | | | After 2010 | | |
|---|---|---|---|---|---|
| Factors | % | Description | Factors | % | Description |
| 1. Market | 10 | Market growth and consumer preferences | 1. Direction | 10 | Purpose of brand, development path |
| 2. Stability | 15 | Customers' loyalty | 2. Alignment | 10 | Organization commitment to brand development |
| 3. Leadership | 25 | Position of the brand in the sector | 3. Empathy | 10 | Relationships with stakeholders |
| 4. Profit trend | 10 | Performance and brand value | 4. Agility | 10 | Market growth |
| 5. Support | 10 | Support the brand has received | 5.Distinctiveness | 10 | Intangible assets which are difficult to imitate |
| 6. Geographic spread | 25 | Strength of the brand in foreign countries | 6. Coherence | 10 | Different customers interactions in terms of brand positioning |
| 7. Protection | 5 | Company's ability to protect the brand | 7. Participation | 10 | Collaboration with stakeholders |
| | | | 8. Presence | 10 | Brand perception by different customers |
| | | | 9. Trust | 10 | Respond to customers' needs |
| | | | 10. Affinity | 10 | Shared values with stakeholders |
| Σ | **100** | | Σ | **100** | |

Thus, factors have been changed in order to take into consideration the real situation as the method was criticized for a lack of a future orientation, due to the usage of historical data (Soto 2008). Nevertheless, there is a subjectivity of the assessment, since it can be determined in different ways.

As a result of calculations, the company receives a brand strength value in the range from 0 to 100 points. Further, the brand multiplier and the brand strength index are correlated on the S chart in order to find the multiplier. At this stage, the brand value is already calculated by multiplying the cash flow from intangible assets by the found brand multiplier. This approach is widely regarded for its ability to address all aspects of branding.

# 3 Methodology of Research

## 3.1 General Background of Research: Description of Sberbank

The method that we use is a case study of Sberbank. The case study method is beneficial because it allows real-time observations (Yin 2004). Moreover, it corresponds to our need to investigate "an event, activity, process, or people" in a particular situation (Creswell 2002).

Sberbank is the biggest bank of the Russian Federation; the main shareholder is represented by the Ministry of Finance of the Russian Federation with the majority (over 50%) shareholding of voting rights. In 2021 Sberbank has a widely spread regional network, 11 regional banks with 14,1 thousand subdivisions in 83 constituent entities of the Russian Federation. The foreign network of Sberbank Group includes subsidiary banks, branches and representative offices in 17 countries around the world. About 3 million clients of Sberbank are located outside of Russia and international business accounts for 5% of total activity.

Moreover, Sberbank is the largest bank in Russia in terms of transactions with individuals and usage of credit cards (71% as of September 2020), in terms of the number of private clients who keep their savings in it (64% as of September 2020). Moreover, Sberbank has become more digital, offering different digital channels. The average increase of Sberbank's web version and mobile applications' utilizers is 35% per annum (Rozhkova et al. 2021).

In July 2019, Sberbank topped the list of the most expensive brands in Russia according to Brand Finance. The value of it for the year increased by 25,6% and amounted to 842,1 billion rubles. Sberbank has topped this rating since 2017, ahead of companies in the oil and gas sector.

In 2020, a rebranding was proposed (from brand "Sberbank" to "Sber"). The main reason is the growth of the bank's ecosystem and the development of new services (in addition to financial services, Sberbank has been developing more than 50 new business areas – from e-commerce and cloud technologies to medicine and telecom).

The brand development strategy introduced also fully embodies the core value of "customer-centric" of this century-old bank, specifically the following points are key to the brand evolution and value (Sberbank 2021):

1. The best customer experience. Establish a customer experience evaluation system to continuously improve customer service quality and evaluate employee performance through this system.
2. Power sinks. Expand the power of the account manager, expand the scope of responsibility of the account manager, and reduce the power of the person closest to

the customer, allowing the account manager to change Sberbank's product functions, sales channels, service and communication methods, etc.

3. Multi-channel sales and service. With vigorous investment and development of digital technology and products, Sberbank's customers can obtain what they need at any time, any place, and through any means (for example, intelligent machines and digital personnel), and realize information exchange and sharing among multiple channels (Rogulenko et al. 2019).

4. Expand the product range. Improve product portfolio capabilities to cover all financial-related needs of customers, gradually include non-financial products, provide intuitive, convenient and reliable services, and become the best partner for clients.

5. Data and analysis. Collect, store and analyze customer information and behavioral data, use big data technology to better understand and serve clients, and provide clients with better services in the most appropriate way.

6. Build an ecosystem around customer needs. Build an ecosystem with partners to bring more added value to customers. Customers can complete housing payments, travel and exit preparation, transportation and shopping through the ecosystem.

With reference to the five elements of internationally recognized brand evaluation: quality, service, technological innovation, tangible and intangible assets, in order to maintain a "customer-centric" organizational operation and practice by all employees, and to avoid being out of touch with clients, slow to respond, becoming a bloated organization, and to much bureaucracy that are all common in large companies, Sberbank continues to carry out large-scale customer-centric changes. A change in 2016 they called the Sbergile project, is directly responsible for the company's senior vice president, customer-oriented optimization and the reengineering of all business processes of the company, to ensure that all business operations of the company will always put customers first.

### 3.2    Interbrand Model

Interbrand is one of the world market leaders in brand valuation. Interbrand annually compiles and publishes the rating of the "100 most valuable brands in the world". It is crucial to emphasize that this top 100 brands include only companies with a global scale of activity with all the necessary information available. Accordingly, companies that operate locally or are held privately cannot be included; as well as such giants as VISA, BBC, Mars and CNN because there is not enough information about these companies in public sources. In addition to the rating of the most expensive brands, Interbrand publishes a rating of brand portfolios owned by companies. During the preparation process, Interbrand relies on data provided by Citigroup. The company also provides its brand multiplier to the Financial World magazine, which calculates the value of brands itself.

This brand value calculation methodology is based on the net present value of the brand. The following four stages are considered:

Stage 1. The economic profit is calculated, in other words, the cash flow that is created by the intangible assets of the company (formula 2):

$$EP = NOPAT - [Capital\ Employed * WACC] \qquad (2)$$

where:

- EP – economic profit (cash flow from intangible assets);
- NOPAT – net operating profit after taxes;
- Capital Employed – the amount of capital required to create a similar non-branded product;
- WACC – weighted average cost of capital.

The main goal is to separate profit generated by intangible assets (including a brand) from profit generated by physical capital. To calculate the amount of capital involved in the creation of a non-branded analogue, the industry average indicator of the ratio of capital employed in the industry to an indicator of income is used. Multiplying this ratio by the volume of sales of the evaluated company, we get the required value. This value is considered "natural" for the production of non-branded products.

Stage 2. The objective is to highlight in the cash flow created by intangible assets, the part created by the brand.

Stage 3. It is necessary to determine the brand multiplier. The basis of the calculation is the analysis of brand strength index while considering ten indicators (presented in the previous chapter).

All parameters consider brand attractiveness from different perspectives. By assessing the brand for each of the ten parameters, the index of brand strength can be formed.

Interbrand has created its own S-curve, which shows the relationship between the brand strength index and the brand multiplier (Rocha 2016). However, the exact equation of the graph is kept secret and still is unknown since the S-curve is considered as intellectual property of the company.

Stage 4. The brand value is calculated. In order to do this, the brand value added, and the brand multiplier are multiplied in order to obtain the brand value.

The main advantage of the Interbrand brand assessment methodology is that it offers a financial assessment of the brand value. That's why it gained worldwide popularity. Before, there were many brands assessment models, but they all represented a non-financial component. The company was not able to express the intangible power of the brand and its effectiveness in monetary terms.

Another advantage is manifested in the fact that the method reflects the past and current results of the company's activities. The framework provides a comprehensive picture of brand potential using quantitative and qualitative metrics.

However, it doesn't take into consideration the brand's future. Because of this, there are strong fluctuations in brand value from year to year, which in practice should not be. Theoretically, the brand-multiplier itself should demonstrate the future capabilities of the brand, but due to the closed formula used for finding the multiplier itself, questions arise about the correctness of the method.

## 4   Interbrand Model: Calculation and Analysis

In this chapter, we have estimated the value of Sberbank's brand using the Interbrand model for 2019–2020:

- it is necessary to calculate the cash flow created by intangible assets;
- it is necessary to find the share of the brand in the cash flow from intangible assets;
- it is necessary to calculate a brand multiplier.

We have chosen these two years as the beginning of 2020 was the start of the rebranding (from Sberbank to Sber), representing an important milestone. Furthermore, an important element of this year is the COVID pandemic as a "black swan event" (Antipova 2021) and the role of a brand through communication strategies in order to link the current COVID-19 situation and people – "a force that is bringing people together" (Sobande 2020).

Calculation of the cash flow created by the intangible assets of Sberbank is presented in Table 2.

**Table 2.** Cash flow generated by the intangible assets of Sberbank (2019–2020).

| Indicator | 2019 | 2020 |
|---|---|---|
| NOPAT, billion rubles | 845 | 760,3 |
| WACC | 8.8 | 8.5 |
| CE Sberbank, billion rubles | 4487 | 5047 |
| Cash flow, billion rubles | 450 | 331 |

Net operating profit after taxes (NOPAT) and capital employed (CE) has been extracted from published Sberbank consolidated financial statements as of 31.12.2020 (Sberbank 2021). We did not use an industry average indicator of ratio of capital employed in the Russian bank industry to an indicator of income due to the absence of official data. As all financial statements use Russian rubles as the reporting currency, we have decided to use Russian rubles and convert the final brand value into dollars (using exchange rate from Sberbank's annual reports).

WACC has been extracted from S&P Global Market Intelligence and https://finbox.com.

Cash flow generated by the intangible assets of Sberbank has been decreased from 450 billion rubles in 2019 to 331 billion rubles in 2020, mostly due to the fact that profit fell. Capital employed has been increased from 4487 billion rubles in 2019 to 5047 billion rubles in 2020.

Table 3 represents the calculation of the brand-multiplier of Sberbank (2019–2020) using 10 parameters.

**Table 3.** Brand-multiplier of Sberbank.

| Factors | % | Description | 2019 | 2020 |
|---|---|---|---|---|
| Direction | 10 | Purpose of brand, development path | 10 | 10 |
| Alignment | 10 | Organization commitment to brand development | 8 | 10 |
| Empathy | 10 | Relationships with stakeholders | 7 | 7 |
| Agility | 10 | Market growth | 7 | 7 |
| Distincti-veness | 10 | Intangible assets which are difficult to imitate | 7 | 7 |
| Coherence | 10 | Different customers interactions in terms of brand positioning | 8 | 8 |
| Participation | 10 | Collaboration with stakeholders | 5 | 5 |
| Presence | 10 | Brand perception by different customers | 7 | 7 |
| Trust | 10 | Respond to customers' needs | 8 | 9 |
| Affinity | 10 | Shared values with stakeholders | 7 | 9 |
| **Total** | **100** | | **74** | **79** |

After that a S-curve that Interbrand created is used (See Fig. 1):

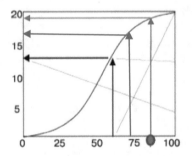

**Fig.1.** S-curve Interbrand.

The horizontal axis is the brand strength (in our case, it is 74 in 2019 and 79 in 2020), and the vertical axis is the brand multiplier. The brand multiplier would be 16 in 2019 and 17 in 2020.

Next, the calculations of the brand value are carried out in Table 4. The share of the brand in the cash flow from intangible assets is assumed to be 10%. The exchange rate has been extracted from published Sberbank consolidated financial statements as of 31.12.2020 (Sberbank 2021).

**Table 4.** Assessment of the Sberbank brand value (2019–2020)

| Indicator | 2019 | 2020 |
|---|---|---|
| Cash flow, billion roubles | 450 | 331 |
| Brand share, % | 10 | 10 |
| CE Sberbank, billion roubles | 4487 | 5047 |
| Cash flow (brand), billion roubles | 45 | 33 |
| Brand-multiplier | 16,0 | 17,0 |
| Brand value, billion roubles | 720 | 563 |
| Exchange rate, roubles for 1 $ | 64,7 | 71,9 |
| Brand value, million dollars | 11 132 | 7 834 |

Finally, the brand value according to this model is 11,132 million dollars in 2019 and 7,834 million dollars in 2020.

## 5  Summary and Conclusions

The main aim and purpose of the work were to identify the concept of the brand, calculate and analyze brand value, using methodologies of the most famous consulting agencies Interbrand and Brand Finance.

Table 5 is built based on the top-4 brand values according to top 500 Global Brands 2019 and 2020. We added information from Interbrand (2019–2020) and calculated the brand value of Sberbank for 2019–2020 using the Interbrand methodology.

**Table 5.** Assessment of the Sberbank brand value (2019–2020).

| Indicator | Amazon | | Google | | Apple | | Microsoft | | Sberbank* | |
|---|---|---|---|---|---|---|---|---|---|---|
| | 2019 | 2020 | 2019 | 2020 | 2019 | 2020 | 2019 | 2020 | 2019 | 2020 |
| Brand Finance | 187905 | 220791 | 142755 | 159722 | 153634 | 140524 | 119 595 | 117 072 | 12361 | 13233 |
| Interbrand | 125263 | 200667 | 167713 | 165444 | 234341 | 322999 | 108 847 | 166 001 | **11132** | **7834** |
| Δ, % | *50,01* | *10,03* | *−14,88* | *−3,46* | *−34,44* | *−56,49* | *9,87* | *−29,48* | *11,04* | *68,91* |
| Δ, mln. $ | *62642* | *20 124* | *−24 958* | *−5 722* | *−80 707* | *−182475* | *10 748* | *−48929* | *1229* | *5 399* |

*Brand Finance Banking 500 (2019–2020), Interbrand approach (calculated by authors)*

Based on Table 5 the following conclusions have been made:

1. Existence of drastic deviation of brand value. For example, for Amazon in 2020 the difference is 20,124 million dollars or approximately 10% (more for the Brand Finance approach). For Microsoft in 2020 the difference is 48,929 million dollars or approximately 30% (more for Interbrand approach).
   For Sberbank the deviation in brand value is around 11% in 2019 and 69% in 2020.

Therefore, as the calculation methods give conflicting results, there is no way to make a reasonable conclusion about a particular brand value if the Brand Finance and Interbrand models are used simultaneously.

2. Existence of different dynamic. For example, Apple's brand value has grown in 2020 according to the Interbrand model, but has fallen according to Brand Finance. For Google and Sberbank the situation is reversed; we observe an annual brand growth if the Brand Finance approach is utilized and an annual brand drop if Interbrand is used.

Therefore, brand dynamic estimates are different as well. While results from Brand Finance show a positive Sberbank's brand value dynamic (growth by 7%), the Interbrand approach's calculations demonstrate a decrease by 30%.

The deviation can be explained by the different methodology of these methods. Thus, brand value estimates are biased; improvements and modifications are needed. In addition, the Interbrand methodology contains approximations and inaccuracies in the structure of the assessment, which makes it ineffective in making marketing management decisions. The brand value calculated using the Interbrand method shows significant fluctuations (compared with Brand Finance), despite the fact that the brand is a fairly stable asset. And all this is due to the subjective factor of assessment. This means that the model cannot assess the brand value correctly, it is necessary to make the model less universal and more versatile for different sectors of the economy.

This research has shown that Interbrand model is applicable in Russian realities, however the results are quite different from results obtained by Brand Finance. It seems that any assessment of brand value will always be subjective and depend on the evaluator, the information provided and the method of valuation. Despite the fact that stakeholders prefer an approach with as little subjectivity as possible (Reyneke et al. 2014), there are many firms in the market that provide services for recognizing the financial component of the brand. Each model has variables that are not published publicly, so the difference in values can be explained by this.

# References

Aaker, D.A.: Managing Brand Equity: Capitalizing on the Value of a Brand Name. Simon and Schuster, New York (2009)

Antipova, T.: Coronavirus pandemic as black swan event. In: Antipova, T. (ed.) ICIS 2020. LNNS, vol. 136, pp. 356–366. Springer, Cham (2021). https://doi.org/10.1007/978-3-030-49264-9_32

Chaudhuri, A., Holbrook, M.B.: The chain of effects from brand trust and brand affect to brand performance: the role of brand loyalty. J. Mark. 65(2), 81–93 (2001)

Creswell, J.W.: Educational Research: Planning, Conducting, and Evaluating Quantitative and Qualitative Research. Merrill Prentice Hall, Upper Saddle River (2002)

Daum, J.H.: Intangible Assets and Value Creation, 1st edn. Wiley, Hoboken (2007)

Firat, D., Badem, A.C.: Brand valuation methods and reflection of brand value in financial statements. J. Account. Finan. 38, 210–219 (2008)

Interbrand. https://www.interbrand.com. Accessed 30 June 2021

Jacoby, J., Chestnut, R.: Brand Loyalty Measurement and Management. Wiley, Hoboken (1978)

Kim, H.B., Kim, W.G., An, J.A.: The effect of consumer-based brand equity on firms' financial performance. J. Consum. Mark. **20**(4), 335–351 (2003)

Knowles, J.: Value-based brand measurement and management. Interact. Mark. **5**(1), 40–50 (2003)

Lomax, W., Raman, A.: Analysis and Evaluation 2007–2008. Routledge, Milton Park (2007)

Murphy, J.J.: Technical Analysis of the Financial Markets a Comprehensive Guide to Trading Methods and Applications, Hardback (1998)

Nikitina, V.S.: Brand essense: elements and added value. E-Management **4**(1), 58–67 (2021). https://doi.org/10.26425/2658-3445-2021-4-1-58-67. (in Russian)

Pessemier, E.A.: A new way to determine buying decisions. J. Mark. **24**(10), 41–46 (1959)

Porter, M.E., Kramer, M.R.: Creating shared value. Harv. Bus. Rev. **89**(1/2), 62–77 (2011)

Raggio, R.D., Leone, R.P.: The theoretical separation of brand equity and brand value: managerial implications for strategic planning. J. Brand Manag. **14**(5), 380–395 (2007)

Reichheld, F.F.: The Loyalty Effect: The Hidden Force Behind Growth, Profits and Lasting Value. Harvard Business School Press, Boston (1996)

Reyneke, J., Abratt, R., Bick, G.: What is your corporate brand worth? A guide to brand valuation approaches. S. Afr. J. Bus. Manag. **45**(4), 1–10 (2014)

Rocha, M.: Financial applications for brand valuation: delivering value beyond the number. Interbrand (2016)

Rogulenko, T.M., Ponomareva, S.V., Krishtaleva, T.I.: Competition between intelligent machines and digital personnel: the coming crisis in the labor market during the transition to the cyber economy. In: Filippov, V.M., Chursin, A.A., Ragulina, J.V., Popkova, E.G. (eds.) The Cyber Economy. CE, pp. 185–194. Springer, Cham (2019). https://doi.org/10.1007/978-3-030-31566-5_20

Rozhkova, D., Rozhkova, N., Tozhihonov, S.: Electronic banking and financial performance of the Russian Bank. In: Antipova, T. (ed.) ICCS 2020. LNNS, vol. 186, pp. 227–237. Springer, Cham (2021). https://doi.org/10.1007/978-3-030-66093-2_22

Sberbank. https://www.sberbank.ru/ru/person. Accessed 30 June 2021

Simon, C.J., Sullivan, M.W.: The measurement and determinants of brand equity. A financial approach. Mark. Sci. **12**(Winter), 28–52 (1993)

Sobande, F.: We're all in this together: commodified notions of connection, care and community in brand responses to COVID-19. Eur. J. Cult. Stud. **23**(6), 1033–1037 (2020)

Soto, T.: Methods for Assessing Brand Value: A Comparison Between the Interbrand Model and the BBDOs Brand Equity Evaluator Model. Diplom. de (2008)

Tarakçi, I.E., Göktaş, B., Baş, M.: Financial brand value vs marketing brand equity. Muhasebe ve finans alanında bilimsel çalışmalar-inci erdoğan tarakçi-bora göktaş (2021)

Voskanyan, Y., Kidalov, F., Shikina, I., Kurdyukov, S., Andreeva, O.: Model of individual human behavior in health care safety management system. In: Antipova, T. (ed.) ICCS 2020. LNNS, vol. 186, pp. 413–423. Springer, Cham (2021). https://doi.org/10.1007/978-3-030-66093-2_40

Wood, L.: Brands and brand equity: definition and management. Management Decision (2000)

Yin, R.K.: The Case Study Anthology. Sage, Thousand Oaks (2004)

# Blockchain Technology Adoption in the South African Financial Service Sector: Perceived Advantages, Challenges and Potential Use Cases

Lorian Barrett and Jean-Paul Van Belle(✉)

University of Cape Town, Cape Town, South Africa
jean-paul.vanbelle@uct.ac.za

**Abstract.** Blockchain Technology (BCT) is a decentralized, trustless system that is set to transform, and possibly revolutionize, the financial services industry worldwide. In South Africa, perceptions concerning BCT are mixed, with some organizations unsure if it is just a hype or if it is indeed as disruptive as predicted. While there have been numerous studies on the use of cryptocurrencies, there is limited research on the alternative uses and intended adoption of the technology within the South African investment industry. This paper explores the intention of South African investment organizations to adopt blockchain and probes into future intended use cases. Semi-structured interviews were conducted with participants from eight financial services organizations. Factors from TOE, TTF, DOI, PCI as well as institutional theory were integrated. The study revealed many different factors influencing BCT adoption intentions within the South African investment industry. Several intended use cases specific to the investment industry were identified with smart contracts, equity post trade and securities registry being the most cited. Finally, the study identified several key factors that either had a positive or negative affect on blockchain adoption. The revised TOETI model can serve as a foundation for future studies on blockchain which in time could live up to its hype to be as fundamental and disruptive as predicted.

**Keywords:** Blockchain Technology (BCT) · Distributed Ledger Technology (DLT) · South African investment organizations · Technology adoption factors · TOE framework · Blockchain use cases

## 1 Introduction

Blockchain was conceptualized by Satoshi Nakamoto in 2008 as a key component of the cryptocurrency bitcoin and combines peer-to-peer technology with public-key cryptology [1]. Across the world Blockchain Technology (BCT) is attracting interest as it is made up of a distributed, tamper resistant, digital ledger that is not regulated by a central authority such as a bank or government [2]. Bitcoin and blockchain are often referred to interchangeably as the success of Bitcoin was attributed to its underlying technology; Blockchain [3]. The potential use of bitcoin and blockchain can be

© The Author(s), under exclusive license to Springer Nature Switzerland AG 2022
T. Antipova (Ed.): DSIC 2021, LNNS 381, pp. 231–243, 2022.
https://doi.org/10.1007/978-3-030-93677-8_20

grouped into 3 categories 1) Blockchain 1.0 relating to its use as a currency; 2) Blockchain 2.0 using the technology for financial contracts such as stocks, bonds and loans; and finally 3) Blockchain 3.0, comprising of all applications beyond currency and contracts [1, 7].

Even though the next era of blockchain looks promising for a number of different applications, the technology is considered new and its potential use cases are often misunderstood. If financial services industries seek to adopt blockchain technology (BCT), they will need to disrupt their own operations, processes, culture and mind-sets [4]. Organizations need to understand what differentiates this distributed database technology from standard database improvements [5] and assess and evaluate the state of this potential product before including it in strategic plans [6].

Existing research on blockchain was primarily focused at a general level around the hype of the technology, possible use cases, proof of concepts as well as the benefits and shortcomings of the technology. There also seemed to be a lack of research around perceptions of intended adoption of the technology from organizations. This study aimed to gain a better understanding regarding the current status and potential intended use of blockchain, specifically within the South African capital market space. This study is not presenting technical arguments but looks at the perceptions about the feasibility of adoption in the South African context. This will equip investment management organizations with required information needed to make informed decisions about the potential future use of the technology. The objective was achieved through answering the following two research questions:

- What are the factors driving the intended adoption of blockchain?
- What does the South African investment industry perceive the potential use cases in blockchain to be?

These questions were addressed through the development of an initial research framework identified through the review of the existing literature. Qualitative interview data was then used to answer the questions within the research framework.

## 2   Literature Review

### 2.1   Blockchain Technology

Blockchain is said to be an architecture for a system of decentralized trustless transactions [7]. A great analogy for blockchain is to think of it as a database that allows people across the world to not only register any kind of asset (property, votes, ideas, reputation, health), but also be able to transact these assets without the assistance of a middleman such as a bank [7]. Blockchain is also commonly referred to as "distributed ledger technology" (DLT) [8] or "distributed ledger system" (DLS) [9].

A blockchain is typically made up of the following components, a distributed shared database, a cryptography application, and a way to determine or validate which is the correct alternative database [8]. These components can be combined to develop a number of different solutions over and above using it as a virtual currency [8].

Blockchains can be categorized as either public/permissionless or private/ permissioned [8]. A permissionless system contains anonymous participants whose identity is unknown [9]. On a public or permissionless blockchain, anyone can join and run their own node in order to process transactions without permission from a specific authority [10]. In this type of network participants are unknown to each other and trust is inherent through game-theoretical incentives [11]. Well-known examples to date of a permissionless blockchain are Bitcoin and Ethereum [11]. Ethereum includes a Turing-completed programming language that could be used to create smart contracts [9].

On a permissioned or private blockchain the network is closed and only participants who have been granted access, or permission to use the network, can interact with the blockchain making it a possible issue for data that requires privacy [11]. On the other hand, by permitting only trusted participants on a specific blockchain, the network operations can be made more flexible, faster and efficient [11]. Examples of this type of network include Ripple, a cryptocurrency exchange and Hyperledger [9].

## 2.2 Blockchain 2.0 (Use Cases and Benefits in the Investment Industry)

There are a growing number of opportunities or potential use cases of blockchain that have been identified in the literature that are not specific to an industry, refer to Table 1. These include the foundational use case of blockchain as a cryptocurrency, asset exchange systems, uses within the Internet of Things (IoT), protection of personal data and content distribution systems.

**Table 1.** Potential use cases of blockchain technology identified in the literature.

| Use case | Use case |
|---|---|
| Asset Exchange System [7] | Internet of Things [14] |
| Automated Compliance [19] | Insurance Industry [19] |
| Banking Industry: interbank payments [27] | Private Securities Trading [18] |
| Blockchain- or Consensus-as-a-Service [9] | Protection of Personal Data [14] |
| Contracts Rights Management [15] | Remittance System [12] |
| Cryptocurrency [13, 20] | Securities Registry (Proxy Voting) [8, 19] |
| Digital Content Distribution System [16] | Self-sovereign Identities [20, 21] |
| Equity Post Trade (Clearing and Settlement) [19] | Smart Contracts [7, 8, 10, 22] |

In the investment industry, blockchain has the ability to cryptographically validate transactions and provides opportunities for equity post trade by enhancing the security of trading and settlement platforms [8, 25]. Other potential use cases relevant to the investment industry included fractional assets, alternative revenue streams as well as self-sovereign Identities [20, 21]. The most cited potential use of blockchain in the literature was usage as a smart contract [7, 8, 10, 22].

## 2.3    Blockchain Challenges/Limitations

Many challenges and limitations that need to be addressed before organizations can start adopting BCTs are highlighted in the literature. Softer issues around hype, perception and understanding of the technology are some of the most frequently occurring limitations [10]. Technical issues around blockchain include scalability – as measured by latency and throughput capacity [17] – and architectural security issues such as a 51% attack [10] or other hacking interventions. Uncertain future governance, privacy issues, ecological impact and excessive energy consumption are further issues [23].

## 2.4    Current State of Blockchain Adoption and Industry Perceptions in the South African Investment Industry

Despite the potential benefits of blockchain, it appears that investment management organizations have chosen not to engage as rapidly as the other financial organizations for a number of reasons such as the perceptions that blockchain will only affect outsourced areas such as settlement or they are not under as much pressure as the banks to reduce costs [4]. However, with the potential use cases in the investment industry discussed above and listed in Table 1, there is a need to perform research in order to understand these perceptions from an investment industry perspective. The PwC FinTech Survey 2017 cited Africa as having the lowest percentage of respondents extremely familiar with blockchain technology [4] which indicates another opportunity to explore this technology further within the African continent. Given that the Johannesburg Stock Exchange is the only Africa-based stock exchange in the top 20 by market capitalization, the South African investment industry is a feasible area for research in BCT.

## 2.5    Theoretical Research Models of IT Innovation Adoption

In this study, a compound model using constructs from the Technology-Organization-Environment (TOE) and Task-Technology-Fit (TTF) frameworks was developed to account for additional factors borrowed from DOI (Diffusions of Innovations), PCI (Perception Characteristics of Innovation), as well as institutional theory. This framework is referred to as the Technology-Organization-Environment-Task-Individual model (TOETI) [24] (Fig. 1).

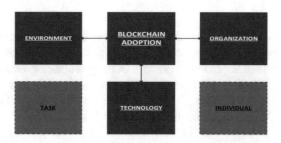

**Fig. 1.**  Technology-Organization-Environment-Task-Individual framework [24].

# 3    Research Methods and Design

The research purpose was both exploratory and descriptive as it examined the research questions in an attempt to understand the adoption of blockchain. Descriptive research is defined as "…making careful observations and detailed documentation of a phenomenon of interest" [25: 7]. Semi-structured interviews were conducted with random participants of a target population. The target population identified were investment management organizations based in South Africa, as the study aimed at understanding their perceptions as well as intended adoption. These organizations needed to be members of the Association for Savings and Investments in South Africa (ASISA). In addition, the target population included specialists and vendors in blockchain technology to provide the research with richer information and to validate some of the factors identified. We initially aimed to interview between eight and twelve participants but ended up with nine as saturation was reached.

Thematic analysis was used for analyzing the information derived from the interviews. The process of thematic analysis that was followed for this research was based on thirteen steps outlined by [26]. Steps 1–4 were performed iteratively to refine and finalize the data. The literature and conceptual TOETI research framework were continually and iteratively returned to in order to ensure the themes identified during analysis were in line with the research questions, objectives and propositions. NVivo software was used for the analysis.

This study followed the UCT guidelines with regards to ethics, privacy and confidentiality. Privacy and anonymity were requested by all participants and was guaranteed and applied when recording the data.

# 4    Findings, Analysis and Discussion

For the semi-structured interviews, 9 interviewees participated. Theme saturation point occurred from the fifth interview onwards as no new themes emerged in the last three interviews. Therefore, the sample size can be deemed to be sufficient.

As the focus of the study was refined to South Africa and investment industries, it was important that participants and the organizations they represent remained anonymous. To achieve this, the participants and the organizations they represent were assigned a unique identifier, e.g. P-A, which was applied when transcribing interviews. Only P-A1 and P-A2 worked in the same organization. Most of the participants held senior positions at the company and were well educated with a significant amount of experience (Table 2).

**Table 2.**  Demographics of the interviewees.

| Inter-viewee | Position | Experience | Company Size | Qualification | Industry |
|---|---|---|---|---|---|
| P-A1 | Senior IT Manager | 16 Years | 501-1000 Employees | Degree | Financial Services |
| P-A2 | Senior Risk Officer | 17 Years | 501-1000 Employees | Business Science Degree, Masters | Financial Services |
| P-B | Head of Information Systems | 21 Years | 201-500 Employees | Degree | Investment Management |
| P-C | Head of Information Systems | 20 Years | 1001-5000 Employees | Bachelor of Commerce Degree, MBA | Financial Services |
| P-D | Head of Information Systems | 34 Years | 51-200 Employees | Unknown | Investment Management |
| P-E | Head of Product Development | 27 Years | 201-500 Employees | CFA | Financial Services |
| P-F | Chief IT Officer | 20 Years | 51-200 Employees | Masters | Financial Services |
| P-G | Managing Director | 14 Years | 2-10 Employees | Degree | Education Management |
| P-H | Head of Information Systems | 20 Years | 501-1000 Employees | Bachelor of Commerce & Business Science Degree | Financial Services |

Using the previously defined TOETI model as a foundation for the major themes, 24 codes or sub-themes emerged from the participant's personal experience, organizational context and perceptions of blockchain. These were sorted in descending order from the most occurring subtheme among participants. As one of the main objectives was to identify factors that affect the adoption of blockchain technology, all subthemes identified had either a positive or negative impact. The findings of these subthemes will be discussed in more detail in the next subsections under their applicable major TOETI themes, Technology, Organization, Environment, Task and Individual (Table 3).

**Table 3.** Factors identified as driving the adoption of BCT by interviewee.

| | P-A1 | P-A2 | P-B | P-C | P-D | P-E | P-F | P-G | P-H | Count |
|---|---|---|---|---|---|---|---|---|---|---|
| **TECHNOLOGY** | | | | | | | | | | |
| Perceived Challenges | X | X | X | X | X | X | X | X | X | 9 |
| Relative Advantage (Perceived Benefits) | X | X | X | X | X | X | X | X | X | 9 |
| Available BCT Technologies | X | 0 | X | 0 | X | 0 | 0 | 0 | 0 | 3 |
| Disruptive Technology | X | 0 | 0 | X | 0 | X | 0 | 0 | 0 | 3 |
| **ORGANIZATION** | | | | | | | | | | |
| Technology Readiness | X | X | X | X | X | X | X | X | X | 9 |
| Change Management & Agility | 0 | X | X | X | X | X | X | 0 | X | 7 |
| Financial Resources | 0 | 0 | X | X | X | X | X | 0 | X | 6 |
| Formal Linking structures (E.g. Board) | 0 | 0 | X | X | X | X | X | 0 | X | 6 |
| Trialability | X | X | 0 | X | X | 0 | X | 0 | X | 6 |
| Compatibility | X | 0 | X | X | 0 | 0 | X | 0 | X | 5 |
| Risk Management | X | 0 | X | 0 | 0 | X | X | 0 | X | 5 |
| Human Resources | 0 | X | X | X | 0 | 0 | 0 | 0 | X | 4 |
| Culture | 0 | 0 | 0 | 0 | X | 0 | 0 | 0 | X | 2 |
| **ENVIRONMENT** | | | | | | | | | | |
| Collaboration | X | X | X | X | X | X | X | X | X | 9 |
| Regulation | X | X | X | X | X | X | X | X | X | 9 |
| Isomorphic Pressures | X | X | X | X | 0 | X | X | X | X | 8 |
| Competitive Advantage | 0 | X | X | X | 0 | 0 | X | X | X | 6 |
| Market Structure | X | X | X | X | 0 | X | 0 | X | 0 | 6 |
| **TASK (Use Cases)** | X | X | X | X | X | X | X | X | X | 9 |
| **INDIVIDUAL** | | | | | | | | | | |
| Complexity or Understanding | X | X | X | X | X | X | X | X | X | 9 |
| Hype | X | X | X | X | X | X | X | X | X | 9 |
| Perceptions | X | X | X | X | X | X | X | X | X | 9 |
| Education | X | X | X | 0 | 0 | 0 | X | X | X | 6 |
| Individual Characteristics (Leadership) | 0 | 0 | 0 | X | 0 | X | X | X | 0 | 4 |

Given the limited space, not all the factors will be discussed. Instead, focus will be given to the identified perceived challenges and benefits. Then, as part of the 'Task" factor, specific use cases of BCT in the investment industry will be identified.

### 4.1 Perceived Challenges

Several challenges, concerns and limitations were highlighted in the literature review such as issues around scalability, hype/perceptions/understanding, security, infrastructure, governance, privacy and wasted resources (Table 4). Two new challenges were raised, first a concern around a loss of jobs and second, as a threat to an organization's value chain. Given the limited space, only some illustrative quotes are given.

A number of concerns were raised by the participants around how the technology would impact their organization's infrastructure and thus would have a negative impact on blockchain adoption. One concern was that some of the main industry players such as the Johannesburg Stock Exchange (JSE) were not ready from an infrastructure point of view to adopt blockchain as mentioned by P-F "…I am concerned that they are not ready for it because we are so dependent on [the JSE] that we will struggle and we will build workarounds around it to enable it." Another concern was the required focus on storage of the blockchain to ensure there is sufficient storage which could have a potential cost impact. P-C raised this "…One of the things that worries me is storage [...] Once you start storing every record in a chain you know, what is that going to do to the cost of storage? How do you maintain that so with things like POPPI, you delete records after a certain period of time, how will the blockchain manage that particular process?".

**Table 4.** Potential challenges and concerns around blockchain technology adoption in finance.

| | P-A1 | P-A2 | P-B | P-C | P-D | P-E | P-F | P-G | P-H | Participant Total |
|---|---|---|---|---|---|---|---|---|---|---|
| Regulation | X | X | X | X | X | X | X | X | X | 9 |
| Complexity or Understanding | X | X | X | X | X | X | X | X | X | 9 |
| Perceptions | X | X | X | X | X | X | X | X | X | 9 |
| Hype | X | X | X | X | X | X | X | X | X | 9 |
| Infrastructure | X | X | X | X | 0 | X | X | X | X | 8 |
| Scalability | X | 0 | 0 | 0 | X | X | X | X | 0 | 5 |
| Loss of Jobs (NEW) | X | X | 0 | 0 | 0 | X | 0 | 0 | X | 4 |
| Value Chain (NEW) | 0 | X | 0 | X | X | X | 0 | 0 | 0 | 4 |
| Privacy | X | 0 | 0 | 0 | X | 0 | 0 | 0 | X | 3 |
| Wasted Resources | X | X | 0 | X | 0 | 0 | 0 | 0 | 0 | 3 |

The issue of scalability was consistently confirmed as a negative factor affecting blockchain adoption during the interviews, with five of the participants citing this as a concern. P-F stated that "…it must just be scalable for us. We cannot as a financial services organization compete with the Azures and the Amazon Web Services and the other companies that have the economies of scale to provide that scalability [...] it should be smart enough to do that. To scale as it sees there is pressure on the infrastructure." On the other hand, some participants had a positive view on the scalability of the technology, citing benefits such as reducing human resources and allowing their organizations to be able to handle future impacts. P-D said "…if we can scale something we do not have to hire additional bodies to do that we can just scale it with the technology…" as well as P-E: "…we would also be looking at if it is future fitting, is it scalable?".

A new and negative factor for the adoption of blockchain was around the loss of jobs, cited by four participants. This was seen to more likely to impact those jobs that involve administration within the investment industry. P-E said that there "…is a big worry across the world about unemployment and the effect technologies like this could have on unemployment…If we are employing ten percent of our people, [...] to confirm trades that the asset managers are making. If blockchain takes it over, there's thirty jobs gone." Due to the manual nature of administrative jobs, blockchain could potentially automate some of the manual tasks and thus organizations may not require as many people to do the job as mentioned by P-A2 "…there are already people that are being employed to do specific jobs and if you go that automated route those people may lose their jobs and hence the apprehension to implement it."

The concern of privacy was raised by a few of the participants and thus would have a negative impact on the adoption of blockchain. P-H "…privacy concerns me a lot, in our business privacy is big." It is important that issues around privacy are resolved before blockchain can be implemented. As mentioned in the literature review, privacy is important in the investment industry especially in the securities market that cannot afford to disclose transactions or large positions publicly as it could result in investors not acting in the best interests of their clients due to the exposure to information in the blockchain [6]. This concern was reiterated with participant P-A1 saying that there will be "…an issue when one guy can see what the other guy is trading and clearly that would not be allowed."

Another new and negative factor that could affect the adoption of blockchain concerned the impact the technology would have on the organization's value chain as well as the industry as a whole. P-A2 "…the biggest challenge that I'm seeing there is that you are killing a whole industry." This concern was even more applicable to those organizations that provide fund administration as a service as blockchain could potentially replace some of these services as stated by participant P-E "…our role within the value chain then gets diminished in that we can't offer [Service X] any-more…" and participant P-G "…How do we implement this without putting ourselves out of the job?" Organizations will need to ensure they are able to understand the technology in order to be able to incorporate it into their strategy as mentioned by P-C "…we are going to have to figure out what our place is in the new world […] yes that's our challenge."

Large amounts of energy are consumed to process the blockchain algorithms [23]. Three of the participants were aware of this negative factor, such as participant P-A2 "…At the moment it requires a lot of electricity […] that is one of the issues." Another participant mentioned that careful consideration needs to be made around the type of consensus algorithm that will be used to ensure the legitimacy of the blockchain as the existing one used for bitcoin wastes a large amount of resources.

## 4.2  Relative Advantages

"Relative advantage", how an innovation increases economic profitability, social status or provides other benefits, was highlighted as the second biggest factor under the technology major theme. Thus, relative advantage of blockchain can be seen as having a positive impact on its adoption. A number of perceived benefits were identified by participants, highlighted in Table 5, with transaction speed, security, transaction cost and transparency being cited the most among participants.

Transaction Speed and Costs together formed the most counted factors among both the media articles and the participants indicating they were two of the more important perceived benefits. An example of this was in the equity post trade process where transactions can take up to seven days to clear, as mentioned by P-B "… in the investment area the settlement is T, T plus one, two, three, five, up to seven and it would be advantageous if it was at a lower cost […] We could get real time settlement and reporting and everything that is associated to it. That is a main benefit."

**Table 5.** Perceived relative advantages or benefits of blockchain technologies.

| | P-A1 | P-A2 | P-B | P-C | P-D | P-E | P-F | P-G | P-H | Participant Total |
|---|---|---|---|---|---|---|---|---|---|---|
| Transaction Speed | X | 0 | X | 0 | X | 0 | X | 0 | 0 | 4 |
| Transaction Costs | 0 | X | 0 | 0 | X | X | X | 0 | 0 | 4 |
| Transparency | X | X | 0 | X | 0 | 0 | 0 | 0 | X | 4 |
| Security | 0 | X | 0 | 0 | X | 0 | 0 | X | 0 | 3 |
| Accessibility | 0 | 0 | 0 | 0 | X | 0 | 0 | 0 | 0 | 1 |
| Disintermediation | 0 | 0 | 0 | 0 | 0 | X | 0 | 0 | 0 | 1 |
| Immutability | 0 | 0 | 0 | 0 | X | 0 | 0 | 0 | 0 | 1 |
| Revenue Generating | 0 | 0 | 0 | 0 | 0 | 0 | X | 0 | 0 | 1 |

### 4.3 Blockchain Use Cases

One of the research objectives was to understand the intended use within the investment industry in South Africa. A combined analysis from the literature review, media articles and interviews revealed 24 potential use cases of blockchain which were not necessarily specific to the investment industry. The most cited use case across the board was as a cryptocurrency and only half of the use cases identified were deemed relevant to the investment industry. A few of the most applicable use cases, are discussed here.

Apart from cryptocurrency, the use cases Equity Post Trade, Securities Registry and Smart Contracts were the most frequent use cases and the ones participants intended using. New ones identified in the interviews that were not identified in the literature review or media articles included, KYC (Know your Customer), Messaging, Script Lending Registry and using blockchain as a Transfer Agency System. On the other hand, despite smart contracts being cited as one of the most important potential uses of blockchain in the literature review they were only mentioned by two of the participants.

All participants had heard of blockchain prior to the interview in some form or another but the majority had heard about it due to bitcoin. P-D said "…the success of bitcoin in itself is testimony to the fact that blockchain actually does work".

Equity post trade includes functions such as settling and clearing of transactions which was consistently cited by a number of the participants "…There are a lot of possible use cases, but I think primarily settlement…" (P-B), "…Settlement definitely…" (P-F), "…Settlements and clearing…" (P-G), "…Settlement obviously comes in as one of these quick wins…" (P-D), "…settlements is definitely something that we should consider…" (P-H).

As mentioned in the literature, a blockchain could be used to administer the ownership of securities allowing for retroactive traceability by retrieving a list of shareholders and invoking specific corporate actions such as dividend payments, stock splits [8] or distributing proxy voting statements [19]. Participants primarily cited proxy voting as an area of interest "…Definitely proxy voting…" (P-E), "…yes proxy voting is a good one for corporate actions and entitlement" (Table 6).

**Table 6.** BCT use cases specific to investment industry as identified by interviewees.

| TASK (Use Cases) | Literature Review | P-A1 | P-A2 | P-B | P-C | P-D | P-E | P-F | P-G | P-H | Frequency Count |
|---|---|---|---|---|---|---|---|---|---|---|---|
| Cryptocurrency | X | X | X | X | X | X | X | X | X | X | 9 |
| Equity Post Trade | X | 0 | X | X | X | X | 0 | X | X | X | 7 |
| Securities Registry | X | X | 0 | 0 | 0 | 0 | X | 0 | X | X | 4 |
| Automated Compliance | X | 0 | 0 | X | X | 0 | 0 | 0 | 0 | X | 3 |
| Blockchain as a Service | 0 | 0 | 0 | 0 | 0 | 0 | 0 | X | X | 0 | 2 |
| Remittance & Payment Systems | 0 | X | 0 | 0 | 0 | 0 | 0 | 0 | X | 0 | 2 |
| Smart Contracts | X | X | X | 0 | 0 | 0 | 0 | 0 | 0 | 0 | 2 |
| Exchange | 0 | 0 | 0 | 0 | 0 | X | 0 | 0 | 0 | 0 | 1 |
| KYC (New) | 0 | 0 | 0 | 0 | X | 0 | 0 | 0 | 0 | 0 | 1 |
| Messaging (New) | 0 | 0 | 0 | 0 | 0 | 0 | 0 | 0 | 0 | X | 1 |
| Private Securities Trading | X | X | 0 | 0 | 0 | 0 | 0 | 0 | 0 | 0 | 1 |
| Scrip Lending Registry (New) | 0 | 0 | X | 0 | 0 | 0 | 0 | 0 | 0 | 0 | 1 |
| Transfer Agency System (New) | 0 | 0 | 0 | 0 | 0 | 0 | X | 0 | 0 | 0 | 1 |

# 5 Conclusion and Future Research

The research study revealed that there was a good understanding of blockchain from participants within the investment industry. However, this could be attributed to their level, experience and position within their organization and could therefore not be generalized across the investment organization or industry. Despite a keen interest in the technology, most perceptions around blockchain were mixed. Several potential use cases were highlighted by participants relevant to the investment industry with Equity Post Trade, Securities Registry and Smart Contracts being the most mentioned. New use cases that were not identified in the literature review included Know your Customer (KYC), Messaging, Script Lending Registry and using blockchain as a Transfer Agency System. It was also identified that there was a consistent association between the cryptocurrency use case bitcoin and blockchain. Finally, the factors that affected the adoption of the technology that were grouped using the TOETI model. A total of 24 new factors were identified that either had a positive or negative affect on blockchain adoption. Most of the factors were identified under the Organization major theme indicating this was dominant on the minds of participants. A significant factor identified under the Environment major theme included collaboration i.e. competitors working together in formal structures like consortiums and working groups. Education and upskilling programmes could contribute towards a better understanding of the technology.

In conclusion: whilst there is a good general understanding of blockchain from individuals, this may not be reflective of their organization. Overall perceptions at the time of study are mixed however these could change with time and as the technology becomes more widely accepted due to actual implementations. Main intended use cases specific to the investment industry were identified with smart contracts, equity post trade and securities registry being the most cited. A revised TOETI model has been developed that can assist organizations in their decision-making process concerning blockchain adoption.

There were several limitations that need to be considered, the first being that while the investment industry is made up of organizations of varying sizes, the study provided a narrower focus on larger South African investment companies. Future research could be conducted across investment organizations of varying sizes and could also be expanded to include the financial services industry within South Africa. Another limitation was that there was a lack of actual use cases that had been implemented within investment organizations and organizations in general in South Africa. Future research should be conducted once actual use cases have been implemented.

Further studies around blockchain adoption should be conducted to test the revised TOETI model as well as to identify new factors that might impact adoption, ideally using a larger sampling frame. Overall, it can be said that the adoption of blockchain in South Africa's investment industry is in its infancy, but once matured, could live up to its hype and be as fundamental and disruptive as predicted.

# References

1. Zhao, J.L., Fan, S., Yan, J.: Overview of business innovations and research opportunities in blockchain and introduction to the special issue. Financ. Innov. **2**(1), 1–7 (2016)
2. Yaga, D., Mell, P., Roby, N., Scarfone, K.: Blockchain technology overview (2019)
3. Ross, E.S.: Nobody puts blockchain in a corner: the disruptive role of blockchain technology in the financial services industry and current regulatory issues. Catholic Univ. J. Law Technol. **25**(2), 7 (2017)
4. Yazdani, D., Weber, G.: PwC: Global FinTech Report (2017). https://www.pwc.com/gx/en/industries/financial-services/assets/pwc-global-fintech-report-2017.pdf. Accessed 20 Sept 2018
5. Collomb, A., Sok, K.: Blockchain/Distributed Ledger Technology (DLT): what impact on the financial sector? DigiWorld Econ. J. **103**(1), 93–111 (2016)
6. Wang, H., Chen, K., Xu, D.: A maturity model for blockchain adoption. Financ. Innov. **2**(1), 1–5 (2016). https://doi.org/10.1186/s40854-016-0031-z
7. Swan, M.: Blockchain thinking: the brain as a decentralized autonomous corporation [commentary]. IEEE Technol. Soc. Mag. **34**(4), 41–52 (2015)

8. Santo, A., et al.: Applicability of Distributed Ledger Technology to Capital Market Infrastructure. Japan Exchange Group (2016). https://blockchainlab.com/pdf/E_JPX_ working_paper_No15.pdf. Accessed 21 Sept 2021
9. Swanson, T.: Consensus-as-a-service: a brief report on the emergence of permissioned, distributed ledger systems (2015). http://www.ofnumbers.com/wp-content/uploads/2015/04/ Permissioned-distributed-ledgers.pdf. Accessed 20 Sept 2018
10. Broby, D., Paul, G.: Blockchain and its use in financial settlements and transactions. J. Chartered Inst. Securities Investment (2017). http://strathprints.strath.ac.uk/59818/. Accessed 20 Sept 2018
11. Mattila, J.: The Blockchain Phenomenon. Berkeley Roundtable of the International Economy, p. 16 (2016). https://brie.berkeley.edu/sites/default/files/juri-mattila-.pdf. Accessed 20 Sept 2018
12. Lee, L.: New kids on the blockchain: how bitcoin's technology could reinvent the stock market. Hastings Bus. LJ **12**, 81 (2015)
13. Nakamoto, S.: Bitcoin: a peer-to-peer electronic cash system. Decentralized Bus. Rev. 21260 (2008)
14. Trautman, L.J.: Is disruptive blockchain technology the future of financial services? Consum. Financ. Law Q Rep. **69**(1), 232–242 (2016)
15. Watanabe, H., Fujimura, S., Nakadaira, A., Miyazaki, Y., Akutsu, A., Kishigami, J.: Blockchain contract: securing a blockchain applied to smart contracts. In: IEEE International Conference on Consumer Electronics (ICCE), pp. 467–468 (2016)
16. Kishigami, J., Fujimura, S., Watanabe, H., Nakadaira, A., Akutsu, A.: The blockchain-based digital content distribution system. In: 2015 IEEE Fifth International Conference on Big Data and Cloud Computing, pp. 187–190 (2015). https://doi.org/10.1109/BDCloud.2015.60
17. Beck, R., Müller-Bloch, C.: Blockchain as radical innovation: a framework for engaging with distributed ledgers as incumbent organization (2017)
18. Crosby, M., Pattanayak, P., Verma, S., Kalyanaraman, V.: Blockchain technology: beyond bitcoin. Appl. Innov. **2**, 6–10 (2016)
19. McWaters, R.J., Galaski, R., Chatterjee, S.: The future of financial infrastructure: an ambitious look at how blockchain can reshape financial services. In: World Economic Forum, vol. 49 (2016)
20. Polyviou, A., Velanas, P., Soldatos, J.: Blockchain technology: financial sector applications beyond cryptocurrencies. In: Multidisciplinary Digital Publishing Institute Proceedings, vol. 28, no. 1, p. 7 (2019)
21. Abraham, A.: Self-sovereign identity. Styria. EGIZ. GV. AT (2017)
22. Mohanta, B.K., Panda, S.S., Jena, D.: An overview of smart contract and use cases in blockchain technology. In: 2018 9th International Conference on Computing, Communication and Networking Technologies (ICCCNT), pp. 1–4. IEEE (2018)
23. Deshpande, A., Stewart, K., Lepetit, L., Gunashekar, S.: Distributed ledger technologies/ blockchain: challenges, opportunities and the prospects for standards (2017)
24. Neill, D., Van Belle, J., Ophoff, J.: Understanding the Adoption of Wearable Technology in South African Organisations. CONF-IRM 2016 Proceedings, Paper 5 (2016)
25. Bhattacherjee, A.: Social Science Research: principles, methods, and practices. Textbooks Collection **3** (2012). https://doi.org/10.1186/1478-4505-9-2
26. Skovdal, M., Cornish, F.: Qualitative Research for Development: A Guide for Practitioners. Practical Action Publishing, Rugby (2015)
27. Grau Miró, J.: Strategic innovation in financial sector: blockchain and the case of Spanish banks (2016)

# An Interaction Design Dome to Guide Software Robot Development in the Banking Sector

Juanita L. Beytell$^{(\boxtimes)}$ ⓘ and Jan H. Kroeze ⓘ

School of Computing, University of South Africa, Pretoria, South Africa
44260121@mylife.ac.za, kroezjh@unisa.ac.za

**Abstract.** Robotics, robotic processes, artificial intelligence and machine learning are concepts that create uncertainty among people. Technological expansion influences the preconceived, adverse feelings of dislike that users are harboring. This uncertainty influences the acceptance of software robots that are implemented in systems used by employees. This may also be the case regarding the interaction of banking employees with automated information systems used to perform their daily work activities. It is against this background that a framework for interaction design is required. Such a framework could guide designers to consider employed users and establish a positive, accepting inter-action. Due to the lack of existing interaction design frameworks aimed at addressing the implementation of software robots, the study aims to answer the question: Which core components will contribute to a framework that will guide the interaction design of a software robotic system in the banking sector? An initial systematic review approach was performed to identify the influence of robotics on user experience. This literature review has identified principles in other industries that were pulled through to financial institutions. This study contributes to filling the gap in academic knowledge by exploring related literature to identify and evaluate applicable principles and integrating these into a new synthesized framework. The result of the study is a new framework called an interaction design dome. The dome can be used to guide software robot design and development from a human-computer interaction viewpoint, with specific reference to the financial sector.

**Keywords:** Interaction design · Framework · User experience · Software robots · Robotics

## 1 Introduction

Robotics is a rapidly growing area of computers that continuously increases in importance [1]. A universal academic definition of this popular field is difficult to come by due to the continuous change of the subject area [2]. An early definition is that robotics concentrates on creating movements through the guidance of computers and is mainly used in industry [3–5]. In later definitions, robotics includes manipulation (through motorized functionality) and navigation (through software programming) [5].

The word 'robot' was popularized by the Czech playwright Karel Capek in 1921 [6]. According to the International Organization of Standardization, a robot is an "actuated mechanism programmable in two or more axes with a degree of autonomy, moving within its environment, to perform intended tasks" [7]. This definition includes only independent mechanical robots that can move. Laschi, Mazzolai and Cianchetti [8] accept this definition of robots in their studies, whereas You and Robert [9] add both virtual- and physical-embodied actions. Robots are considered as devices that perform the "three Ds": **d**irty, **d**angerous and **d**ull, characteristics which are often used to describe the jobs that robots are most suited for [10].

Robots are increasingly replacing humans in the workforce due to the various functions they can perform [11]. As organizations are embracing and integrating robots in the workplace alongside their human employees, user acceptance and user experience need to be considered [12]. To achieve a positive user experience, designers should utilize the multidisciplinary characteristics of interactive design [13]. Interactive design is the design of systems that assist users in communicating with each other and with computers [13].

The literature in the field of robotics and design for a better user experience presents findings that confirm that the appropriate implementation of a robot can enhance user experience. Most of the research was done based on mechanical robots implemented in industry. Information is also available on the topics of interaction design for humans interacting with robots, social robots and computer hardware. In the financial environment, users are more likely to interact with software robotics within the systems that they use, rather than mechanical robots. Developers of software robotics are unlikely to consider the findings of research based on mechanical robotics and need studies focused on software robotics. No framework for interaction design that enhances the use of software robotics in the financial environment has been found in the literature.

Therefore, there is a need to integrate and amend existing research relating to interaction design for mechanical robots in the industrial environment to fill the gap in the literature and be relevant and applicable for software robots in the financial industry.

## 2   Methodology

The research question of this initial literature review is: Which core components will contribute to a framework that will guide the interaction design of a software robotic system in the banking sector? This question will be answered by identifying elements of software robotics that influence user experience. An initial systematic literature review (SLR) approach was performed to identify the influence of robotics on user experience. Through an SLR as the methodology, we can identify and evaluate the existing research that is relevant to the specific research question [14]. This methodology has been selected due to its nature of clear boundaries that explicitly addresses the 'how' and 'why' questions [15]. These questions will highlight the components needed for a positive user experience in interacting with robotics in a banking institution. The qualitative, interpretive study used the keywords robot and interaction design to search the digital libraries of the ACM Digital library, IEEE Xplore, ResearchGate, Elsevier and Springer Link for articles that were relevant to the topic. The perusal of the abstract and conclusion of the found articles resulted in discarded articles where the focus was not relevant to the study. The remaining sources were explored to identify and evaluate possible concepts until conceptual maturity was reached. The review aims to identify relevant design principles, the relationship between them and to use this information to develop theoretical guidelines for developing software robots. These guidelines are integrated and visualized in this paper as a 'dome' resting on and connecting 'pillars' carved from the building blocks that were deemed applicable. The new construct (i.e., the interaction design dome) is a conceptual framework that synthesizes existing principles into an innovative construct.

Robotics, robotic processes, artificial intelligence and machine learning are concepts that create uncertainty among people [16]. Robotics is defined as a computer or computer program that is built to perform human tasks [8, 17]. A positive user experience is achieved when people can interact with robots in the same way that they interact with each other [18]. Alenljung et al. [20] concluded that user experience is influenced by the responsiveness, robustness and trickiness of the interaction with a robot. Concepts that influence user experience design has been identified are reflected in Table 1.

**Table 1.** Concepts that influence user experience design.

| No | Concepts | Description | Authors |
|----|----------|-------------|---------|
| 1 | Behavior | How the system (or robot) responds to the user | [19–26] |
| 2 | Capabilities | The skills of both the user and system | [19, 23, 27] |
| 3 | Experience | The way users feel about interacting with the system | [22–24, 28–31] |
| 4 | Interaction | The activity where user and system work together | [19, 21, 22, 24, 27, 32, 33] |
| 5 | Level of autonomy | How independent the system can work | [27, 34] |
| 6 | Motivation | The goals and purpose of using the system | [19, 27, 31, 33, 35–37] |

The framework to guide the interaction design of a robotic system should include these concepts to ensure acceptance by users and a positive user experience. In the context of financial institutions, users are considered to be the employees using banking systems in their daily activities. The sources reviewed did not yield a previously established framework to guide the development of software robots that will enhance the user experience of its consumers in the banking sector.

## 3 Literature Review

The literature review assists in discussing the topics of interaction design, user experience, robots and robotics. Understanding the concepts in terms of software robotics leads to the identification of the contributing factors in the conceptual framework.

The first step of **interaction design** is to define what interaction is. "Interaction is the essence of all user experiences" [38]. Interaction is described as the dialogue or language used between the user and the system and explain that users who are not engaged will withdraw from the interaction [38, 39]. Interaction is also defined to "give life" to a system [31]. Four types of interaction are defined as instruction, conversing, manipulating and exploring [13]. System developers will define the required type of interaction that will guide the design of the user interface. Interaction design is defined as the development of "interactive products to support the way people communicate and interact in their everyday and working lives" [13]. Furthermore, interaction design can improve the user's work life by intentionally constructing user experiences that improve the user's work experience [13]. Another definition considers that the main motivation of interaction design is the impact of the system on the user [40]. In this paper, interaction design is seen as the creation of a link between the system and the user in human-computer interaction [38]. Although several authors consider interaction design as a sub-component of user experience [31, 35, 38, 39, 41], the motion that interaction design is an umbrella term for all actions relating to the design of the system is accepted in this paper [13]. Interaction design is considered to include all design aspects, including user experience design.

A positive **user experience** contributes towards employees being motivated and performing their duties within appropriate work practices [42]. The definition of user experience can guide the decision towards which attributes can facilitate a positive user experience. The most accepted definition of user experience is that of the ISO: "[a] user's perceptions and responses that result from the use and/or anticipated use of a system, product or service" [43]. Several authors discuss this definition in depth. The ISO standard results from focusing on the use of a product or system developed and on a user's thoughts and responses [36]. Hellman and Rönkkö [44] discuss this and state that the ISO measures of efficiency, effectiveness and user satisfaction lead to only functionality and usability being measured. They propose the inclusion of aspects of pragmatic and business cost aspects when assessing user experience [44]. While the way an experience evolves through experience (actual interaction) should be added to the definition [45], the inclusion of hedonic aspects (such as fun, beauty and pleasure) would provide for an all-inclusive user experience [42, 44]. This study will be based on the opinion that user experience subsumes usability, and usability is, therefore, an attribute of user experience. The definition against which user experience will be tested is a user's overall sense of fulfilment about the computer system being used to perform his or her daily activities.

Only a few academic authors have defined **robots.** The definition supplied by the ISO is used as a starting point. An all-inclusive definition for robots is challenging to find [46]. The definitions by scholars are presented below (see Table 2).

The definition found in the Merriam-Webster dictionary is considered the foundation on which this research is based: a robot is "a device that automatically performs complicated, often repetitive tasks" [49]. The different types of robots are intricately linked to the definition of robots and the age of the definition. The definitions of robots in the 1960s refer to types of mechanical robots and still form part of the classification by the ISO. The more modern definitions that include computers that function in virtual environments, also include classifications linked to their function or purpose. Three

**Table 2.** Robot definitions.

| No | Author | Definition |
|----|--------|-----------|
| 1 | [7] | "An actuated mechanism programmable in two or more axes with a degree of autonomy, moving within its environment, to perform intended tasks." |
| 2 | [12] | "Technologies that include computer-simulated and physical actions." |
| 3 | [47] | "A robot is an inherently active agent that interacts with the real world, and often operates in uncontrolled or detrimental conditions." |
| 4 | [48] | Robots are "collections of robotic primitives" which are combined into new constructs with "sensing and actuation power" |

such categories are attended, unattended and free robots. The definition of unattended robots is most relevant to this study.

As with the definition of robots, there is also no universal definition for **robotics**. This could be due to the continual changes in the character of robotics due to the ongoing maturation of technology [4].

Table 3 reflects the academic definitions found in the literature. Robotics involves the study and design of computer systems and processes that have been designed to control robots through a set of rules [50, 51]. Another informal definition found on the Internet is that robotics is the technology used for robot design, production, process and use of robots [51]. This study focuses on software robotics as a field of investigation dealing with scenarios where a virtual computer performs a set of detailed, repetitive steps in the same way as what humans would have done [21, 52].

**Table 3.** Robotics definitions.

| No | Author | Definition |
|----|--------|-----------|
| 1 | [3] | Robotics involves the movement of physical objects by a computer |
| 2 | [4] | Robotics became defined as the study of the intelligent connection of sensing to actuation |
| 3 | [5] | Robotics includes three areas of computer influence: navigation, manipulation and interaction |
| 4 | [53] | Robotics refers to a science or art that involves artificial intelligence and mechanical engineering |
| 5 | [54] | Robotics can be either a science or engineering. It is considered science when it is used for explorative studies and engineering when used for innovation and new system discovery |

The next section discusses the main elements (the pillars of the dome) in more detail, and Sect. 5 integrates them into a new construct (or interaction design dome).

# 4    Results of Initial Literature Review

The definition of interaction design indicates that the design of a system affects users and can improve how they communicate by designing for a positive user experience [13]. The influence of a positive user experience on the motivation of employees has been established [42] and the attributes or concepts of the interaction design can guide the user experience design [13].

In this study, user experience is a user's overall sense of fulfilment about the computer system being used to perform his or her daily activities. The framework should thus be emphasizing aspects that influence user experience that should be considered when a robot is implemented in the system used.

Concepts identified in the literature are behavior, capabilities, experience, interaction, the level of autonomy and motivation as described in Table 1.

The **behavior** of the system or how the system responds to the user is the first identified concept. Four behavioral characteristics are highlighted in the literature: anthropomorphic behavior, adaptive behavior, repetitive, rule-based steps and expected behavior. Systems that perform anthropomorphically with feedback of human nature are more accepted by users than a nonhuman reply [19, 20, 22]. Sachs [22] and Comsa [25] raised strong opinions that technology should have an adaptive nature and be able to respond to users. Users should not have to adapt to the technology [22, 25]. Another behavioral concept that authors agree to is the type of function that robots can perform. High-volume, repetitive functions where technology behaves in the same continuous manner are ideal tasks for a robot [21, 23, 26]. The last behavioral characteristic that is needed for a great interaction design is expected behavior. A system should be clear about what is happening in the background so that users know exactly what to expect [25]. Technology that includes these behaviors as part of the interaction design will lead to a positive user experience.

The **capabilities** or skills of both the user and the system are important concepts identified in the literature. This involves more than just the ability of the human to use the system and the system to be able to react to the user's input. It also includes the skills required by the developer of the system. The required skills include the ability to understand the function being automated and programming the actions within the limitations of the technology available [27]. The abilities and limitations of the users and the technology should be a consideration in every interaction design [19, 23]. The consideration of the capabilities of all three role players will result in a positive user experience.

The real **experience** of the user using the system is a concept that cannot be overlooked. Experience can be classified into three questions. First, what is the user expecting to experience? Then, what was the user's previous experience with similar technology and lastly, how does the user feel when using this system? Although these three concepts are alike, they are also different. Systems are unintelligent and are programmed (by a human) to act in a specific manner [24]. Users often unrealistically expect the system to understand the outcomes it produces [24]. Designers should manage the expectations of users by identifying the correct processes to automate that will satisfy the needs of the users [22, 23]. Research into the experience of users with similar technology and whether they have liked using it can guide interaction design [31]. The motions and feelings that a user goes through when interacting with a system are the last experience factor identified. Users often fear the unknown and the interaction design should guide them into a successful interaction [24]. The user's operation of the system will influence their opinion of the success of the system as a whole [29]. A great program with poor interaction will lead to users rejecting the system. This also links to the expected experience; the design of a system is often criticized where the expectation is higher than the reality [22]. Interaction design as part of the system development life cycle will often go back and look at the user's need and real experience and improve upon it.

**Interacting** with a system is about more than just how a user operates the system. Goodrich and Schultz [20] define interaction as "the process of working together to accomplish a goal." This definition establishes the foundation for this concept where the interaction between user and system needs to be realistic and achievable [27]. Interaction can take three forms. Fully automated, collaborative and modifiable. Fully automated systems are still interacting with input, output and other systems where the system has pre-programmed rules [34]. These systems quite often function in a manner where programs and applications are used, but not changed [26]. More familiar systems are those that require user collaboration where user decision making forms part of a system [33]. These interactions between user input and system output create an interdependent collaboration [32]. A scarce interaction is where a user can modify the system when system changes or business requirements change [21]. Designers should establish the type of interaction that is required so that the interaction design complements the need of the users.

The **level of autonomy** of the system might be similar to the interaction, but what we are looking at here is specifically how independent the system can work. When designing a system that interacts with people, there is always a chance for error. If a system is not autonomous, higher user abilities are required [27]. Interfaces need to be designed with consideration to the ease of use by designing a simple and logical path [31]. Systems with a hundred percent autonomy are not always the right answer and might not be resolving the client's needs [34]. The level of autonomy needed for an effective interaction design should be determined before the design starts.

The last concept is probably the most important one as the **motivation**, goals and purpose of user interaction with the system will weigh heavily on the interaction design. To understand the users' needs, the designers must understand why they use the system, what their preferences are to reach their goal and how important their goals are to them [31]. Two characterizations of motivation are prominent in the literature. A system is supposed to ease your workload and users still want some form of control. The purpose of a computerized system is to deliver the correct results quickly and inexpensively [26]. These functions should not add to a user's workload but ease it. Samuels and Haapasalo [33] describe the ease with which mathematical collaboration can be achieved through the correct interaction platform. Previous studies have proven that users prefer to remain in control of a process [27, 37]. Children that were allowed to control robots were more motivated to use the systems [27]. Control can be in the forms of input parameters, schedule times, remote control or partial delegation [37]. The aspects that motivate users to interact with a system form part of the interaction design.

# 5   Discussion of a Proposed Framework

An initial conceptual framework has been designed using the concepts extracted from the literature. The identified concepts have been applied as user experience attributes of robotics, resulting in a framework that could be applied in a financial environment. Figure 1 shows the six surrounding pillars on which the "dome" of interaction design of software robots rests (behavior, motivation, capabilities, autonomy levels, interaction and experience). In turn, each foundational pillar is constituted of various principles that have been discussed in the narrative sections above.

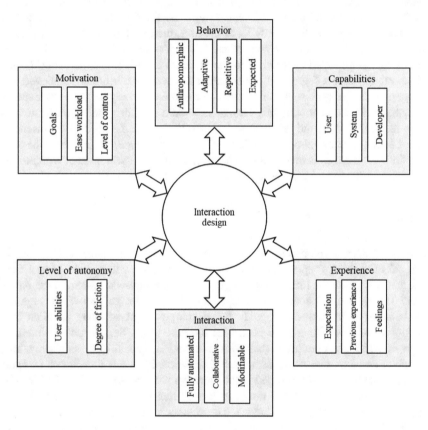

**Fig. 1.** An interaction design dome to serve as a conceptual framework to guide the development of software robotics in the banking sector.

The appropriateness of the framework for the banking sector can be linked back to the literature. **Behavior** intentions have been found to adapt to the technology used in banking systems [55]. The **motivation** behind the use of banking systems is to use the system without interruptions [56]. The ease of use of a system links to the **capabilities** of the system as well as the users and their **interaction** [55]. The autonomy of a banking system has been highlighted during the recent rapidly changing situation in the world and the need to support the financial system [57]. The experience of the bank's employees represents "business management know-how" [58].

The pillars identified as seen in Fig. 1 are not independent of each other. They all interact with and influence each other. One interaction example is between **motivation** and **behavior**. The way a system behaves can influence how a user feels about the system. A system that behaves as expected will enhance the users' motivation to continue using the system. A system that performs repetitive tasks in a predictive manner will reduce the required level of control desired by the user.

While robots are expected to behave in a human manner that is adaptive, the programmed repetitive, rule-based steps still need to be followed. Robots that include adaptive behaviors will be easily accepted by users. Robots need to be designed with cognizance of the capabilities or skills of the user, the developer and the infrastructure available. These considerations will reduce frustration and termination of the process. Comparing the expected experience to the physical experience while using the robot will reflect the user's feelings and determine the sense of fulfilment achieved when interacting with the system. Robotic interaction should not only be fully automated but also collaborative and modifiable. Designers should establish the type of interaction that is required so that the interaction design complement the need of the users. The level of autonomy of a robot and the level of human exchange will influence the degree of friction that the user is involved in. The motivation why users interact with the robot has a great influence on their acceptance of a robotic solution to a problem.

## 6   Conclusion

Software robotics development is a relatively new technology that is expanding. This paper asks the research question: Which core components will contribute to a framework that will guide the interaction design of a software robotic system in the banking sector? This question is answered through an interaction design "dome" that can serve as a conceptual framework to guide the development of software robotics in the banking sector. Six pillars have been identified on which the "dome" of interaction design of software robot rests (behavior, motivation, capabilities, autonomy levels, interaction and experience). In turn, each foundational pillar is constituted of various principles that have been discussed in the narrative sections above. A limitation of this study is that the framework has not been tested empirically. In future work, the conceptual framework should be validated. The authors plan to implement the framework in a practical banking environment and evaluate and improve the interaction design dome during and after the process. Subsequent research could determine if the framework could be generalized and applied in more areas of the financial sector and other industries.

**Acknowledgments.** The work is based on research supported in part by the National Research Foundation (NRF) of South Africa (Grant Number 132180) and the University of South Africa's Research Professor Support Programme. The grant holder acknowledges that opinions, findings and conclusions or recommendations expressed are that of the authors and that the NRF/Unisa accepts no liability whatsoever in this regard. We would like to thank the editors, programme chairs and reviewers for their feedback that guided us to improve the article.

# References

1. Ben-Ari, M., Mondada, F.: Elements of Robotics. Springer, Heidelberg (2017). https://doi.org/10.1007/978-3-319-62533-1
2. Mataric, M.J.: The Robotics Primer. The MIT Press, London (2007)
3. Halperin, D., Kavraki, L., Latombe, J.: Robotics. In: Handbook of Discrete and Computational Geometry, pp. 1–24 (1997)
4. Siciliano, B., Khatib, O. (eds.): Springer Handbook of Robotics. Springer, Cham (2016). https://doi.org/10.1007/978-3-319-32552-1
5. Kanda, T., Ishiguro, H.: Human-Robot Interaction in Social Robotics. CRC Press, Boca Raton (2013). https://doi.org/10.1201/b13004
6. Gates, B.: A robot in every home. Sci. Am. **296**, 58–67 (2007). https://doi.org/10.1038/scientificamerican0107-58
7. ISO 8373:2012: Robots and robotic devices—Vocabulary
8. Laschi, C., Mazzolai, B., Cianchetti, M.: Soft robotics: Technologies and systems pushing the boundaries of robot abilities. Sci. Robot. **1**, 1–12 (2016). https://doi.org/10.1126/scirobotics.aah3690
9. You, S., Robert, L.P.: Facilitating employee intention to work with robots. In: Proceedings of the Diffusion Interest Group in Information Technology, pp. 1–5. AIS Electronic Library, Seoul (2017)
10. Fahmideh, M., Lammers, T.: A study of influential factors in designing self-reconfigurable robots for green manufacturing. In: ACIS 2018 - 29th Australasian Conference on Information Systems, pp. 1–8. AIS Electronic Library, Sydney (2018). https://doi.org/10.5130/acis2018.bg
11. Ahn, J., Ahn, J., Oh, S.: The rise of the machines: Robotis, a frontier in educational and industrial robots in Korea. In: Proceedings of the International Conference on Electronic Business (ICEB), pp. 449–455. Atlantis Press, Hong Kong (2015)
12. You, S., Robert, L.P.: Emotional attachment, performance, and viability in teams collaborating with embodied physical action (EPA) robots. J. Assoc. Inf. Syst. **19**, 377–407 (2018). https://doi.org/10.17705/1jais.00496
13. Sharp, H., Rogers, Y., Preece, J.: Interaction Design: Beyond Human-Computer Interaction. Wiley, Indianapolis (2019)
14. Santos, J.S., Andrade, W.L., Brunet, J., Araujo Melo, M.R.: A systematic literature review of methodology of learning evaluation based on item response theory in the context of programming teaching. In: 2020 IEEE Frontiers in Education Conference, pp. 1–9 (2020). https://doi.org/10.1109/FIE44824.2020.9274068
15. Saunders, M.N.K., Lewis, P., Thornhill, A.: Research Methods for Business Students. Pearson, Harlow (2019)
16. Beck, C.: The Rise of the Robots. Oneworld Publications, London (2016)
17. Reichardt, M., Föhst, T., Berns, K.: On software quality-motivated design of a real-time framework for complex robot control systems. Electron. Commun. EASST. **60**, 1–20 (2013). https://doi.org/10.14279/tuj.eceasst.60.855.849
18. Alenljung, B., Lindblom, J., Andreasson, R., Ziemke, T.: User experience in social human-robot interaction. Int. J. Ambient Comput. Intell. **8**, 12–31 (2017). https://doi.org/10.4018/IJACI.2017040102

19. Goodrich, M.A., Schultz, A.C.: Human–robot interaction: a survey, foundation and trends. Human-Comput. Interact. **1**, 203–275 (2007). https://doi.org/10.1561/1100000005Human-Robo
20. Foster, M.E., Deshmukh, A., Janarthanam, S., Lim, M.Y., Hastie, H., Aylett, R.: Influencing the learning experience through affective agent feedback in a real-world treasure hunt (Extended Abstract). In: Proceedings of the 14th International Conference on Autonomous Agents and Multiagent Systems. pp. 1711–1712. ACM Press, Istanbul (2015). https://doi.org/10.5555/2772879.2773398
21. Asatiani, A., Penttinen, E.: Turning robotic process automation into commercial success - Case OpusCapita. J. Inf. Technol. Teach. Cases. **6**, 67–74 (2016). https://doi.org/10.1057/jittc.2016.5
22. Sachs, B.: The crucial role of UX in the design of software robots
23. UX for robotic process automation: UX for robotic process automation: The future of emerging technologies efficiency and productivity
24. Desai, N.: The invisible UX of RPA – An insight. https://www.virtusa.com/perspective/the-invisible-ux-of-rpa-an-insight/%0AThe
25. Comsa, A.: User experience (UX) design with robotic process automation (RPA). https://medium.com/accesa/user-experience-ux-design-with-robotic-process-automation-rpa-8162dfaeef25. Accessed 08 Apr 2021
26. Willcocks, L., Lacity, M., Craig, A.: The IT function and robotic process automation (2015)
27. Encarnação, P., et al.: Virtual robot and virtual environments for cognitive skills assessment. In: Assistive Technology Research Series, pp. 508–516, Maastricht (2011). https://doi.org/10.3233/978-1-60750-814-4-508
28. Wei, C., Yu, Z., Fong, S.: How to build a chatbot: Chatbot framework and its capabilities. In: Proceedings of the 10th International Conference on Machine Learning and Computing, pp. 369–373. ACM Press, Macau (2018). https://doi.org/10.1145/3195106.3195169
29. Yabe, H., Ono, D., Horikawa, T.: Space fusion: context-aware interaction using 3D scene parsing. In: Proceedings of 2018 International Conference on Computer Graphics Interaction Technology Virtual Augment, Reality, pp. 3–4 (2018). https://doi.org/10.1145/3275495.3275498
30. Skjuve, M., Haugstveit, I.M., Følstad, A., Brandtzaeg, P.B.: Help! Is my chatbot falling into the uncanny valley? An empirical study of user experience in human-chatbot interaction. Hum. Technol. **15**, 30–54 (2019). https://doi.org/10.17011/ht/urn.201902201607
31. Riddle, R.: What is interaction design: the practical framework. https://www.uxpin.com/studio/blog/what-is-interaction-design-the-practical-framework/
32. Schmeil, A., Eppler, M.J., De Freitas, S.: A structured approach for designing collaboration experiences for virtual worlds. J. Assoc. Inf. Syst. **13**, 836–860 (2012). https://doi.org/10.17705/1jais.00309
33. Samuels, P., Haapasalo, L.: Real and virtual robotics in mathematics education at the school-university transition. Int. J. Math. Educ. Sci. Technol. **43**, 285–301 (2012). https://doi.org/10.1080/0020739X.2011.618548
34. Goodrich, M.A., Schultz, A.C.: Human–Robot Interaction: A Survey. Now Publishers, Hanover (2007)
35. Adikari, S., McDonald, C., Campbell, J.: A design science framework for designing and assessing user experience. In: Jacko, J.A. (ed.) HCI 2011. LNCS, vol. 6761, pp. 25–34. Springer, Heidelberg (2011). https://doi.org/10.1007/978-3-642-21602-2_3
36. Jokinen, K., Wilcock, G.: User experience in human-robot interactions. In: Proceedings of the 4th International Workshop on Perceptual Quality of Systems, Vienna, pp. 125–130 (2013). https://doi.org/10.21437/pqs.2013-24

37. Rühr, A., Berger, B., Hess, T.: Can I control my robo-advisor? Trade-offs in automation and user control in (digital) investment management. In: Proceedings of the 25th Americas Conference on Information Systems, pp. 1–10. AIS Electronic Library, Cancun (2019)
38. Cao, J., Zieba, K., Ellis, M.: Interaction Design Best Practices: Mastering the Intangibles. UXPin, San Francisco (2015)
39. Kolko, J.: Thoughts on Interaction Design. Elsevier, Amsterdam (2011). https://doi.org/10.1016/C2009-0-61347-7
40. Dix, A., Finlay, J., Abowd, G.D., Beale, R.: Human-Computer Interaction. Pearson, Harlow (2004)
41. Hassenzahl, M.: User experience and experience design (2011)
42. Savioja, P., Norros, L.: Systems usability framework for evaluating tools in safety-critical work. Cogn. Technol. Work. **15**, 255–275 (2013). https://doi.org/10.1007/s10111-012-0224-9
43. ISO 9241-11:2018(en): Ergonomics of human-system interaction—Part 11: Usability: definitions and concepts. International Organization for Standardization (ISO), Switzerland (2018)
44. Hellman, M., Rönkkö, K.: Is user experience supported effectively in existing software development processes? In: Law, E.L., Bevan, N., Christou, G., Springett, M., and Lárusdóttir, M. (eds.) Proceedings of the International Workshop on Meaningful Measures: Valid Useful User Experience Measurement, pp. 32–37. Institute of Research in Informatics of Toulouse (IRIT), Reykjavik (2011)
45. Bevan, N.: What is the difference between usability and user experience? In: Proceedings of Workshop in UXEM, 4 p. (2009)
46. Isleifsdottir, J., Larusdottir, M.: Measuring the user experience of a task oriented software. In: Law, E.L.-C., Bevan, N., Christou, G., Springett, M., and Lárusdóttir, M. (eds.) Proceedings of the International Workshop on Meaningful Measures: Valid Useful User Experience Measurement, pp. 97–102. Institute of Research in Informatics of Toulouse (IRIT), Reykjavik (2011). https://doi.org/10.1145/2070481.2070498
47. Sünderhauf, N., et al.: The limits and potentials of deep learning for robotics. Int. J. Rob. Res. **37**, 405–420 (2018). https://doi.org/10.1177/0278364918770733
48. O'Kane, J.M., LaValle, S.M.: On comparing the power of mobile robots. Robot. Sci. Syst. **2**, 65–72 (2007). https://doi.org/10.15607/rss.2006.ii.009
49. Merriam-Webster Dictionary: Robot
50. Robots.net: Robotics in 2020: Types of robots that we use
51. Kym: What is robotics: How robots benefit mankind. Robots.net. 1–6 (2019)
52. Tripathi, A.M.: Learning Robotic Process Automation: Create Software Robots and Automate Business Processes with the Leading RPA Tool - UiPath. Packt Publishing, Birmingham (2018)
53. Clarke, R.: Asimov's laws of robotics: Implications for information technology. Computer (Long. Beach. Calif). **27**, 57–66 (1994). https://doi.org/10.1109/2.248881
54. Bonsignorio, F.P., Hallam, J., Del Pobil, A.P.: Defining the requisites of a replicable robotics experiment. In: Workshop on Good Experimental Methodologies in Robotics, pp. 1–7 (2009)
55. Hossain, S.A., Bao, Y., Hasan, N., Islam, M.F.: Perception and prediction of intention to use online banking systems. Int. J. Res. Bus. Soc. Sci. **9**, 112–126 (2020). https://doi.org/10.20525/ijrbs.v9i1.591
56. Haralayya, B.: Core banking technology and its top 6 implementation challenges. J. Adv. Res. Oper. Mark. Manag. **4**, 25–27 (2021)

57. Demirguc-Kunt, A., Pedraza, A., Ruiz-Ortega, C.: Banking sector performance during the COVID-19 drisis. J. Bank. Financ. 106305 (2021). https://doi.org/10.1016/j.jbankfin.2021.106305
58. Savchenko, T., Kovacs, L.: Trust in the banking sector: EU experience and evidence from Ukraine. Financ. Mark. Institutions Risks. **1**, 29–42 (2017). https://doi.org/10.21272/fmir.1(1).29-42.2017

# Methodology for Innovative Projects' Financing in IT Business

Gennady Ross[1] ⓘ, Tatiana Antipova[2(✉)] ⓘ,
and Valery Konyavsky[1] ⓘ

[1] Plekhanov Russian University of Economics, Moscow, Russia
[2] Institute of Certified Specialists, Perm, Russia

**Abstract.** The comprehensive approach to solving the problem of the modeling of commercial risks of IT business innovative projects is considered. The model of efficient management organization for innovative project within the technology park are proposed. The system of logical connections between individual projects and their stages as well as general goals and objectives of the technology park were taken into account. This study considers the outcomes of innovative project stages, general resource restrictions for all projects performed in the technology park. An example of applying the modeling technique using "Combinatorics" module of "Decision" tool system is given.

**Keywords:** Information technologies (IT) · Evolutionary simulation model · Decision system · IT business · Technology park (techno park) · Innovative project

## 1 Introduction

In the context of the rapid development of information technologies (IT), various sectors of the Russian economy are showing great interest in participating in the financing of innovative projects, the implementation of which will improve the efficiency of production and the quality of products [1–3]. However, the analysis of statistics on innovative projects shows that successfully completed projects account for about 17%, troubled projects account for 43%, and failed projects account for up to 40%. Such a state of affairs is associated with commercial risks in the IT sphere, in particular, with the risks of making decisions on the forms, amount and timelines of state support for the IT business, which multiply both in the case of successful decisions and in the case of unsuccessful decisions [4]. This creates specific threats for the economic component of information security both on the scale of the IT industry and on the scale of the entire national economy.

The existing measures of state management of the economy are mainly limited to taxation and the placement of state orders, and there are also various specific methods of state support for certain IT spheres. In particular, the establishment of innovation centers (techno parks), the major task of which is to create favorable conditions for small and medium-sized innovative businesses, is of great importance for the IT business. The IT business is a highly efficient business on the one hand, and a highly risky business on the other hand. Upon that, the main source of risks is the uncertainty generated by numerous and fundamentally irremovable sources, such as commercial prospects for scientific and technical innovations, the efficiency of innovation management, the preferences of potential consumers, etc.

For above reasons, each innovative project and each of its stage has a low probability of a favorable outcome. The essence of the above problem is determined by a specific combination of the following peculiarities [5–9]:

- firstly, the already mentioned specifically meaningful role of state support for ICT in the form of appropriate techno parks;
- secondly, high investment risks (each innovative project and each of its stages has a low probability of a favorable outcome assuming that the greatest effect is achieved at the final stage depending on the previous ones), which, in their turn, generate the following sequence of interrelated risks: industry lag - national economic costs;
- thirdly, the risk of insufficient support for innovations in the IT business since the successful completion of at least small portion of innovative projects makes the entire innovative project effective, while the lack of support leads to technological stagnation, which generates the same risks and threats as inefficient investments in state support of the IT business.

Thus, the problem of reducing the risks and threats of the IT business appears as a problem of efficient management of innovative project portfolio within the techno park. This problem is non-trivial from a scientific point of view, as it is associated with the need to comprehensively consider the probabilities of outcomes of the innovation project (IPr) stages, common resource constraints for all projects implemented in the technology park, a complex system of logical links between individual projects and their stages, as well as common goals and objectives of the technology park [10–14].

## 2 Mathematical Formulation of the Problem

Innovative projects (IPr) are usually performed by specialized funds or techno parks. These projects have a number of specific peculiarities. In particular, the probability of successful completion of each individual IPr is not high, but the successful completion of at least one of them can recoup all costs, including the costs of unfinished activities or activities that have turned out to be the dead-end ones.

Another important peculiarity of the innovative project is that each IPr has a number of successively implemented stages, and almost each stage is an activity being an IPr that can bring a certain income and become the final one. In addition, the alternative options of implementation of stages are possible. Taking into account all these conditions, it is necessary to draw up the most cost-efficient comprehensive innovative project.

Let us consider the main stages of IPr:

$$FR \rightarrow AR \rightarrow D \rightarrow P_r \rightarrow C \rightarrow MA \rightarrow IP \rightarrow M \rightarrow Sa,$$

where: FR stands for fundamental (theoretical) research; AR stands for applied researches; D stands for development; $P_r$ stands for projecting; C stands for construction; MA stands for mastering; IP stands for industrial production; M stands for marketing; Sa stands for sale.

The alternative options of implementation of any of these stages may differ in the composition of contractors, the materials and components used, the cooperation relations, the workflow management methods, the scope of work, the deadlines or the methods of financing.

For ease of presentation, let us introduce the following definitions: an IPr stage or additional measures for IPr stage implementation shall be called an activity. Each activity shall be characterized by the amount of costs, profit and reliability. Upon that, the probability that the activity will be completed within the target time frame and with the expected result shall be called the reliability of the activity. The requirement for the sequence of implementation of activities, their alternativeness, the obligatory implementation or approval of the independence of the activity from the rest shall be called logical connections between the activities (Table 1). The activity shall be called major if in addition to it there are additional activities at this IPr stage and if the stage cannot be successfully completed without this activity.

**Table 1.** Logical connections

| Logical connection | Record mode | Comment |
|---|---|---|
| "Mutually exclusive: …,…, …" | […,…,…] | The numbers of activities are listed in square brackets separated by commas |
| "At least one of the activities shall be present: …,…,…"; | {…,…,…,} | The numbers of activities are listed in curly brackets separated by commas |
| "If …,…,… then …,…,…"; | (…,…, …,) → (…, …,…,) | The numbers of "activities being causes" and "activities being effects" are listed in round brackets and are connected by an arrow |
| "Independent: …,…,…"; | … | The activity number is recorded on a separate line |

As a rule, the technopark activities require the state support. Let us consider the task of the most effective use of state subsidies. Let us assume that $M_{i,j,1}$ is the activity being j stage of IPr number i. Figure one in the index stands for "major" activity. $P_{i,j,1}$ stands for the reliability of $M_{i,j,1}$ activity. Let us assume that $M_{i,j,k}$, $k = 2,\ldots,K(i,j)$ are "additional" activities, each of which increases the reliability of $M_{i,j,1}$ major activity by $P_{i,j,k}$, $k = 2,\ldots,K(i,j)$ values correspondingly.

The reliability of the stage itself, regardless of the previous stages, is equal to the reliability of major activity increased due to the implementation of additional activities, that is:

$$P_{i,j} = \begin{cases} \sum_{k=1}^{K(i,j)} P_{i,j,k}, & \text{if } \sum_{k=1}^{K(i,j)} P_{i,j,k} \leq 1 \\ 1, & \text{otherwise} \end{cases} \qquad (1)$$

and the probability of unfavorable outcome of j stage of i IPr is equal to $1-P_{i,j}$. If each of the additional activities can ensure a favorable outcome of the stage, the probability of unfavorable outcome of j stage of i IPr is equal to the probability of unfavorable outcome of all activities $M_{i,j,k}$, k = 1,...,K(i,j) simultaneously: The probability that all these activities will have an unfavorable outcome is equal to $\prod_{k=1}^{K(i,j)}(1 - P_{i,j,k})$. The probability of unfavorable outcome of j stage of i IPr (let us denote it as $Q_{i,j}$) is equal to:

$$Q_{i,j} = \begin{cases} 1 - P_{i,j}, \text{major and additional activities} \\ \prod_{k=1}^{K(i,j)}\left(1 - P_{i,j,k}\right), \text{mutually replaceable but not alternative activities} \end{cases}, \forall i, \forall j$$

(2)

Let us assume that $Z_{i,j,k}$ is the amount of required financing for $M_{i,j,k}$ activity, and $\alpha_{i,j,k}$ is Boolean variable, which takes on a value of 1 if $M_{i,j,k}$ activity is accepted for implementation (included in the plan), or a value of 0 if it is rejected:

$$\alpha_{i,j,k} = \begin{cases} 0 \\ 1 \end{cases}, \forall i, \forall j, \forall k$$

(3)

Let us assume that V is the limit of financing (on the part of the technopark) in the considered period of time. The limit of financing is as follows:

$$\sum_{i=1}^{N}\sum_{j=1}^{9}\sum_{k=1}^{K(i,j)} Z_{i,j,k} * \alpha_{i,j,k} \leq V$$

(4)

The reliability of the stage with the reliability of the previous stages being taken into account, which is to say, the probability that IPr number i will not be stopped at r stage (let us denote it as $H_{ir}$) is equal to:

$$H_{i,r} = \prod_{j=1}^{r}(1 - Q_{i,j}), \forall i, \forall r$$

(5)

IPr number i can be set not only at the last stage, but at stage j = 9, when the production of some new goods and services containing know-how is carried out. Even the initial stage, namely the stage of fundamental research, can yield certain dividends through the sale of licenses.

The reliability of $M_{i,j,k}$ activity with its logical connections being taken into account (let us denote it as $h_{i,j,k}$) is defined both by the reliability of stage $H_{i,r}$, and by the reliability of activity $P_{i,j,k}$:

$$h_{i,j,k} = 1 - \left(1 - H_{i,j-1}\right) * \left(1 - P_{i,j,k}\right), \forall i, \forall j, \forall k$$

(6)

If we denote the estimated profit from $M_{i,j,k}$ activity by $R_{i,j,k}$, where k = 1,...,K(i,j), the expected profit from $M_{i,j,k}$ activity is equal to the following product: $R_{i,j,k}*h_{i,j,k}$.

The goal of IPr management in the technopark for a sufficiently long period of time is making profit:

$$\sum_{i=1}^{N}\sum_{j=1}^{9}\sum_{k=1}^{K(i,j)} R_{i,j,k} * h_{i,j,k} * \alpha_{i,j,k} \rightarrow max \qquad (7)$$

The requirement that the stages of any IPr be carried out in the required sequence is expressed by the following inequalities:

$$\alpha_{i,j+1,k} \geq \alpha_{i,j,k}\ldots\forall i, \forall j, \forall k \qquad (8)$$

If there is a set of G measures, among which at least one shall be performed (for example, for the implementation of a certain technical policy in the technopark), then this requirement is expressed by the condition:

$$\sum_{(i,j,k)\in G}\alpha_{i,j,k} \geq 1 \qquad (9)$$

Among the activities, in particular, among the activities being additional to the same major activity, there may be alternative, mutually exclusive ones (for technical, organizational or other reasons). Let us assume that W is the set of alternative activities. The requirement of alternativeness is expressed by the following condition:

$$\sum_{(i,j,k)\in W}\alpha_{i,j,k} \leq 1 \qquad (10)$$

The existence of logically independent activities is also possible. In particular, an IPr consisting of one activity is possible.

The system of relations (1)–(10) is a mathematically correct formulation of the problem of management of innovative project portfolio in the form of a Boolean programming problem.

## 3   The Forms of Collection of Initial Data for All Activities and Their Logical Connections

We have three forms of collection of initial data for all activities and their logical connections: Form 1 (Table 2) contains information about a specific activity, Form 2 (Table 3) contains data on the names of innovative projects, and Form 3 (Table 4) contains a complete list of activities and their indicators.

**Table 2.** Form 1: Activity

| Innovative project | | | Stage | Event option | |
|---|---|---|---|---|---|
| Name | | No. (i) | No. (j) | Description | No. (k) |
| | | | | | |
| The volume of financing, c.u. $(Z_{i,j,k})$ | Reliability, unit fraction $(P_{i,j,k})$ | Estimated profit, c.u. $(R_{i,j,k})$ | Execution period | | |
| | | | | Beginning | Completion |
| | | | | | |

**Table 3.** Form 2: The limit of financing

| Technopark: | | | The list of IP | |
|---|---|---|---|---|
| | | | No. | Name |
| Period | | | | |
| Beginning | Completion | | | |
| | | | | |
| The volume of financing, c.u.: | | | | |
| | | | | |

**Table 4.** Form 3: The list of activities of _____ ... _____ techno park for the period (name) from ... to ...

| Item No. | Code | The volume of financing, c.u. $(Z_{i,j,k})$ | The share of state (technopark) participation | | Reliability, $(h_{i,j,k})$, unit fraction | The expected profit, c.u. $(h_{i,j,k} * R_{i,j,k})$ |
|---|---|---|---|---|---|---|
| | | | % | c.u. | | |
| 1 | | | | | | |
| 2 | | | | | | |
| ... | | | | | | |

The code in Form 3 is a set of numbers separated by commas. Upon that, the first number is the IPr number, the second number is the stage number, the third number is the activity number. If the activity is conditional, then the code contains more than 3 numbers. Wherein the 3rd number and the subsequent numbers are the numbers of activities included in the conditional activity. IPr – the name and number of the innovation project. Stage – the name and number of the stage. Activity – the name and number of the activity. If the activity is conditional, the names and numbers of all activities, being included in it, shall be indicated.

The methods of the construction of logical connection network and the calculation of reliabilities of activities. The cases when the alternativeness of activities generates branching alternative chains are possible. The methods of the calculation of reliabilities of activities using (1), (4) and (5) formulas depend on logical connections between the activities. Some typical situations are illustrated in Fig. 1.

In Fig. 1(A), $M_{1,1,1}$ activity is major while $M_{1,2,1}$, $M_{1,2,2}$ and $M_{1,2,3}$ activities are the options of additional non-alternative activities performed at the 2nd stage. In Fig. 1 (Б), $M_{2,1,1}$ activity is major while $M_{2,2,1}$, $M_{2,2,2}$ and $M_{2,2,3}$ activities are alternative. Figure 1(B) shows a certain example illustrating the assertion that more complex situations can be reduced to the previous two ones. Factually, the probability of an unfavorable outcome of $M_{3,4,1}$ depends on the probability of unfavorable outcomes of $M_{3,2,1}$ and $M_{3,3,1}$, and it is calculated according to the option of alternative activities. Upon that, the latter does not depend on the method by which the probability of unfavorable outcomes of $M_{3,2,1}$ and $M_{3,3,1}$ alternative is calculated.

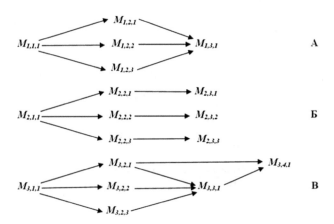

**Fig. 1.** The options of connected subnetworks at multiple activities. Source: Authors' elaboration.

In order to solve the problem formulated above, "Combinatorics" module has been developed within the framework of Decision tool system.

*Example.* To develop a comprehensive innovation project consisting of 3 financing projects, the first of which is connected with the creation of new volumetric media for the RAM of servers, the second one is connected with the development of special software for mobile communication facilities, and the third one is connected with the creation of new software and hardware for information protection.

The activities of innovation project are connected by the following network of logical connections (Table 1):

**IPr 1:** $(M_{1,1\text{-}3,1}) \rightarrow (M_{1,4\text{-}6,1})$; $(M_{1,1\text{-}3,1}) \rightarrow (M_{1,4\text{-}6,2})$; $(M_{1,4\text{-}6,1},M_{1,4\text{-}6,2}) \rightarrow (M_{1,7\text{-}9,1})$;
$(M_{1,4\text{-}6,1},M_{1,4\text{-}6,2}) \rightarrow (M_{1,7\text{-}9,2})$.
**IPr 2:** $(M_{2,2,1}) \rightarrow (M_{2,(3,8,9),1},M_{2,(3,8,9),2})$.
**IPr 3:** $M_{3,3\text{-}9,1}$.

The graphic interpretation of these IPRs is shown in Fig. 2.

Let us suppose that Form 1 is filled in for each activity that appears in the above list of logical connections. Based on these data, a summary of the activity indicators shown in Table 2 is compiled. The total volume of financing for all activities shall not exceed 1,000 c.u.

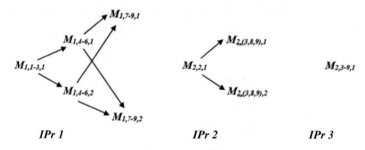

**Fig. 2.** The logical network of activity connections.

**Table 5.** Activity indicators.

| No (i) | Stage (j) | Option (k) | The volume of financing, c.u.($Z_{i,j,k}$) | Reliability, unit fraction ($P_{i,j,k}$) | The estimated profit, c.u.($R_{i,j,k}$) |
|---|---|---|---|---|---|
| 1 | 1-3 | 1 | 1.35 | 0.9 | 235 |
| 1 | 4-6 | 1 | 117 | 0.75 | 0 |
| 1 | 4-6 | 2 | 97 | 0.65 | 0 |
| 1 | 7-9 | 1 | 317 | 0.65 | 755 |
| 1 | 7-9 | 2 | 517 | 0.5 | 1,325 |
| 2 | 2 | 1 | 35 | 0.95 | 0 |
| 2 | 3,8,9 | 1 | 97 | 0.55 | 211 |
| 2 | 3,8,9 | 2 | 33 | 0.75 | 65 |
| 3 | 3-9 | 1 | 25 | 0.6 | 35 |

Source: authors' compilation based on their experience

The reliabilities of activities with their logical connections $h_{i,j,k}$ being taken into account calculated using formulas (1), (2), (5), (6) in accordance with the logical connections between the activities. In particular, $h_{i,j,k} = P_{i,j,k}$ equation holds true for the following activities: M1,1-3,1, M2,2,1, M3,3-9,1. $Q_{i,j} = 1\text{-}P_{i,j,1}$ equation holds true for the following activities: M1,4-6,1, M1,4-6,2, M2,(3,8,9),1, M2,(3,8,9),2. Formula (2) holds true for M1,7-9,1 and M 1,7-8,2 activities with "major and additional activities" option.

**Table 6.** The reliabilities of activities

| The activity code | $P_{i,j,k}$ | $Q_{i,j}$ | $H_{i,r}$ | $h_{i,j,k}$ |
|---|---|---|---|---|
| 1,1-3,1 | 0.90 | | 0.90*0.75≈0.68 | 0.90 |
| 1,4-6,1 | 0.75 | 0.25 | 0.90*0.65≈0.59 | 0.08 |
| 1,4-6,2 | 0.65 | 0.35 | 1.00 | 0.15 |
| 1,7-9,1 | 0.65 | 1-0.65-0.50: = 0.00 | 1.00 | 0.65 |
| 1,7-9,2 | 0.50 | 1-0.65-0.50: = 0.00 | | 0.50 |
| 2,2,1 | 0.95 | | 0.95*0.55≈0.52 | 0.95 |
| 2,(3,8,9),1 | 0.55 | 0.45 | 0.95*0.75≈0.71 | 0.21 |
| 2,(3,8,9),2 | 0.75 | 0.25 | | 0.07 |
| 3,3-9,1 | 0.60 | | | 0.60 |

Table 7 provides the list of activities and their indicators.

**Table 7.** The list of techno park activities

| Item No | The activity code | The volume of financing, c.u. ($Z_{i,j,k}$) | Profit, c.u ($R_{i,j,k}$) | Actual activities ($h_{i,j,k}$), unit fraction | The expected profit, c.u ($h_{i,j,k}$* $R_{i,j,k}$) |
|---|---|---|---|---|---|
| 2 | 1,1-3,1 | 135 | 235 | 0.90 | 211.5 |
| 3 | 1,4-6,1 | 117 | 0 | 0.08 | 0 |
| 4 | 1,4-6,2 | 97 | 0 | 0.15 | 0 |
| 5 | 1,7-9,1 | 317 | 755 | 0.65 | 490.75 |
| 6 | 1,7-9,2 | 517 | 1,325 | 0.50 | 662.5 |
| 8 | 2,2,1 | 35 | 0 | 0.95 | 0 |
| 9 | 2,(3,8,9),1 | 97 | 211 | 0.21 | 44.31 |
| 10 | 2,(3,8,9),2 | 33 | 65 | 0.07 | 4.55 |
| 14 | 3,3-9,1 | 25 | 35 | 0.60 | 21 |

Table 8 shows the initial information entered into Combinatorics module of Decision tool system [15, 16].

**Table 8.** Initial data on the activities

| 1 | Activities | Costs | Profit |
|---|---|---|---|
| 2 | 1,1-3,1 | 135 | 211.5 |
| 3 | 1,4-6,1 | 117 | 0 |
| 4 | 1,4-6,2 | 97 | 0 |
| 5 | 1,7-9,1 | 317 | 490.75 |
| 6 | 1,7-9,2 | 517 | 662.5 |
| 7 | 2,2,1 | 35 | 0 |
| 8 | 2(3,8,9),1 | 97 | 44.31 |
| 9 | 2(3,8,9),2 | 33 | 4.55 |
| 10 | 3,3-9,1 | 25 | 21 |

The introduced logical connections between the activities constructed in accordance with Table 5:

(1,1-3,1) → (1,4-6,1); (1,1-3,1) → (1,4-6,2); ((1,4-6,1), (1,4-6,2)) → (1,7-9,1); ((1,4-6,1), (1,4-6,2)) → (1,7-9,2); (2,2,1) → ((2(3,8,9),1), (2(3,8,9),2)); 3,3-9,1. Table 9 shows the result of the optimization calculation.

**Table 9.** The result of the optimization calculation

| 1 | Costs | Profit | TP/RR | The activities selected for implementation | | | | | | | |
|---|---|---|---|---|---|---|---|---|---|---|---|
| 2 | 25.00 | 21.00 | 0.84 | 3,3-9,1 | | | | | | | |
| 3 | 93.00 | 25.56 | 0.27 | 2,2,1 | 2(3,8,9),2 | 3,3-9,1 | | | | | |
| 4 | 132.00 | 44.31 | 0.34 | 2,2,1 | 2(3,8,9),1 | | | | | | |
| 5 | 135.00 | 211.50 | 1.57 | 1,1-3,1 | | | | | | | |
| 6 | 160.00 | 232.50 | 1.45 | 1,1-3,1 | 3,3-9,1 | | | | | | |
| 7 | 228.00 | 237.05 | 1.04 | 1,1-3,1 | 2,2,1 | 2,(3,8,9),2 | 3,3-9,1 | | | | |
| 8 | 267.00 | 255.81 | 0.96 | 1,1-3,1 | 2,2,1 | 2,(3,8,9),1 | | | | | |
| 9 | 292.00 | 276.81 | 0.95 | 1,1-3,1 | 2,2,1 | 2,(3,8,9),1 | 3,3-9,1 | | | | |
| 10 | 325.00 | 281.36 | 0.87 | 1,1-3,1 | 2,2,1 | 2,(3,8,9),1 | 2,(3,8,9),2 | 3,3-9,1 | | | |
| 11 | 666.00 | 702.25 | 1.05 | 1,1-3,1 | 1,4-6,1 | 1,4-6,2 | 1,7-9,1 | | | | |
| 12 | 691.00 | 723.25 | 1.05 | 1,1-3,1 | 1,4-6,1 | 1,4-6,2 | 1,7-9,1 | 3,3-9,1 | | | |
| 13 | 759.00 | 727.80 | 0.96 | 1,1-3,1 | 1,4-6,1 | 1,4-6,2 | 1,7-9,1 | 2,2,1 | 2,(3,8,9),2 | 3,3-9,1 | |
| 14 | 798.00 | 746.56 | 0.94 | 1,1-3,1 | 1,4-6,1 | 1,4-6,2 | 1,7-9,1 | 2,2,1 | 2,(3,8,9),1 | | |
| 15 | 823.00 | 767.56 | 0.93 | 1,1-3,1 | 1,4-6,1 | 1,4-6,2 | 1,7-9,1 | 2,2,1 | 2,(3,8,9),1 | 3,3-9,1 | |
| 16 | 856.00 | 772.11 | 0.90 | 1,1-3,1 | 1,4-6,1 | 1,4-6,2 | 1,7-9,1 | 2,2,1 | 2,(3,8,9),1 | 2,(3,8,9),2 | 3,3-9,1 |
| 17 | 866.00 | 874.00 | 1.01 | 1,1-3,1 | 1,4-6,1 | 1,4-6,2 | 1,7-9,2 | | | | |
| 18 | 891.00 | 895.00 | 1.00 | 1,1-3,1 | 1,4-6,1 | 1,4-6,2 | 1,7-9,2 | 3,3-9,1 | | | |
| 19 | 959.00 | 899.55 | 0.94 | 1,1-3,1 | 1,4-6,1 | 1,4-6,2 | 1,7-9,2 | 2,2,1 | 2,(3,8,9),2 | 3,3-9,1 | |
| 20 | 998.00 | 918.31 | 0.92 | 1,1-3,1 | 1,4-6,1 | 1,4-6,2 | 1,7-9,2 | 2,2,1 | 2,(3,8,9),1 | | |

In Table 9, the TP/RR indicator expresses the profitability of state financing, that is the ratio of total profit to the volume of financing. Let us explain the foregoing using the example of an option with the optimum set of activities that delivers maximum return on financing. As we can see in the TP/RR column, the maximum ratio of the target parameter (profit) to the restricted resource (state financing) equal to 1.57 is achieved with state financing of 135 c.u. This amount ensures the ability of performing activities No. 5 and provides a total profit in the amount of 211.50 c.u.

Table 6 shows that return reaches its peak at a low level of financing, then it declines and further, after reaching the level of financing from 135 to 228 c.u., it remains almost constant and indicates the presence of a small profit. However, it stands to reason that it is expedient to use both the available state financing as much as possible and to invest the techno park's own funds in financing the same innovative projects.

# 4   Conclusions

The above formulation of the problem as well as the optimization calculation are aimed at the search of a set of options of the activities that ensure maximum profit at various levels of state financial support. In addition, other options of optimization calculations are of interest, in particular for maximum profit at various levels of financing from the techno park's own resources; maximum profit at different levels of aggregate financing; maximum profit upon several resource restrictions, in particular there may be restrictions on labor, material and energy resources in addition to financial restrictions; calculation for minimum state financing, provided that a given profit is ensured; calculation for minimum aggregate financing, provided that a given level of return on aggregate financing is ensured; calculation for minimum state financing, provided that the required profitability of own financing is ensured; calculation for minimum state financing upon several resource restrictions, etc.

# References

1. Ogoleva, L.N. (ed.): Innovation Management: Textbook. INFRA-M, 238 p. (2001)
2. Ganina, G.E., Klementyeva, S.V.: Innovative Project Management, 36 p. BMSTU Publishing House, Moscow (2014)
3. Bykovsky, V.V., Mshchenko, E.S., et al.: Innovative Project and Program Management, 104 p. TSTU Publishing House, Tambov (2011)
4. Sklyarova, E.E.: Statistical analysis of the innovative sector of the economy in Russia. Bull. Perm Univ. 3(14), 50–57 (2012)
5. Ross, G., Fedorov, Yu.P.: Methodology for diversifying innovative risks in IT technology parks. Information Technologies of Management in Socio-Economic Systems. M.: Federal Agency for Information Technologies "Rosinformtechnologii", pp. 94–105 (2007)
6. Lichtenstein, V., Ross, G.: Information technologies in business. practicum: the application of decision system in solving of applied economic problems. M.: Finance and Statistics, 560 p. (2009)
7. Ross, G., Konyavsky, V.: Modeling of investment in It-business. In: Antipova, T. (ed.) ICADS 2021. AISC, vol. 1352, pp. 377–389. Springer, Cham (2021). https://doi.org/10.1007/978-3-030-71782-7_33
8. Ross, G., Konyavsky, V.: Financial costs optimization for maintaining critical information infrastructures. In: Antipova, T. (ed.) ICIS 2020. LNNS, vol. 136, pp. 188–200. Springer, Cham (2021). https://doi.org/10.1007/978-3-030-49264-9_17
9. Mingaleva, Z., Gayfutdinova, O., Podgornova, E.: Forming of institutional mechanism of region's innovative development. World Acad. Sci. Eng. Technol. 58, 1041–1051 (2009)
10. Mingaleva, Z., Mirskikh, I.: The problems of legal regulation of intellectual property rights in innovation activities in Russia (institutional approach). World Acad. Sci. Eng. Technol. 29, 464–476 (2009)
11. Akatov, N., Mingaleva, Z., Klačková, I., Galieva, G., Shaidurova, N.: Expert technology for risk management in the implementation of QRM in a high-tech industrial enterprise. Manag. Syst. Prod. Eng. 27(4), 250–254 (2019). https://doi.org/10.1515/mspe-2019-0039
12. Mingaleva, Z., Mirskikh, I.: On innovation and knowledge economy in Russia. World Acad. Sci. Eng. Technol. 42, 1018–1027 (2010)

13. Mingaleva, Z., Shaidurova, N., Prajová, V.: The role of technoparks in technological upgrading of the economy. The example of agricultural production. Manag. Syst. Prod. Eng. **26**(4), pp. 241–245 (2018)
14. Mingaleva, Z., Deputatova, L., Starkov, Y.: Values and norms in the modern organization as the basis for innovative development. Int. J. Appl. Bus. Econ. Res. **14**(9), 5799–5808 (2016)
15. Liechtenstein, V., Ross G. Patent: A way to optimally control an equilibrium random process # 2557483 on 25.06.2015. https://new.fips.ru/registers-doc-view/fips_servlet?DB=RUPAT&rn=9052&DocNumber=2557483&TypeFile=html. Accessed 04 Oct 2017
16. Liechtenstein, V., Ross G. Patent: A way of selecting values of environmental parameters consistent with optimal control of an equilibrium random process # 2558251 on 27.07.2015. https://new.fips.ru/registers-doc-view/fips_servlet?DB=RUPAT&rn=9284&DocNumber=2558251&TypeFile=html. Accessed 21 Oct 2017

# The Role of Digital Integrated Reporting Disclosure to Firm Value (Evidence in Indonesia-Singapore)

St. Dwiarso Utomo$^{(\boxtimes)}$ ⓘ, Zaky Machmuddah ⓘ,
and Dian Indriana Hapsari

Universitas Dian Nuswantoro, Semarang, Indonesia
dwiarso.utomo@dsn.dinus.ac.id

**Abstract.** Empirically proving the application of digital integrated reporting disclosure to the firmvalue is the aim of the research. Signal theory explains that companies offer better information can influence investors' economic decisions and attract them to contracts with more significant benefits than other companies that offer worse information. Manufacturing companies listed on the Indonesia Stock Exchange (IDX) and the Singapore Stock Exchange (SGX) are research objects. Observations were carried out for five years starting from 2016–2020 so that as many as 300-panel observations were obtained. WarpPLS 5.0 was used as a research analysis tool. The results showed that firm value was influenced by digital integrated reporting disclosure, firm value was influenced by firm size, and firm value was influenced by leverage. The research findings theoretically support the signal theory. The signal given by the company to its stakeholders influences on the value of the company. The implication of this research is the importance of companies implementing digital integrated reporting disclosures to increase firm value so can achieve that company sustainability.

**Keywords:** Digital integrated reporting · Firm size · Leverage · Firm value

## 1 Introduction

The phenomenon that occurred in June 2020 in manufacturing companies showed that operating conditions in the entire Asian manufacturing sector continued to deteriorate. This is reflected in the IHS Markit report, which states that the purchasing managers' index in June 2020 was 43.7 or up 8.2 points from 35.3 in May 2020. Despite the increase in the number, it remained in the contraction area and even indicated a deterioration. As is known, purchasing managers' index above 50 indicates that manufacturing is expanding, while below 50 indicates that manufacturing is in recession. Indonesia recorded further contraction, although the rate of decline was much reduced from the previous survey period, with the purchasing managers' index rising to 39.1. Likewise, conditions in Singapore, conditions in Singapore also worsened, with the most striking decline from each survey participating countries. The purchasing managers' index of 38.8 signaled a sharp contraction, albeit up 12.4 points from May. This information is one of the signals that can influence investors in making investment decisions.

A good company should send a credible signal about its quality to the capital market to distinguish itself from bad companies [1]. This is in line with [2], which states that the disclosure of quality reports will influence the value of the company. In this era, a new reporting paradigm has been considered. The economic, social, and environmental activities of the company are integrated into one to provide a more holistic view of company performance and ensure that ethical responsibility is at the forefront of business activities [3]. Managers can create corporate value by implementing holistic reporting called integrated reporting. Integrated reporting can be understood as a convergence of sustainability reports and financial reports aimed primarily at investors. Top management provides its views on how sustainability issues and initiatives are expected to contribute to a long-term strategy for business growth [4]. Research [2, 5–9] show that integrated reporting disclosures positively influence firm value. However, [4, 10] show that integrated reporting does not affect on firm value.

Another factor that can affect the value of the company is the size of the company. The size of the company can be measured and known through several measures. When the company gets bigger, the disclosure of information will usually be more significant and broader, and this is a signal given by the company so that it can affect the value of the company. This explanation is by research from [11–13], which proves that firm value is positively influenced by firm size. However, this is not in line with [14–16], which provide evidence that firm value is not influenced by firm size.

In addition to the size of the company, leverage can also affect the value of the company. Leverage is the company's ability to pay off its debts. The number of loans used by the company will later be charged when carrying out industrial operations. This ratio aims to manage how effective the company's investment structure is so that it can affect the value of the company. This is in line with [17–19], which state that firm value is positively influenced by leverage. However, this is not the same as the research findings from [20], which states that firm value is not affected by leverage.

Based on the phenomena described and the variations in research findings, this research is still interesting to be reviewed. The originality of this research is to use the object of manufacturing companies from two Asian countries, namely Indonesia and Singapore because based on the phenomenon of manufacturing companies in Indonesia and Singapore, the purchasing managers' index has decreased and is in a critical position.

## 2 Literature Review and Hypotheses Development

### 2.1 Signal Theory

[1] states that good companies can differentiate themselves from bad by sending credible signals about their quality to the capital market. [21] shows that companies with high debt can signal that the company is more optimistic and of good quality compared to companies that have low debt. In addition, signal theory explains that company insiders generally know more information about to the company's prospects than external parties. Signal theory is fundamentally concerned with reducing

information asymmetry between two parties [22]. To reduce information asymmetry, managers (insiders) are advised to provide the information needed by investors or potential investors [23]. Companies that offer better information can influence investors' economic decisions and attract them to contracts with more significant benefits than other companies that offer worse information [24].

## 2.2 The Effect of Digital Integrated Reporting Disclosure on Firm Value

A good company can differentiate itself from a bad company by sending a credible signal about its quality to the capital market [1]. This is in line with [22] which states that fundamentally signal theory reduces information asymmetry between two parties. For this reason, managers (insiders) are advised to provide the information needed by investors or potential investors [23]. The credible signal given by the company can be in the form of integrated reporting disclosure because it can provide added value for the company so that this will affect the value of the company in the eyes of investors or potential investors. This means that the more integrated reporting disclosures submitted by the company to investors or potential investors, the better the company's value in the eyes of investors or potential investors. This is supported by [2, 5–9], which state that there is a strong positive impact of integrated reporting on firm value. The first hypothesis can be formulated as follows:

**H1.** Firm value is positively influenced by digital integrated reporting disclosure.

## 2.3 Effect of Firm Size on Firm Value

The company's size can be measured and known through several measures so that companies that send quality credible signals to the capital market can distinguish themselves from bad companies [1]. When the company gets bigger, the disclosure of information will usually be more significant and broader. This is a signal given by the company so that it can affect the value of the company. This explanation is by research from [11–13], which proves that firm value is positively influenced by firm size. The second hypothesis can be formulated as follows:

**H2.** Firm value is positively influenced by firm size.

## 2.4 The Effect of Leverage on Firm Value

Leverage is the company's ability to pay off its debts. The number of loans used by the company will later be charged when carrying out industrial operations. This ratio aims to manage how effective the company's investment structure is. [21] explains that companies with high debt can signal that the company is more optimistic and of good quality compared to companies that have low debt. This means that the more debt the company has, the higher the value of the company, and the lower the debt owed by the company, the lower the company's value. Research findings [17–19] are in line with the previous explanation, which states that firm value is positively influenced by leverage. The third hypothesis can be formulated as follows:

**H3.** The firm value is positively influenced by leverage.

## 3   Research Method

This research is a quantitative research, secondary data in the form of information in integrated reporting is used as a source of information. Manufacturing companies listed on IDX and SGX are used as research objects. Based on the specified sample criteria, 30 companies were obtained as samples from Indonesia and Singapore. The sample criteria include manufacturing companies listed on IDX and SGX, publishing integrated reporting, having complete data according to the observation period from 2016–2020. Samples were obtained from the consumer goods industry sub-sector, basic chemical industry, and various industries. The observation period is for five years, starting from 2016 to 2020, so that a total of 300 panel data are obtained. Secondary data in the form of financial reports, annual reports and sustainability reports are used as sources of data information.

Measurement of the firm value variable in the proxy using Tobin's Q [25]:

$$\text{Tobin's } Q = (MVS + D)/TA \tag{1}$$

Where:

– MVS = market value of all outstanding shares, i.e., company share price * extraordinary shares.
– TA = Company assets, such as; cash, accounts receivable, inventory, and book value of land.
– D = debt

The measurement of integrated reporting disclosure uses the integrated reporting index, which is calculated by dividing the total items disclosed by the total disclosure items (8 items) [2]. Meanwhile, company size is proxied by the natural logarithm of total assets [26, 27]. Leverage is measured by dividing total debt by total assets [28–30]. The analytical tool used to test the hypothesis is Partial Least Squares (PLS) - Structural Equation Modeling (SEM) using the WarpPLS version 5.0 program. Variant-based structural equation analysis (SEM) which can simultaneously test the measurement model as well as test the structural model. The measurement model is used to test the validity and reliability, while the structural model is used to test causality (testing hypotheses with predictive models). The advantages of PLS include: PLS is able to model many dependent and independent variables (complex models), the results remain solid even though there are abnormal data, can be used on reflective and formative constructs, can be used on small samples, does not require data to be distributed evenly. normal, can be used on data with different types of scales, namely: nominal, ordinal, and continuous.

## 4   Results and Discussion

See Fig. 1 and Table 1.

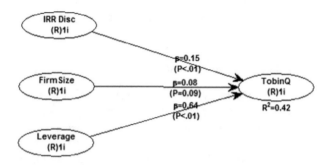

**Fig. 1.** Path analysis charts

**Table 1.**   Research result

| Variable | Index | p-value | Criteria | Annotation |
|---|---|---|---|---|
| APC | 0.286 | P < 0.001 | P < 0.05 | Accepted |
| ARS | 0.422 | P < 0.001 | P < 0.05 | Accepted |
| AVIF | 1.076 | | ≤ 3.3 | Accepted |

### 4.1   The Effect of Digital Integrated Reporting Disclosure on Firm Value

The estimation results of the study show that the disclosure of digital integrated reporting elements has a positive effect on firm value. The study results imply that the increased disclosure of the company's digital integrated reporting elements will provide a positive perception for investors so that the market will rate the company higher. The research results align with signaling theory which states that good companies can differentiate themselves from bad companies by sending credible signals about their quality to the capital market [1]. These results also support previous empirical studies which concluded that higher disclosure of integrated reporting could positively affect firm value [2, 5–9]. But it is not in line with research from [4, 10], which shows that integrated reporting does not affect on firm value.

### 4.2   Firm Size Effect on Firm Value

The estimation results of the study indicate that firm size has a positive effect on firm value with a significance level of 0.1. The study results imply that the larger the company's size will provide a positive perception for investors or potential investors because when the company gets bigger, the disclosure of information will usually be more significant and broader, so the market will value the company higher. The research results align with signaling theory which states that good companies can differentiate themselves from bad companies by sending credible signals about their quality to the capital market [1]. These results also support previous empirical studies which concluded that firm size could positively affect firm value [11–13]. However, the findings of this study are not the same as those of [14–16], which provides evidence that firm value is not influenced by firm size.

### 4.3   The Effect of Leverage on Firm Value

The estimation results of the study show that leverage has a positive effect on firm value. The study results imply that the greater the leverage will give a positive perception for investors or potential investors because companies that have high debt can be a signal that the company is more optimistic and of good quality compared to companies that have low debt [21]. So the market will value the company higher. The research results align with signaling theory which states that good companies can differentiate themselves from bad companies by sending credible signals about their quality to the capital market [1]. These results also support previous empirical studies which concluded that leverage could positively affect firm value [17–19]. However, the findings of this study are not the same as the findings of [20], which provides evidence that firm value is not influenced by leverage.

## 5   Conclusion and Recommendation

The conclusion from the findings of this study is that firm value is significantly affected by the disclosure of integrated reporting, firm size, and leverage. The limitation of this study is the R square value of 42%, so suggestions for future research are to modify the model by adding mediating or moderating variables such as corporate governance to find varied results. The implication of this research is the importance of companies implementing integrated reporting disclosures to increase firm value so can achieve that company sustainability.

## References

1. Spence, M.: Job market signaling. Q. J. Econ. **87**(3), 355–374 (1973)
2. Lee, K.-W., Yeo, G.-H.: The association between integrated reporting and firm valuation. Rev. Quant. Financ. Acc. **47**(4), 1221–1250 (2015). https://doi.org/10.1007/s11156-015-0536-y
3. Lodhia, S.: Exploring the transition to integrated reporting through a practice lens: an Australian customer owned bank perspective. J. Bus. Ethics **129**, 585–598 (2015). https://doi.org/10.1007/s10551-014-2194-8
4. Churet, C.R., Sam, Eccles, R.G.: Integrated reporting, quality of management, and financial performance. J. Appl. Corpor. Finan. **26**(1), 8–16 (2014)
5. El Deeb, M.S.: The impact of integrated reporting on firm value and performance: evidence from Egypt. Alexandria J. Account. Res. **3**(2), 1–50 (2019)

6. Cosma, S., Soana, M.G., Venturelli, A.: Does the market reward integrated report quality? Afr. J. Bus. Manag. **12**(4), 78–91 (2018)
7. Barth, M.E., Cahan, S.F., Chen, L., Venter, E.R.: The economic consequences associated with integrated reporting quality: capital market and real effects. Accounting, organizations and society, Working paper, vol. 3546, pp. 1–52 (2017)
8. Martinez, C.: Effects of integrated reporting on the firm's value: evidence from voluntary adopters of the IIRC's framework. SSRN Electron. J. (2016). https://ssrn.com/abstract=2876145. https://doi.org/10.2139/ssrn.2876145
9. Mervelskemper, L., Streit, D.: Enhancing market valuation of ESG performance: is integrated reporting keeping its promise? Bus. Strateg. Environ. **26**(4), 536–549 (2016)
10. Suttipun, M.: The effect of integrated reporting on corporate financial performance: evidence from Thailand. Corp. Ownersh. Control. **15**(1), 133–142 (2017)
11. Nursetya, R.P., Hidayati, L.N.: How does firm size and capital structure affect firm value? J. Manag. Entrepr. Res. **01**(2), 67–76 (2020)
12. Hardinis, M.: Capital structure and firm size on firm value moderated by profitability. Int. J. Econ. Bus. Adm. **VII**(1), 174–191 (2019)
13. Berger, A., Petti, E.B.: Capital structure and firm performance: a new approach to testing packing order theory and an application to banking industry. J. Bank. Finan. **30**(4), 1065–1102 (2006)
14. Oktaviani, R.F.: Effect of firm growth and firm size on company value with earning per share as moderation. Econ. Account. J. **3**(3), 219–227 (2020)
15. Setiadharma, S., Mechali, M.: The effect of asset structure and firm size on firm value with capital structure as intervening variable. J. Bus. Finan. Affairs **6**(4), 1–5 (2017)
16. Mule, R.K., Mukras, M.S., Nzioka, O.M.: Corporate size, profitability and market value: an econometrics panel analysis of listed firm in Kenya. Eur. Sci. J. **11**(13), 376–396 (2015)
17. Jihadi, M., Vilantika, E., Hashemi, S.M., Arifin, Z., Bachtiar, Y., Sholichah, F.: The effect of liquidity, leverage and profitability on firm value: empirical evidence from Indonesia. J. Asian Finan. Econ. Bus. **8**(3), 0423–0431 (2021)
18. Febriyanto, F.C.: The effect of leverage, sales growth and liquidity to the firm value of real estate and property sector in Indonesia stock exchange. Econ. Account. J. **1**(3), 198–2005 (2018)
19. Cheng, M.C., Tzeng, Z.C.: The effect of leverage on firm value and how the firm financial quality influence on this effect. World J. Manag. **3**(2), 30–53 (2011)
20. Tahu, G.P., Susilo, D.D.B.: Effect of liquidity, leverage and profitability to firm value (dividend policy as moderating variable) in manufacturing company of Indonesia stock exchange. Res. J. Finan. Account. **8**(18), 89–98 (2017)
21. Ross, S.A.: The determination of financial structure: the incentive signalling approach. Bell J. Econ. **8**(1), 23–40 (1977)
22. Spence, M.: Signaling in retrospect and the informational structure of markets. Am. Econ. Rev. **92**, 434–459 (2002)
23. Dainelli, F., Bini, L., Giunta, F.: Signaling strategies in annual reports: evidence from the disclosure of performance indicators. Adv. Account. Incorporating Adv. Int. Account. **29**, 267–277 (2013)
24. Grossman, S.J., Stiglitz, J.E.: Stockholder unanimity in making production and financial decisions. Q. J. Econ. **94**(3), 543–566 (1980)
25. Lindenberg, E.B., Ross, S.A.: Tobin's q ratio and industrial organization. J. Bus. **54**(1), 1–32 (1981)
26. Basuroy, S., Gleason, K.C., Kannan, Y.H.: CEO compensation, customer satisfaction, and firm value. Rev. Account. Finan. **13**(4), 326–352 (2014)

27. Feng, M., Wang, X.A., Saini, J.S.: Monetary compensation, workforce oriented corporate social responsibility, and firm performance. Am. J. Bus. **30**(3), 196–215 (2015)
28. Desoky, A.M., Mousa, G.A.: An empirical investigation of the influence of ownership concentration and identity on firm performance of Egyptian listed companies. J. Account. Emerg. Econ. **3**(2), 164–188 (2013)
29. Lahouel, B.B., Peretti, J.M., Autissier, D.: Stakeholder power and corporate social performance. Corpor. Gov. **14**(3), 363–381 (2014)
30. Zou, H., Zeng, S.X., Lin, H., Xie, X.M.: Top executives' compensation, industrial competition, and corporate environmental performance evidence from China. Manag. Decis. **53**(9), 2036–2059 (2015)

# Information Support for the Sustainable Development of Small and Medium-Sized Businesses in the Real Estate Development Sector

Svetlana Grishkina[1] , Vera Sidneva[1(✉)] , Yulia Shcherbinina[1] ,
Valentina Berezyuk[2] , and Khizir Pliev[1]

[1] Financial University under the Government of the Russian Federation,
Leningrad Prospect, 49, 125167 Moscow, Russia
[2] Karaganda Economic University of Kazpotrebsoyuz, Akademicheskaya 9 Str.,
100009 Karaganda, Kazakhstan

**Abstract.** This article aims to research the information support for the sustainable development of small and medium-sized businesses in the real estate development sector of the construction industry of Russia and substantiated the directions of its progress in the context of digitalization. In order to substantiate the proposals, features of sustainable development of the real estate development business were identified, the leading international initiatives on sustainable development of the construction industry were analyzed, the role of small and medium-sized businesses in the real estate development sector of the construction industry was shown. Also, the factors of sustainable development of SMEs and large businesses were compared, including the perspective of providing information about these factors to stakeholders. Recommendations for the transformation of accounting and analytical systems in small and medium-sized businesses in the real estate development sector, aimed at increasing the transparency of small and medium-sized businesses and information support for sustainable development by using the digital economy's advantages.

**Keywords:** Small and medium-sized enterprises (SMEs) · Sustainable development · Global Reporting Initiative (GRI) · Real estate development · Sustainable development reporting

## 1 Introduction

Despite the ups and downs experienced by the construction industry during various periods of time depending on the current state of the Russian economy, construction remains one of the most important industries. Thus, ensuring the achievement of the sustainable development goals of this industry is one of the most crucial tasks for the country, which has also been reflected in the actively elaborating strategy for the development of the construction industry of the Russian Federation until 2030 [1].

One of the major sectors of the construction industry is the sector of construction of residential and non-residential buildings – the real estate development sector. Over the

© The Author(s), under exclusive license to Springer Nature Switzerland AG 2022
T. Antipova (Ed.): DSIC 2021, LNNS 381, pp. 277–286, 2022.
https://doi.org/10.1007/978-3-030-93677-8_24

past five years, the share of the real estate development sector in the construction industry has remained almost unchanged and accounts for about 35% of the total volume of work in the industry [2]. The very concept of real estate development could be defined as "improvement, development, and construction of the real estate." In general, development is understood as an entrepreneurial activity associated with the creation, reconstruction, or change of real estate objects, including buildings or land plots, leading to an increase in their value. In a narrow sense, development is often understood as commercial construction [3].

On the one hand, in terms of disclosure of non-financial information on sustainable development targeted at a wide range of stakeholders, the real estate development sector is one of the most transparent and open in the construction industry of Russia. On the other hand, in the real estate development sector, sustainable development is traditionally understood only in the context of large companies. A low level of information transparency characterizes small and medium-sized businesses; thus, there is a need for elaborating the rules for the formation of financial information to assess the sustainable development of SMEs. It is also necessary to adapt non-financial reporting standards for small and medium-sized businesses since small and medium-sized enterprises do not disclose information about their activities' social and environmental aspects.

To assess the information disclosure of the sustainable development of small and medium-sized businesses in the real estate development sector, we used the data provided by the international sustainability reporting database (GRI's Sustainability Disclosure Database) [4], as well as data from sample surveys of small and medium-sized businesses.

In carrying out this research, a deductive approach has been applied, complemented by the use of scientific methods such as analysis, generalization, grouping, and comparison. In order to conduct empirical research, there also have been used methods of financial analysis and content analysis.

## 2    Sustainable Development of the Construction Industry

To analyze the role of the construction industry in the activity of the Russian economy, have been used the indicator of gross value added (GVA) since this indicator allows to assess the contribution of a particular industry to the economy more clearly than the indicator of gross domestic product (GDP). GVA differs from GDP by leveling the impact of taxes and taking into account subsidies on the final product of the industry (see Fig. 1).

Despite the decline in the period from 2016 to 2019, from 6.4% to 5.5%, respectively, in 2019, the share of GVA of the construction industry in the country's economy has stabilized. In 2020, despite the pandemic, there was a slight increase in the construction industry share in the GVA of Russia, up to 5.7%. In the period from 2015 to 2020, the volume of the added value of the construction industry has increased by 16% and reached 5.5 trillion rubles.

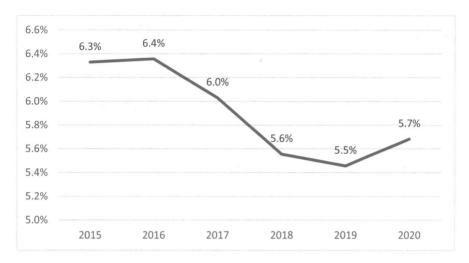

**Fig. 1.** Share of construction industry in the GVA of Russia for 2015–2020. Source: compiled by the authors based on materials [5].

According to the traditional models of SD: the triune concept of SD and the concept of ESG-factors, there are four main groups of SD factors: economic, environmental, social, and corporate governance [6, 7].

Many standards and initiatives for SD management in the construction industry are currently in force in the world. Some of them have been developed by the state control bodies of the construction sector; others are the result of cooperation between commercial and non-commercial organizations, which are guided by the goal of standardization issues of accounting for SD factors in the construction industry.

The most common ones are presented in Table 1.

Recent years, there has been a tendency in Russia for more widespread application of international initiatives and standards for SD of the construction industry. So, in 2016, by order of the Federal Agency for Technical Regulation and Metrology dated November 21, 2016, N 1722-st-1725-st, standard GOST R 57274–2016 "Sustainable Development in Construction" was put into effect on the territory of the Russian Federation. This standard is identical to the European standards, such as the EN 15643 series "Sustainable development in construction" given in Table 1. This document regulates the basic principles for assessing indicators of SD in the construction sector [11–14]. Despite the advantages of this document, as in fact the first regulatory document in Russia dedicated to the problem of SD of the construction industry, it needs to be improved and adapted to the Russian economy.

However, with all the variety of standards and initiatives for the SD of the construction industry, including the disclosure of non-financial information by construction organizations, the issue of adapting SD practices and disclosing information on SD in small and medium-sized businesses has not been entirely covered. The main barrier to introducing procedures for the disclosure of the SD information in small and medium-

**Table 1.** International initiatives to mainstream sustainable development (SD) issues in the construction industry.

| Initiative | Year | Basic principles and content |
|---|---|---|
| ISO standards: 11 standards developed by ISO / TC 59 / SC 17 Sustainability in buildings and civil engineering works | 2014–2019 | Guides the application of sustainability principles in construction work and real estate development in particular. The standards demonstrate to various construction stakeholders how to incorporate these principles into their decision-making processes to maximize the contribution of construction activities to SD |
| GRESB | 2009–2020 | GRESB is an abbreviation for Global Real Estate Sustainability Benchmark For Commercial Real Estate Enterprises (CRE), GRESB is a tool for providing investors with information on their ESG data. The organization has become an industry organization with over 250 members who use this data to manage the risk of their real estate investments. GRESB is the leading ESG benchmark for real estate and infrastructure investments worldwide |
| European standards developed by the European Committee for Standardization: EN 15643-1:2011 EN 15643-2:2011 EN 15643-3:2012 EN 15643-4:2012 | 2011–2012 | Contains general principles and requirements, expressed in a series of standards, for evaluating construction projects in terms of environmental, social, and economic performance, considering technical characteristics and functionality. That helps to quantify the contribution of construction work to SD. Applies to all types of constructed buildings and relates to the assessment of the SD performance of new buildings throughout their entire life cycle |
| ASTM Standards for Sustainable Construction | 2015 | ASTM International is an international standards organization that develops and publishes a wide range of voluntary harmonized technical standards. Provides a model that includes sustainability measures for the entire construction project and its site, from design and construction to commissioning and during the exploitation |

Source: compiled by the authors based on materials [3, 8–10].

sized businesses is the complexity of the standards and the uncertainty of the concept of the information support for SD of the small and medium-sized enterprises.

## 3   Small and Medium-Sized Businesses in the Sustainable Development of the Real Estate Development Sector of the Construction Industry

Construction activity, and, in particular, development, is primarily a project activity. Furthermore, in many ways, the level of sustainable development of a company depends on the degree to which the construction project meets the requirements of sustainable development.

The development project implementation model consists of three stages: pre-development, development, and post-development (Fig. 2).

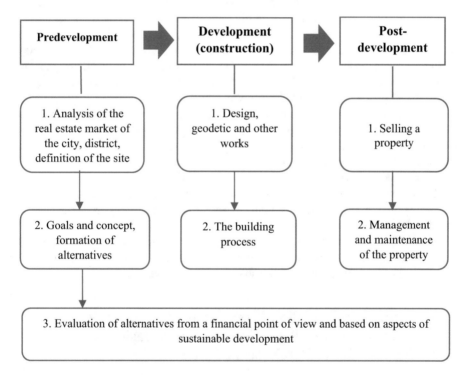

**Fig. 2.**  Stages of a sustainable real estate development project. Source: compiled by the authors based on [3].

New solutions for the most efficient organization of production processes in development organizations offer digital products and technologies, examples of which at various stages of a development project are presented in Table 2.

**Table 2.** Using digital technologies in the development process

| Stage | Proposed solution | Impact |
|---|---|---|
| Predevelopment and Development | BIM - Building Information Modeling System | Virtual modeling systems that allows to consider and predict various scenarios for the implementation of a development project |
| Post-development | CRM, digital mortgage and online sales, systems "Smart home" | Designed to automate the collection and analysis of information on the use of the finished real estate project |

Source: authors development.

The construction industry in most countries, including the Russian Federation, structurally consists of a small number of large companies and a large number of small and medium-sized enterprises [2]. At the same time, there is no uniform definition of small and medium-sized enterprises in different jurisdictions. In scientific research of Western scientists, the definition adopted in the European Union is usually used [15]. In the Russian Federation, this issue is regulated by the Federal Law of July 24, 2007, N 209-FZ (as amended on June 8, 2020) "On the development of small and medium-sized businesses in the Russian Federation." In general, these are companies with a staff of up to 250 people and a turnover of up to 2 billion rubles.

In 2020, up to 95% of all companies operating in the construction industry were SMEs, while their share in the total volume of work performed was more than 50%. It is evident that despite the dominance of large companies in the industry, to achieve sustainable development of the entire construction industry, it is necessary to involve small and medium-sized enterprises in the sustainability processes.

Let us consider the main existing differences between small and medium-sized businesses in the real estate development sector and large ones in the context of sustainable development factors (see Table 3).

Thus, it is evident that small and medium-sized companies, due to the complexity and cost of using sustainable development practices, lack of financial and labor resources, pay less attention to sustainable development factors.

## 4   Problems of Information Presentation of Sustainable Development of Small and Medium-Sized Businesses in the Real Estate Development Sector

As already noted, the construction industry, including the development sector in Russia, is based on small and medium-sized enterprises. Construction of residential and non-residential buildings is the most popular activity among small and medium-sized enterprises, with a share of 26% among the ten most popular activities listed in the Russian classifier of economic activities. In large cities, the real estate development sector is dominated by large businesses, while in small cities, the leading role in construction is played by small and medium-sized businesses [16].

**Table 3.** Key differences in the context of sustainable development between SMEs and big business in the construction industry

| Characteristics | Small and medium-sized businesses | Large businesses |
|---|---|---|
| Corporate governance | | |
| Organizational structure | One or more owners; usually managed by an owner | Several or more owners; usually running by hired managers |
| Long-term development | Usually do not have a thorough long-term strategy | There is a long-term strategy, reputation and experience are highly valued |
| Social responsibility | | |
| Relations with the state authorities | Insignificant for authorities | Can and usually must cooperate with authorities |
| Investments in human capital | Not interested in investing in human capital | Ready to invest in human capital development |
| Corporate culture | Unformalized | Formalized |
| Economic efficiency | | |
| The main goal of the business | Profit maximization | Value maximization |
| KPIs, including for investors | Financial and operational | Integrated, taking into account risk factors and long-term factors |
| Environmental sustainability | | |
| Enforcement of environmental standards | Usually neglected | Implement environmental responsibility and safety practices |

Source: authors development.

The analysis of small and medium-sized enterprises in Russia's real estate development sector showed that the subjects under study do not disclose information on sustainable development in their reports, which allows us to conclude that their management is not sufficiently interested in the factors of sustainable development.

This is due to the following reasons:

- shortage of highly qualified personnel who can understand and apply green technologies, as well as introduce new best practices in construction processes;
- closedness of small and medium-sized businesses, that is, the absence of any relevant and transparent reporting information, which makes it possible for stakeholders to assess the sustainability of small and medium-sized businesses;
- a high mortality rate of SMEs.

The solution of these problems is a prerequisite for the development of sustainable development practices in small and medium-sized construction organizations and ensuring the proper level of information presentation of sustainable development.

Features of small and medium-sized businesses predetermine the feasibility of adapting existing standards to form and disclose financial and non-financial information on sustainable development for SMEs. The analysis showed that the management, accounting, and analytical mechanisms of sustainable business development widely used in large companies are not practiced at small and medium-sized enterprises. In small and medium-sized businesses, accounting financial statements are formed in a simplified manner, which reduces the quality of accounting information for assessing the economic factors of sustainable development, and non-financial information on the sustainable development of enterprises is not generated and is not given to stakeholders.

The introduction of digital technologies in management processes by economic entities, including in the accounting and analytical sphere, opens up new opportunities for small and medium-sized enterprises to update their accounting and analytical systems. The use of cloud technologies and the digitalization of the tax system make it possible to abandon most of the simplifications in accounting and preparation of financial statements, which lead to the information "closeness" of small and medium-sized businesses.

For an informational presentation of sustainable development of small and medium-sized enterprises in the real estate development sector, non-financial information is needed; proposals for its formation and presentation have been formulated in the various studies of many economists, including in the authors' works [17–21]. However, small and medium-sized businesses need to adapt the existing non-financial reporting standards to the conditions of their activities, and, above all, to present clear and precise rules for the formation of such information, which, in our opinion, should be formalized in the form of an appropriate standard for SMEs.

## 5   Conclusion

The digital transformation of the construction industry, including the real estate development sector, contributes to implementing sustainable development practices in economic entities. Digitalization is based on the use of information and communication technologies to solve managerial and production problems and the accumulation and analysis of big data with their help for planning purposes, building an effective development strategy, and optimizing business processes.

The real estate development sector of the construction industry in Russia is based on small and medium-sized enterprises, and therefore their sustainable development is an indispensable condition for the sustainable development of the entire sector.

Sustainable development of small and medium-sized enterprises in the real estate development sector has its characteristics compared to large businesses. It is primarily ensured by the effectiveness of accounting and analytical systems designed to generate information for disclosing sustainable development factors and contribute to the development of an optimal development strategy for the development business.

To improve the efficiency of accounting and analytical systems in small and medium-sized businesses in the real estate development sector, necessary to update accounting rules, including the rejection of some simplifications that negatively affect the quality of accounting information and are not advisable in conditions of computer processing of information. In addition, it is necessary to develop separate standards for the formation of financial and non-financial information, taking into account the specifics of these entities.

A practical accounting and analytical system will also be helpful at all stages of implementing a real estate development project, from the design and construction stage to the stage of real estate exploitation.

Taking into account the factors of SD by small and medium-sized companies of the real estate development sector allows: - to create a favorable investment image of the company; - to satisfy the demand of the society for greater transparency of business; - to create the basis for the long-term development of the company, by improving stability and competitiveness of the organization.

# References

1. The strategy for the development of the construction industry of the Russian Federation until 2030 (Draft, 2019/12/06). http://stroystrategy.ru/docs/%D0%9F%D0%A0%D0%9E%D0% 95%D0%9A%D0%A2%20%D1%80%D0%B0%D1%81%D0%BF%D0%BE%D1%80% D1%8F%D0%B6%D0%B5%D0%BD%D0%B8%D1%8F%20%D0%A1%D0%A2%D0% A0%D0%90%D0%A2%D0%95%D0%93%D0%98%D0%98-2030+.pdf. Accessed 17 July 2021
2. Construction in Russia, Statistical Book Rosstat (2020). https://rosstat.gov.ru/storage/ mediabank/tASKTSkO/Stroitelstvo_2020.pdf    https://www.rbc.ru/economics/05/02/2019/ 5c5948c59a794758389cfdf7. Accessed 17 July 2021
3. GRESB Real estate appraisal. GRESB. Text: electronic. https://gresb.com. Accessed 17 July 2021
4. Sustainability disclosure Database (2020). https://database.globalreporting.org/search. Accessed 30 Apr 2021
5. Gross value added by industry, Federal State Statistics Service (2021). https://rosstat.gov.ru/ accounts. https://www.rbc.ru/economics/05/02/2019/5c5948c59a794758389cfdf7. Accessed 18 July 2021
6. Grishkina, S.N.: Information support for sustainable development of the agricultural sector of the economy. In: Theory, Methodology, Practice: Monograph. 224 p. Ruscience, Moscow (2018)
7. Barilenko, V.I., Efimova, O.V., Nikiforova, E.V., et al.: Information and Analytical Support of Sustainable Development of Economic Entities: Monograph, 160 p. Ruscience (2015)
8. ASTM. Standards for Sustainable Building/ASTM (2015). Text: electronic. https://www. astm.org/GLOBAL/MDCP_sector_overview.pdf. Accessed 07 July 2021
9. CEN - European Committee for Standardization. https://standards.iteh.ai/catalog/standards/ cen. Accessed 07 July 2021
10. ISO TC59 SC 17 Sustainability of buildings and construction works. - Text: electronic. https://www.iso.org/committee/322621/x/catalog/p/1/u/0/w/0/d/0. Accessed 07 July 2021
11. Order of the Federal Agency for Technical Regulation and Metrology of November 21, 2016, No. 1722-st (revised on November 21, 2016) "On approval of the national standard of the Russian Federation" "Sustainable development in construction. Part 1. General provisions"//SPS "Garant"
12. Order of the Federal Agency for Technical Regulation and Metrology of November 21, 2016 No. 1723-st (as amended on November 21, 2016) "On approval of the national standard of

the Russian Federation" "Sustainable development in construction. Part 2. Principles for assessing environmental performance"//SPS "Garant"

13. Order of the Federal Agency for Technical Regulation and Metrology of November 21, 2016 No. 1724-st (as amended on November 21, 2016) "On approval of the national standard of the Russian Federation" "Sustainable development in construction. Part 3. Principles for assessing social indicators."//SPS "Garant"

14. Order of the Federal Agency for Technical Regulation and Metrology dated November 21, 2016 No. 1725-st (as amended on November 21, 2016) "On approval of the national standard of the Russian Federation" "Sustainable development in construction. Part 4. Principles for evaluating economic indicators." //SPS "Garant"

15. Annual report on European SMEs 2020/2021. The SME performance review, the European Commission (2021). https://ec.europa.eu/growth/smes/sme-strategy/performance-review_en. Accessed 21 July 2021

16. The minor victory of the construction over trade. RBK. 145 (2021). https://www.rbc.ru/newspaper/2021/09/30/615324889a7947b07d9887ec. Accessed 21 July 2021

17. Grishkina, S.N., Sidneva, V.P.: Modern approaches to the preparation and presentation of reporting in the field of sustainable business development. Econ. Bus. Banks 4(42), 86–96 (2020)

18. Grishkina, S.N., Sidneva, V.P., Dubinina, G.A.: Non-financial reporting in the system of increasing corporate transparency of Russian business. Adv. Soc. Sci. Educ. Human. Res. 252, 82–86 (2018)

19. Hassan, A., Elater, A.A., Lodh, S., Roberts, L., Nandy, M.: The future of non-financial businesses reporting: learning from the Covid-19 pandemic. Corp. Soc. Responsib. Environ. Manag. 28(4), 1231–1240 (2021)

20. Efimova, O.V.: Formation of reporting on sustainable development: stages and procedures of preparation. Account. Anal. Audit. 5(3), 40–53 (2018)

21. Adams C.A.: The sustainable development goals, integrated thinking and the integrated report (2017). http://integratedreporting.org/wp-content/uploads/2017/09/SDGs-and-the-integrated-report_full17.pdf. Accessed 15 Mar 2020

# Cryptocurrencies as a Financial Asset: An Evidence from an Institutional Investors Perspective

Miloš Grujić$^{(\boxtimes)}$ 

Independent University of Banja Luka, Banja Luka 78000, Republic of Srpska

**Abstract.** The study provides a systematic analysis and review of the relevant empirical literature based on the fundamental topics that have been associated with the cryptocurrencies market. The research was carried out by using the method of description, literature analysis, and research conducted. The study investigates the empirical verification of the efficacy of investment diversification using the main stock exchange indices in the Eurozone countries and Bitcoin. The aim of the research is to examine whether it is justified and to what extent to include Bitcoin in the portfolio of an institutional investor. The analysis includes data on the daily movement of selected action indices as well as the movement of Bitcoin. The methodology involves the analysis of high-frequency data, given that daily trading data were used. The results show that it is justified to include Bitcoin or any other cryptocurrency in the portfolio structure. Also, the results show which share of Bitcoin in the portfolio is justified from the aspect of institutional investors. The data used in the analysis cover the period from 2020. The results of the research show that Bitcoin is a good source of diversification in a portfolio that contains traditional financial instruments, both for an investor who is not prone to risk, and for those investors who have a greater appetite for risk. The conclusion is that the rational behavior of institutional investors requires consideration of investing in Bitcoin using the Markowitz model.

**Keywords:** Blockchain · Financial markets · Cryptocurrencies · Digital Currencies · Bitcoin · Portfolio optimization · Institutional investors

## 1 Introduction

Cryptocurrencies are digital financial assets, for which ownership and transfers of ownership are guaranteed by a cryptographic decentralized technology [1]. Therefore, cryptocurrency is a currency and hence it is an asset. Cryptocurrencies exist only in digital form. Cryptocurrencies can be transferred completely only between digital addresses. Cryptocurrencies like Bitcoin have matured from being associated exclusively with computer technicians to being considered by central banks as a technology to implement digital money. Cryptocurrencies, as an asset, exist only in digital form but cryptocurrencies can be transferred between digital addresses.

Cryptocurrencies are not covered by some assets as a background, nor can they be used to pay taxes. They have value in the exchange among the users of these

T. Antipova (Ed.): DSIC 2021, LNNS 381, pp. 287–299, 2022.
https://doi.org/10.1007/978-3-030-93677-8_25

currencies. Accordingly, they represent private and decree (proclaimed or fiat) money. Given the difficult prediction of cryptocurrency price movements, their instability is a guarantee that rational market participants will not easily engage in international cryptocurrency transactions.

The emergence of cryptocurrencies was first treated by the general public as one of the technical innovations, with the sign "risky", "unreliable" and "short-lived". In this regard, plastic cards have gone through a similar path, and today they are generally accepted and regulated. Thanks to the flexibility and customization of users, cryptocurrencies are becoming widespread and accepted as payment methods, which allows liquidity and flexibility in payment.

The aim of this research is to examine and determine whether, and to what extent, it is advisable to include a certain share of Bitcoin in a portfolio containing stock indices. It is assumed that the cryptocurrency Bitcoin is one of the sources for portfolio diversification [2] For the purposes of the study, the methodology or technology of modern portfolio optimization was used, which shows the optimal share of traditional financial instruments and Bitcoin in the portfolio. In order to optimize the portfolio, Markowitz's portfolio theory was used in the research.

The problem of the research is the definition and creation of guidelines and the adoption of a framework for making investment decisions of the participants, ie. institutional investors in the international financial market. The subject of the research is the daily movements of stock indices and Bitcoin. The objectives of the research are to show that inclusion in the portfolio of risk instruments can reduce the risk of the overall portfolio but also increase the transfer. The study investigates the scope and limitations of such an approach. The motives and mission of the research are to try to explain the reasons and consequences of combining Bitcoin and traditional financial instruments in the portfolio and to point out the application of this knowledge in theory and practice. The basic hypothesis of the research is: "Observing at data on a daily basis, if we look at a portfolio that includes traditional financial instruments, the cryptocurrency Bitcoin is a desirable source of portfolio diversification, ie it can reduce portfolio risk and increase yield."

The study is made up of several parts. The first part is an introduction. The second part describes previous research and the most important conclusions on this topic. The third part describes the methodology and assumptions used. The fourth part presents the data and descriptive statistics used in the research. After that, the most important research results are shown. Finally, the conclusion of the research is given in the recommendations for further research.

## 2    Literature Review

### 2.1    Cryptocurrencies

Over the years, financial markets have undergone tremendous changes. One of them is the reduction of barriers and access to foreign portfolio flows from one market to another [3], but also the emergence of new forms of financial assets such as cryptocurrencies.

Unlike other currencies, cryptocurrencies are not supported by the rule of law, but by technology. The transfer is simple, there are no intermediaries and the third party cannot prevent or correct the transactions. Assuming that all legal systems collapse, cryptocurrencies would continue to exist with the existence of the Internet and people willing to use them. Other important characteristics of cryptocurrencies are ownership provided by strong cryptography, transactions are visible but users are anonymous, the person who sends cryptocurrencies, unlike the recipient, must be connected to the Internet and except for supply and demand nothing guarantees its value.

Today, cryptocurrencies have entered the flow of real economies, and economists, lawyers, IT experts, hackers, central banks and security services deal with their phenomenon. One of the reasons that, at first, attracted supporters of technology, the Internet and investors is the fact that cryptocurrencies are not subject to control by central banks or government agencies but their value is determined by a multitude of computers. Namely, they are protected from inflation by a mathematical function that prevents their amount from rising above a predetermined limit. Another feature of cryptocurrencies, recognized as an advantage, is reflected in the elimination of intermediaries, which makes transactions cheaper - especially in international payments. As advantages of cryptocurrencies, it is pointed out that they are based on a decentralized system without the existence of regulatory authority, while decentralization, user anonymity and the absence of a regulatory agency prove to be the main disadvantages [4].

## 2.2   Blockchain

Blockchain is the decentralized managing technique, designed, inter alia, for issuing and transferring money for the users of Bitcoin. Described technique can support the public ledger of all Bitcoin transactions that have ever been executed - without any control of a third-party organization [5].

The system described so far is a decentralized database that cannot threaten or change the attack of a large number of users at the same time [5]. Thus, the security of information stored in the system is greater than in a centralized database. Therefore, with certain corrections, the described blockchain system can be used to store and process different types of data. Consequently, this technology can also reform the financial markets, but it can be used to verify the authenticity of different types of documents: vehicle registration, health card certification, voting in elections, record keeping, or copyright [6]. There are numerous opportunities or potential use cases of blockchain that have been identified in the literature: asset exchange system [5, 7], banking industry [17], blockchain or consensus-as-a-service [10], clearing & settlement [15], cryptocurrency [4], identity management [12], insurance industry [16], internet of things [14], private securities trading [11], remittance system [7], securities registry (proxy voting) [13], smart contracts [8, 9], voting system [18] etc. A certain transaction, together with all other changes, is recorded on all computers and servers, which means that manipulations are almost excluded. Given the gained popularity as cryptocurrencies and accompanying blockchain technologies, numerous studies examining the speculative nature of Bitcoin [7–10], Yermack [7] and Kristoufek [8] after examining the nature of Bitcoin concluded that it cannot be explained with the help of modern economic theories because it shows more speculative features compared to other financial instruments and currencies. Other authors try to explain whether this cryptocurrency is a currency or a good commodity [11–14]. In addition, Dyhrberg [10, 11] and Šoja & Chamil [1] point out that Bitcoin has certain characteristics of other asset classes or financial instruments such as currencies and gold. Therefore, it enables portfolio diversification. There are also conflicting findings. Baur et al. [15] demonstrated that the characteristics of Bitcoin are different from that of other assets such as gold and the US dollar.

Other research suggests that the cryptocurrency Bitcoin may be a desirable diversification tool [2]. This claim has been tested precisely in the portfolio that consists of traditional financial instruments such as gold, bonds, and stocks, both for conservative investors and for those investors who prefer risky investments [16, 17].

The main motive for attracting institutional investors to invest in Bitcoin is the diversification of placements and the ease of investing capital [3]. Unlike other financial instruments, Bitcoin does not have a basis that would guarantee its value and sustainability. Therefore, institutional investors approach investing in this instrument with special caution. The importance of institutional investors is manifested through the improvement of the efficiency of the financial system. Considering the weak or almost no connection of Bitcoin with other assets, this financial instrument is recognized as an instrument for hedging or "hedging" in the portfolio [2]. Using monthly return data, Chan et al. [19] prove that Bitcoin can be used by the administration as hedging against the well-known stock indexes Nikkei, S&P 500, Euro STOXX, Shanghai A-Share, and TSX Index.

Ram [20] in an extensive analysis using Sharp's ratio showed that Bitcoin provides risk-weighted returns higher compared to other financial instruments or asset classes. Trautman and Dorman [21] conducted a correlation analysis between different financial instruments such as stocks, real estate, gold, and bonds. They proved that the benefits of applying diversification to these financial instruments are. In this regard, there are various studies that examine the risks associated with investing in Bitcoin. For example, Eisl et al. [22] in their analysis start from conditional VaR with the aim of examining whether investing in Bitcoin can really optimize the portfolio. Their findings suggest that, despite the fact that Bitcoin increases VaR, the additional risk is offset by the high return that Bitcoin rejects.

Briere et al. [15] using data on weekly changes during the period from 2010 to 2013 analyzed the investment in Bitcoin from the position of an institutional investor from the United States investing in traditional financial instruments (stocks, bonds, and major currencies) and alternative investments (goods and real estate) They concluded that, during the observed period, investing in Bitcoin is characterized by an extremely high level of return but also risk. In addition, they note that investing in Bitcoin provides benefits from inclusion in such a portfolio. Carpenter [23] insisting on variance and at the expected value showed that Bitcoin can be a desirable instrument for diversification. The inclusion of Bitcoin in the portfolio of traditional financial instruments is in line with the tests of diversification by investing in traditional financial instruments. Empirical findings of our study suggest that throughout the period, investors will benefit from diversification by combining investments in companies from developed and new economies [4]. Also, previous research shows that this strategy gives excellent results when applied to instruments with similar characteristics and levels of risk, such as government bonds [5].

## 3   Methodology

In order to explore the optimal portfolio diverzification using Bitcoin for a globalinvestment portfolio, a few stocke Exchange indexes were chosen. Portfolio optimization involves several steps. The first step is the selection of instruments to be included in the analysis, the second step is the calculation of returns, the next is the calculation and expression of covariance among financial instruments and the last step is the construction of an optimal portfolio.According to this theory, a higher return does not necessarily mean a higher risk. According to this theory, a higher return does not necessarily mean a higher risk. The assumptions of portfolio theory that Markowitz implies are [24]:

- the investor takes into account any alternative for investments that is represented by the distribution of the probability of expected returns in the observed period;
- the investor assesses portfolio risk based on the variability of expected portfolio returns;
- investor decisions are based only on the expected return and risk, so that their indifference curves are a function of the expected return and the expected variance (or standard deviation) of the return;
- investors maximize the expected benefit and their indifference curves demonstrate a diminishing marginal utility of wealth;
- for a given level of risk, investors prefer higher returns than lower ones and vice versa. For a given level of expected returns, investors prefer less risk than the higher one.

In this study, the indices are considered as separate securities. In practice, the closest to this are ex-change traded funds that replicate index movements.

Volatile financial instruments have a high standard deviation. On the other hand, the standard deviation for relatively stable financial instruments is low to low. In this regard, the standard deviation expresses uncertainty as a risk, even in the case where there are above-average returns that are desirable for the investor. Value-at-risk (VaR) is a methodology that institutional investors use to express, ie. calculation of potential losses on a particular investment. The same methodology makes it possible to express the probability of such losses occurring. The use of the VaR model shows the worst-case scenario that an individual investor could encounter when investing in a specific asset. The VaR estimate takes into account the observed period, the confidence level, and the sum of all losses expressed as a percentage.

$$VaR_{x,a} = -z_a \sigma S \qquad (1)$$

In this formula for makes a quantile order and $\alpha$ is a standardized random variable while S is the value of the position. In order to calculate the VaR of the portfolio using the Makrowitz methodology, it is necessary to express the covariance and correlation between the observed financial instruments, the expected return and risk of each individual financial instrument included in the portfolio

$$E_{r(p)} = \sum_{i=1}^{n} W_1 E_{i(r)} = W_1 E_{1(r)} + W_2 E_{2(r)} + \ldots + W_N E_{N(r)} \tag{2}$$

In this case, Er (p) represents the expected return r of the portfolio, p and W represent the weight or percentage share or share of the financial instrument in the portfolio p. In order to gain insight into how the values of financial instruments included in the portfolio move, it is necessary to calculate the correlation between them.

Regardless of the fact that there are indications that investing in Bitcoin really reduces the risk of the entire portfolio, it is desirable to consider VaR in the analysis. Namely, rational institutional investors will certainly examine and express the level of risk expressed through daily changes in the value of a particular financial instrument.

$$Covariance(A, B) = \frac{\sum(r_A - \bar{r}_A)\sum(r_B - \bar{r}_B)}{N} \tag{3}$$

Where $r_A$ represents the return of asset A, and from the investor's point of view, $r_B$ represents the return of asset B while N represents the battle of observations.

The correlation between assets A and B in the portfolio is expressed in the usual way.

$$Correlation(p) = \frac{cov(A, B)}{\sigma_A \sigma_B} \tag{4}$$

In this statement, Cov (A, B) shows the covariance between financial instruments A and B, while σA and σB represent the standard deviation of financial instruments A and B.

## 4  Data

The portfolio that is being structured includes stock exchange indices from the euro area countries, which represent an approximation of the stock markets of selected countries. The sampled countries are the euro area countries. They were sampled with the aim of ignoring currency risk and arbitrage in the calculations and with the aim of focusing the research on the effects of diversification. In the analysis, the indices were viewed as separate securities. In practice, the closest to this are exchange traded funds that replicate index movements. Namely, in developed markets there are numerous "exchange traded funds" that replicate investment in a particular index or sector. To determine the risk of individual positions in which investments are made, a standard deviation, ie variance, is needed, because it measures how far individual amounts, in this case returns, are from the average. When determining the optimal portfolio that includes the cryptocurrency Bitcoin, the study observes daily changes in stock exchange indices from the position of an institutional investor investing in Eurozone countries.

Initially, daily returns are reported for each financial instrument that will be considered for inclusion in the optimization process. After that, using the modern theory of portfolio optimization, an efficient portfolio was calculated, ie. "Optimal portfolio", which is characterized by the fact that it rejects the least risk - expressed by standard deviation. Certain limits are set when calculating the optimal portfolio. First, the share of each instrument can be 0% and go up to 100%. In other words, during the simulation, an investment in a stock exchange index can be completely rejected, but the entire amount can only be invested in one. Further, the goal of optimization is to find a portfolio with minimal risk, minimal standard deviation that includes Bitcoin, and a portfolio that does not include Bitcoin. Third, no brief passes are provided. In the end, the costs of transactions and portfolio adjustments are ignored or neglected.

Descriptive statistics for the analyzed period include data on average return, standard deviation, frequency of negative return and VaR for the analyzed period. Daily returns for the analyzed instruments show that the return for traditional instruments is generally modest, while this data is the highest for Bitcoin and amounts to 0.60% per day. However, the standard deviation, as a measure for expressing risk, is the largest in Bitcoin, which can be considered a confirmation that this instrument is really the most risky (Table 1).

**Table 1.** Descriptive statistics for the analyzed period include

| Index | Mean | StDev | Freq < 0 | Parametric VaR | Empirical VaR | Parametric CVaR | Empirical CVaR |
|---|---|---|---|---|---|---|---|
| Austria | −0.03% | 2.33% | 51.78% | −3.85% | −3.51% | −4.82% | −6.01% |
| Belgium | −0.01% | 2.12% | 47.43% | −3.50% | −3.80% | −4.38% | −5.61% |
| Cyprus | −0.05% | 1.24% | 51.78% | −2.09% | −1.62% | −2.60% | −3.35% |
| Finland | 0.06% | 1.77% | 48.62% | −2.86% | −3.03% | −3.60% | −4.58% |
| France | −0.01% | 2.05% | 48.62% | −3.38% | −3.52% | −4.24% | −5.32% |
| Germany | 0.03% | 2.08% | 50.20% | −3.39% | −3.69% | −4.26% | −5.19% |
| Greece | 0.00% | 2.61% | 47.43% | −4.30% | −4.00% | −5.39% | −7.14% |
| Ireland | 0.03% | 1.95% | 46.64% | −3.18% | −2.88% | −4.00% | −5.18% |
| Italy | 0.00% | 2.21% | 45.85% | −3.63% | −3.34% | −4.55% | −5.76% |
| Netherlands | 0.03% | 1.79% | 46.64% | −2.92% | −3.06% | −3.67% | −4.54% |
| Portugal | −0.01% | 1.68% | 48.22% | −2.78% | −2.27% | −3.48% | −4.22% |
| Slovakia | −0.41% | 6.42% | 34.78% | −10.96% | −1.99% | −13.64% | −10.56% |
| Spain | −0.04% | 2.14% | 49.01% | −3.56% | −3.28% | −4.45% | −5.18% |
| Slovenia | 0.00% | 1.33% | 42.29% | −2.19% | −1.87% | −2.75% | −3.61% |
| Bitcoin | 0.60% | 2.84% | 39.53% | −4.08% | −3.51% | −5.27% | −5.72% |

If the risk of a particular financial instrument varies widely during the observed period, as is the case during the "pandemic" 2020, the standard deviation offers less reliable estimates of the actual risk. Therefore, it is useful to apply the VaR method for risk assessment. In this way, it is expressed how much an institutional investor can expect to lose by investing in a financial instrument, under the same market conditions as they were in the observed period. For example, banks and regulators use the VaR measure with the intention of stating which part of the portfolio must be covered by a possible loss at a certain level of risk.

In this study, parametric and empirical VaR and CvaR are presented, with a confidence interval of 95%. The results show that the standard deviation as a measure of risk as well as the VaR for Bitcoin do not deviate significantly from other instruments. This means that Bitcoin had lower volatility during the observed period. The reasons for that are different, but some of the probable ones are the fact that many companies have started to accept Bitcoin as a payment instrument (Tesla company), while payment card providers have pointed out that they are ready to adapt their platforms to cryptocurrencies (Mastercard and Visa). Such circumstances have boosted confidence in Bitcoin and certainly diminished its daily volatility.

## 5   Results and Discussion

Modern portfolio theory leads to the conclusion that portfolio diversification must include those instruments that are not highly correlated with each other. Based on this condition, a correlation was expressed for the observed financial instruments (Fig. 1).

| | Austria | Belgium | Cyprus | Finland | France | Germany | Greece | Ireland | Italy | Netherlands | Portugal | Slovakia | Spain | Slovenia | Bitcoin |
|---|---|---|---|---|---|---|---|---|---|---|---|---|---|---|---|
| Austria | 1 | | | | | | | | | | | | | | |
| Belgium | -0.063 | 1 | | | | | | | | | | | | | |
| Cyprus | 0.140 | 0.049 | 1 | | | | | | | | | | | | |
| Finland | 0.706 | 0.044 | 0.084 | 1 | | | | | | | | | | | |
| France | 0.023 | 0.901 | 0.041 | 0.112 | 1 | | | | | | | | | | |
| Germany | 0.110 | 0.118 | -0.152 | 0.170 | 0.092 | 1 | | | | | | | | | |
| Greece | 0.211 | -0.115 | 0.121 | 0.129 | -0.202 | 0.327 | 1 | | | | | | | | |
| Ireland | 0.074 | 0.809 | 0.037 | 0.184 | 0.875 | 0.091 | -0.201 | 1 | | | | | | | |
| Italy | 0.247 | 0.168 | 0.023 | 0.255 | 0.216 | 0.216 | 0.092 | 0.270 | 1 | | | | | | |
| Netherlands | 0.053 | 0.853 | 0.059 | 0.149 | 0.933 | 0.084 | -0.211 | 0.836 | 0.196 | 1 | | | | | |
| Portugal | 0.062 | 0.789 | 0.052 | 0.140 | 0.867 | 0.101 | -0.185 | 0.789 | 0.211 | 0.844 | 1 | | | | |
| Slovakia | 0.021 | 0.023 | -0.002 | -0.036 | 0.048 | -0.007 | 0.003 | 0.055 | 0.006 | 0.024 | 0.038 | 1 | | | |
| Spain | -0.017 | 0.884 | 0.078 | 0.084 | 0.929 | 0.074 | -0.160 | 0.820 | 0.261 | 0.847 | 0.854 | 0.074 | 1 | | |
| Slovenia | 0.568 | 0.009 | 0.165 | 0.630 | 0.076 | 0.062 | 0.094 | 0.111 | 0.062 | 0.127 | 0.113 | 0.003 | 0.027 | 1 | |
| Bitcoin | -0.071 | 0.049 | 0.056 | -0.031 | 0.038 | -0.034 | 0.123 | 0.015 | -0.096 | 0.049 | 0.008 | -0.064 | 0.026 | 0.001 | 1 |

**Fig. 1.**  Matrix of correlation

The results show that there is a low correlation between the observed financial instruments. More precisely, it almost does not exist. This result shows that an optimally diversified portfolio can be created from the analyzed financial instruments. To examine the role and contribution of Bitcoin in the portfolio, two portfolios were analyzed. Both portfolios aim to minimize investor risk, ie create a portfolio that minimizes risk and offers an appropriate degree of return (Table 2).

**Table 2.** Optimal portfolio with investing in Bitcoin without investing in it

|  | Minimal risk, without Bitcoin | Minimal risk, with Bitcoin |
|---|---|---|
| Average return | 0,02% | −0,02% |
| Standard deviation | 0,75% | 0,78% |
| Slope | 2,60% | −2,80% |
| Cyprus | 33,74% | 35,67% |
| Germany | 12,10% | 6,37% |
| Greece | 2,62% | 6,28% |
| Italy | 6,22% | 6,70% |
| Netherlands | 2,17% | 5,69% |
| Portugal | 12,20% | 12,56% |
| Slovakia | 1,48% | 1,37% |

The results show that a portfolio that does not include Bitcoin carries a negative daily return of −0.022%, while a portfolio that includes Bitcoin carries a daily return of 0.019%. It is noticed that the standard deviation of the portfolio that includes Bitcoin is lower than the standard deviation of the portfolio that does not include this instrument, which suggests that from the aspect of risk it is justified to include Bitcoin in the structure of the portfolio. Specifically, the results suggest that the inclusion of Bitcoin in a share of about 6.7% increases the risk-return ratio, ie makes the portal more efficient.

The results show that, if Bitcoin is included in the portfolio, it is possible to achieve a higher return with a lower risk, if the risk is measured by a standard deviation. The results show that the Shape ratio is positive for a portfolio that includes Bitcoin, which is another measure that confirms the usefulness of including Bitcoin in the investment portfolio. Also, diversification and its application to the investment portfolio is originally a theoretical concept but has a wide application in practice. In addition, this concept is a widely accepted instrument for the academic community but also for practitioners and institutional investors. These results are in line with research conducted by Liu and Tsyvinski in 2018, which showed that it is justified to include about 6% of Bitcoin in the portfolio in order to achieve an optimal portfolio. According to this author, even those who are strongly skeptical about Bitcoin should keep at least 4% of Bitcoin in their portfolio. The results of the research he conducted in this study confirm the thesis presented by Tsyvinski [24]. Likewise, earlier studies [2] have shown that it is justified to include about 2% of Bitcoin's portfolio structure. In this case, moving all the data from 2020, it was shown that it is justified to increase that share. Graph No. 1 shows an efficient set for a portfolio without Bitcoin for the analyzed period, ie for 2020, daily returns.

It is obvious that an efficient set that includes Bitcoin brings a positive return, while an efficient set that represents the portfolio structure without Bitcoin generally brings negative daily returns. The results show that a portfolio that includes Bitcoin is a portfolio that provides a better risk-return ratio, which is acceptable from the point of view of a rational investor.

The optimal portfolio, ie the portfolio that carries minimal risk, except for Bitcoin, includes stock indices of Cyprus (33.74%), Germany (12.1%), Greece (2.6%), Italy (6.2%), the Netherlands (2.17%), Portugal (12.2%), Slovakia (1.47%) and Bitcoin (6.7%). On the other hand, a portfolio without Bitcoin includes the same instruments with an increase in their share in the overall portfolio structure.

Given that other similar cryptocurrencies have a high correlation (Table 3) and a similar Var as a Bitcoin. For example Litecoin has 7.27%, Ripple 7.80, Ethereum −6.41% and Dogecoin 5.59. According to that we can claim that we would get similar conclusions by combining the observed indices and other cryptocurrencies.

**Table 3.** Correlation between cryptocurrencies

|      | BTC      | LTC      | XRP      | ETH      | DOGE |
|------|----------|----------|----------|----------|------|
| BTC  | 1        |          |          |          |      |
| LTC  | 0.852603 | 1        |          |          |      |
| XRP  | 0.574493 | 0.654042 | 1        |          |      |
| ETH  | 0.856685 | 0.863682 | 0.665725 | 1        |      |
| DOGE | 0.575929 | 0.568864 | 0.601548 | 0.581072 | 1    |

# 6  Conclusion

The study considers the scope and limitations of the application of modern portfolio theory in the diversification of portfolios for institutional investors in the Eurozone financial markets. The results of the research give new perspectives on investments in cryptocurrencies. A large space for further research has been identified. First, there is insufficient research in the existing literature on the applicability of cryptocurrencies and blockchain technology. We identified studies that considered usability from a user perspective and from a developer perspective. However, in addition to the use of cryptocurrencies in financial markets, there is still a lot of room for the development and use of blockchain technology in the economy. In this regard, most current research relates to Bitcoin, and few have focused on other cryptocurrencies. Further research should focus on smart contracts and increase knowledge beyond cryptocurrencies. Although blockchain is presented in a cryptocurrency environment, this idea can be used on various occasions. Further room for research exists in the fact that there are not enough high-quality publications on the use of cryptocurrencies and blockchains in the financial markets. In terms of implications, this study contributes to the literature on the scope and limitations of the application of modern portfolio theory in the diversification of portfolios for institutional investors in the Eurozone financial markets. The results of the research give new perspectives on investments in cryptocurrencies.

There were a number of limitations that need to be considered. Future studies could examine development and use of blockchain technology in the economy. In this regard, most current research relates to Bitcoin, and few have focused on other cryptocurrencies. Although blockchain is presented in a cryptocurrency environment, this idea can be used on various other occasions. Further room for research exists in the fact that there are not enough high-quality publications in the use of cryptocurrencies and blockchains in the financial markets.

In comparison with other assets, cryptocurrencies are an extremely volatile instrument. Given that it is a speculative and highly volatile financial instrument, investors have different views on Bitcoin. First in terms of defining this cryptocurrency and then in terms of including this instrument in the investment portfolio. By including Bitcoin in the investment portfolio, the goal of diversification has been achieved. This is to reduce the risk of the institutional investor to a minimum. In practice, this means that it is possible to create a portfolio that carries an acceptable level of risk with the desired level of return. Given that Bitcoin is an extremely volatile and consequently - risky instrument, the expected return is also - high. The results of the research show that the cryptocurrency Bitcoin can serve as a desirable instrument for diversification of the investment portfolio when looking at a portfolio that includes stock indices. These results confirm the initial research hypothesis. Also, the results suggest that it is desirable to include in the structure of the portfolio a certain share of Bitcoin, about 6%. Also, if the investor wants to be additionally exposed to risk, he can create a portfolio in which there will be significant participation of Bitcoin and which can reject a higher daily return but also a significantly higher level of risk. However, before making a decision on investing in Bitcoin, investors should make an investment policy and goals, which will result in a risk appetite and target. Expected return, and to decide accordingly the share of bitcoins they want to include in the portfolio. Ultimately, the inclusion of Bitcoin in the portfolio enhances portfolio diversification, creating an efficient portfolio acceptable to a rational investor. When interpreting the obtained results, it is necessary to keep in mind the assumptions of portfolio optimization. The institutional investor will consider each investment alternative that is represented by the probability distribution of the expected returns in the observed period. It will then express the risk based on the variability of the observed financial instruments. Therefore, its decisions are based solely on the expected return and risk, so for the same level of risk, investors want to achieve higher returns and vice versa. In addition, it is assumed that transaction costs are neglected in the model. Finally, the study opens up space for other research. The first, in the direction of combining other cryptocurrencies with different financial instruments and, the second, in the direction of analyzing in the direction of portfolio diversification using cryptocurrencies, but from the aspect of investors with different risk aversion.

We recommend for future research the use of other research models for diversification based on the other theories. Also, further research can go in the direction of testing the model in different periods. Furthermore, future directions of research could be pointed towards the introduction of different transaction cost models and analysis of the impact of different criteria on the selection of shares in the portfolio, as well as different lengths of time series on the basis of which estimates are obtained. Also, future research can be directed towards the improvement of the optimization process by introducing the maximum allowed number of transactions and costs for portfolio optimization, which will facilitate institutional investors' decisions for further trading.

Finally, further research may go in the direction of comparing the main indices on regional stock exchanges in order to investigate whether there is a statistically significant correlation between the returns of different indices.

# References

1. Giudici, G., Alistair, M., Vinogradov, D.: Cryptocurrencies: market analysis and perspectives. J. Ind. Bus. Econ. **47**(1), 1–18 (2020). https://doi.org/10.1007/s40812-019-00138-6
2. Šoja, T., Chamil, S.: Bitcoin in portfolio diversification: the perspective of a global investor. Bankarstvo **48**(4), 44–63 (2019)
3. Muguto, T., Rupande, H., Muzindutsi, P.-F.: Investor sentiment and foreign financial flows: evidence from South Africa. Zbornik Radova Ekonomskog Fakulteta U Rijeci: Časopis Za Ekonomsku Teoriju I Praksu **37**(2), 473–498 (2019)
4. White, L.: The market for cryptocurrencies. Cato J. **35**(2), 383 (2015)
5. Swan, M.: Blockchain thinking: the brain as a decentralized autonomous corporation. IEEE Technol. Soc. Mag. **34**(4), 41–52 (2015)
6. Yli-Huumo, J., Ko, D., Choi, S., Park, S., Smolander, K.: Where is current research on blockchain technology?—a systematic review, **11**(1) (2016)
7. Yermack, D.: Is bitcoin a real currency? An economic appraisal. In: Handbook of Digital Currency, pp. 31–43 (2015)
8. Kristoufek, L.: What are the main drivers of the bitcoin price? Evidence from wavelet coherence analysis. Plos One **10**(4), e0123923 (2015)
9. Baek, C., Elbeck, M.: Bitcoins as an investment or speculative vehicle? A first look. Appl. Econ. Lett. **22**(1), 30–34 (2015)
10. Dyhrberg, A.H.: Hedging capabilities of bitcoin. Is it the virtual gold? Finan. Res. Lett. **16**, 139–144 (2016)
11. Dyhrberg, A.H.: Bitcoin, gold and the dollar–a garch volatility analysis. Financ. Res. Lett. **16**, 85–92 (2016)
12. Katsiampa, P.: Volatility estimation for bitcoin: a comparison of garch models. Econ. Lett. **158**, 3–6 (2017)
13. Pieters, G., Vivanco, S.: Financial regulations and price inconsistencies across bitcoin markets. Inf. Econ. Pol. **39**, 1–14 (2017)
14. Radivojac, G., Grujić, M.: Domains and limitations of the utilization of cryptocurrencies and blockchain technology in international business and financial markets. Acta Economica **16** (29), 79–102 (2018)
15. Baur, D., Hong, K., Lee, A.: Bitcoin: medium of exchange or speculative assets? J. Int. Finan. Mark. Inst. Money **54**, 177–189 (2018)
16. Guesmi, K., et al.: Portfolio diversification with virtual currency: evidence from bitcoin. Int. Rev. Financ. Anal. **63**, 431–437 (2019)
17. Abid, F., et al.: International diversification versus domestic diversification: mean-variance portfolio optimization and stochastic dominance approaches. J. Risk Finan. Manag. **7**(2), 45–66 (2014)
18. Platanakis, E., Andrew, U.: Portfolio management with cryptocurrencies: the role of estimation risk. Econ. Lett. **177**, 76–80 (2019)
19. Chan, W.H., Le, M., Wu, Y.W.: Holding bitcoin longer: the dynamic hedging abilities of bitcoin. Q. Rev. Econ. Finan. **71**, 107–113 (2019)
20. Ram, A.J.: Bitcoin as a new asset class. Meditari, Account. Res. **21**(1), 147–168 (2019)
21. Trautman, L., Dorman, T.: Bitcoin as asset class (2018)

22. Eisl, A., Gasser, S., Weinmaye, K.: Caveat emptor: does bitcoin improve portfolio diversification? Exchange Rates & Currency (2015)
23. Carpenter, A.: Portfolio diversifiication with bitcoin. J. Undergraduate Res. Finan. **6**(1), 1–27 (2016)
24. Liu, Y., Tsyvinski, A.: Risks and returns of cryptocurrency. National Bureau of Economic Research (2018)
25. Markowitz, H.: Portfolio selection. J. Finan. **7**(1), 77–91 (1952)

# Cryptocurrency as a Vector for the Digitalisation of the World Economy

Luka Ilich(⊠) 📵

Moscow State Institute of International Relations, Vernadsky av. 76,
119454 Moscow, Russia

**Abstract.** In the digital economy, all participants in the payment market are not required to simply provide a payment service, but there is a requirement to provide the maximum range of customer requests in the provision of such services. Thus, the content of the payment service, its elemental composition under the conditions of digitalization become much more complex. Today, virtually every month, new payment services enter the financial market that are able to respond to such needs to a greater or lesser degree of completeness. Most of them use surrogates instead of real funds in customers' bank accounts for mutual settlements. Thus, the customer is forced to incur costs in order to convert their money into a convenient means of payment for these applications. In addition, the existence of many different applications without the ability to integrate them significantly complicates the life of a modern person looking for satisfying payment needs in one place and preferably at once, without wasting time. In this regard, the hybrid, integrative nature of payment services should be recognised as one of the most important features influencing the transformation of their content and structure in the digital economy. The development and implementation of financial technology in a digitalised environment is transforming traditional areas and ways of providing financial and other services, and new products and services are emerging.

**Keywords:** Payment Services · Legal Regulations · Blockchain

## 1 Introduction

At the present stage of the economy it seems correct to talk not about payment services in general and their transformation, namely, innovative payment services as a driver for the development of the entire financial market. [1] Thus, innovation should be recognized as a driver for the development and updating of payment services in a digital economy.

The development and implementation of financial technologies in digitalization transforms traditional directions and methods of providing financial and other services, new products and services appear.

T. Antipova (Ed.): DSIC 2021, LNNS 381, pp. 300–310, 2022.
https://doi.org/10.1007/978-3-030-93677-8_26

In the fall of 2017. The President of the Russian Federation gave an order until July 2018. legally regulate the ICO procedure (attracting money) and determine what "cryptocurrency", "Mining", etc. The preparation of the bill "On Digital Financial Assets" was carried out by the Bank of Russia and the Ministry of Finance of Russia. 2017 is remembered as a period of active up-trend of virtual currencies, ICO - the release of new tokens and the development of cryptospace infrastructure.

The aim of the work is to consider cryptocurrency as a vector of digitalization of the global economy. To achieve this goal, the following tasks were developed:

- to study the technological basis of the operation of cryptocurrency;
- identify the principle of operation of cryptocurrencies using the example of Bitcoin;
- identify the problems of international legal regulation of cryptocurrency;
- consider the prospects for the functioning of cryptocurrencies based on the implementation of blockchain technologies;
- assess the impact of the economic consequences of the pandemic;
- coronavirus infection COVID-19 on the cryptocurrency market.

During the study, the following methods were used: literature analysis; analysis of regulatory documentation on the topic of work, as well as analysis of documents; study and synthesis of domestic and foreign practices; review of economic and financial resources, articles; observation.

The object of research is the development of the cryptocurrency market. The subject of the study is the factors of digitalization of the economy.

The structure of the work is determined by the tasks, purpose and logic of the study. The work consists of an introduction, three main bodies, a conclusion and references.

## 2    Cryptocurrency as an Innovative Tool for the Global Economy

In the context of this study, several distinctive characteristics of cryptocurrencies need to be differentiated:

- volume of output (limited and unlimited);
- centralised (centralised and decentralised currencies);
- hybridity and the ability to change its type: the type of token can be changed in the course of project development or in connection with the so-called SAFT agreement (Simple Agreement for Future Tokens), a contract between developers and investors regarding the future of the tokens to be issued;
- nativity and nonnativity of the functioning mechanism (nativity operates on its own blockchain, nonnativity operates on a borrowed blockchain);
- specific consensus algorithms (currently more than 11) to validate transactions (see Fig. 1 as an example):

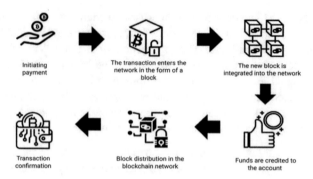

**Fig. 1.** Consensus blockchain mechanism Proof-of-Work Source: Author's interpretation based on [2].

Figure 1 shows the Proof-of-Work consensus blockchain mechanism on which bitcoin operates, but it is not the only version of blockchain. There are other consensus variants, the most common being Proof-of-Stake ("proof of share" - Nxt, Ethereum in transition) and Proof-of-Importance ("proof of importance" - NEM Foundaition). However, in the now classic Proof-of-Work (PoW) consensus, there are technological variations where so-called masternodes - master nodes with a special status compared to regular nodes - are added. In the Dash network, for example, they intermingle transactions (PrivateSend technology) and allow transactions to be sent more quickly (InstantSend technology). North America and Europe have the highest number of active cryptocurrency wallet users - a combined 30%. 32% of active wallets use proprietary software, while the remaining 68% use open source software. Mobile apps are the most common, at 65% of the total (Fig. 2).

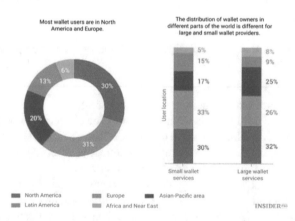

**Fig. 2.** Territorial distribution of international cryptocurrency market participants, Source: Authors's compilation from [3].

More than half of miners rate their influence on the development of the bitcoin protocol as strong or very strong - this is due to the conflict surrounding the SegWit update. 58% of major mining pools are in China, so many believe that the PRC is the most important player when it comes to mining and the direction of the protocol. The US ranks second with 16%. Mining, despite the rising value of digital currencies, is growing unevenly. Total bitcoin revenues for the year (blockchain creation fees + transaction fees), when immediately converted to dollars, were markedly lower in 2016 ($563 million) than in 2014 ($786 million). Certain cryptocurrencies have a specific usage pattern that provides additional opportunities for users (Fig. 3).

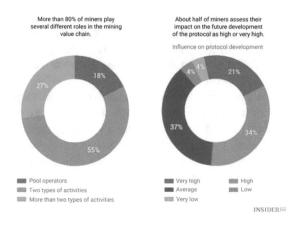

**Fig. 3.** Impact of miners on cryptocurrency market development, Source: Authors's compilation from [3].

The use of computer networks is driven by different kinds of tasks. Local area networks such as BAN (Body Area Network), PAN (Personal Area Network), LAN (LAN, Local Area Network), CAN (Campus Area Network), MAN (Metropolitan Area Network) have a specific business purpose. It is this purpose that determines their architecture, access as well as the relationships of network users [4]. Therefore, an analogy with cloud service models is appropriate here: Platform-as-a-Service, Infrastructure-as-a-Service (IaaS). Cloud technology and blockchain in these models are designed to solve certain narrowly targeted tasks, such as deployment of developed applications on a platform provided by a provider (PaaS) or organisation of workflow in an enterprise. Blockchain used in such computer networks is therefore a technological property of pre-existing and defined relationships between users interacting within that network. As such, the use of blockchain in these networks will receive less attention.

## 3   The Role of Cryptocurrencies in Today's Global Economy

### 3.1   How Cryptocurrencies Work, Using Bitcoin as an Example

The main feature of bitcoin is that it is not pegged to any existing state currency, but can be exchanged for dollars or euros. For blockchain technology to be applied to financial and payment services on a large scale, a gradual shift in trust from today's efficient but costly centralised counterparty services to a distributed model is needed. Without effectively addressing the "trust issue", many players, or even the entire industry, will not be able to realize the promising benefits that this technology brings to the payment business. Here are some of the challenges that will need to be addressed in the near term.

– establish an unambiguous or hybrid status of cryptocurrency at the international level: commodity, medium of payment or exchange, etc., which would then be projected into national legislation as well;
– come to an understanding of whether a public registry can be subject to a hacker attack and create the conditions, the tools to prevent them;
– understand whether bitcoin's negative reputation can be overcome and how steps in this direction can be implemented;
– overcome the regulatory complexities associated with the introduction of block-chain technology. For example, even though the validation procedure during a blockchain transaction is performed by almost all participants in the network at the same time, if most of the participants forming the consensus network model collude to commit fraud, the registry could be manipulated. This could be a problem for a payment service provider with a relatively small network that does not have proper pre-clearance procedures.

The number of new cryptocurrencies is constantly increasing, and bitcoin's market share has been gradually declining over the past few years.[5] While it used to be possible to buy bitcoin and have a high probability of making a huge profit after a short time, the situation has now changed. Most banks are exploring the use of Blockchain technology to streamline business processes and reduce costs [6]. Figure 4 shows the statistics of the leading players in the international arena among the registration of cryptocurrency startups.

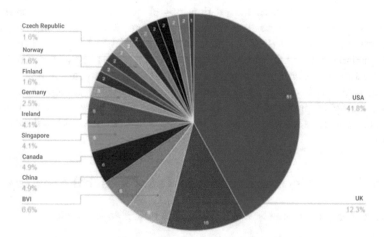

**Fig. 4.** Leading countries in terms of cryptocurrency penetration in the world, Source: Authors's compilation from [7].

Once they are adopted and a stable legal practice is established, we can talk about a qualitatively new stage in the attitude of Russian regulators towards cryptocurrencies as a means of payment. In general, the consensus algorithm used in blockchain and the features of the network organisation affect three critical factors in the provision of payment services: speed, cost and security. Security in the context of blockchain refers to its decentralised nature. This term, already explained earlier in the paper, remains rather ambiguous in practice in terms of content due to the lack of standards and other regulations defining its regulation and nature.

## 3.2  Problems of International Legal Regulation of Cryptocurrencies

The evolving digitalisation of the economy in today's environment is an irreversible process, encompassing all areas of economic activity. According to researchers, digitalisation processes are now most clearly reflected in the financial sector as the dominant segment of any country's economy. Today, the financial sector around the world is undergoing digital transformation, that is, a radical and complex change in the way business thinks in the new digital economy. Leading participants in the financial and payment services markets as part of the financial infrastructure, the so-called "digital elite", are seeking to combine a proactive, innovative, ahead of the curve digital activity and strong management (including risk management in the new environment), continuing the transition from using information technology as a means of business automation to transforming it on a massive scale in a digitalized environment. With the digitalisation of the global economy, every country is now looking to outpace innovation in the financial sector.

Digitalisation processes are bringing many innovations to the financial sector, transforming the market for financial and payment services: they promote new financial assets, new types of services, new distribution channels, new tools for shaping contract terms (such as "big data"), etc. Given the insufficient legal and regulatory framework and lagging legislation, today's users of electronic payment systems, as well as service providers, are not sufficiently aware of their rights and obligations. Despite the fact that this type of payments is growing in popularity among consumers, neither they nor service providers have full information about how to protect themselves from fraud in this area. This protection is also not facilitated by the lack of a legal framework for digital technology in financial and payment services. Amidst the clear prospects, the most significant problem in the area under study is the lack of legal regulation and legal framework for blockchain and cryptocurrency technology, both at national and international levels. It should be noted that in the legal dimension, and in many respects even in conceptual terms, there is no clear and unambiguous idea of who will be responsible for the credibility of information when using this technology: all participants or any of them.

Other significant problems in the development and application of the technology in question also include the slow speed of the network compared to current solutions, the need for high computer power, the shortage of qualified specialists, and hence the high cost of implementing and maintaining the technology. Regarding the legal aspect among the problems identified above, it is worth noting that the approach to regulating cryptocurrencies is currently very different worldwide, with a general trend towards state regulation rather than prohibition of cryptocurrency circulation, despite the very controversial debate about its fate and current situation in financial circles and at the governmental level in different countries, as well as globally, as shown in the data systematised below (see Table 1), This demonstrates the prospects for the use of cryptocurrency as a means of payment, if the scientific, methodological and regulatory framework for its functioning is carefully developed. At the same time, cryptocurrency has already obtained official status in this area in a number of countries around the world (see Table 1). The table shows that cryptocurrencies are not currently banned in the world's largest economies.

**Table 1.** Legal regulation of cryptocurrencies in countries around the world

| Approach to legal regulation cryptocurrency | Countries |
|---|---|
| Cryptocurrency transactions banned at state level | Algeria, Bolivia, Ecuador, Indonesia, Kyrgyzstan, Lebanon, Morocco, Namibia, Nepal, Pakistan, Viet Nam |
| The state is wary of cryptocurrencies and some restrictions have been introduced, but citizens are not prosecuted for their use | Armenia, Brazil, China, Colombia, Dominican Republic, Estonia, France, Iran, Jordan, Kazakhstan, Lithuania, Nigeria, Russia, South Korea, Thailand, UAE |
| Cryptocurrencies have an official status (commodity, financial asset, means of payment, etc.), there is state control of the market, or their use is not restricted in any way by the authorities. | Argentina, Australia, Belarus, Bulgaria, Canada, Denmark, EU, Finland, Georgia, Germany, Hong Kong, Israel, India, Japan, Kenya, Luxembourg, Malaysia, Mexico, New Zealand, Norway, Philippines, Poland, Saudi Arabia, Senegal, Singapore, Slovakia, Slovenia, South Africa, Spain, Sweden, Switzerland, Taiwan, Tunisia, Turkey, USA, Venezuela, Zimbabwe |

Source: Author's compilation based on and updated from [7]

In Russia, a ban on accepting payment in cryptocurrency and exchanging it into roubles has been legally established. At the moment, cryptocurrency cannot be a legal means of payment in Russia, i.e. the right to own cryptocurrency is allowed, but the right to make transactions is excluded.

The importance of cybersecurity for institutions and their regulators in the financial and payment sphere will only grow in the future. The main challenge remains to maintain a balance between security and customer convenience. For full-service providers, offering a varied range of services and seeking to maintain a presence across multiple channels, the challenge will also continue to grow. But digitalisation can also create a new environment for financial institutions in which threats can be more effectively identified and prioritised, risks can be quickly identified and mitigated, and security concerns can be understood [8]. As part of the study of international experience in the development of payment systems in the digital environment and their regulation, the increasing role of standardization should be noted. It is important to note that standards have traditionally served not only as an important regulatory tool, but also as a means of accumulating and disseminating best practices in the financial sector, including in methodological terms, as well as for improving the efficiency of economic operators.

# 4  Prospects for Cryptocurrencies in Today's Global Economy

## 4.1  Prospects for Cryptocurrencies Based on the Implementation of Blockchain Technology

The key to the rapid development of cryptocurrency and the distributed ledger (blockchain) technology that supports it is that they do not impose significant transaction costs (unlike the classical currency market), provide transparency and irrevocability of transactions and eliminate the intermediary as a verifier of transactions. Cryptocurrencies are much more volatile than stocks, commodities and indices [9]. There is a clear trend in the cryptocurrency market to diversify, with bitcoin's share declining. Investors are getting to know the industry and investing:

- to favourites from the top 10 capitalisation rankings according to Coinmarketcap;
- promising coins from the top 50;
- assets little known to the general public. The share of the which has more than tripled in a year, from 7 to 25%. Of the individual cryptocurrencies, Ethereum is the closest to bitcoin. However, investments in ETH are extremely volatile - since early 2018, there have been times when BTC and ETH capitalisations have differed between them by 70% and more than 300% [10].

The digitalisation of the financial sector is putting new demands on the consumer, including an understanding of 'financial hygiene', cybersecurity fundamentals and how to handle personal data. However, the speed and scale of digitalisation is creating multiple risks for different groups of consumers of financial services (market, regulatory, behavioural, technological), its sources and the response of the financial sector and consumers themselves. The same can be said for payment and financial services providers. With the development of the digital economy and the active introduction of digital technologies in the financial sector, it is planned that payment service providers, which include payment system operators, will be able to be not only banks and non-bank financial institutions (NBFIs), but also other business entities, whose activities are currently regulated rather ambiguously under national legislation in many countries, including Russia. The status of foreign payment service providers, payment aggregators and mobile application developers providing payment services integrated into their functionality, including those operating globally in the global financial market, remains unclear in many countries. In this context, for example, in the domestic context, the Bank of Russia announced in 2019 the intention to strictly regulate and control payment applications and payment aggregators, especially foreign ones; in addition, federal legislation also established in 2019 a restriction for electronic money operators operating in Russia, prescribing that only Russian infrastructure shall be used for these purposes; several draft laws were introduced whose contents oblige banks to participate It is planned to establish legal functioning of the notions of "payment aggregator", "payment application provider" in the economic turnover, as well as to set the terms of their involvement by the banks and procedure of control over their activity.

If the full package of proposed amendments is finally adopted and enshrined in law, foreign payment service providers (e.g., AliPay, WeChat) will be able to operate in

Russia only if they are included in a special register of the Bank of Russia. Foreign payment systems (MoneyGram, Westernpay, Uniteller, Intel Express) will have to open a representative office or branch. In order to improve efficiency in a rapidly changing regulatory environment, new technology solutions are required, which are provided by the so-called Regtech and Suptex companies. The introduction of digital technologies for financial institutions to comply with regulatory requirements and improve the efficiency of regulators themselves (Regulatory Technology and Supervisory Technology) helps to simultaneously reduce the regulatory burden and improve the quality of reporting information, ensure transparency of the financial institution or institution providing financial and payment services.

The current inefficiencies in bank credit information systems mainly occur for the following reasons: firstly, the scarcity and poor quality of data makes the credit information system unable to effectively assess the perceived quality of the credit provided; the second difficulty is the exchange of inter-institutional data; the third aspect is the lack of privacy, the possibility of hackers stealing user data, which leads to difficulties in users providing their personal data from.

Although solving these problems requires the cooperation and involvement of various stakeholders, Blockchain technology can provide some assistance in solving these problems. In addition, this technology can be unprecedentedly effective in data verification. Blockchain technology can encrypt data, allowing users to control their own data and establish ownership of it. It can also ensure that the information is authentic and reliable, as well as reducing the cost of data collection for credit agencies. With the use of Blockchain technology, Big data could become credit resources with the character of privacy, and even form the basis of future credit systems.

Having access to an open, transparent register of bank transactions would also be useful for regulators. This would allow governments to deal with tax fraud. Blockchain technology, which is a type of underlying technology that provides ample opportunity to develop tailored and flexible business processes in banking transactions. That is, Blockchain delivers capabilities that can be applied to business processes with different scenarios. It can digitise assets and accurately reflect the amount of market value, thereby regulating and optimising the financial infrastructure. This dramatically increases the efficiency of the process of valuing and buying and selling financial assets while reducing costs.

### 4.2   Impact of the Economic Impact of the Pandemic

The cryptocurrency market has long been independent of events related to the spread of the coronavirus. The Chinese government first reported cases of coronavirus infection in December 2019, with the official announcement of an outbreak only coming in the second half of February 2020. In March 2020, a single bitcoin was offered for around $6,500, and in February 2021 the cryptocurrency was worth around $56,000. But the COVID-19 pandemic has led to a serious change in currency markets and stock indices [11]. Millions of investors around the world had huge financial problems amid the pandemic - so-called margin calls. All the while, bitcoin, as a major cryptocurrency, has risen rapidly, surpassing the $55,000 mark. Along with bitcoin, major altcoins have also grown. For example, the ETH token peaked at $126 in May 2021. Investors were actively buying bitcoin in December, expecting the price to rise, and when it did, they started selling aggressively. That and the panic factor. The market fear index peaked in March. Investors tried to sell assets that could still be sold. The first to sell off were bitcoins and other cryptocurrencies that were held in an investor's portfolio.

Following decisions by several central banks, the US Federal Reserve and the ECB to maintain or set minimum interest rates and to launch dollar liquidity programs, the dollar began to strengthen markedly against other world currencies. With a sharp drop in production around the world, closed borders and a slowing global economy, cryptocurrencies, which have traditionally been seen as a way to store savings in difficult times, have also begun to suffer. The drop in their prices was contributed to by a mass exodus from the market by institutional investors, who suddenly found themselves in a difficult environment and began selling off their cryptocurrency assets in an attempt to build up their liquidity reserves. Investors are again choosing the US currency as the safest way to save assets.

## 5   Conclusion

The immutable nature of blockchain systems is a guarantee of the authenticity of information. That is, once a piece of information enters the system, it cannot be altered. This eliminates the subsequent problems of fraud, but it also implies that prior verification of information must be more thorough. Thus, mechanisms for accessing information must be established as sufficiently stringent regulations, and the data at each node must be verifiable to ensure that fraud is then ruled out. Once a transaction is initiated, it cannot be reversed. Thus, its authenticity and reliability must be verified in order to avoid accidental losses. The need for special attention to improve the regulation of financial and payment services is due to the highly complex and sophisticated nature of this market in an evolving digital environment. The modern system of regulation of the financial and payment services market includes the relevant policies of economic agents, regulations and standards of regulatory bodies at the national and international level, as well as legislation (national and international legal frameworks), which act as elements of state and international financial policies aimed at ensuring the security of the world's financial system and subordination of digital economy actors to existing international, national and international.

The study of international experience in the development of payment systems in the digital environment and their regulation has led to the conclusion that there is an increasing role for standardization in this system. The development of the digital economy requires changes in domestic legislation in accordance with the new legal relations that arise between the subjects of the financial and payment market and new, non-traditional objects in this area.

There is a situation where each functioning ESS actually forms not a regulatory but almost a "legal field" at the level of its economic entity, using for this purpose approaches and instruments that are unacceptable from a legal point of view. This paper reflects the need to reach uniformity and (possibly at the level of international standards) determine how to interpret digital technologies in general, electronic payment systems and new digital payment and financial technologies in the current legal framework, which will then allow their formulation to be projected at the national level.

The main conclusion of this work is cryptocurrencies, although they have existed on the market since 2011, however, over 10 years many legal issues have not been resolved, there is also a strong dependence of the cryptocurrency rate on exogenous factors. The cryptocurrency market shows a tendency to strengthen and has brought serious changes to payment services, which in turn at the level of the leading countries of the world gave rise to discussions about the creation of digital currencies (stablecoin and digital ruble). Nevertheless, today cryptocurrencies are not reliable tools for storing or investing capital for the long term.

In continuation of this work, it is proposed to analyze the possible consequences of creating digital currencies, following the example of stable coin (the digital currency of which is tied to the dollar exchange rate).

# References

1. Juffa, D.A.: The impact of cryptocurrency on the Russian economy. Young Sci. **20**, 257 (2019)
2. Lewis, A.: The Basics of Bitcoins and Blockchains: An Introduction to Cryptocurrencies and the Technology that Powers Them (Cryptography, Crypto Trading, Digital Assets, NFT), Lewis A. Mango, London (2018)
3. Insider Cryptocurrency. https://markets.businessinsider.com/cryptocurrencies. Accessed 13 May 2021
4. Bondarchuk, D.: Cryptocurrency, mining and smart contracts to be legalised. How will it affect legal work. EJ Lawyer **6**, 1–2 (2018)
5. Antipova, T.: Is it worth investing in cryptocurrency? In: MATEC Web of Conferences, vol. 342, p. 08007. Universitaria SIMPRO 4-5 (2021). https://doi.org/10.1051/matecconf/202134208007
6. FINMA: Guidelines for Enquiries Regarding the Regulatory Framework for Initial Coin Offerings (2018)
7. Comply Advntage Cryptocurrency Regulations Around. https://complyadvantage.com/blog/cryptocurrency-regulations-around-world/. Accessed 09 May 2021
8. Kucherov, I.I.: Cryptocurrency as a legal category. Finan. Law **5**, 3–8 (2018)
9. Antipova, T.: Is it worth investing in cryptocurrency? In: MATEC Web of Conferences, vol. 342, p. 08007. Universitaria SIMPRO, 2 (2021). https://doi.org/10.1051/matecconf/202134208007
10. Binance ETH/BTC. https://www.binance.com/en/trade/ETH_BTC. Accessed 20 May 2021
11. Antipova, T.: Digital view on COVID-19 impact. In: Antipova, T. (ed.) ICCS 2020. LNNS, vol. 186, pp. 155–164. Springer, Cham (2021). https://doi.org/10.1007/978-3-030-66093-2_15

# Role of the Banking Sector in the Financial System and Its Macroeconomic Functions in Russia

Luka Ilich[(✉)] [ID]

MGIMO University, Vernadsky av. 76, 119454 Moscow, Russia

**Abstract.** The banking system of Russia has been functioning in a challenging environment in recent years. Changes to international market conditions and the macroeconomic situation, approaches to banking regulations and the rehabilitation policy have been affecting proficiency, with which banks have been discharging their functions in the Russian economy.

Importance of the banking system for social and economic development of the Russian Federation shall be currently increasing in the setting of damped growth of the world economy, contraction of external demand, persisting sanctions, consequences of containment measures.

This thesis consists of five chapters. The second chapter addresses essence, structure and main functions of the banking sector. The third chapter provides analysis of the current condition and peculiarities of development of the banking sector of the Russian economy. The fourth one addresses current issues and development potential of the Russian banking sector.

**Keywords:** Russian Federation · Bank assets · Monetary policy

## 1 Introduction

The financial system is one of the most important subsystems of the economic system and performs important functions in the economy. In the expanded presentation of the financial system that is used in this work, it ensures the fulfillment of the socio-economic functions of the state, serves as a clearing house for all other elements of the economic system, provides the economy with financial resources, supports the work of the mechanism for transforming savings into investments.

The banking system of Russia in recent years has been operating in difficult external conditions. Changes in the situation of foreign markets and the macroeconomic situation, approaches to regulating the activities of banks, recovery policies - all this influenced the efficiency of banks in fulfilling their functions in the Russian economy.

The main goal of this work is to analyze the role that the banking sector plays in the financial system and economy of Russia. To achieve this goal, the following main tasks are set in the work:

- consider the essence and structure of the financial system, as well as the macroeconomic role of the banking sector as one of its elements;
- describe the features of the structure of the banking sector of the Russian economy;

T. Antipova (Ed.): DSIC 2021, LNNS 381, pp. 311–319, 2022.
https://doi.org/10.1007/978-3-030-93677-8_27

– to analyze the main trends in the development of the banking sector of the Russian Federation;
– identify key problems and show the main directions of development of the banking sector of Russia.

The object of the study is the financial system of the national economy, the subject is the banking sector as one of the structural elements of the financial system.

When conducting the study, general methods of scientific research were used, in particular, analysis and synthesis, comparison. In addition, methods of calculation and graphic analysis of the development indicators of the Russian economy and its banking sector were used.

When writing the work, the works of the book of the monograph of leading experts of economists of financiers were used: Abramova M. A., Aganbegyana A. G., Berdysheva A.V., Borisova E. R., and others, regulations: federal laws, instructions and instructions of the Bank of Russia, Internet resources: http://base.garant.ru/, http://www.cbr.ru/banking_sector/, https://proficomment.ru/gosudarstvennye-banki-rossii/, https://www.gks.ru/ank.

## 2   Analysis of the Current Condition and Development Peculiarities of the Banking Sector of Economy of the Russian Federation

Deposit and lending transactions are the main transaction groups for the banking sector of the Russian Federation. It is explicitly seen from the analysis of structure of assets and obligations of the banking sector (Figs. 1 and 2).

Russian banks form the resource base by attracting funds from non-financial enterprises and population to deposits and other accounts. The share of enterprises' funds in the liabilities of the banking sector has ranged from 33 to 39%, in those of the population - from 32 to 39% of the total bank liabilities in the last 5 years [1].

In general, 64 to 74% of the resource base of the Russian banks were formed out of attracted funds during the period from 01.01.2016 to 01.01.2021 (Fig. 1).

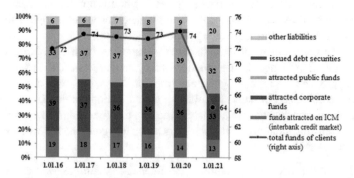

**Fig. 1.** Resource base of the Russian banking sector, relative share in the volume of bank liabilities, %. Sources based on: [2]

The assets of the banking sector of the Russian Federation are primarily formed out of lending transactions (Fig. 2).

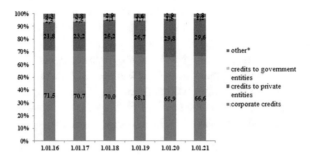

**Fig. 2.** Structure of assets of the Russian banking sector, relative share in the total bank assets, %. Source based on [2]

Moreover, as can be seen from Fig. 2, the share of corporate credits in assets was 2–2.5 times higher than the share of retail credits over the past 5 years. However, the relative share of credits provided to enterprises was decreasing in favor of credits to private entities.

Runtime trends as regards deposit transactions of the banking sector are shown in Fig. 3.

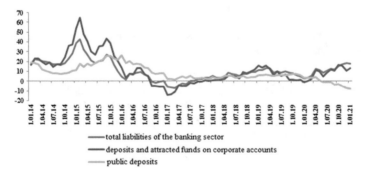

**Fig. 3.** Rates of growth of total liabilities and deposits of the banking sector by the previous year, %. Sources based on: [2]

The Figure distinguishes two main periods – the period of deferred dynamics of deposits and resource base of banks from the beginning of 2015 to the beginning of 2017. The boundaries of these periods are primarily subject to macroeconomic reasons.

In 2015–2016, there was an economic slowdown in Russia caused, first of all, by a rather sharp drop in oil prices. Since the early 1990s, hydrocarbons and other raw commodities have been the basis of Russia's exports, and federal budgeting, the exchange rate dynamics and, in general, the growth rate of the Russian economy largely depend on exports.

The slowdown of dynamics and a successive GDP drop were followed by an incoming drop in actual disposable public income, deterioration of the financial status of enterprises. It could not but affect the volume of deposit transactions in the banking sector. And vice versa – some improvement in the situation in 2017–2019 due to resumed growth in oil prices had a positive impact on attraction of customer funds by the banking sector.

The same factors influenced the bank crediting market in Russia in the sense that crediting dynamics largely depends on creditworthiness of borrowers, and this function is driven by their income and financial condition.

There are two main periods in the credit market development – the decline in 2015–2016 and the growth recovery since 2017, accordingly (Fig. 4).

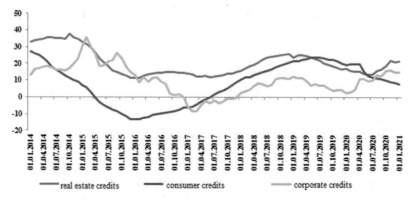

**Fig. 4.** Growth rate of debt on credits provided by the banking sector to population and enterprises, year-on-year, %. Source: [2]

Herewith, a change in the interest rates of the credit market, which followed changes to the key rate of the Bank of Russia, as illustrated in Fig. 5, was another important factor that ascertained demand for bank credits.

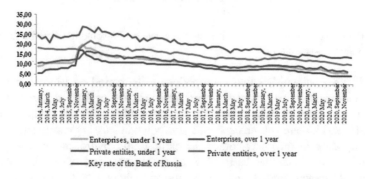

**Fig. 5.** Weighted average rates on ruble-denominated corporate and retail credits with maturities under and over 1 year, key rate, %. Source: [3]

Firstly, the Fig. 5 demonstrates a direct correlation between changes in the key rate and corporate and retail credit rates. Secondly, the Figure shows that the overall rates were gradually decreasing after the growth thereof in 2014 and 2015.

However, as is discussed in the following chapter of this paper, credits are inaccessible for most enterprises in the real sector in Russia giving rise to a number of economic issues disrupting operation of the savings-to-investment mechanism.

## 3   Current Issues and Development Prospects in the Banking Sector of the Russian Federation

### 3.1   Current Issues in the Banking Sector

According to the author of this paper, the main current issues in the banking sector include, firstly, the consequences of the rehabilitation policy that the Bank of Russia has been pursuing since 2013. Secondly, dysfunction of the savings-to-investment mechanism and insufficient participation of banks in financing the Russian economy development.

The policy of rehabilitation of the banking sector pursued in the recent years suggested withdrawing of the banks, which were potentially unstable in financial terms, according to the Bank of Russia, or were suspected of participation in illegal or shady transactions, from the market. Such banks were deprived of their licenses, and they were subject to a bankruptcy or rehabilitation procedure with the following change of control or reorganization [4].

This policy has primarily led to a rather significant reduction of the amount of banks in Russia. According to Fig. 6, the total amount of the banking services provided on the Russian market had reduced by over 2.5 times, the multi-branch banking had gone down by over four times in the period from 01.01.2013 to 01.01.2021.

It had not impacted the stability of the banking sector; however, the competitive environment on the banking services market had not improved, but had worsened as a result of a concentration increase caused by a flow of the client base to large systemically important banks.

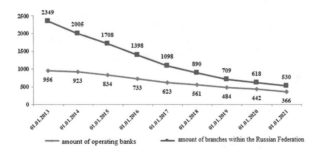

**Fig. 6.** Amount of banks and their branches in the Russian Federation, pcs. Source based on: [5]

The last statement can be confirmed by Fig. 7, which, according to the Bank of Russia, shows change in the share of state, private and foreign banks in the main segments of the banking services market.

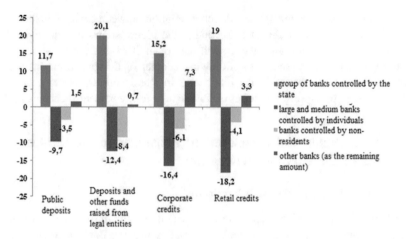

**Fig. 7.** Growth of share of the main bank groups in the most important segments of the banking market in Russia over the period from 01.01.2013 to 01.01.2019, %. Sources based on: [4]

As is seen from the above, state banks have significantly raised credit and deposit market shares by reducing market shares of all other banks.

Against the macroeconomic background, the total amount of banks on the market and the degree of competitiveness of this market are not so important as effectiveness exercised by the banking sector when fulfilling the savings-to-investment transformation function. Besides, in this case, this refers to the actual but not financial investments as was previously described herein.

In the recent years, effective functioning of this mechanism becomes challenging in Russia, which can be illustrated by Fig. 8.

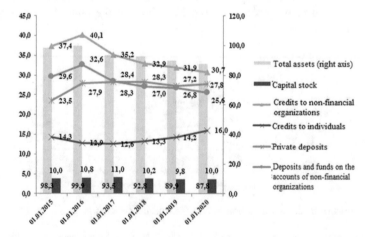

**Fig. 8.** Macroeconomic indicators of the banking sector of the Russian Federation, % of GDP. Sources based on: [6]

Figure 8 shows a pronounced trend towards a reduction in corporate credit debt since the beginning of 2016. Moreover, this was apparent against the background of no significant problems with the capital of the banking sector and with the resource base of the banks. Meanwhile, over half of the circulating capital of enterprises in Russia is financed by credit, and a significant part of investments in fixed assets should be financed by it as well.

### 3.2 Lines of Development of the Russian Banking Sector in the Midterm

Academic A.G. Aganbegyan and V.V. Ivanter, Doctor of Economics M.V. Yershov mention in their articles of 2019–2020 that the recent-year recession in the Russian economy was caused by lack of investment into it. This is primarily due to inaccessibility of bank credits to enterprises, which is mostly a result of monetary and supervisory policy of the Bank of Russia [7].

It shows that, despite the reduction of the key rate and the credit market rates, the level of interest rates for corporate credits exceeded the asset profitability rates in most branches of the Russian economy.

In such situation, credit start being serviced not from income but from circulating assets, in particular, from the depreciation and wages funds. Firstly, this reduces credit demand. Secondly, this creates obstacles for growth of investments into capital assets and wages on a macroeconomic level.

Moreover, the Bank of Russia abandoned direct control of the ruble exchange rate from January 2015, which has led to no ruble devaluation only in 2017 in the period from 2013 (refer to Fig. 2).

On one hand, it makes an adverse impact on the living standards through the rise in prices for imports and import components of the cost of Russian goods. This means that it adversely affects domestic demand, which is vital for economic growth. On the other hand, it disrupts investment mechanisms, since it is simply impossible to plan investment projects in conditions of high volatility of the ruble exchange rate.

Furthermore, switching to the Basel III standards initiated by the Bank of Russia significantly imposed stricter economic standards on the banks and stiffened the requirements to the techniques of assessment of risks, in particular, of the credit and deposit ones [8]. Other conditions being equal, this increases surcharge to credit interest rates that the banking sector sets for risk and, accordingly, lifts the level of these interest rates. Thus, despite the decrease of the key rate, the issues of credit accessibility for the real sector persist in Russia.

Since a simple decrease in the key rate does not provide a solution to this problem, for the reasons discussed above, it is necessary to use specialized instruments for bank refinancing in this case, whereas the methods of such instruments have already been worked out by the Bank of Russia.

Secondly, an issue of volatility of the ruble exchange rate is discussed in most of these publications. It is suggested that these issues may be solved through tightened foreign exchange regulation and control, through new measures, which are quite multiple, required for limiting currency speculations, "carry trade" transactions with participation of non-residents and large Russian banks, through limited outflow of capital from Russia.

These are just high priority measures. The general list of the current proposals as regards improvement of significance of the banking system in stimulating the Russian economy is much wider and requires focused consideration [9].

Today, the Central Bank prefers to adhere to traditional methods of doing business sometimes failing to minimize risks that could destabilize country's economy.

The issues discussed above are still relevant in the main development lines of the financial market, and development of the banking sector is associated with digitalization, improvement of competitive environment, with subsequent improvement of the banking supervision procedures. It is associated with the issues that have no immediate relevance to any changes in macroeconomic situation in Russia, which remains challenging.

## 4  Conclusion

The paper shows that the financial system in its broad sense includes such basic constituents as state finance, corporate and household finance, as well as securities markets and the financial sector of economy.

The banks included in the banking sector of the Russian economy are predominantly multi-function, are divided into groups by their system-related significance, regional coverage of the branch network and by whether the main owners belong to a state sector, private sector of by a non-resident sector.

The main lines of activity of the banking sector in Russia include deposit operations, which mainly form the resource base of banks, and crediting, which forms over 2/3 of the total assets of the banking sector.

The paper specifies significant reduction of the amount of banks due to the rehabilitation policy pursued by the bank of Russia since 2013, firstly, and deterioration of the competitive environment and increasing of the bank service market concentration due to the same policy, secondly, as the main trends of development of the banking sector.

Thirdly, two main periods as regards key bank transactions are specified in the paper: a period of decline on deposit and credit markets in 2015–2016 and a period of recovered dynamics of attracted client's funds and credit indebtedness starting from 2017.

According to the author of the paper, the main factors that defined the dynamics of indicators in the banking sector within the period discussed in the paper included:

- changes in the macroeconomic situation that affected financial standing of the clients;
- monetary policy of the Bank of Russia defining the level of interest rates on the credit and deposit markets;
- and switching to the recommendations contained in Basel III as regards banking regulation that made supervision standards much stricter.

Decreased significance of banks in ensuring savings-to-investment transformation through credits provided to the economy, which has been slowing down since the beginning of 2014, is the main problem in the development of the banking sector, which is highlighted in the paper. The issue of availability of bank credits to the real sector. Providing a solution to this issue, improvement of significance of the banking sector in the economy development financing are, according to the author of this paper and many Russian scientists, a key line of development of the banking sector in Russia in the midterm.

It is an open secret that complex mechanisms and systems (such as the banking ones) have been a fruit of hard work since the ancient times and, as pertinent to any complex "mechanism", banking system also requires constant modernization. Based on my research, we can conclude that the lion's share of the issues described in the research process are far from new and were inherent in the banking system of the Russian Federation before.

# References

1. Information on operating credit institutions involving non-resident, 01 Jan 2020. https://www.cbr.ru/Collection/Collection/File/27599/PUB_200101.pdf. Accessed 23 June 2021
2. CBR time series of indicators of individual tables of the collection "Statistical indicators of the banking sector of the Russian Federation". http://www.cbr.ru/statistics/bank_sector/review/#highlight=o??op%7C?a??o?c?o?o%7Cce??opa%7Co??opa. Accessed 23 June 2021
3. CBR Interest rates on loans and deposits and structure of loans and deposits by maturity. http://www.cbr.ru/statistics/bank_sector/int_rat/#highlight=?pe???a?%7C?pe???o?%7Cc?e?e???%7Cpy???x%7C?o??apax%7Ce?po. Accessed 25 June 2021
4. CBR Report on development of Banking Sector and Banking Supervision. http://www.cbr.ru/about_br/publ/nadzor/. Accessed 26 June 2021
5. CBR Quantitative characteristics of the banking sector of the Russian Federation. http://www.cbr.ru/statistics/bank_sector/lic/. Accessed 26 June 2021
6. CBR macroeconomic indicators of the banking sector of the Russian Federation, tab. 1 / Overview of the banking sector of the Russian Federation. http://www.cbr.ru/statistics/bank_sector/review/#highlight=o??op%7C?a??o?c?o?o%7Cce??opa%7Co??opa. Accessed 27 June 2021
7. Aganbegyan, A.G.: How to resume social and economic growth in Russia. Sci. Works Liberal Econ. Soc. Russia **222**(2), 164–182 (2020)
8. Larionova, I.V.: On bringing banking regulation in line with the standards of the Basel Committee on Banking Supervision (Basel III) in an unstable economic situation: monograph. Eds Larionova I.V., KNORUS, Moscow, p. 190 (2018)
9. CBR The main directions of the unified state monetary policy. http://www.cbr.ru/about_br/publ/ondkp/. 28 June 2021

# Analysis of the Innovations' Trends in the United Kingdom and Some Proposals for Russia

Svetlana Tsvirko[(⊠)] [iD]

Financial University under the Government of the Russian Federation,
Leningradskiy prospekt 49, 125993 Moscow, Russian Federation
STsvirko@fa.ru

**Abstract.** The article focuses on the features of the United Kingdom innovation policy. The primary method of research is SWOT analysis. The study's objective was to analyze strengths, opportunities, weaknesses, and threats connected with innovations' activities in the UK and to suggest some proposals for the development of innovations in Russia. The main peculiarities of the innovations' development in the UK revealed in the paper are as follows: significant participation of the private sector in innovations; striving for the optimal share and forms of state participation in financing the innovation sphere; great attention to fundamental research at universities and ensuring the commercialization of scientific developments; active cooperation with international partners; attracting foreign capital to the innovation sphere; active implementation of innovations in the public sector. The most serious problems of the innovations' development that are identified in the paper include: unresolved issues related to the Brexit conditions; problems with productivity growth, especially in some regions of the country; a relatively small share of government spending on R&D so far; the negative consequences of the COVID-19 pandemic. The analysis provides a basis for decision-makers to exploit opportunities and minimize the drawbacks in innovations' policy.

**Keywords:** Innovations · Rankings · Innovations policy · Innovations inputs · Innovations outputs · Research · Development · Technologies · SWOT analysis

## 1 Introduction

The study of issues related to the innovations is of current interest. The importance of innovations can be explained by their necessity for socio-economic development in different types of countries – both developed and developing. Innovative development is one of the trends of our time, but it manifests itself in various countries in different ways. The most innovative economies are considered to be developed countries and newly industrialized countries. Russia also continues to introduce new technologies, but so far, it is still in 45th place in the Global Innovation Ranking. In these conditions, it becomes necessary to study foreign experience in the formation and development of innovativeness [1].

© The Author(s), under exclusive license to Springer Nature Switzerland AG 2022
T. Antipova (Ed.): DSIC 2021, LNNS 381, pp. 320–331, 2022.
https://doi.org/10.1007/978-3-030-93677-8_28

The theoretical foundations for studying the topic of innovation were laid by Joseph A. Schumpeter [2], who introduced the term «innovation»; S.S. Kuznets [3], A. Kleinknecht [4], who further developed the concept of waives in innovations; R.N. Foster [5], P.F. Drucker [6], who researched innovation concerning the phenomenon of entrepreneurship, and many other scientists.

Among Russian contemporary authors who deal with the problems of the innovative development, we can name publications by Glazyev S. Y. [7], Sukharev O. S. [8–10]; country and sectoral features of innovation are considered in the works of Alexandrova, E., Poddubnaya, M. [11], Chernomorova T. V. [12], Ekimova N. A. [13], Mironova V.N., Starodubtseva E.B. [1], Melnik M., Antipova T. [14], Ross G., Konyavsky V. [15], Rozhkova D., Rozhkova N., Blinova U. [16] and others. Several international organizations have issued their working papers devoted to the innovations' policy and the results of the innovations' development.

However, not all aspects of the topic are disclosed, particularly the development of tools to stimulate innovative activity. In addition, some rethinking of innovative activity in the actions of individual states is necessary in connection with the impact of the COVID-19 pandemic. As far as we know, there are no papers with SWOT analysis of the innovations' system in the United Kingdom that can contribute to a more integrated assessment of the peculiarities of the innovations' policy in this country.

In this context, this study aims to analyze possibilities and drawbacks in the innovations' policy. For achieving the aim of the research, it is necessary to explain briefly the methodology of research in the form of SWOT analysis, to show some examples of the realization of innovations' policy in practice in the United Kingdom, clarify the findings in the framework of SWOT analysis and finally reveal conclusions of research and potential for future studies.

## 2   Methodology of Research

The methodological basis of the research are system approach, fundamental provisions of management, methods of comparative analysis, expert assessments. The comparative analysis method was used to determine the differences between innovations systems' developed in various countries and to reveal the successful cases.

Analysis of a specific country's experience with the innovation policy can produce relevant data and provide some understanding of the necessary measures to implement in the innovations' sphere. We will use SWOT analysis for this research.

SWOT analysis is a method of strategic planning consisting in identifying factors of the internal and external environment for the system. Factors are divided into four categories: Strengths (S), Weaknesses (W), Opportunities (O), Threats (T). The objective of the SWOT analysis is to provide a structured description of the situation regarding which a decision needs to be made. This is a universal method that is applicable in a wide variety of areas of economics and management.

As for input data, we used several rankings characterizing innovative activity and publications of the international organizations, such as the United Nations Educational, Scientific and Cultural Organization (UNESCO), World Intellectual Property Organization (WIPO), European Commission.

## 3    The UK's Place in the World's Leading Rankings of Innovations

The UK is one of the leading countries in innovation. As the experience of this country shows, a successful innovation policy allows ensuring competitiveness, investment attractiveness and stable economic development. Great Britain is high in various rankings characterizing innovative activity.

In the Global Innovation Index 2021, the UK ranked 4th after Switzerland, Sweden, and the United States [18]. The specified Global Innovation Index is calculated based on more than 80 variables and presents the complete set of the innovative development indicators across the world. The index is calculated as a weighted sum of scores from two groups of indicators:

1. disposable resources and conditions for innovation (Innovation Inputs), which include: institutions, human capital and research, infrastructure, development of the domestic market, business development;
2. obtained practical results of the implementation of innovations (Innovation Outputs), including: development of technologies and the knowledge economy, results of creative activity.

The final Global Innovation Index can also be considered in terms of the ratio of the cost of innovation development and the achieved effect. According to the results of the ranking for 2021, the UK is in 7th place in terms of indicators characterizing available resources and conditions for conducting innovations, and in 6rd place in terms of achieved results of implementation of innovations. More detailed information on the positions in the Global Innovation Index in recent years is presented in Table 1.

**Table 1.** Positions in the Global Innovation Index 2018–2021

|                       | 2018 | 2019 | 2020 | 2021 |
|-----------------------|------|------|------|------|
| GII of the UK         | 4    | 5    | 4    | 4    |
| - Innovation Inputs   | 4    | 6    | 6    | 7    |
| - Innovation Outputs  | 6    | 4    | 3    | 6    |
| GII of Russia         | 46   | 46   | 47   | 45   |
| - Innovation Inputs   | 43   | 41   | 42   | 43   |
| - Innovation Outputs  | 56   | 59   | 58   | 52   |

Source: Compiled by the author based on [17, 18]

Based on the ratio of available resources, as well as the conditions for innovation and the practical results obtained, the UK is highly effective in terms of efforts to develop innovation. Russia not only lags significantly behind the UK in this rating, but also has a different ratio of innovation inputs and innovation outputs. As it can be seen from Table 1, Russia has considerably lower rankings of innovation outputs then rankings of innovation inputs.

The UK has traditionally ranked high in the European Innovation Scoreboard. In its 2020 edition, the UK was not included in the European Union for the first time, so the

EU average was compiled for the 27 member countries. This has led to a decline in the EU average for innovation performance, as the UK has consistently performed above the EU28 average in previous years [19].

The UK's position in the separately identified indicator characterizing eco-innovation in Europe is also quite high. Thus, the United Kingdom rose in the corresponding ranking from 11th place in 2017 to 7th place in 2019. In 2019, the country received 118 points, which is higher than the EU average. In the ranking, the UK follows Germany, Austria, Sweden, Finland, Denmark and Luxembourg. Among the indicators included in the Eco-Innovation Index, the UK shows high results in terms of resource efficiency and in 2019 was ranked 3rd among the 28 EU countries [20].

Great Britain is distinguished by the high positions of universities. In the World University Rankings for 2021, Oxford takes the first place among the world's universities, and the University of Cambridge is also included in the top ten. The top 100 universities in The World University Rankings 2021 include 11 universities in the UK [21]. The country is recognized as a world leader in the quality of scientific publications.

Thus, the above results confirm that the UK is among the leading countries in the development of innovation.

## 4 UK Innovations' Organization System

For several decades, the UK's strategic planning system has been incorporated into the European Union's programming system and was linked to pan-European innovation development programs [12, p.92].

Until the early 2000s there was no officially announced innovation policy in the UK. In 2003, the UK Department of Trade and Industry published the Government's Technology Development Strategy, and in 2004 a Technology Strategy Board was established to invest in new technologies, support their development, and commercialize. In 2004, the UK government developed a strategy for state support for the development of high technologies, the main principles of which were presented in the report "Global Economic Challenges in the Long Term and Opportunities for the UK". Also, the Program of Investments in Science and Innovative Technologies for 2004–2014 was developed, which implied joint activities of various ministries and departments.

In the period from 2005 to 2008, in the structure of the national innovation system of Great Britain, reforms were carried out aimed at increasing the efficiency of the innovation management mechanism. In 2008, the Innovative Nation program was published, highlighting the value of an open innovation model, in which organizations partner with universities, other companies, and suppliers. Consumers are also agents of innovation - independently, jointly with businesses, or as co-producers of government services [12, p.91].

In December 2011, the Innovation and Research Strategy for Growth (IRSG) was presented. An important step was the identification in 2012 of priority research areas ("8 great technologies"), which received significant support in the UK:

1. big data and energy-efficient computing;
2. satellites and commercial use of space;
3. robotics and autonomous systems;
4. synthetic biology;
5. regenerative medicine;
6. agronomy;
7. modern materials and nanotechnology;
8. energy and its storage.

Each of these technologies creates long-term benefits for society, has the prerequisites for widespread use in various industries and sectors to create new products and services, as well as a high potential for entering international markets.

In 2018, UK Research and Innovation (UKRI), a nongovernmental organization for Research and Innovation (UKRI), was founded, it is funded by the Department of Business, Energy and Industrial Strategy. The UKRI structure includes:

1) Art and Humanities Research Council (AHRC);
2) Biotechnology and Biological Sciences Research Council (BBSRC);
3) Engineering and Physical Sciences Research Council (EPSRC);
4) Economic and Social Research Council, formerly the Economic and Social Research Council (ESRC);
5) Medical Research Council (MRC);
6) Natural Environment Research Council (NRC);
7) Science and Technology Facilities Council (STFC);
8) Innovate UK (innovation agency);
9) Research England division.

The first eight of these nine UKRI units have a UK-wide funding mandate, while Research England provides funding only to England.

Innovate UK supports UK organizations interested in international cooperation by providing access to international experience and practice, thus keeping their growth and development. The tasks of the UKRI include: identifying the achievements of science and technology that will stimulate economic growth; educating British innovators on innovative ideas in various fields and connecting them with partners that are necessary for further growth; assistance in starting, building, and developing a business.

UKRI's activities are quite successful. In its annual report for 2019–2020, UKRI notes that the supported research and innovation is of great economic importance to the country. Companies participating in research projects funded by UKRI increased their turnover and employment by 5.8–6% more in the three years after the project and 22.5–28% more in six years after the project than similar firms that did not receive support [22, p.10]. According to the recent Annual report 2020–2021 by UKRI, since establishment, they have supported more than 52800 awards with a combined value of more than £ 22 billion [23, p.21].

## 5   Features of Financing Innovation in the UK. Directions and Tools for Innovative Development in the UK

The share of public investment in total R&D investment in the UK is relatively small compared to other countries. So, according to the latest available statistics, in 2018 the share of R&D expenditures in GDP was 1.7%, and according to this indicator, the UK was not included in the list of 15 countries with the highest corresponding indicators (these include, for example, South Korea - 4.3%, Japan - 3.4%, Finland - 3.2%, Switzerland - 3.2%, Austria - 3.1%, Sweden - 3.1%, etc.). According to the Industrial Strategy adopted in the UK in 2017, it is planned to increase the share of R&D in GDP to 2.4% by 2027 [24, p.11] and to 3% in the long term. This is consistent with the ambitious goal of becoming the most innovative country in the world by 2030 [24, p.63].

As for the absolute indicators of R&D expenditures, according to the latest available information for international comparisons, the UK is in 8th place with $ 44 billion at PPP (after the USA, China, Japan, Germany, South Korea, France, India) [25]. The structure of the R&D expenditures is as follows: business – 65.15%, universities – 25.78%, government – 7.26%, private noncommercial organizations – 1.81%. In the UK, a favorable environment has been created that allows combining research with business, stimulating the innovative activity of companies, using public-private partnership tools to solve the problem of increasing the innovativeness and competitiveness of the economy.

In 2020, in the course of the pandemic crisis, the UK Government announced a significant increase in public investment in R&D to £ 22 billion annually by 2024–2025 [23]. For comparison, in 2017, public investment in R&D for the fiscal year 2021/22 was planned only in the amount of £ 12.5 billion [24]. It is likely that the conditions for recovery from the coronavirus crisis will require more government participation in financing investments than before.

The UK has an effective ecosystem of clusters, accelerators, technology parks, etc. The country has four of the 100 largest science and technology clusters in the world: London (15th), Cambridge (57th), Oxford (71st), and Manchester (93rd). Cambridge and Oxford, according to the compilers of the Global Innovation Index, are the most high-tech clusters in the world [17]. For instance, since 2010, the Tech Nation platform has been operating, which is a catalyst for change, accelerating the entire ecosystem, bringing together a network of founders and philanthropists, enterprises and politicians, foundations and institutions to collaborate in creating technological innovation. Such platforms perform important functions of coordinating the activities of clusters, accelerators, technology parks, etc.

Let's mention some examples of various funds and programs that contribute to innovative development in the UK. The Industrial Strategy Challenge Fund (ISCF) is the UKRI's premier innovation program to address challenges. Around £2.6 billion in public funds and £3 billion in negotiated funding from the private sector have been invested in projects that bring together researchers and businesses to tackle the social and industrial challenges of our time [26]. Projects financed from this fund must comply with one of the four areas of the UK Government's industrial strategy: "net growth"; "aging society"; "the future of mobility"; "artificial intelligence and data economics".

«Smart» is an innovative UK open grant funding program to support effective and commercially viable ideas in a range of technology areas. Projects can relate to any area of technology and economics, including arts, design and media, creative industries, and science and technology.

«Creative Clusters» and «Audience of the Future», funded by UKRI, are two of the world's largest R&D programs for creative industries, using new digital and immersive technologies to create new markets, services, and products in the field of fashion, cinema, games, museums, sports theaters.

An important program is the Knowledge Transfer Partnerships (KTP). It is a partially government-funded program to promote collaboration between businesses and universities in the United Kingdom. The Knowledge Transfer Partnership was launched in 2003, replacing the Learning Companies Program (TCS), which has existed since 1975. The program is funded by several public sector organizations and is led by Innovate UK, an executive government agency reporting to the Department of Business, Energy and Industrial Strategy. The partnership includes three participants: 1) company or agency; 2) knowledge base (university or other higher education institution, research organization); 3) newly qualified graduate. The objectives of each KTP program are to foster the transfer of knowledge and technology and the diffusion of technical and business skills within the company, stimulate and expand relevant business research and training conducted through the knowledge base, and enhance the business and special skills of the newly qualified graduate. Thus, this Partnership promotes cooperation between enterprises, research organizations and young scientists.

Regional Development Agency can be an example of the development institution used in the UK. As N.A. Ekimova revealed, these institutions were quite effective, primarily in the least developed regions [13]. Currently, the Regional Development Agency operates only in Scotland. The i54 South Staffordshire Technology Business Park is an example of a successful business park that has been revitalized with funds from the Regional Development Agency. The park houses Jaguar Land Rover Engine Factory, Food Testing Laboratory and feed Eurofins, the American space company Moog research centers, etc. In 2019, it was decided to expand the business park, which can help create an additional 1,700 jobs to the existing 2,700 jobs and attract up to £ 300 million in investments [27].

A distinctive feature of innovation in the UK is active collaboration with international partners. The UK is a member of EUREKA (European Research Coordination Agency). As part of their relationship with EUREKA, UK organizations involved in international projects are eligible for funding through programs such as Network Projects, Globalstars, Eurostars and Clusters.

In the UK, 17% of R&D investments are financed from overseas; about half of the research and development in the business sector was carried out by foreign companies [24]. In addition, a significant part of scientific publications was written in co-authorship with foreign scientists.

The UK is active in promoting eco-innovation, circular economy, and new business models with environmental benefits. A significant number of organizations in the country have been involved in promoting systemic eco-innovation by providing financial support, advice and networking opportunities to innovative SMEs, as well as government bodies at the regional and national level.

## 6  Results of SWOT Analysis of the Innovation System in the United Kingdom

Results of SWOT analysis of the innovation economy of the United Kingdom are presented in Table 2.

**Table 2.** SWOT analysis of the innovations system in the United Kingdom.

| Strengths | Weaknesses |
|---|---|
| - significant private sector participation in innovation,<br>- striving for an optimal form of state participation in financing the innovation sphere,<br>- developed infrastructure, including information and communication technologies' infrastructure,<br>- market sophistication, including competition and market scale,<br>- active cooperation with international partners,<br>- significant share of funding from abroad,<br>- hosting of the world's leading science and technology clusters,<br>- high quality of scientific publications,<br>- great attention to fundamental research and ensuring the commercialization of scientific developments | - relatively small share of government spending on R&D so far,<br>- problems with the growth of labor productivity, especially in some regions of the country,<br>- curtailment of the regional development agencies (except for Scotland) |
| Opportunities | Threats |
| - further increase in attracting foreign capital to the innovation environment,<br>- maintaining the reputation and further development of international cooperation,<br>- active innovation in the public sector,<br>- strengthening cooperation between universities and business,<br>- further development of the education system, including continuous education | - unresolved issues related to the Brexit conditions, which have an impact on the relationship with the EU related to cooperation in the innovation sphere and funding opportunities from European funds,<br>- increased international competition,<br>- negative consequences of the COVID-19 pandemic |

Source: Summarized by the author

## 7  Discussion

To sum up, the main strong points of the innovations' system in the United Kingdom are as follows: active participation of the private sector in innovation; striving for the optimal share and forms of state participation in financing the innovation sphere; great attention to fundamental research at universities and ensuring the commercialization of scientific developments; cooperation with international partners; attracting foreign capital to the innovation sphere; active implementation of innovations in the public sector.

The problems related to the stimulation of innovative activity include the unresolved issues related to the Brexit conditions. Opportunities for free labor migration, which previously allowed the UK to attract and retain researchers and innovators both in the university environment and in business, are likely to be limited. The UK's status in some programs is questionable, including the European Union's research and technology development program Horizon (Horizon 2020), which ran from 2014 to 2020, and was the largest EU framework program with a budget of 80 billion euros at 2011 prices. It will continue under the name Horizon Europe in the period 2021–2027 with the budget increased to 100 billion euros. Without a "deal" with the EU, the UK will be considered a "third country" in the Horizon Europe program. It should be clarified that the participants in this program are divided into EU member states, associate members, and «third countries». Organizations from «third countries» can participate in the research and technology development program, but are not eligible for automatic financial support from the program budget. Financing of projects for organizations from «third countries» is possible, but subject to some conditions.

The program Scale-Up Europe, actively discussed currently, is an initiative started by French President Emmanuel Macron in June 2021, is dominated by France, especially taking into consideration of France's presidency of the European Union in the first half of 2022; so the future of the UK participation in it is not clear.

The difficulties faced by the UK also include the problem of low productivity growth after the global financial and economic crisis of 2008–2009. In the period from 2011 to 2017 there was a stagnation in the performance indicator. In terms of productivity growth in 2016, the UK lagged behind such G7 countries as Germany, France, USA, Italy [28]. Since 2017, there has been a slight increase in productivity. Year 2020 metrics are volatile and inconsistent. The pandemic situation poses major challenges to performance measurement. It will likely be difficult to identify sustainable, long-term performance trends for some time.

The coronavirus pandemic has had a significant impact on the UK economy, as in other countries, becoming a test of the resilience of health care systems, social welfare, and, in general, the economy and society. Research and innovation have become an essential part of a coordinated European response to the pandemic crisis and will be vital to support a sustainable and comprehensive recovery in Europe.

As for the recommendations that can be done for Russia in accordance with the findings about innovation sphere of the United Kingdom, to our mind, they are as follows:

- it is necessary to have thoughtful, officially announced innovation policy; strategy of state support for innovation;
- prioritization is essential (similar to "8 great technologies", chosen in the UK), and the system of the supported projects' selection should be improved;
- in conditions of a lack of investment, an optimal mix of public and private investment is needed;
- Russia is famous for some deep breakthrough research, but it is necessary to use the tools to commercialize the achievements;
- universities, private sector and public sector should improve knowledge-sharing networks to increase awareness of successful innovations;
- it is recommended to use the variety of instruments (different funds, tax credits, subsidies) and to create an effective ecosystem of clusters, accelerators, technology parks, etc.;
- the development of human capital is relevant, which ultimately leads to creating new products and services in modern industries and spheres of the economy and innovative development in general.

# 8 Conclusion

In this paper, we have revealed the features of the innovations policy in the United Kingdom. We can conclude that the UK uses a wide range of mechanisms and instruments to stimulate innovative development. The following strong points of innovation policy in the UK can be noted: significant participation of the private sector in innovation; striving for the optimal share and form of state participation in financing the innovation sphere; great attention to fundamental research at universities and ensuring the commercialization of scientific developments; active cooperation with international partners; attracting foreign capital to the innovation sphere; active implementation of innovations in the public sector.

The problems and challenges in the field of innovative development in the UK include: unresolved issues related to the Brexit conditions, which has an impact on relations with the EU; problems with productivity growth, especially in some regions of the country; a relatively small share of government spending on R&D so far; the negative consequences of the COVID-19 pandemic.

Research and innovation in the UK, in line with current conditions, aims to improve the resilience of manufacturing sectors, the competitiveness of the economy, and society's digital and environmental transformation.

Analysis of the experience of the United Kingdom in the sphere of innovations can be used as a base for suggestions how to increase efficiency of the innovations policy in Russia. The strengths and opportunities, as well as threats and weaknesses of the innovations system in the United Kingdom, highlighted in the research are to be noted for economic policy considerations. Suggestions for future research are connected with practical recommendations for more efficient innovations' policy.

# References

1. Mironova, V.N., Starodubtseva, E.B. (eds.): Trends in innovative development of the world economy: monograph. Knorus, Moscow (2021)
2. Schumpeter, J.A.: The theory of economic development. An inquiry into profits, capital, credit, interest, and the business cycle. Harvard Econ. Stud. (1934)
3. Kuznets, S.S.: Schumpeter's business cycles. Am. Econ. Rev. **30**(2), 257–271 (1940)
4. Kleinknecht, A.: Innovation, accumulation, and crisis: waves in economic development, review (Fernand Braudel Center) IV, Spring, 687–711 (1981)
5. Foster, R.N.: Innovation: The Attacker's Advantage. Summit Books (1986)
6. Drucker, P.F.: Innovation and Entrepreneurship. Harper Business (1993)
7. Glazyev, S.Y., Lvov, D.S.: Economic Theory of Technical Development: Monograph. Science, Moscow (1990)
8. Sukharev, O.S.: Economic Policy. Introduction into the crises and growth theory. Finance and Statistics, Moscow (2011)
9. Sukharev, O.S.: Management of economy policy and industrial development. Finance and Statistics, Moscow (2012)
10. Sukharev, O.S.: Structural analysis of the economy. Finance and Statistics, Moscow (2012)
11. Alexandrova, E., Poddubnaya, M.: Digital technologies development in industry sectors and areas of activity. In: Antipova, T. (ed.) ICIS 2020. LNNS, vol. 136, pp. 112–124. Springer, Cham (2021). https://doi.org/10.1007/978-3-030-49264-9_10
12. Chernomorova, T.V.: Great Britain: innovation policy and methods of its implementation. Act. Prob. of Europe **1**, 89–116 (2013). (in Russian)
13. Ekimova, N.A.: Regional development institutions: UK experience. J. Inst. Stud. **12**(3), 42–59 (2020). https://doi.org/10.17835/2076-6297.2020.12.3.042-059. (in Russian)
14. Melnik, M., Antipova, T.: Organizational aspects of digital economics management. In: Antipova, T. (ed.) ICIS 2019. LNNS, vol. 78, pp. 148–162. Springer, Cham (2020). https://doi.org/10.1007/978-3-030-22493-6_14
15. Ross, G., Konyavsky, V.: Modeling of investment in it-business. In: Antipova, T. (ed.) ICADS 2021. AISC, vol. 1352, pp. 377–389. Springer, Cham (2021). https://doi.org/10.1007/978-3-030-71782-7_33
16. Rozhkova, D., Rozhkova, N., Blinova, U.: Development of the e-government in the context of the 2020 pandemics. In: Antipova, T. (ed.) ICADS 2021. AISC, vol. 1352, pp. 465–476. Springer, Cham (2021). https://doi.org/10.1007/978-3-030-71782-7_41
17. Global Innovation Index 2020: Who Will Finance Innovation? (2020). https://www.wipo.int/global_innovation_index/en/2020/. Accessed 12 Aug 2020
18. Global Innovation Index 2021: Tracking Innovation through the COVID-19 Crisis (2021). https://www.wipo.int/edocs/pubdocs/en/wipo_pub_gii_2021.pdf. Accessed 23 Sept 2021
19. European innovation scoreboard 2020 - main report. European Commission. https://ec.europa.eu/docsroom/documents/42981. Accessed 12 Dec 2020
20. Eco-innovation. United Kingdom. https://ec.europa.eu/environment/ecoap/united-kingdom_en. Accessed 10 Dec 2020
21. The Times Higher Education World University Rankings 2021. https://www.timeshighereducation.com/student/best-universities/best-universities-uk. Accessed 10 Dec 2020
22. UK Research and Innovation. Annual Report and Accounts 2019-20. https://www.ukri.org/about-us/what-we-do/annual-report-and-accounts/. Accessed 10 Dec 2020

23. UK Research and Innovation. Annual Report and Accounts 2020-2021. https://www.ukri.org/wp-content/uploads/2021/07/UKRI-200721-AnnualReport2020-2021.pdf. Accessed 19 Aug 2021
24. Department for Business, Energy & Industrial Strategy, Industrial Strategy: building a Britain fit for the future (2017). https://assets.publishing.service.gov.uk/government/uploads/system/uploads/attachment_data/file/664563/industrial-strategy-white-paper-web-ready-version.pdf. Accessed 22 Nov 2020
25. How much your country invest in R&D? UNESCO Institute for Statistics. http://uis.unesco.org/apps/visualisations/research-and-development-spending/. Accessed 12 Aug 2021
26. What is the Industrial Strategy Challenge Fund. https://www.ukri.org/our-work/our-main-funds/industrial-strategy-challenge-fund/what-is-the-industrial-strategy-challenge-fund/. Accessed 10 Dec 2020
27. New era for I54 as work starts on western extension. 27.06.2019. https://www.i54online.com/latest-news/new-era-for-i54-as-work-starts-on-western-extension. Accessed 10 Dec 2020
28. Productivity: Key Economic Indicators. https://commonslibrary.parliament.uk/research-briefings/sn02791/. Accessed 10 Dec 2020

# The Organization of Analytical Procedure for Assessing the Tax Efficiency of the Company According to the Financial Statements for Digitalization

Liudmyla Lakhtionova[1]([envelope]) [iD], Svitlana Kalabukhova[2] [iD],
Oksana Isai[2] [iD], and Olena Chuk[2] [iD]

[1] National Aviation University, 1, Liubomyr Husar Avenue,
Kyiv 03067, Ukraine
Ludmilala@i.ua
[2] Kyiv National Economic University named after Vadym Hetman,
54/1, Peremohy Avenue, Kyiv 03057, Ukraine

**Abstract.** The article examines the organization of the analytical procedure for assessing the tax efficiency of the company according to the financial statements for its digitalization. It is proposed to organize an analytical procedure for studying the tax efficiency of the company, using financial statements. According to the annual financial report of the Joint Stock Company "Southern Mining and Processing Plant" shows an example of the proposed organization of the analytical procedure for disclosing the tax efficiency of the company. The results of the study are useful for users for further processing in the context of digitalization.

**Keywords:** Unification · Analytical procedure · Tax efficiency

## 1 Introduction

The difficult economic situation in the world, associated with quarantine conditions, requires the adaptation of companies in the long run of the COVID-19 pandemic.

Ukrainian enterprises work especially in difficult conditions. The heavy long-term anti-terrorist operation in the east of the country and the COVID-19 pandemic have led to a difficult economic situation in the country: most companies are not working or working part-time, living standards continue to fall, the number of unemployed and low-income.

Bankruptcy and business collapse is becoming a hallmark of most countries.

And all this is happening against the background of the fourth industrial revolution, the introduction into human life of digitalization, attempts to use artificial intelligence.

The difficult economic situation in the world requires the adaptation of tax systems of different countries to modern force majeure conditions of companies.

An important lever for regulating the distribution between the state and enterprises of all forms of profit as the main source of investment and innovation activities of

T. Antipova (Ed.): DSIC 2021, LNNS 381, pp. 332–341, 2022.
https://doi.org/10.1007/978-3-030-93677-8_29

enterprises is the corporate income tax, which is part of the tax system of any country in the world.

The mechanism of tax policy of the enterprise makes it possible to maximize the net profit of management staff with optimal parameters of the tax burden.

The need for different groups of stakeholders to assess the impact of the tax system on the financial results of economic entities has led to the emergence of tax analysis.

Many world scientists consider the construction and improvement of the tax system, tax law, types of taxes and fees, resolution of tax disputes, optimization of the tax burden: 1) Australia: C. Coleman [1], G. Hart [1], B. Bondfield [1], M. Mckerchar [1], J. McLaren [1], K. Sadiq [1], A. Ting [1]; 2) Canada: G.C. Ruggeri (2018) [2], Carole Vincent (2018) [2]; 3) Great Britain: Eduardo Baistrocchi (2018) [3]; 4) India: Karthik Sundaram (2021) [4]; 5) Jordan: F. Alasfour (2016), M. Samy (2016), R. Bampton (2016) [5]; 6) USA and the Netherlands: CCH Tax Law Editors (2020) [6]; (2018) [7]; 7) Russia: KS Grigorieva (2019) [8], AM Grinkevich (2019) [9] and others.

However, in the thematic literature, not everyone understands the organization of the analytical procedure for disclosing the tax efficiency of the company according to the financial statements for its digitalization.

The urgency of these issues is growing in the context of economic globalization against the background of the fourth industrial revolution, which accelerates the development of digital technologies in the methodology and organization of tax analysis and tax management at the micro and macro levels.

Existing world experience and review of professional literature has shown that in general the key task of tax analysis is to substantiate economic decisions aimed at optimizing the tax costs of enterprise profits.

At the same time, a number of issues related to the analytical procedure remain unresolved:

- to understand external users according to the financial statements of the nature of taxation of financial results;
- assessments of the impact of the tax environment on financial results;
- determining the tax efficiency of the business entity;
- clarifying the intentions of management to preserve the capital of owners and investors

All these problematic issues need to be addressed in the context of the fourth industrial revolution and the digitalization of the economy, in real attempts to use artificial intelligence in human life.

**The purpose of the article** is to develop recommendations for the organization of an analytical procedure for disclosing the tax efficiency of the company according to the financial statements for its digitalization.

**Development and testing of proposals** for the organization of an analytical procedure for assessing the tax efficiency of the company according to the financial statements for digitalization will be carried out on the example of the Joint Stock Company "Southern Mining and Processing Plant") [10].

## 2  Research Method

The study is used: absolute, relative and average values, comparison, grouping, tabular, modeling, time series, detailing, visualization, generalization.

## 3  Literature Review

An interesting textbook on taxes, tax law, legal analysis, tax analysis was prepared by Australian scientists Cynthia Coleman (2013), Brett Bondfield (2013), Margaret Mckerchar (2013), John McLaren (2013), Kerrie Sadiq (2013), Antony Ting (2013) [1]. Authors use specific examples to consider various tax problems and ways to solve them.

Issues of tax reform and the tax burden on taxpayers are being studied by various scholars. For example, Canadian scholars Ruger (2018) and Vincent (2018) analyze various proposals for tax reform to create a discourse on debunking the myths associated with a fixed tax. This book proposes progressive and comprehensive tax reforms that simplify the tax system for the vast majority of taxpayers. At the same time, reforming the tax system does not hinder economic growth [2]. The authors reveal the problems and prospects of tax reform.

A prominent place in the tax academic literature is occupied by a unique and triumphant edition of the University of Cambridge - the book "Global analysis of disputes under tax treaties" edited by Eduardo Baistrocchi (2017), London School of Economics and Political Science [3].

This two-volume set offers an in-depth analysis of the leading tax treaty disputes in the G20 and beyond within the first century of international tax law. Including country-by-country and thematic analyzes, the study is structured around a novel global taxonomy of tax treaty disputes and includes an unprecedented dataset with over 1500 leading tax treaty cases. By adopting a contextual approach the local expertise of the contributors allows for a thorough and transparent analysis. This set is an important reference tool for anyone implementing or studying international tax regulations and will facilitate the work of courts, tax administrations and practitioners around the world. It is designed to complement model conventions such as the OECD Model Tax Convention on Income and on Capital. Together with Resolving Transfer Pricing Disputes (2012), it is a comprehensive addition to the current debate on the international tax law regime [3].

Analysis of the evolution of tax law in India was studied by Indian scientist Karthik Sundaram (2021) [4].

F. Alasfour (2016), M. Samy (2016), R. Bampton (2016) (Jordan) study how people determine their level of tax morality and decisions about compliance with tax requirements [5].

With the publication of Coronavirus (COVID-19) Tax Relief: Law, Explanation & Analysis, Wolters Kluwer (2020) is providing practitioners with a single integrated source for law and explanation of the tax provisions of the Families First Coronavirus Response Act and the Coronavirus Aid, Relief, and Economic Security (CARES) Act as well as other important guidance issued by the IRS. This volume includes the text of amended sections of the Internal Revenue Code, Congressional reports, and detailed explanations, complete with analysis and practitioner comments covering the tax provisions of these two massive Acts (USA and Netherlands) [6].

With the publication of Tax Cuts and Jobs Act: Law, Explanation and Analysis, Wolters Kluwer (2018) is providing practitioners with a single integrated source for law and explanation of the tax provisions of important 2017 tax legislation. The 2017 LEA includes the complete text of amended laws, Congressional reports, and detailed explanations, complete with analysis and practitioner comments covering the tax provisions and landmark Tax Reform legislation USA and Netherlands) [7].

Russian scientists KS Grigoriev (2019) and AM Grinkevich (2019) analyze the interests of the main participants in tax relations, their contradictions and ways of harmonization in the direction of forming an effective and fair tax system. The authors substantiate the use of a functional approach to assessing the effectiveness of the tax system, offer indicators for assessing the effectiveness of the tax system [8].

The monograph of the Russian scientist O. Ordynska (2017) is devoted to topical issues of tax system reform in different countries. The book examines the experience of tax reforms in a number of European countries, the United States, China and some other foreign countries, as well as in the Russian Federation. A feature of the monograph is a detailed analysis of the similarities and differences of the main directions of tax reforms carried out in different countries, as well as the impact of these reforms on the level of tax burden on different groups of taxpayers [9].

A number of issues related to the analytical procedure for external users to understand the nature of taxation of financial results, assess the impact of the tax environment on financial results, determine the tax efficiency of the entity and clarify the intentions of management to save capital remain unresolved. owners and investors.

## 4    Research Results

The main regulatory document that regulates the issues related to the determination of the costs of corporate income tax is IAS 12 "Taxes for profit "[11]. IAS 12 "Income Taxes" uses the following meanings: "accounting profit is the profit or loss for the period before tax expense is deducted; taxable profit (tax loss) - profit (loss) for the period determined in accordance with the rules established by the tax authorities; tax expenses (tax income) - the total amount included in the determination of profit or loss for the period in accordance with current and deferred taxes; current tax - the amount of income taxes payable (refundable) in respect of taxable income (tax loss) for the period; deferred tax liabilities - the amount of income taxes payable in future periods; deferred tax assets - the amount of income taxes that are reimbursable in future periods "[11].

As can be seen from the above provisions, tax analysis is based on the use of two independent sources of information: data generated by accounting rules (accounting profit, expenses (income) from income tax) and data generated by tax accounting rules (tax income, current income tax, income tax rate). The difference between the accounting profit (loss) and tax profit (loss) of the reporting period, formed as a result of the application of different rules of recognition of income and expenses, which are established in the regulations on accounting and tax legislation, is the basis for recognition in accounting and reporting deferred tax assets and deferred tax liabilities.

The vast majority of external users of financial statements are interested in the question at what real rate the corporate income tax was paid for the reporting period, as such information is useful for comparing it with industry average values and understanding the tax efficiency of the entity. IAS 12 "Income Taxes" when disclosing information in the financial statements recommends calculating the average tax rate as tax expense (income) divided by accounting profit [11, paragraph 86]. In the foreign scientific professional literature, the average corporate income tax rate is interpreted as an indicator of "effective tax rate" (ETR) and calculated by formula (1) [12, p. 224]:

$$ETR = (Current\ Tax + Deferred\ Tax)/Net\ Incom\ Befor\ Tax \qquad (1)$$

Using the Ukrainian practice, the amount consisting of the current income tax (Current Tax) taking into account the deferred tax liability (VPZ) and the deferred asset (VPA), ie the deferred income tax (Deferred Tax), is proposed to be interpreted as "expenses from income tax "and formula 1 for the purposes of tax analysis is recommended to be modified as follows (formula 2):

$$ETR = income\ tax\ expense/accounting\ profit \qquad (2)$$

In order to identify taxpayers who are at risk of non-payment of taxes and fees, it is recommended to define the level of income tax payment as the ratio of the amount of income tax stated in the declaration (hence the current income tax) to income from any activity (less indirect taxes), determined by the rules of accounting, which is taken into account when determining the object of taxation.

This indicator characterizes the tax burden of the company on income tax and can be useful for external users in the process of analyzing the financial statements of the company to compare with the industry average and monitor the likelihood of evasion of income tax.

As can be seen from formulas 1, 2, the effective income tax rate is affected by deferred tax, which is formed from indicators of deferred tax assets and deferred tax liabilities, which are present in the balance sheet and, accordingly, affect the calculation of various financial ratios. Obviously, the movement of deferred tax assets refers to the investment activities of the enterprise, as it is associated with the inflow and outflow of non-current assets, and the movement of deferred tax liabilities relates to the financial activities of the enterprise, as it leads to changes in equity and debt. Therefore, the excess of the amount of deferred tax liabilities over the amount of deferred tax assets (passive balance) can be characterized as a kind of additional source of financing.

It is proposed to organize the analytical procedure of assessing the company's tax efficiency according to the financial statements for digitalization in four stages and start with a coefficient analysis of tax efficiency by calculating the effective income tax rate and the level of tax burden on income tax. To do this, it is recommended to organize the composition, structure and movement of information for its digitalization in the following way (Table 1).

Previous author's developments were used for clarity of calculations [13].

**Table 1.** Calculation of the effective income tax rate and the level of tax burden on corporate income tax (stage 1)

| Indexes | For the previous period, (0) | For the reporting period, (1) | | Absolute deviation, Δ | Progress (T),% |
|---|---|---|---|---|---|
| 1. Accounting profit (before tax) (OP) | 9 922 237 | 2 409 262 | | +12 487 025 | 225,8 |
| 2. Income tax expense (income) (BPP) | 1 779 648 | 3 996 985 | | +2 217 337 | 224,6 |
| 3. Current income tax (PPP) | 2 001 791 | 4 151 036 | | +2 149 245 | 207,4 |
| 4. Total income from activities (D) | 23 402 666 | 35 638 713 | | +12 236 047 | 152,2 |
| Analytical operations | | | | | |
| Calculation of the effective income tax rate (ETR): (p.2/p.1 × 100%) | 17,9 | 17,8 | | 17,8% <18% (there is tax efficiency) 17,8–17,9 = −0,1% (there is a positive trend) | |
| Determining the level of tax burden of the enterprise on income tax: (p.3/p.4 × 100%) | 8,6 | 11,6 | | 11,6–8,6 = +3% (there is a negative trend) | |

(Source: author's elaboration).

Table 1 shows that in the reporting period the effective income tax rate is 17,8% and is lower than the base for Ukraine (18%), which indicates an effective tax policy of management personnel of the enterprise; At the same time, there is an increase in the total tax burden of the enterprise on income tax (+3%), which indicates a negative trend and increasing impact on the financial results of its activities tax environment.

In the second stage, it is proposed to carry out a factor analysis of income tax costs and it is recommended to organize the composition, structure and movement of information for its digitalization in this way (Table 2).

As can be seen from Table 2, income tax expenses increased by UAH 2 217337 thousand, which amounted to 224,6% of the growth rate, while a significant negative impact on their change was caused by an increase in the amount of current income tax (+2 149 245 thousand), as well as a decrease in the benefits of deferred income tax by 68 092 thousand UAH, which indicates a negative trend in the preservation of financial capital for owners and investors.

In the third stage, it is proposed to conduct a structural analysis of tax efficiency by determining the share of deferred income tax in accounting profit and diagnosing the presence of an additional free source of financing due to deferred income tax. To do this, it is recommended to organize the composition, structure and movement of information for its digitalization in the following way (Table 3).

As can be seen from Table 3, the company has reduced the share of deferred income tax in accounting profit to 0,7%, but can use deferred tax liabilities as a free additional source of financing in the amount of UAH 534792 thousand.

**Table 2.** Factor analysis of the costs of corporate income tax (stage 2)

| Indexes | For the previous period, (0) | For the reporting period, (1) | Abs. deviation, Δ | Progress (T),% |
|---|---|---|---|---|
| 1. Costs (income) from income tax (BPP) | 1 779 648 | 3 996 985 | +2 217 337 | 224,6 |
| 2. Current income tax (PPP) | 2 001 791 | 4 151 036 | +2 149 245 | 207,4 |
| 3. Deferred income tax (MPP) (item 1 - item 2) | (222 143) | (154 051) | −68 092 | 69,3 |

| Analytical operations | |
|---|---|
| Construction of a 2-factor additive income tax expense model (BPP) | BPP = PPP + MPP<br>BPP (0) = 1 779 648 =<br>= 2 001 791 + (−222 143)<br>BPP (1) = 3 996 985 =<br>= 4 151 036 + (−154 051) |
| Calculation of the impact of the 1st factor - changes in the amount of current income tax (F1):<br>conditions. BPP (F1) = PPP (1) + MPP (0);<br>ΔBPP (F1) = conditions. BPP (F1) - BPP (0) | conditions. BPP (F1) =<br>4 151 036 + (−222 143) = 3 928 893<br>ΔBPP (Φ1) =<br>= 3 928 893 − 1 779 648 =<br>= +2 149 245 |
| Calculation of the impact of the 2nd factor - changes in the amount of deferred income tax (F2):<br>conditions. BPP (F2) = BPP (1) + MPP (1) = BPP (1);<br>ΔBPP (F2) = BPP (1) - conditions. BPP (F1) | conditions. BPP (F2) =<br>= 3 996 985<br>ΔBPP (Φ2) =<br>= 3 996 985 − 3 928 893 =<br>= +68 092 |

(Source: author's elaboration).

In the fourth stage, it is proposed to assess the risk of non-payment of income tax by the company and the risk of reduction of retained earnings in the future due to deferral of income tax in the reporting period. To do this, it is recommended to organize the composition, structure and movement of information for its digitalization in the following way (Table 4).

Table 4 shows that the company has no risk of non-payment of income tax and non-fulfillment of its tax obligations to the state, but in future periods due to deferral of income tax in the reporting period, the company may lose 1.9% of retained earnings.

**Table 3.** Structural analysis of deferred corporate income tax (stage 3)

| Indexes | At the beginning. period, (0) | At the end. report. period, (1) | Abs. deviation, $\Delta$ | Progress (T),% |
|---|---|---|---|---|
| 1. Accounting profit (before tax) (OP) | 9 922 237 | 22 409 262 | +12 487 025 | 225,8 |
| 2. Deferred tax liabilities (VPZ) | 0 | 534 792 | +534 792 | 1000 |
| 3. Deferred tax assets (VPA) | 163 791 | 0 | −163791 | 0 |
| 4. Deferred income tax (deferred income tax) (MPP) | (222 143) | (154 051) | −68 092 | 69,3 |
| Analytical operations | | | | |
| Determining the share of deferred income tax in accounting profit: % MPP = NPP/OP × 100% | 2,2 | 0,7 | 0,7 – 2,2 = −1,5% (negative trend) | |
| Finding out if there is an additional free source of funding ($\varPi K^{B\varPi 3}$) due to deferred income tax: $\varPi K^{B\varPi 3}$ = VPZ − VPA | $\varPi K^{B\varPi 3}$ = 534 792 − 0 = + 534 792 thousand UAH (there is a free additional source of funding in the balance sheet) | | | |

(Source: author's elaboration).

**Table 4.** Assessment of risks of loss of tax efficiency of the company (stage 4)

| Indexes | At the beginning. period, (0) | At the end. report. period, (1) | Abs. deviation, $\Delta$ | Progress (T),% |
|---|---|---|---|---|
| 1. Retained earnings (NP) | 9 394 667 | 28 141 355 | +18 746 688 | 299,5 |
| 2. Deferred tax liabilities (VPZ) | 0 | 534 792 | +534 792 | 1000 |
| 3. Deferred tax assets (VPA) | 163 791 | 0 | −163791 | 0 |
| 4. Current income tax (PPP) | 2 001 791 | 4 151 036 | +2 149 245 | 207,4 |
| 5. Total income from activities (D) | 23 402 666 | 35 638 713 | +12 236 047 | 152,2 |
| Analytical operations | | | | |
| Testing for the risk of non-payment of income tax: PPP < D | 207,4 > 152,2 (no risk) | | | |
| Assessment of the risk of decrease in retained earnings in the future due to deferred income tax in the reporting period (RNP): (VPZ − VPA) / NP × 100% | (534 792 – 0)/28 141 355 × 100% = 1,9% | | | |

(Source: author's elaboration).

## 5   Conclusion

In modern conditions of digitalization of the financial reporting process, analytical support for stakeholder's decision-making in practice has a low information culture of documenting the analytical process, which does not contribute to the digitalization of the analytical process in the context of facilitating the formulation of significant analytical conclusions by different stakeholder groups.

Since the vast majority of financial capital providers are interested in the question at what real rate was paid corporate income tax for the reporting period, the creation of uniform forms of analytical documents to provide such information is the main direction of development of analytical processes.

Modeling of accounting and analytical information creates an experimental basis for understanding the process of managing an economic entity as a technological process in the digital age. Therefore, modeling of analytical procedures with information of financial statements in the form of analytical tables should be considered an important step in the organization of relevant analytical support of the process of management of economic entities in the information environment.

The proposed forms and content of analytical tables to assess the tax efficiency of the company according to the financial statements contribute to the unification of the text of the analytical document in this subject area, when constant information is presented in the form of common names of indicators, analysis tools, analytical operations, entered in the appropriate columns and variable information such as specific data and conclusions will be entered in the columns in accordance with the specific financial report.

It will greatly simplify and facilitate the preparation of forms of analytical documents for implementation in economic software products in the field of accounting and taxation, increase the information culture of documenting the analytical process and transparency of information on the formation of the effective income tax rate and the level of tax burden on the income tax of the company selected for investment or business partnership; will strengthen the clarity and informativeness of analytical conclusions about the results and prospects of business continuity, will create an innovative basis for the digitalization of the analytical process in a digital economy.

## References

1. Ruggeri, G.C., Carole, V.: An Economic Analysis of Income Tax Reforms. London (2018)
2. Coleman, C., et al.: Australian Tax Analysis: Cases, Commentary. Commercial Applications and Questions. Thomson Reuters, Pyrmont (2013)
3. Baistrocchi, E. (ed.): A Global Analysis of Tax Treaty Disputes. Cambridge University Press, Cambridge (2017)
4. Karthik, S.: Tax, Constitution and the Supreme Court: Analysing the evolution of taxation law in India. Oak Bridge Publications, Bangalore (2021)
5. Alasfour, F., Samy, M., Bampton, R.: The Determinants of Tax Morale and Tax Compliance: Evidence from Jordan. In: Advances in Taxation, pp. 125–171, vol. 23. Emerald Group Publishing Limited, Bingley (2016)

6. Coronavirus (COVID-19) Tax Relief: Law, Explanation & Analysis / CCH Tax Law Editors. Wolters Kluwer, Alphen aan den Rijn, Netherlands (2020)
7. Tax Cuts and Jobs Act: Law, Explanation and Analysis / CCH Tax Law Editors. Wolters Kluwer, Alphen aan den Rijn, Netherlands (2018)
8. Grigor'yeva K.S., Grinkevich A.M.: Otsenka effektivnosti nalogovoy sistemy Rossiyskoy Federatsii (Evaluation of the efficiency of the tax system of the Russian Federation). Natsional'nyy issledovatel'skiy Tomskiy gosudarstvennyy universitet, Tomsk (2019)
9. Ordynskaya Ye.: Transformatsiya nalogovoy sistemy v period krizisa. Rossiyskiy i zarubezhnyy opyt (Transformation of the tax system during the crisis. Russian and foreign experience). Prospekt, Moscow (2017)
10. Aktsionerne tovarystvo «Pivdennyy hirnycho-zbahachuval'nyy kombinat». Finansova zvitnist' vidpovidno do Mizhnarodnykh standartiv finansovoyi zvitnosti ta zvit nezalezhnoho audytora 31 hrudnya 2020 r. (Joint-stock company "Southern Mining and Processing Plant". Financial statements in accordance with International Financial Reporting Standards and the report of the independent auditor as of December 31, 2020). http://www.ugok.com.ua/upload/iblock/c2f/%D0%A4%D1%96%D0%BD%D0%B0%D0%BD%D1%81%D0%BE%D0%B2%D0%B0%20%D0%B7%D0%B2%D1%96%D1%82%D0%BD%D1%96%D1%81%D1%82%D1%8C%20%D0%90%D0%A2%20%D0%9F%D0%86%D0%92%D0%94%D0%93%D0%97%D0%9A%20%D0%B7%D0%B0%202020%20%D1%80%D1%96%D0%BA%20%D1%80%D0%B0%D0%B7%D0%BE%D0%BC%20%D0%B7%D1%96%20%D0%B7%D0%B2%D1%96%D1%82%D0%BE%D0%BC%20%D0%B0%D1%83%D0%B4%D0%B8%D1%82%D0%BE%D1%80%D0%B0.pdf. Accessed 23 Aug 2021
11. Mizhnarodnyy standart bukhhalters'koho obliku 12 «Podatky na prybutok» (International Accounting Standard 12 «Income Taxes». http://zakon.rada.gov.ua/laws/show/929_012. Accessed 23 Aug 2021
12. Scholes, M., Wolfson, M., Erickson, M., Hanlon, M., Maydew, E., Shevlin, T.: Taxes and Business Strategy: A Planning Approach 5th Ed. 528 p. Prentice Hall, Upper Saddle River (2015)
13. Lakhtionova, L., Muranova, N., Bugaiov, O., Ozeran, A., Kalabukhova, S.: Balance sheet (statement of financial position) transformation in the light of new digital technology: ukrainian Experience. In: Antipova, T. (ed.) Integrated Science in Digital Age 2020, pp. 25–41. Springer, Cham (2021). https://doi.org/10.1007/978-3-030-49264-9_3

# Use of Digital Channels by Micro Loan Customers in India

Kadambelil Paul Thomas[(✉)]

ESAF Small Finance Bank, Building No. VII/83/8, ESAF Bhavan,
Mannuthy, Thrissur, Kerala 680651, India
paul@esafbank.com

**Abstract.** Microfinance has played a significant role in financial inclusion in India. Formerly, the commercial banks have excluded a large section of the population in the country which is low-income and lack credit worthiness for formal credit. They have not been able to reach the poor; a considerable segment of that excluded class is covered by access to microfinance. The inaccessibility of credit at the grass root level is one of the reasons for rampant under-development of rural India. As the formal credit institutions were rather incapable of dealing with the financial requirements of the poor, microfinance emerged as an alternate credit system. Microfinance has further gained momentum by leveraging technology to provide financial services to the poor. This paper outlines how digitalization has influenced financial inclusion and what are the various challenges that must be tackled with to make it more effective. The paper also emphasizes how both the lenders and borrowers can use digital media to reap more benefits of the various advances in financial inclusion. It gives insight into how the impact of micro-lending can be augmented using technology.

**Keywords:** Microfinance · Digitalization · Financial inclusion · Customer-centric

## 1 Introduction

Financial inclusion is the foundation of any developing economy. Like elsewhere, microfinance has a key role in the development of rural and urban India. It act as an anti-poverty vaccine for the people living in rural areas. It aims at assisting communities of the economically excluded to achieve greater level of asset creation and income security at the household and community level. The utmost significance is that it dispenses easy access to capital for small entrepreneurs. The Indian microfinance institutions provide loans, insurance and access to savings accounts. A lot of commendable novel experiments and approaches have happened in India, especially in easing the financial exclusion of the small and marginal clients, different face-to-face methods in microfinance branch banking and off late, use of digital channels [1].

This work defends how the lenders and borrowers can augment the micro-lending process using digital channels. The paper also focuses on the positioning of time and cost-saving means to benefit the needy stakeholders with a blend of conventional and digitally aided financial services viz., in-branch communication with customers, digital

T. Antipova (Ed.): DSIC 2021, LNNS 381, pp. 342–357, 2022.
https://doi.org/10.1007/978-3-030-93677-8_30

transformation of the microfinance branches, customer-centric initiatives, and the consequent redefined role of field and branch staff. Thus, this work judgmentally delivers the range, path, progress, adoptable and adaptable sequential strategies for using digital technology in the domain of Indian microfinance.

## 1.1 The Research Question

India has been keen in bringing down the divide in banking finance, micro credit, financial exclusion and the popular 'digital divide'. The increasing use of digital tools has had an impact across all sectors of the economy. The financial sector has also seen itself driven into the new digital era, and microfinance is not an exception. As digital tools are introduced by microfinance institutions (MFIs), the key question is how far the various digital approaches have been helpful; whether the new information and communications technologies and digitized tools can serve as strong counterpart in making the micro financing providers user-friendly and especially in India.

## 1.2 Relevance of the Topic

In India, digital channels have been evangelized by the banks with Government's infrastructural support of developing an ecosystem of national level identification, credit infrastructure, receipt and payment infrastructure such as engagement of Business Correspondents (BCs) and agents for last mile delivery; and upsurge of FinTech entities which leverage technology to offer suitable financial products and easier processes. The Government-led initiative of Jan Dhan Accounts, Aadhaar biometric ID (Unique Identification numbers (UID)) and Mobile (JAM Trinity) which used biometric identification system, mobile technology and a bank account accelerated the adoption of various digital channels. Moreover, popularization of retail payment system products such as CTS, AEPS, NACH, UPI, IMPS operated by National Payments Council of India (NPCI) have caused digital transactions to increase manifolds. These systems have been governed with proper regulations that ensure a level the playing field, interoperability and allow for effective regulation and supervision. Bharat Bill Payment System (BBPS), payments via feature phones or smart phones or WhatsApp are evolving to be few digital channels helping the micro loan customers.

## 1.3 Research Gap and Pros and Cons of Employing Digital Channels in Microfinance Sector of India

Microfinance digitalization: is a menace or chance? A clear answer to this question is the research gap envisaged by the researcher. The digital revolution of the MFI front is a necessity for the survival of the sector, as viewed by many. This paper discusses the challenges and risks of this necessary transition for microfinance.

### Possible Benefits Due To Digitalization of Microfinancing

1. It allows MFIs to increase revenues and reduce costs.
2. It will bring the opportunity to leverage the technology continuance theory (TCT) for a special type of relationship banking.

3. It will help MFIs learn and practice the new type of personalized customer experience.
4. The technology ambience allows the provider and modern client to link clean financial services to the real-world micro financing

**Possible Risks Due of Microfinancing**

1. Being a novice element in MFIs, the myth around digital transformation is always there. A strong argumentation is: "Digitalization does not solve the finer aspects of organizational problems., and it may not truly match all micro credit establishments, loan providers and customers. To fit apt technology, serious research is needed - location specific, scale specific and client specific"
2. In the transformation process, MFIs may lose social and people-specific
3. Redundant creation of products and services may happen
4. A total digitalization may cause the valuable human interactions

The present paper has summarized the above opportunities and threats and put forth a via media in the conclusion paragraph.

## 1.4    Research Methodology

The widely used qualitative research technique namely content analysis and integrated reviewing method was followed for this study. The three approaches viz., conventional, directed, and summative were used according to the reviews, incidences and cases analyzed [15]. A meaningful sample of initiatives, cases, reports, research papers were purposively selected for the study according to the objectives intended. Personal interviews with experts in microfinance and information and communication technologies in the financial sector were also done.

## 1.5    Microfinance – Global History and Milestones

Microfinance has had a stretched and diverse history beginning with the first pawn-shops founded by Franciscan monks in the 15th century to today's use of crowd funding and peer to peer lending [3]. In the mid-1800s, Lysander Spooner (1870) in his book- No Treason: The Constitution of No Authority, wrote about the benefits of minor loans to small businesses as an assistance to escaping poverty, and Friedrich Wilhelm Raiffeisen founded the first cooperative lending bank to support farmers in rural Germany. On April 3, 1948, President Truman signed the Economic Recovery Act of 1948. It became known as the Marshall Plan, named after the Secretary of State George Marshall, who in 1947 proposed that the United States provide economic assistance to restore the economic infrastructure of post war Europe.

From its origins as an action-research project in 1976, the Grameen Bank conceptualized by the Nobel laureate Prof. Mohammed Yunis of Bangladesh has grown to provide collateral-free loans to 7.5 million clients in more than 82,072 villages in Bangladesh of which 97 per cent are women. Basically, microfinance, with its roots in the 1970s in organizations like Grameen Bank of Bangladesh, facilitated accessibility of financial services including savings and current accounts, insurance funds and microcredit to micro and small enterprises typically owned and run by poor and low-income faction. These sections of the society are either unserved or underserved by financial aids. The small businesses may be formal (registered) or informal (non-registered) units.

Today, the World Bank estimates that more than 16 million are served by some 7,000 microfinance institutions all over the world. In a gathering at a Microcredit Summit in Washington DC the goal was reaching 100 million of the world's poorest people by credit from the world leaders and major financial institutions. The year 2005 was proclaimed as the International Year of Microcredit by the Economic and Social Council of the United Nations to stimulate the spirit of micro and small entrepreneurship across the world [2].

## 1.6    Indian Microfinance Sector on the Threshold of a Second Revolution

It has become increasingly evident that the next leap in India should come from the Information and Communication Technology (ICT) digital channels. Though the digitalization of microfinance segment has come up well, majority of the customers are not aware of it and its mode of use. At the same time, it is difficult for the micro credit agent to attend to all the customers in an area. With the recent trend of minimizing the number of field workers, there is a need to use the latest technologies for delivering micro credit services. The telecom network and mobile phone service even in rural India is quite impressive [25]. This network can be used for delivering the micro credit facilities to the stakeholders. Thus digital channels play a significant role in providing speedy and reliable micro loan service. Among the states of India, Kerala lies in the top of the e-readiness pyramid, which positions the state in the leader category among e-ready population [29].

Information and Communication Technologies (ICTs) are emerging as an important tool for development world-wide. ICTs are no more confined to assist research and development; the new technologies have made substantial expansions in the life-styles and the efficiency-levels of all sectors of economy. The Agriculture sector is also gearing itself to make optimal use of the new information and communication technologies [34].

The products, services and all the functions of the microfinance sector heavily rely on the element called 'customer'. Customer-centricity has developed over a period wherein the needs of the client are to be addressed rapidly in this competitive field to build and nurture batches of loyal customers.

The evolution of technology can be attributed to the ever-increasing demand of the customer aspirations for better quality of service, ease of getting things and availing product or service at the very moment of time at one's convenient place of choice. Data has become the new oil, a prerequisite for delivering any kind of service. Storage size has shrunk, and speed of data has increased, both manifolds.

In such a scenario, delivery of customer-centric products and high-quality services through diverse channels and devices has become the primary requirement for an organization to survive. The customer needs to be given his choice of channels which keep facing him, be available in his pocket all the times conveniently and he will be always in total command to the access of product and services at all times.

Digitalization of financial services has enabled access of financial services and credit facilities to vulnerable groups by reducing the cost. This has helped leaps and bounds to achieve financial inclusion world over. Financial inclusion means that individuals and businesses have access to useful and affordable financial products and services that meet their needs–transactions, payments, savings, credit and insurance, all delivered in a responsible and sustainable way [36].

Since two decades, huge investments have been made by banks in technology to reduce their cost and improve customer's experience. Banks operate digital banking channels like ATM, Internet banking, mobile banking and digital banking kiosks to deliver the best quality services to customers with the hope of bringing down operational costs so as to enhance profitability [28]. It has been experienced that banks' costs reduce by shifting of a mass of customers to modern banking channels [16]. Despite such efforts, due to the slow pace of customers getting acclimatized to the modern technology aided digital banking channels the ambitious cost reduction in operating expenses has not been accomplished [18]. Such a situation is much precarious for an emerging economy like India as only around 16 per cent of the rural population practice the Internet for making digital payments [22].

Still, there have been incidences of hope in the banking sector. The success of the 'Digital India' campaign gives rays of hope that more than a billion Indian citizens have a digital identity with 560 million Internet connections [26]. The purpose of digitization is to bring disconnected rural remote regions into the formal financial sector through electronic banking channels which in turn will contribute to economic development. Digital banking channels help connect the disconnected folk with mainstream banking system by offering various innovative banking services. The modern mobile banking apps also enable customers to use non-financial services. However, due to the lack of awareness and knowledge, these services have not been fully utilized by customers [20]. Undoubtedly, there is a stringent need to absolutely influence customers through awareness campaigns on advantages of technology aided digital channels to make the banking and microfinance facilities pleasant.

## 1.7   Recent Developments in Banking and MFI Digitization in India

As a big leap, in 2008, the National Payments Corporation of India (NPCI) India was initiated as an umbrella organization promoted by the Reserve Bank of India for all retail payment systems, with a vision to enable citizens to have access to e-payment services at anyplace and anytime. As a Unified Payments Interface, it meant to move India to a cashless society. It has successfully completed the development of a domestic card payment network called RuPay, reducing the dependency on international card schemes. The retail payment system products such as Bharat Bill Payment System (BBPS); National Financial Switch (NFS), the largest network of shared Automated Teller Machines (ATMs); Immediate Payment Service (IMPS), an instant real-time inter-bank electronic funds transfer system; National Automated Clearing House (NACH), a web-based clearing service that could ease the work of financial institutions by removing any geographical barriers in efficient banking; Cheque Truncation System (CTS) or Image–based Clearing System (ICS) for faster clearing of cheques;

Aadhaar enabled Payment System (AePS), a bank led model which allows online interoperable financial inclusion transaction at PoS (MicroATM) through the Business correspondent of any bank using the Aadhaar authentication [21]. Among them the BBPS allows payments through feature phones or smart phones or WhatsApp is a unique digital channel helping the micro loan customers.

## 2 Objectives of This Review Research

This research paper analyzes the importance of trust in microfinance, and how this trust has helped reduce barriers to financial transactions. Trust has assisted in reducing information asymmetries and in creating compatible incentives, which have helped MFIs to succeed in the provision of financial services, especially credit, to the low-income sectors of the population. It also analyzes the importance of information in efficient credit decisions and the delivery of other services. In microfinance, information has been collected and trust has been built through human contact. This research effort also analyses how the new digital tools may interfere in the personal relationships MFIs have built with their clients over the years; evaluates if these tools may help or hinder in the development of trust. A blend of personalized and digital approach could be a via media approach, at least to begin with.

Thanks to several interviews conducted with experts in microfinance and with experts who are closely working with information and communication technologies in the financial sector, the paper aims to assess the opportunities, challenges and threats for MFIs and their clients in this digitalization process.

## 3 Defining Digitalization in the Context of Financial Inclusion

Digitalization is creating products and services available which can give enhanced customer service with the aid technologies, which are interoperable.

Introduction digital banking has revolutionized the banking sector and facilitates large volume of Government to Person (G2P) transactions and vice versa (P2G), directly through bank accounts, facilitated cashless transactions, changed the entire procedure bank transfers, facilitated the purchasers to see their account details, pay online bills and transfer money from one account to the opposite in a faster way. This has helped the end-user enjoy a methodical financial life, further embracing hassle-free online banking.

## 4 Digitalization Infrastructure in India

The transformation of India into a digital society and a knowledge economy has bettered the ease of citizens' living. The combination of the Government-led infrastructure such as the JAM Trinity and India Stack forms the bedrock of the evolution of digital channels across the board.

The JAM trinity, a government-led initiative comprising the Pradhan Mantri Jan Dhan Yojana (PMJDY) scheme, Aadhaar biometric identification, and mobile technology, is a key driver of financial inclusion in the country.

The PMJDY scheme, introduced in 2014, opened accounts for hundreds of millions of previously unbanked adults. A PMJDY bank account is a no-frills, low-cost, zero balance account aimed at bringing traditionally disadvantaged population groups, such as women, rural and below-poverty individuals, into the financial inclusion fold.

Especially in a country like India, where the Governments steps in to give subsidies, grants and freebies to the needy, unserved, underserved sections of the society and tax, charge or collect dues by Tax, octroi, tolls, fees from those communities who can remit a part of their income back to the Government, digital channels come handy. Because of the population, the transactions are many and the largest democratic country can carry out both G2P and P2G transaction without spillage and pilferage [4].

The "India Stack"–a government-led digital infrastructure, based on a set of APIs (Application Programming Interface)–has served as a foundation for the growth of the digital ecosystem, enabling presence-less, paperless and cashless digital payments.

Other supply infrastructure such as storage on cloud, local and hybrid servers also rapidly expanded. Telecom and data connectivity was boosted through more tower and fibre optic broadband infrastructure by the Government and private players. Government felt the need for Fibre to home connections and hence BSNL and other telecom companies joined the party by including various households to the network. Thus from 2G, 3G, Indian customers graduated to 4G, enabling financial inclusion across board.

## 5    Evolution of Digital Channels in Indian Banking Sector: Speed, Storage and Connectivity

The evolution of digital channels was expedited by the Indian telecom operators using Aadhaar as their base for onboarding customers–to conduct their Know Your Customer [KYC] verification. The data connectivity improved and the telecom services were taken up by more and more bottom of the pyramid (BOP) customers. These customers began using data through YouTube, WhatsApp and other online channels on their phones (Fig. 1).

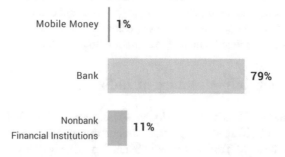

**Fig. 1.** Financial inclusion breakdown by accounts Source: Findex Report 2017, World Bank

Banks witnessing this evolution started extending their channels apart from SMS to mobile banking. Also, non-banks could enter the market through differentiated banking licenses. Though non-banks can offer e-money, it was never targeted at the unbanked. Money deposited into e-money accounts could be used only for making payments and cannot be withdrawn in cash at agents or at ATMs.

## 6   Integration of Financial Services in India

Providing access to financial services and products at affordable rates to the economically weaker section of the society has been an agenda of successive governments in India. The Jan Dhan–Aadhaar–Mobile (JAM) trinity has a positive impact on the banking sector and financial inclusion in the country. With the launch of JAM services, there has been a significant improvement in terms of targeted and accurate payments. They have also helped in weeding out duplication of entries, and bringing down the reliance on cash mode of payments.

Since its pan-India roll out in 2014, 488 schemes and services from 63 ministries have been brought under the DBT (DBTM 2019a). Similarly, until March 2020, the total number of beneficiaries under Aadhaar-enabled services had been 436.98 crores [DBTM 2019b] (Fig. 2).

## 7   Role of Digital Channels in Financial Inclusion

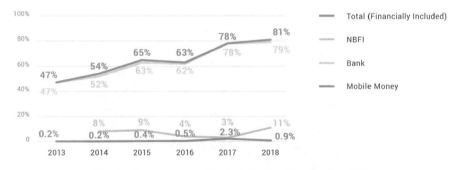

**Fig. 2.**  Account ownership in India: by Year Source: RBI

In the recent years, Government of India and RBI have undertaken several initiatives to expand the financial outreach to the unbanked people of India. For example, RBI introduced no-frill account in the year of 2005 for providing basic banking services to the unbanked people and to increase financial inclusion in India. No-frill account scheme allows the customers to open their bank accounts with minimum balance or no balance. Similarly, other major schemes such as Know your Customer, Electronic Benefit Transfer, Unique Identification Authority of India, Direct Benefit Transfer, PMJDY and Mudra banks have been introduced in India for the promotion of financial inclusion in India.

Apart from these major schemes, the Reserve Bank of India has also introduced several financial products and services for promoting financial inclusion in India. Some of the financial services has extremely benefited to the unbanked people in India. For example, through the use of Kisan Credit Card, the bank is providing small credits or loans to the poor farmers to cover their cost of cultivation, harvest and farm maintenance (Fig. 3).

Similarly, the introduction of BCs has helped the banks to provide the basic banking services to the poor people at a low and an affordable cost (Fig. 4).

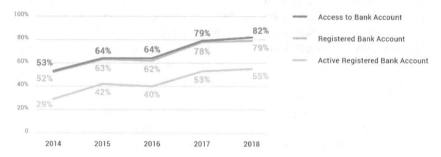

**Fig. 3.** Increase in bank account holders in India Source: RBI

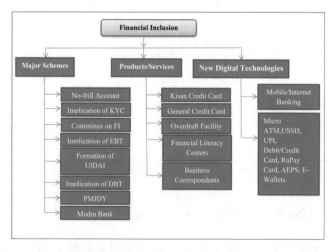

**Fig. 4.** Recent development in financial inclusion (FI) Source: RBI- National Strategy for Financial Inclusion 2019–2024

## 8   Cash and Digital Modes

Off late, instead of putting brakes on the cash system, banks attempt to circulate cash in the community rather than killing physical cash and land it into electronic cash. Banks are creating Village Level Entrepreneurs (VLEs) who helps farmers, micro customers

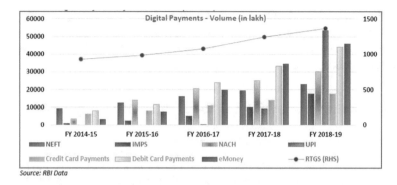

Source: RBI Data

**Fig. 5.** Volume of digital transactions in India Source: RBI

to reuse the cash in the community instead of remitting it in the bank. Agency banking, assisted channel, VLEs, Agri-clinics, Milk societies and other local entities and small businesses help to recirculate cash saving time and energy for all the stakeholders in the system (Fig. 5).

Banks promote this cash and digital mode system to thrive as all the participants in the eco system gets benefited some way or other. Banks saves a lot of resources if otherwise have to deploy to make the services available. Cash holding, handling, risk associated with it all come down helping reduction of cost and increase productivity. Cash rich and cash starved system's interplay helped to remove friction, reduce cost, increase productivity and becomes planet friendly. Direct-to-Consumer (D2C) businesses also witnessed a growth of 87% during a period of eight months of the national lockdown [14].

## 9  Role of FinTechs in Furthering Digitalization

There were an estimated 1,500 FinTechs in India by 2017. One of the earliest Indian FinTechs, Financial Innovation and Network Operations (FINO) was formed in 2017 as a service provider offering end-to-end IT and service solutions to financial institutions to enable them to reach out to unbanked geographies. IFC was an early equity investor. FINO went on to offer business correspondent (agent) solutions to banks, microfinance institutions, insurance companies and government agencies serving rural areas of India. FINO payments bank has 265 branches and over 170,000 service points providing deposits, savings, loans, and insurance services (through partners), as well as over one million monthly remittances, cost-effectively delivered to rural recipients.

More recently, FinTech start-ups have proliferated in India. Some of these companies have scaled at a very fast pace, and now dominate their respective categories. Prominent applications include mobile wallets, alternative lending and insurance. Mobile wallets have proliferated, and a few have acquired meaningful scale and usage. Paytm has >200M users of its mobile wallet, allowing customers to make payments from/to any account at no fee, and enabling over eight million merchants to accept electronic payments.

Online insurance marketplaces such as Policy Bazaar, Coverfox, and Turtlement, have simplified and digitalized the process of purchasing an insurance policy and help customer make informed decisions. These are regulated under Insurance Regulatory and Development Authority (IRDA) as agents or brokers.

## 10 Differentiated Banking for the Bottom of the Pyramid

The essential components of digital financial inclusion:

1. Digital transactional platforms enable customers to make or receive payments and transfers and to store value electronically through the use of devices that transmit and receive transaction data and connect to a bank or non-bank permitted to store electronic value
2. Devices used by the customers can either be digital devices (mobile phones, etc.) that transmit information or instruments (payment cards, etc.) that connect to a digital device such as a point-of-sale (POS) terminal.
3. Retail agents that have a digital device connected to communications infrastructure to transmit and receive transaction details enable customers to convert cash into electronically stored value ('cash-in') and to transform stored value back into cash ('cash-out') [32].
4. Additional financial services via the digital transactional platform may be offered by banks and non-banks to the financially excluded and underserved—credit, savings, insurance, and even securities—often relying on digital data to target customers and manage risk.

## 11 Bottlenecks to Financial Inclusion

There is no doubt that India has significantly improved the financial inclusion of the marginalized sections, and Digital India has turned out to be an important intervention. However, the digital divide is too wide and there exist many bottlenecks and challenges which need immediate attention.

To begin with, High Level Principles for Digital Financial Inclusion, published by the G20 under the rubric of Global Partnership for Financial Inclusion in 2016, provides useful insights to address the issues impeding financial inclusion, and how digital technologies can help in the process [36]. These include the promotion of financial services as a national plan, the need to balance innovation and risk, providing legal and regulatory frameworks, and expanding the digital financial ecosystems, among others.

The most common barriers to the digital financial inclusion include the non-availability of suitable financial products, lack of skills among the stakeholders to use digital services, infrastructural issues, teething problems between various systems, and low-income consumers who are not able to afford the technology required to access digital services [30].

Another challenge to digital financial inclusion arises from the attitude of the stakeholders. For instance, take the case of Jan Dhan bank accounts. When the scheme was launched in 2015, banks were given ambitious targets to open accounts for the marginalized. This has resulted in the opening of many dormant accounts which never saw actual banking transactions.

Another major bottleneck faced by Digital India, with respect to financial inclusion, is the heavily dominated cash economy in the country. The data from RBI reveals that cash circulation has increased in 2018 after demonetization. As per a report of the International Labour Organization (ILO), about 81% of the employed persons in India work in the informal sector [30]. The combination of a huge informal sector along with a high dependence on cash mode of transaction poses an impediment to digital financial inclusion.

There is also a gender dimension to financial inclusion in the country. According to the 2017 Global Findex database, 83% of males above 15 years of age in India held accounts at a financial institution in 2017 compared to 77% females [36]. This is attributed to socio-economic factors, including the availability of mobile handset and internet data facility being higher among men than women.

## 12   Last Mile Challenges and Their Impact – The Indian Scene

Some last mile challenges in India's digital financial inclusion journey are as follows:

### 12.1   Digital Exclusions

KPMG report (2020) found that India's Internet usage is the lowest of the BRICS nations. Similarly, the Digital Quality of Life (DQL) Index 2020 reflects the dismal performance of India in digital parameters. Moreover, digital illiteracy, innumeracy, and unfamiliarity with technology hinder the digital product's full uptake.

### 12.2   Unfamiliarity with Direct Benefit Transfer (DBT)

The DBT program initiated in 2013 by Government of India to change the mechanism of transferring subsidies. This program aims to transfer benefits/subsidies from various welfare schemes, directly to the people through their bank accounts. But in several cases Beneficiaries of DBT still do not know what to do when their payments get rejected, often due to technical reasons such as incorrect account numbers and incorrect Aadhaar mapping with bank accounts. More importantly, the workers/beneficiaries have rarely been consulted regarding their preferred mode of transacting [18].

### 12.3   Breeding of Corruption

The digital exclusion and unfamiliarity of beneficiaries with DBT, has created new forms of corruption. This was evidenced in the massive scholarship scam in Jharkhand, where many poor students were deprived of their scholarships owing to a nexus of intermediaries, government officials, banking correspondents and others.

## 12.4     Inadequate Rural Banking

There are just 14.6 bank branches per 1 lakh adults in India. It is sparser in rural India. Moreover, rural banks are short-staffed and tend to get overcrowded. Accessing banks in rural areas leads to loss of one wage day for the worker. Also, people have to spend money on transportation to access the bank to withdraw their payments/subsidies.

## 12.5     Accountability Issue

The lack of accountability and absence of grievance redressal continue to impact all DBT programmes.

# 13     Digitalization - Trends and Opportunities

The technology improved manifold, enabling emerging Artificial Intelligence and Machine Learning AI/ML to help the bottom of the pyramid (BOP) customers to use the voice bots and chat bots in communicating the requirement. BOP customers can put across the requirement by choosing the visual icons on a mobile application and interchangeably use voice to navigate in the app for availing credit.

Reading has been a challenge for the micro-loan customers, whereas speaking in vernacular was never been a challenge. Hence Natural Language processing technology has helped to speed up financial inclusion by dovetailing AI/ML with NLP to provide voice bot, which can comprehend a request for a product or service by a customer to the digitalized system to trigger a series of operations/actions from the back end. Thus, the front-end demand has been satisfied by the back-end supply system [22].

Finger printer and facial IDs which are unique to the gullible customer have been leveraged for service delivery at the required speed, right time more accurate. Apart from the biometrics which the customer has always with him, voice tags are developing to support this trend.

Another trend is helping the customer to browse or recall the transaction that he has done over the time. The electronic passbook system in the applications has driven the customer to use the digital channels more, as the customer need not keep any hard copies of the transactions. Planet friendly, records available electronically and can be produced with authenticity on either side of transaction.

# 14     Furthering Inclusion - Suggestions

In order to achieve the objective of providing universal access to financial services, it is important to provide a robust and efficient digital network infrastructure to all the financial service outlets/touch points for seamless delivery of the financial services. It is also recommended to extend the digital financial infrastructure to co-operative banks and other specialized banks (Payments Banks, Small Finance Banks) as well as other non-bank entities such as fertilizer shops, Office of the Local Government bodies/Panchayats, fair price shops, common service centres, educational institutions etc. to promote efficiency and transparency in the services offered to customers. Banks may endeavor to sort out issues relating to remuneration, need for furnishing cash-based collaterals and cash retention limits, etc. to strengthen the BC network.

## 15 Summary and Conclusion

The present qualitative study had a multi-faceted specific focus on the origin, history, milestones, evolutions and positive revolutions, quick takes off, significant outcomes, contributions as well as setbacks in the banking, micro financing and digitalization to address the financial and banking exclusion and inclusion in India. A decent number of research papers, articles, initiatives and cases were reviewed. For the success of digital inclusion of the poor, there has to be a multidimensional approach through which existing digital platforms, infrastructure, human resources, and policy frameworks are strengthened [27]. More importantly, human resources should be leveraged by skilling and positively engaging with them to achieve the last-mile connectivity of financial institutions.

Given the prominence of digitalization, is there a strategy for integration of prominent digital tools in an appropriate and operative way in the microfinance sector? Surely, MFIs need a comprehensive, integrated strategy for digitalization – and then to segregate to adoptable and adaptable segments – so that digital transformation becomes a progressive expedition. Just focusing on efficiency processes or channels will not be enough. Once the overall strategy is defined, break it in to sub-strategies. The next approach would be the digitization of the products and services. The upsurge and elasticity of mobile banking, for example warrants that MFIs adapt their product to meet this new demand. New digital products should be developed in a sequential manner so as to respond to customers' preferences. Then comes digitalization of the channels which involves using tech platforms to improve customer acquisition and user experience. The advent of digital platforms and alternate channels have greatly influenced a good faction of Indian customers do their banking. Clients now a days prefer self-service platforms allowing them liberty, choice, and control. MFIs, must come forward to face the challenges put forth by systems like FinTechs and design new unorthodox financial services with efficiency and effect to reach the rural populations of India [20].

The digital revolution offers the chance to deliver rapid, responsive, responsible and unique financial and social services to the low-income people of India with a sprit that the country has not been able to do in the past. In this context, MFIs really have an added-value, since they know very well their clients and the locations they operate. The Indian MFI's future ought to use their ability banking on their experience and expertise, to prove that digital revolution is both high tech and high touch.

If remedial measures are resorted to tide over the existing constraints like digital divide, MFI's have the potential to strengthen the benefits of economic growth to the poor. On one side it reduces the banking costs it would also address safety concerns and accuracy of the data involved in financial transactions.

So the present paper establishes that MFIs must embrace digital transformation. They must strive through their branch staff and field agents to provide the human touch and assistance, for which they are famous. They must harness the potential of blending their legacy and credibility, experience, relationships and the digital technology know-how. In short, MFI's must deliver a blend of personalized and digitally-enabled services.

# References

1. Convergences.org. (2021). https://www.convergences.org/wp-content/uploads/2019/09/Microfinance-Barometer-2019_web-1.pdf. Accessed 23 Sept 2021
2. Digital India: Where Knowledge is Strength. The Daily Guardian (2021). https://thedailyguardian.com/digital-india-where-knowledge-is-strength/. Accessed 23 Sept 2021
3. How Technology and Data Leaders Can Redefine Financial Inclusion. Crunchbase (2021). https://about.crunchbase.com/blog/how-technology-and-data-leaders-can-redefine-financial-inclusion/. Accessed 23 Sept 2021
4. Kaur, S., Ali, L., Hassan, M., Al-Emran, M.: Adoption of digital banking channels in an emerging economy: exploring the role of in-branch efforts. J. Finan. Serv. Mark. **26**(2), 107–121 (2021). https://doi.org/10.1057/s41264-020-00082-w
5. Mastermicrofinance.com. (2021). https://www.mastermicrofinance.com/Investigacion/WP1_2016-Baldeh_Tena-Microfinance_Goes_Digital_p.pdf. Accessed 23 Sept 2021
6. Simran Jit Kaur—Federation University Australia - Academia.edu. Federation-au.academia.edu. (2021). https://federation-au.academia.edu/simranJitKaur. Accessed 23 Sept 2021
7. Uttamchandani, M.: Consumer Risks in Fintech - New Manifestations of Consumer Risks and Emerging Regulatory Approaches: Policy Research Paper (English) (2021). https://www.worldbank.org/en/topic/financialinclusion. Accessed 23 Sept 2021
8. Welcome to OurDocuments.gov. Ourdocuments.gov. (2021). https://www.ourdocuments.gov/doc.php?doc=82#:~:text=Congress%20overwhelmingly%20passed%20the%20Economic,known%20as%20the%20Marshall%20Plan. Accessed 23 Sept 2021
9. When microfinance goes digital: opportunities, challenges and dangers for microfinance institutions and their clients - Progreso. Progreso. (2021). http://fundacionmicrofinanzasbbva.org/revistaprogreso/en/when-microfinance-goes-digital-opportunities-challenges-and-dangers-for-microfinance-institutions-and-their-clients/. Accessed 23 Sept 2021
10. When microfinance goes digital: opportunities, challenges and dangers for microfinance institutions and their clients - Progreso. Progreso. (2021). http://www.fundacionmicrofinanzasbbva.org/revistaprogreso/en/when-microfinance-goes-digital-opportunities-challenges-and-dangers-for-microfinance-institutions-and-their-clients/. Accessed 23 Sept 2021
11. When Microfinance goes digital: opportunities, challenges and dangers for the microfinance institutions and their clients. Linkedin.com (2021). https://www.linkedin.com/pulse/when-microfinance-goes-digital-opportunities-dangers. Accessed 23 Sept 2021
12. When Microfinance goes digital: opportunities, challenges and dangers for the microfinance institutions and their clients. Es.linkedin.com (2021). https://es.linkedin.com/pulse/when-microfinance-goes-digital-opportunities-dangers. Accessed 23 Sept 2021
13. Wilcox, M.: What is an example of microfinance? Colors-newyork.com (2021). https://colors-newyork.com/what-is-an-example-of-microfinance/. Accessed 16 Sept 2021
14. Yunus, M.: Grameen Foundation. Grameenfoundation.org (2021). https://grameenfoundation.org/about-us/leadership/muhammad-yunus. Accessed 23 Sept 2021
15. McKenzie, D.: Comments Made at IPA/FAI Microfinance Conference Oct. 17 2008. Philanthropy Action (2008). Accessed 17 Oct 2008

16. Bruton, G.D., Chavez, H., Khavul, S.: Microlending in emerging economies: building a new line of inquiry from the ground up. J. Int. Bus. Stud. **42**(5), 718–739 (2011). https://doi.org/10.1057/jibs.2010.58.S2CID167672472

17. Bee, B.: Gender, solidarity and the paradox of microfinance: reflections from Bolivia. Gend. Place Cult. **18**(1), 23–43 (2011). https://doi.org/10.1080/0966369X.2011.535298.S2CID53696094

18. Ly, P., Mason, G.: Individual preferences over development projects: evidence from microlending on kiva. VOLUNTAS: Int. J. Volunt. Nonprofit Organ. **23**(4), 1036–1055 (2012). https://doi.org/10.1007/s11266-011-9255-8

19. Allison, T.H., Davis, B.C., Short, J.C., Webb, J.W.: Crowdfunding in a prosocial microlending environment: examining the role of intrinsic versus extrinsic cues. Entrepreneurship **39**(1), 53–73 (2015)

20. Pazarbasioglu, C., Mora, A.G., Uttamchandani, M., Natarajan, H., Feyen, E., Saal, M.: Digital Financial Services, World Bank Group (2020)

21. Chattopadhyay, S.K.: Financial inclusion in India: a case-study of West Bengal. Department of Economic and Policy Research, Munich (2011)

22. Goldberg, P.: Is Inclusive Growth an Oxymoron? World Bank Blogs (2019)

23. ILO (International Labour Organization): Women and men in the informal economy: a statistical picture (third edition) (2018). https://www.ilo.org›publication›wcms_626831. Accessed 24 Aug 2021

24. Jeník, I., Flaming, M., Salman, A.: Inclusive digital banking: emerging markets case studies. Working Paper. Washington, D.C (2020)

25. Joshi, A.: Mobile phones and economic sustainability: perspectives from India. In: Benyon, D., Surahmanyan, S. (eds.) Proceedings of the First International Conference on Expressive Interactions for Sustainability and Empowerment, 29–30 Oct. 2009, Swindon, UK, pp. 11–19. BCS Learning and Development Ltd. (2009)

26. Karuna, C.H.: Financial Literacy in India. Int. J. Commer. Manag. **4**(5), 76–80 (2018). http://www.managejournal.com/download/703/4-5-26-977.pdf

27. Yong, K.J.: Digital financial inclusion. The World Bank Group (2014). https://www.worldbank.org›topic›publication. Accessed 11 Aug 2021

28. Klapper, L., Lusardi, A., van Oudheusden, P.: Financial literacy around the world. Standard & Poor's Financial Services (2014)

29. Misra, D., Hiremath, B.N.: Livelihood perspective of rural information infrastructure and e-governance readiness in India: a case based study (2015). https://www.researchgate.net/publication/280096530. Accessed 19 Aug 2021

30. Niranjan, J.N.: A case study of barriers to digital financial inclusion of auto-rickshaw drivers in Viman Nagar, Pune, Maharashtra. J. Polit. Sci. Public Affairs **5**(3) (2017)

31. Niti Aayog (Govt. of India) 2021. Digital India – Where Knowledge is strength. https://www.niti.gov.in/. Accessed 11 Aug 2021

32. Zetterli, P., Jeník, I., Salman, A.: How digital banking models are changing inclusion. Presentation, CGAP, Washington, D.C. (2020)

33. 1RBI (Reserve Bank of India). Credit delivery and financial inclusion (2020). https://m.rbi.org.in/scripts/AnnualReportPublications.aspx?Id=1288. Accessed 25 Aug 2021

34. Koshy, S.M., Kumarn, K.: Agricultural information support service vis-à-vis Kisan Call Centre: a performance auditing Ph.D. (Ag.) Thesis. Kerala Agricultural University, Thrissur, India, 152 p. (2013)

35. Spooner, L.: No Treason: The Constitution of No Authority. Rampart Journal. 1 (1). Introduction by Martin, James J. (Spring/Fall 1965). Rampart Journal. **1**(3) (1870)

36. World Bank: India Wave 6 Report Sixth Annual FII Tracker Survey, Financial Inclusion Insights report (2019). https://www.worldbank.org/en/topic. Accessed 25 Aug 2021

# Digital Information Integrated Reporting and Its Impact on Firm Valuation (Evidence in Indonesia)

Zaky Machmuddah[1,2]([envelope]) [iD], Abdul Rohman[1], Anis Chariri[1] [iD], and Agung Juliarto[1]

[1] Universitas Diponegoro, Semarang, Indonesia
Zaky.machmuddah@dsn.dinus.ac.id
[2] Universitas Dian Nuswantoro, Semarang, Indonesia

**Abstract.** Proving the association of digital integrated reporting <IR> information provided directly or indirectly by the company through the company's website on firm valuation and the moderating role of CEO compensation on this association is the purpose of this study. The analysis uses WarpPLS version 7.0 with a sample of 150 observational data, from manufacturing sector listed on the IDX. The results show that digital IR information provided directly or indirectly by companies is positively related to firm valuation, this shows how digitized information is a means for companies to increase firm valuation. In addition, CEO compensation moderates the association, meaning that the compensation received by the CEO moderates the association between digital IR and firm valuation. The implication of this research is the importance of disseminating digital information either directly or indirectly to be used as a company signal to potential investors or investors in making investment decisions that have a positive impact on firm valuation. The contribution of this research is to give a signal to companies about the importance of digital IR disclosures to increase firm valuation, and as a material for government considerations regarding IR disclosures for companies in Indonesia.

**Keywords:** Digital information · Integrated reporting · CEO compensation · Firm valuation

## 1 Introduction

The COVID-19 pandemic has had an impact on all sectors, not only the health sector but also the economic sector. This can be seen in the deteriorating operating conditions of the Asean manufacturing sector that occurred in 2020. Indonesia is one of the Asean countries and Indonesia has experienced this, as shown in 2020 the composite stock price index (CSPI) of manufacturing companies decreased at 31.25%. This incident can affect the interest of potential investors or investors to invest in the manufacturing sector.

To improve this information, companies should provide more credible information about their quality in the capital market, so that they are different from bad companies [1]. Traditional financial reporting models seem to fail to capture the economic implications of business innovation and economic change in a timely manner [2]. In the last decade, digitalization has received the attention of both professionals and academics. Investors are increasingly considering information about company digitization in their decision making [3]. This is in line with [4] which states that the disclosure of quality reports will affect firm valuation.

In recent years, shareholders are worried about financial statements, they think that financial statements will be more complex, less relevant and messy [5]. Although corporate social responsibility reports and sustainability reports provide non-financial information, it seems that financial and non-financial information that is integrated in order to increase shareholder understanding has not been provided by the company [6]. In this era a new reporting paradigm has been considered, in which the economic, social and environmental activities of companies are integrated into one to provide a more holistic view of company performance and ensure that ethical responsibility is at the forefront of business activities [7]. Managers can create corporate value by implementing holistic reporting called integrated reporting. The purpose of integrated reporting as communication in creating short-term, medium-term and long-term company values that contain related company performance, corporate governance, organizational strategy and organizational prospects in the external environment [8]. This explanation is in line with research [4, 9–13] which proves that integrated reporting disclosures affect firm valuation. But this is different from [14, 15] which proves that integrated reporting has no effect on firm value.

To strengthen this explanation, another factor emerged, namely the chief executive officer (CEO). Because in the management board, the CEO is the most important person in influencing the company's strategic decisions [16]. This means that the CEO has power over the relationship between digital integrated reporting of information disclosure on firm valuation. The power of the CEO here is dynamic with the compensation received by the CEO. This is in line with [17–19] which states that the relationship between CSR and financial performance is positive, more prominent by CEO power. Similarly, [20] which explains the moderating impact of the CEO power model on the link between CSR and financial performance, found positive results. Nevertheless [21] found a negative effect of CEO power on the association of CSR and FP.

Based on the phenomena and explanations above, the purpose of this study is to analyze the relationship of digital IR information provided directly or indirectly by the company through the company's website on firm valuation and the moderating role of CEO compensation on this association.

## 2 Literature Review and Hypotheses Development

### 2.1 Signal Theory and Managerial Power Theory

Sending credible signals regarding the quality of companies in the capital market can be used to distinguish between good and bad companies [1]. In line with this [22] explains that fundamentally signal theory is concerned with reducing information asymmetry between two parties. To reduce information asymmetry, managers (insiders) are advised to provide information needed by investors or potential investors [23]. Companies that provide better information can influence investors' economic decisions and allow them to be attracted to contracts with greater benefits than other companies that provide worse information [24].

In addition to signal theory, managerial power theory is also used in this study. [25] argues that structurally and socio-psychologically the CEO is powerful and influential in board-level decision-making related to executive compensation, so that the board is less involved in fair transactions. This mechanism creates little incentive for directors to challenge compensation arrangements that are more in the interests of executives than shareholders, i.e. higher levels of compensation and compensation that is less sensitive to performance.

## 2.2 The Association Between Digital Integrated Reporting Information Disclosure and Firm Valuation

Sending credible signals regarding the quality of companies in the capital market can be used to distinguish between good and bad companies [1]. This is in line with [21] which states that fundamentally signal theory is concerned with reducing information asymmetry between two parties. For this reason, managers (insiders) are advised to provide information needed by investors or potential investors [23]. Firm valuation will be positive if the IR provides benefits for shareholders. The opinion of the proponents states that the quality of information increases after the disclosure of IR because it will be more efficient and the allocation of capital becomes more productive. In addition, articulation in linking organizational strategy, business model and creating value is the goal of IR. So that quality IR disclosures can reduce information processing costs for investors, this should speed up and provide specific information related to asset prices [2, 26, 27]. This means that the more IR disclosures submitted by the company to investors or potential investors, the better the firm valuation in the eyes of investors or potential investors. This is supported by [4, 9–13] which state that the strong positive impact of IR on firm value, so that the first hypothesis is:

**H1.** Firm valuation is positively associated with digital IR information

## 2.3 The Effect of CEO Compensation on the Association Between Digital Integrated Reporting Information Disclosure and Firm Valuation

The influence of the CEO is very large in the company's strategic decision making because the CEO is an important person on the management board [16]. Managerial power theory states that CEO power is entrenched and has an adverse effect on management behavior [28]. The power possessed by a CEO influences the decision-making process of the board to fulfill its own interests that are different from the interests of stakeholders. However, if incentive-based compensation is implemented it is possible to contribute positively to stakeholder demands. On the other hand, CEO tenure and CEO ownership are used as complementary variables [29], this will lead to increased CEO power. So that it will affect non-financial disclosures which have an impact on improving performance. Previous research shows that there is no indication of the effect of one item compared to the other two variables. Thus, increased salary deductions, number of CEO holdings and longer tenure will moderate financial and ESG performance. In line with the arguments and theories discussed as well as the findings of previous research, we assume a moderating effect on the CEO, so that the two hypotheses are:

**H2.** The positive association between digital IR information and firm valuation is more pronounced for CEO Compensation

# 3   Research Method

The sample of this study is manufacturing companies listed on the Indonesia Stock Exchange for the consumer goods industry sub-sector, basic chemical industry, and various industries. The observation period was for five years starting from 2016 to 2020 so that 150 observation data were obtained, this is quantitative research. The source of data is obtained from IR which is presented from the selected sample. Tobin's Q [30] is used to measure firm valuation, with the following formula:

$$\text{Tobin's Q} = (\text{MNS} + \text{D})/\text{TA} \tag{1}$$

Annotation:

MVS = market value of all outstanding shares, i.e., company share price * extraordinary shares.
TA = Company assets, such as; cash, accounts receivable, inventory, and book value of land.
D = debt

Measurement of digital IR information disclosure using the IR index, which is calculated by dividing the total items disclosed by the total disclosure items (8 items) [4]. And for CEO compensation is measured using the natural logarithm of CEO compensation received [31, 32]. The analytical tool used to test the hypothesis is Partial Least Squares (PLS) - Structural Equation Modeling (SEM) using the WarpPLS version 7.0 program.

# 4   Results and Discussion

## 4.1   The Association Between Digital Integrated Reporting Information Disclosure and Firm Valuation

Based on Fig. 1, it shows that the results of the study have succeeded in proving the hypothesis that was built, namely the disclosure of digital IR information is positively related to firm valuation. This means that digital information IR improves the quality of information for financial capital providers to enable more efficiency and productivity, so that this will provide a positive perception for potential investors or investors who are ultimately related to firm valuation. The findings of this study are in line with signal theory. The findings of this study are in line with [4, 9–13] which state that high IR disclosures can positively affect firm valuation. However, this finding is not the same as [14, 15] which states otherwise (Table 1).

**Table 1.** Goodnes of fit structural model

| Criteria | Parameter |
|---|---|
| APC | 0.257, P < 0.001 |
| AVIF | 1.024 |
| AFVIF | 1.643 |
| GoF | 0.399 |
| RSCR | 0.913 |
| SSR | 1.000 |

Source: secondary data processed, 2021

## 4.2 The Effect of CEO Compensation on the Association Between Digital Integrated Reporting Information Disclosure and Firm Valuation

Figure 1 shows that the results of the study strengthen the hypothesis, with a significance level of 0.1 CEO compensation moderating the relationship between digital IR and firm valuation. This means that a strong CEO can influence the decision-making process within the board for their own personal gain in contrast to the interests of stakeholders, because the CEO is the most important person on the management board, he has a great influence on strategic decisions in the company [16]. This means that the CEO has power over the relationship between digital IR of information disclosure on firm valuation. The power of the CEO here is dynamic with the compensation received by the CEO. The findings of this study are in line with managerial power theory. The findings of this study are also in line with [17–19] which states that the positive CSR and financial performance relationship is more pronounced by CEO power. Similarly, [20] who explained the moderating impact of the CEO power model on the link between CSR and financial performance found positive results. Nevertheless [21] found a negative effect of CEO power on the association of CSR and FP.

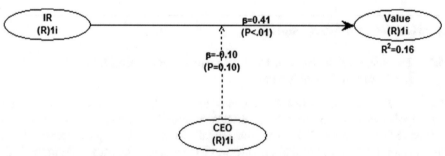

\*\*\*, \*\*, \* denotes significance levels at 0.001, 0.05 and 0.1, respectively.
Source: secondary data processed, 2021

**Fig. 1.** Research result

# 5  Conclusion and Recommendation

The conclusion from the findings of this study is that digital information through IR provided directly or indirectly by companies is positively related to firm valuation, this shows how information about digitization is a means for companies to increase firm valuation. In addition, CEO compensation moderates the relationship, meaning that the compensation received by the CEO will moderate the digital information relationship through IR on firm valuation. The implication of this research is the importance of disseminating digital information either directly or indirectly to be used as a company signal to potential investors or investors in making investment decisions that have a positive impact on firm valuation. The limitation of this study is the R square value of 16%, so suggestions for future research are to modify the model, by adding other independent variables such as CSR costs, intellectual capital, and so on. Giving a signal to companies about the importance of digital IR disclosures to increase firm valuation, and as a consideration for the government regarding IR disclosures for companies in Indonesia is the contribution of this research.

# References

1. Spence, M.: Job market signaling. Q. J. Econ. **87**(3), 355–374 (1973)
2. Healy, P.M., Palepu, K.G.: Information asimmetry, corporate disclosure, and the capital markets: a review of the empirical disclosure literature. J. Acc. Econ. **31**(1–3), 405–440 (2001)
3. Salvi, A., Vitolla, F., Rubino, M., Giakoumelou, A., Raimo, N.: Online information on digitalisation processes and its impact on firm value. J. Bus. Res. **124**, 437–444 (2021)
4. Lee, K.-W., Yeo, G.-H.: The association between integrated reporting and firm valuation. Rev. Quant. Financ. Acc. **47**(4), 1221–1250 (2015). https://doi.org/10.1007/s11156-015-0536-y
5. Financial reporting council (FRC) cutting clutter. Combating clutter in annual reports (2011)
6. KPMG and financial executives research foundation. Disclosure overload and complexity: hidden in plan sight (2011)
7. Lodhia, S.: Exploring the transition to integrated reporting through a practice lens: an Australian customer owned bank perspective. J. Bus. Ethics **129**, 585–598 (2015)
8. International integrated reporting coucil (IIRS). International integrated reporting framework (2013)
9. Mervelskemper, L., Streit, D.: Enhancing market valuation of ESG performance: is integrated reporting keeping its promise? Bus. Strateg. Environ. **26**(4), 536–549 (2016)
10. Martinez, C.: Effects of integrated reporting on the firm's value: evidence from voluntary adopters of the IIRC's framework. SSRN Electron. J. (2016). SSRN: https://ssrn.com/abstract=2876145 or https://doi.org/10.2139/ssrn.2876145
11. Barth, M.E., Cahan, S.F., Chen, L., Venter, E.R.: The economic consequences associated with integrated reporting quality: capital market and real effects. Accounting, Organizations and Society, Working paper 3546, pp. 1–52 (2017)
12. Cosma, S., Soana, M.G., Venturelli, A.: Does the market reward integrated report quality? Afr. J. Bus. Manage. **12**(4), 78–91 (2018)
13. El Deeb, M.S.: The impact of integrated reporting on firm value and performance: evidence from Egypt. Alex. J. Acc. Res. **3**(2), 1–50 (2019)

14. Churet, C., Eccles, R.G.: Integrated reporting, quality of management, and financial performance. J. Appl. Corp. Financ. **26**(1), 8–16 (2014)
15. Suttipun, M.: The effect of integrated reporting on corporate financial performance: evidence from Thailand. Corp. Ownersh. Control **15**(1), 133–142 (2017)
16. Busenbark, J.R., Krause, R., Boivie, S., Graffin, S.D.: Toward a configurational perspective on the CEO. J. Manag. **42**(1), 234–268 (2016)
17. Li, Y., Gong, M., Zhang, X., Koh, L.: The impact of environmental, social, and governance disclosure on firm value. Br. Account. Rev. **50**(1), 60–75 (2018)
18. Javeed, S.A., Lefen, L.: An analysis of corporate social responsibility and firm performance with moderating effects of CEO power and ownership structure. Sustainability **11**(1), 1–25 (2019)
19. Velte, P., Stawigona, M.: Integrated reporting: the current state of empirical research, limitations and future research implications. J. Manag. Control **28**(3), 275–320 (2017)
20. Veprauskaite, E., Adam, M.: Do powerful chief executives influence the financial performance of UK firms. Br. Account. Rev. **45**, 229–241 (2013)
21. Li, F., Li, T., Minor, D.: CEO power, corporate social responsibility and firm value. Int. J. Manag. Financ. **12**(5), 611–628 (2016)
22. Spence, M.: Signaling in retrospect and the informational structure of markets. Am. Econ. Rev. **92**, 434–459 (2002)
23. Dainelli, F., Bini, L., Giunta, F.: Signaling strategies in annual reports: evidence from the disclosure of performance indicators. Adv. Account. Incorp. Adv. Int. Account. **29**, 267–277 (2013)
24. Grossman, S.J., Stiglitz, J.E.: Stockholder unanimity in making production and financial decisions. Q. J. Econ. **94**(3), 543–566 (1980)
25. Bebchuk, L., Fried, J.: Pay without performance: overview of the issues. Acad. Manage. Perspect. **20**(1), 5–24 (2006)
26. Sims, C.A.: Rational inattention: beyond the linier-quadratic case. Am. Econ. Rev. **96**(2), 158–163 (2006)
27. Veldkamp, L.: Information market and the co-movement of asset prices. Rev. Econ. Stud. **70**(3), 823–845 (2006)
28. Bebchuk, L.: The case against board veto in corporate takeovers. Univ. Chicago Law Rev. **69**(3), 973–1035 (2002)
29. Sheikh, S.: An examination of the dimensions of CEO power and corporate social responsibility. Rev. Account. Financ. **18**(2), 221–244 (2019)
30. Lindenberg, E.B., Ross, S.A.: Tobin's q ratio and industrial organization. J. Bus. **54**(1), 1–32 (1981)
31. Shim, E.D., Kim, E.: An empirical examination of the relationship between top executive compensation and firm performance in the post sarbanes-oxley period. Adv. Manage. Account. **25**, 207–228 (2015)
32. Al-Shaer, H., Zaman, M.: CEO compensation and sustainability reporting assurance: evidence from the UK. J. Bus. Ethics **158**, 233–252 (2017)

# Digital Health Care, Hospitals and Rehabilitation

# A Security Review of a Portuguese Hospital Using the Cyber Security Framework: A Case Study

Bruno Pereira[1], João Pavão[1], Dario Carreira[2], Victor Costa[3], and Nelson Pacheco Rocha[4(✉)] (ORCID)

[1] INESC-TEC, Science and Technology School, University of Trás-os-Montes and Alto Douro, Vila Real, Portugal
jpavao@utad.pt
[2] Instituto Universitário da Maia, Maia, Portugal
dariocarreira@netcabo.pt
[3] Centro Hospitalar de Trás-os-Montes e Alto Douro, Vila Real, Portugal
vcosta@chtmad.min-saude.pt
[4] IEETA, Department of Medical Sciences, University of Aveiro, Aveiro, Portugal
npr@ua.pt

**Abstract.** The objective of the study reported by the present paper was to assess the cyber resilience state of a Portuguese hospital. For the study implementation, the Cyber Security Framework (CSF), proposed by the National Institute of Standards and Technology (NIST), was used in conjunction with the Cyber Resilience Review (CRR) tool. The results point to satisfactory levels of cyber resilience of the healthcare entity being studied, but several features need to be optimized. The results also show that the use of CSF and CRR report, generates a large quantity of objective information, which provides an exhaustive identification of aspects that should be improved.

**Keywords:** Information security · Cyber resilience · Cyber security · Risk management framework

## 1 Introduction

With the growing number of devices connected to the communication networks, and also with the increased capabilities for attackers to use these devices to turn a threat into an attack, emerged the necessity of not only using protection techniques (e.g., anti-virus), but also using techniques to mitigate the risks and these threats, using the capabilities (i.e., people, processes and technologies) from the organizations.

Cyber resilience includes not only the definition of standards to prevent attacks, but also the development of methods to identify, detect, protect, respond and recover efficiently and quickly from these attacks.

The study reported by this paper aimed to assess the state of the cyber resilience of a Portuguese hospital. For the study implementation, the Cyber Security Framework

T. Antipova (Ed.): DSIC 2021, LNNS 381, pp. 367–378, 2022.
https://doi.org/10.1007/978-3-030-93677-8_32

(CSF) v1.1 [1] proposed by the National Institute of Standards and Technology (NIST) was used in conjunction with the Cyber Resilience Review (CRR) tool [2].

The paper is outlined as follows. The next Section (Related Work) presents relevant published research that supported the present study. Section 3 (Materials and Methods) describes the two important instruments of this study (i.e., CSF and CRR), and details the study design. Results are presented in Sect. 4. Finally, Sect. 5 presents the discussion of the results and a conclusion.

## 2    Related Work

### 2.1    From Cybersecurity to Cyber Resilience

The complexity of cyberspace is the most effective adversary weapon in the 21st century, anchored in the speed and increasing sophistication of the technological evolution of organizations, which thus increase the attack surface and potential exploitation vulnerabilities.

Small and medium organizations, and in particular healthcare organizations with their communities of users and clients, today constitute a set of interdependent human-cyber-physical systems, where Information and Communication Technologies (ICT) represent the cyber infrastructure that detects, acts and assesses the state of healthcare systems, and help sustain the basic services they offer to citizens. Therefore, cyber threats are ongoing, increasingly complex and sophisticated with some (quite) certainly, resulting in successful attacks originated from multiple sources: Intentional Human (internal or external), Accidental Human, Structural and Environmental [3].

Cybersecurity incidents, namely the cybercrime, has recently shifted from attacking big corporations to smaller industries, also addressing the healthcare sector in general and hospitals in particular. Specifically in this area the trend is rising, with hackers targeting patient healthcare devices that may be reachable from the internet. Most cases include stealing patient information and encrypting it for ransom money.

The healthcare sector has lagged behind other sectors in protecting its main stakeholder (e.g., patients), and now hospitals must invest considerable capital and effort in protecting their systems. It is therefore understandable the European Union's effort on its Horizon 2020 programme, among which the SecureHospitals project seeking to "Raising Awareness on Cybersecurity in Hospitals across Europe and Boosting Training Initiatives Driven by an Online Information Hub" [4].

Reducing susceptibility to cyber threats requires a multidimensional system engineering strategy. However, regarding the healthcare sector and, particularly, the hospitals this is easier said than done, because hospitals are extraordinarily technology-saturated, complex organizations with high end-point intricacy, as well as internal politics and regulatory pressures.

Traditionally, as still much today, cybersecurity is concerned to ensure a secure perimeter and is based on the presumption that all logical (and physical) access points to the environment that is within such perimeter are protected. Therefore, if any breach or violation is detected, cybersecurity recommends a reaction to the event to remedy the situation, usually resorting to technological controls or countermeasures.

However, as the protection perimeter, even in the case of a small organization, can involve numerous access points, electronic, physical and human, such a task requires a huge commitment of resources to be minimally successful, which turns the traditional cybersecurity model to be virtually impracticable [5].

On the other hand, such an approach means that cybersecurity is treated 'bottom up' rather than 'top down', from the outside rather than from inside, reactive rather than proactive and at the tactical level, instead of strategic - the tree management instead of the forest management. Indeed, while the inevitability of attacks is certain, the adoption of only reactive and technological countermeasures in protecting the access points of the surface/perimeter of organizations, according to the traditional concept of cyber-security, are not enough, even ineffective and innocuous, given the extent of that surface/perimeter and the capacity and sophistication of today's highly advanced and persistent threats.

In the absence of such highly effective ('infallible') techniques and the consequent real impossibility of fully protecting cyberspace, the need for a paradigm shift is now recognized. Thus, based on the assumption that, sooner or later, an attack will occur, a more adequate concept emerges - cyber resilience. In fact, accepting the inevitability of not being able to protect the entire perimeter, we are conceding that possible breaches can occur, so, in addition, it is decided to defend in depth, at critical and important levels, according to the assessment of the most critical assets that will need to be preserved at all costs, as opposed to less critical ones, thus mitigating the risks and their consequences [6].

In this context we can define cyber resilience as the ability of a system, organization, mission, or business process to anticipate, withstand, recover from, and adapt capabilities in the face of adversary conditions, stresses, or attacks on the cyber resources it needs to function. From an organizational perspective, cyber resilience is defined as "the ability to continuously deliver the intended outcome despite adverse cyber events", and this definition is systematically described and justified [7].

Decisively, at the base of cyber resilience is cybersecurity, but now added with risk management and, as such, it is not just about technology, but about the organizations' business, depending a lot on people, processes and procedures, in addition to technology. The implementation of a cyber resilience strategy in an organization implies the active involvement of all people, from leaders to employees, and it is necessary that the essential steps are defined in the internal management and governance processes themselves.

## 2.2  Cyber Resilience Management

The development, implementation and management of a resilient cyber system in an organization is complex and difficult, as it faces a wide variety of approaches in terms of resources, technology, processes and procedures, including stored data, data communication, human factors, physical security of assets and privacy, and legal, regulatory and contractual concerns. In addition, there is a continuing need to maintain high confidence in the face of evolving ICT systems, relationships with third parties, staff turnover, changes in physical configuration and the constantly evolving threat landscape.

The only way to ensure survival and improve sustainable resilience is through risk management in this complex digital landscape, albeit a difficult task, based more on art than science, so any attempt to develop an *ad hoc* and autonomous approach to face the situation, it is an invitation to failure. The good news is that a large amount of thought, experimentation and implementation experiences have emerged, providing various elements and reference documents, which represent consensus on the best practices and standards in use, 'de jure' or 'de facto', for planning and implementing cyber resilience strategies. It is therefore convenient to take advantage, not least because its use is allowing economies of scale that drive the development of effective products, services and practices that meet the needs identified in the global digital market [8].

This is how the cyber resilience management process is called, as a continuous effort to seek the best practices and standards that guide decision-makers, for the allocation of resources and implementation of an effective cyber resilience structure. The process is an iterative cycle with four main activities: i) identification of assets and risk assessment, considering the threats and vulnerabilities associated with these assets that affect critical services; ii) election and implementation of controls/countermeasures: selection of controls, allocation of resources, roles and responsibilities; iii) monitoring, reviewing and communicating risks and iv) updating and improving controls.

This repetitive cycle is driven not only by the evolving ecosystem of cyberspace, but also by the evolution of standards and best practices [3].

## 2.3   Standards and Best Practices in Use

There are literally hundreds if not thousands of documented standards that cover topics from accounting and finance to privacy rights, from healthcare records to security risks in critical infrastructure, like energy distribution, transport, defence, healthcare, among others.

In cybersecurity, a set of best practices based on standards is considered relevant, such as ISO, which revised and consolidated several of different standards under the ISO 27000 series; Information Systems Audit and Control Association - Control Objectives for Information and Related Technologies (ISACA) framework; the Standard of Good Practices for Information Security Forum (SGP ISP); the Center for Internet Security Critical Security Controls (CIS CSC); the Computer Emergency Readiness Team-Resilience Management Model (CERT-RMM) of Carnegie Mellon University [9], and the CSF proposed by NIST.

Of these frameworks, the CSF stands out, because it presents a different and innovative approach, such as, instead of being 'one more', it is rather based on a coordinated and integrated summary of the best of other models and references, while remaining fully open to incorporating new references and components, according to circumstances and needs. The CSF consists of five concurrent and continuous functions that constitute the cybersecurity lifecycle for any organization: Identify, Protect, Detect, Respond, and Recover [1].

On the other hand, the main CSF's document is a small and concise one (about 50 pages) on the planning, implementation and assessment of the organizations' cyber resilience capacity, through the use of five high-level functions, 23 categories and 108 clearly defined subcategories that map to other standards. Other NIST documents are also important, such as NIST SP.800-53 r4 and NIST.SP.800-30 r1 among others, with the added advantage of being freely accessible [10].

Currently, the use of the CSF is widespread, both nationally in the USA and internationally, including Portugal[1], in various sectors of activity, in a public or private environment. A confirmation of the importance and interest of CSF as a framer and integrator of reference in cyber resilience, concerning different sectors and organizations, is that they have opted for direct mapping of their objectives and practices, in order to comply with CSF categories and subcategories, as is the case, for example, of the healthcare sector with Health Information Trust Alliance (HITRUST). HITRUST is a United States private organization that exists to ensure that information security becomes a central pillar, rather than an obstacle, to the widespread adoption of healthcare information systems [11].

With the number of cyber-attacks focusing on healthcare for purposes ranging from cause a real damage to the live of patients, the destruction of data, impersonating a physician or other clinician, or just taking data, it is truer than ever that a successful attack is only a matter of time.

## 3   Materials and Methods

### 3.1   CSF and CRR

The CSF provides a comprehensive assessment and continuous improvement based on three essential components [1, 12]: i) Core; ii) Profile; and iii) Implementation Tiers.

The Core aims to present a high-level overview of the cybersecurity situation of any organisation. In turn, the Profile results from the organization needs and represents adjustments and priorities to increase its cybersecurity readiness, while the Implementation Tiers aim to help the organization to measure where it is positioned within the CSF.

The core component is divided into five risk management functions: i) Identity (i.e., complete knowledge about the cyber environment, particularly systems, assets, data, and capabilities); ii) Protect (i.e., appropriate deployment to limit potential cybersecurity risks); iii) Detection (i.e., appropriate activities to identify cybersecurity events); iv) Respond (i.e., appropriate activities to avoid the unwanted impact of cybersecurity events); and v) Recovery (i.e., recovery activities to maintain resilience plans and restore capabilities that may be compromised by a cybersecurity incident). These five functions are divided into 23 categories and 108 subcategories (Table 1). Based on the CSF functions, categories and subcategories is possible a systematic approach to ascertain the organisations cyber security risk management, practices and processes.

The CSF defines seven steps for implementing continuous improvement processes of organization's cyber resilience. An important aspect of these processes is the determination of the organization's current Profile, namely which categories and subcategories of the CSF are conformed and which one's present flaws.

---

[1] In 2019, The National Cybersecurity Center formalized the "QNRCS-Quadro Nacional de Referência de Cibersegurança", based on CSF of NIST.

**Table 1.** CSF core component.

| Functions | Category | Number of subcategories |
|---|---|---|
| Identity | Asset management | 6 |
| | Business environment | 5 |
| | Governance | 4 |
| | Risk assessment | 6 |
| | Risk management strategy | 3 |
| | Supply chain risk management | 5 |
| Protect | Identity management and access control | 7 |
| | Awareness and training | 5 |
| | Data security | 8 |
| | Information protection processes and procedures | 12 |
| | Maintenance | 2 |
| | Protective technology | 5 |
| Detect | Anomalies | 5 |
| | Security continuous monitoring | 8 |
| | Detection processes | 5 |
| Respond | Response planning | 1 |
| | Communication | 5 |
| | Analysis | 5 |
| | Mitigation | 3 |
| | Improvements | 2 |
| Recover | Recovery planning | 1 |
| | Improvements | 2 |
| | Communication | 3 |

There are many tools that help assess the current state of cyber resilience goals and practices, as well as identify areas for improvement (comparing profiles) as well as select and prioritize gaps. We chose to use the CRR, which is a simple and customized tool and which, although based on the CERT-RMM Model, older and slightly different from the CSF, includes a self-assessment kit that guides you through a questionnaire, helping each organization to identify its cyber resilience and risk management practices. Since the CRR seeks to understand the cybersecurity management of services, and their associated assets, that are critical for an organization's mission, it might be used to determine the current state of the institution [2]. It is an interview-based assessment, consisting of 299 questions distributed by ten domains (Table 2).

The answer to each CRR question can be Yes (if what is asked is fully implemented), No (if what is asked is not implemented) or Incomplete (if what is asked is partially implemented). To map the results of the CRR in the functions, categories and subcategories defined by CSF, crosswalk rules are defined [13]. As crosswalk examples, Table 3 indicates the CCR questions that should be used to determine the results of the first four subcategories of the Asset management category of the Identity function.

**Table 2.** CCR domain composition [2].

| Domains | Goals | Goal practices | Maturity indicator level practices |
|---|---|---|---|
| Asset management | 7 | 30 | 13 |
| Controls management | 4 | 16 | 13 |
| Configuration and change management | 3 | 23 | 13 |
| Vulnerability management | 4 | 15 | 13 |
| Incident management | 5 | 23 | 13 |
| Service continuity management | 4 | 16 | 13 |
| Risk management | 5 | 13 | 13 |
| External dependencies management | 5 | 14 | 13 |
| Training and awareness | 2 | 11 | 13 |
| Situational awareness | 3 | 8 | 13 |

**Table 3.** Some examples of the CRR to CSF crosswalk [14].

| CSF subcategories | Maturity indicator level practices |
|---|---|
| ID.AM-1: Physical devices and systems within the organization are inventoried | Are the assets that directly support the critical service inventoried (technology includes hardware, software, and external information systems)? |
| | Do asset descriptions include protection and sustainment requirements? |
| | Are the physical locations of assets (both within and outside the organization) documented in the asset inventory? |
| ID.AM-2: Software platforms and applications within the organization are inventoried | Are the assets that directly support the critical service inventoried (technology includes hardware, software, and external information systems)? |
| | Do asset descriptions include protection and sustainment requirements? |
| | Are the physical locations of assets (both within and outside the organization) documented in the asset inventory? |
| ID.AM-3: Organizational communication and data flows are mapped | Are organizational communications and data flows mapped and documented in the asset inventory? |
| ID.AM-4: External information systems are catalogued | Are the assets that directly support the critical service inventoried (technology includes hardware, software, and external information systems)? |

## 3.2   Study Design

The study reported by this paper took place in a hospital of Portugal. The organization has currently implemented several international security standards as part of its activities related to the management of the information technologies platforms.

The first procedure of the study was the identification of the scope of the organization assessment. For that, the researchers in conjunction with the high-level hospital management identified the: i) Critical services; ii) Stakeholders of the critical service delivery; and iii) Assets (i.e., people, technology, information, and facilities) that are required for this delivery. In parallel, relevant organizational documentation was collected.

Once identified the clinical services and the respective stakeholders and assets, it was identified the group of participants to be interviewed in order to complete the CRR. Moreover, a plan was prepared to interview these participants. The interviews were made individually to the participants. At the end, the team found the mathematic mode among all the answers to the same question. Therefore, the final result to each question was the one that collected more answers with same value.

Based on the analyse of the interviews and the organizational documentation the CRR assessment was consolidated, and a report was created and submitted to the hospital high-level management.

# 4   Results

## 4.1   Scoping of the Organization Assessment

At this stage, the most important areas of the organization were identified and categorized according to the threats that could occur and the vulnerabilities that could be detected. In addition to the administration and information technologies management, four medical services were selected: i) Surgery; ii) Cardiology; iii) Pulmonology; and iv) Paediatrics.

Moreover, relevant organizational documentation was collected, and an inventory was prepared containing the stakeholders and the most important assets of the identified services. Among the collected documentation were hospital politics about: information and security, print and copy, removable data support, internet and e-mail access, videoconferencing, medical devices, user, and patient access.

## 4.2   Participants and Interviews

The interviews were made according to different professional profiles in a total of 12. The distribution of the participants was: seven doctors, four management staff and one informatic specialist, all with more of ten years of service at the hospital.

On average, each interview took around 2 h:30 min with up to two short breaks.

## 4.3   Completing the Assessment

The CRR consolidation was the basis for the application of the crosswalk rules that translate it's results about domains, goals and practices, in categories and subcategories of the CSF. Table 4 results from this crosswalk and represents the number of affirmative, negative, and incomplete answers in the CRR, for each CSF category.

**Table 4.** Summary of the assessment

| Functions (CSF) | Category (CSF) | Number of answers (CRR) | | |
|---|---|---|---|---|
| | | Y | N | Inc. |
| Identity | Asset management | 18 | 0 | 0 |
| | Business environment | 25 | 0 | 3 |
| | Governance | 26 | 2 | 9 |
| | Risk assessment | 16 | 2 | 5 |
| | Risk management strategy | 5 | 3 | 0 |
| | Supply chain risk management | 22 | 0 | 2 |
| Protect | Identity management and access control | 10 | 0 | 0 |
| | Awareness and training | 11 | 1 | 4 |
| | Data security | 8 | 1 | 0 |
| | Information protection processes and procedures | 71 | 10 | 20 |
| | Maintenance | 30 | 0 | 0 |
| | Protective technology | 3 | 1 | 0 |
| Detect | Anomalies | 8 | 0 | 3 |
| | Security continuous monitoring | 8 | 0 | 0 |
| | Detection processes | 9 | 0 | 3 |
| Respond | Response planning | 0 | 0 | 1 |
| | Communication | 9 | 0 | 0 |
| | Analysis | 4 | 0 | 1 |
| | Mitigation | 3 | 0 | 1 |
| | Improvements | 2 | 0 | 0 |
| Recover | Recovery planning | 1 | 0 | 0 |
| | Improvements | 2 | 0 | 0 |
| | Communication | 5 | 1 | 0 |

**Table 5.** List of subcategories with flaws.

| **Identity: business environment** | |
|---|---|
| ID.BE-1 | The organization's role in the supply chain is identified and communicated |
| ID.BE-5 | Resilience requirements to support delivery of critical services are established for all operating states |
| **Identity: governance** | |
| ID.GV-1 | Organizational cybersecurity policy is established and communicated |
| ID.GV-3 | Legal and regulatory requirements regarding cybersecurity, including privacy and civil liberties obligations, are understood and managed |
| ID.GV-4 | Governance and risk management processes address cybersecurity risks |
| **Identity: risk assessment** | |
| ID.RA-6 | Risk responses are identified and prioritized |
| **Identity: risk management strategy** | |
| ID.RM-1 | Risk management processes are established, managed, and agreed to by organizational stakeholders |

(*continued*)

**Table 5.** (*continued*)

| ID.RM-2 | Organizational risk tolerance is determined and clearly expressed |
|---|---|
| ID.RM-3 | The organization's determination of risk tolerance is informed by its role in critical infrastructure and sector specific risk analysis |

**Identity: supply chain risk management**

| ID.SC-3 | Contracts with suppliers and third-party partners are used to implement appropriate measures designed to meet the objectives of an organization's cyber-security program and Cyber Supply Chain Risk Management Plan |
|---|---|
| ID.SC-4 | Suppliers and third-party partners are routinely assessed using audits, test results, or other forms of evaluations to confirm they are meeting their contractual obligations |

**Protect: awareness and training**

| PR.AT-1 | All users are informed and trained |
|---|---|
| PR.AT-2 | Privileged users understand their roles and responsibilities |
| PR.AT-4 | Senior executives understand their roles and responsibilities |
| PR.AT-5 | Physical and cybersecurity personnel understand their roles and responsibilities |

**Protect: data security**

| PR.DS-3 | Assets are formally managed throughout removal, transfers, and disposition |
|---|---|

**Protect: information protection processes and procedures**

| PR.IP-2 | System development life cycle to manage systems is implemented |
|---|---|
| PR.IP-6 | Data is destroyed according to policy |
| PR.IP-7 | Protection processes are improved |
| PR.IP-8 | Effectiveness of protection technologies is shared |
| PR.IP-9 | Response plans and recovery plans are in place and managed |
| PR.IP-11 | Cybersecurity is included in human resources practices |
| PR.IP-12 | A vulnerability management plan is developed and implemented |

**Protect: protective technology**

| PR.PT-1 | Audit/log records are determined, documented, implemented, and reviewed in accordance with policy |
|---|---|

**Detect: anomalies and events**

| DE.AE-2 | Detected events are analysed to understand attack targets and methods |
|---|---|
| DE.AE-3 | Event data are collected and correlated from multiple sources and sensors |

**Detect: detection processes**

| DE.DP-2 | Detection activities comply with all applicable requirements |
|---|---|
| DE.DP-3 | Detection processes are tested |
| DE.DP-5 | Detection processes are continuously improved |

**Response: response planning**

| RS.RP-1 | Response plan is executed during or after an incident |
|---|---|

**Response: analysis**

| RS.AN-4 | Incidents are categorized consistent with response plans |
|---|---|

**Response: mitigation**

| RS.MI-1 | Incidents are contained |
|---|---|

**Response: communications**

| RC.CO-2 | Reputation is repaired after an incident |
|---|---|

### 4.4    Generating the Report

The implementation of about 75 subcategories was completed. Particularly, seven categories, which corresponds to 31 subcategories, were identified without flaws: i) Asset management from the Identity function; ii) Identity management and access control from the Protect function; iii) Maintenance, also from the Protect function; iv) Security continuous monitoring from the Detect function; v) Communication from the Respond function; vi) Improvements, also from the Respond function; vii) Recovery planning from the Recover function; and viii) Improvements, also from the Recover function. In turn, 13 subcategories were not implemented, and 20 subcategories were incomplete. Table 5 presents all the 33 subcategories with flaws. Please note that each CSF subcategory has associated several related questions in CRR. Therefore, each flawed subcategory has, at least, one negative or incomplete CRR answer.

## 5    Discussion and Conclusion

The results point to satisfactory levels of cyber resilience of the healthcare entity being studied, but several features need to be optimized.

In general, the strength of the assessed entity is the fact that about 75 CSF subcategories were conformed. In particular all the subcategories of seven CSF categories were conformed.

On the other hand, as weakness it should be pointed the fact that they were identified flaws in 33 subcategories, which were either not implemented (i.e., 13 subcategories) or were incomplete (i.e., 20 subcategories).

The CSF assessment supported on the CRR report, generates a large quantity of objective information, which provides clear indications about vulnerabilities and how to improve the cyber resilience mechanisms. In addition, since the categories are generally organized in subcategories that take the form of positive questions, the assessment method facilitates the reasoning to resolve the failure.

Definitevily, assuming that complete cybersecurity in the healthcare sector is unachievable, and would exceed financial resources there are still vital steps that can be taken to minimize the risk of cyber- attacks against healthcare facilities. Around 85 percent of targeted cyber-attacks would be preventable if basic prevention rules and protection protocols would be established so, prevention is the cyberdefence for hospitals [14].

Facing the CRR report and Table 5, it appears that 24 of the subcategories flaws (73%) are within of the scope of prevention component, with 11 flaws (33%) in ID function and 13 flaws (40%) in PR function, which is something of major concern to be corrected. Thus, some attention should be drawn to the organizational budget on cybersecurity associated to some specific sectors, to keep it straitly related and mainly based on an understanding of the current and predicted risk environment and risk tolerance.

In Portugal, like in the European Union, the goal is to strengthen its cybersecurity rules in order to tackle the increasing threat posed by cyber-attacks as well as to take advantage of the opportunities of the new digital age. Faced with ever-increasing cybersecurity challenges, the European Union needs to improve awareness of, and response to, cyber-attacks aimed at institutions, both public and private, including in the healthcare sector [15].

The generated report exhaustively identified the organizational flaws, which might help to stablish priorities to the continuous improvement of the cyber resilience state of the assessed organizations.

As future work, the results will be used to study the consequences of the identified vulnerabilities, and to plan and prioritise interventions according to the defined CSF procedures.

## References

1. Sedgewick, A.: Framework for improving critical infrastructure cybersecurity, version 1.0. https://www.us-cert.gov/ccubedvp/self-service-crr. Accessed 29 July 2020
2. Assessments: Cyber Resilience Review (CRR). United States Computer Emergency Readiness Team (US-CERT). https://www.us-cert.gov/ccubedvp/self-service-crr. Accessed 29 July 2020
3. Stallings, W.: Effective Cybersecurity - Understanding and Using Standards and Best Practices. Pearson Education (2019)
4. European Comission: Programme | H2020 | European Commission (europa.eu)-call 'Raising awareness and developing training schemes on cybersecurity in hospitals'. https://project. securehospitals.eu/. Accessed 29 July 2020
5. Brooks, C., Grow, C.: Cybersecurity Essentials. Sybex (2018)
6. Shoemaker, D., Kohnke, A., Sigler, K.: How to Build a Cyber-Resilient Organization. CRC Press, Boca Raton (2019)
7. NIST: Cyber Resiliency. https://csrc.nist.gov/glossary/term/cyber_resiliency. Accessed 29 July 2020
8. Santos, O.: Developing Cybersecurity Programs and Policies. Pearson (2019)
9. CERT: CERT® Resilience Management Model, Version 1.2. https://www.sei.cmu.edu/. Accessed 29 July 2020
10. NIST: NIST Special Publications (SP). https://www.nist.gov/pml/weights-and-measures/ publications/nist-special-publications. Accessed 29 July 2020
11. HITRUST Alliance: Information Risk Management and Compliance. https://hitrustalliance. net. Accessed 29 July 2020
12. Sulistyowati, D., Handayani, F., Suryanto, Y.: Comparative analysis and design of cybersecurity maturity assessment methodology using NIST CSF, COBIT, ISO/IEC 27002 and PCI DSS. JOIV Int. J. Inform. Vis. **4**(4), 225–230 (2020)
13. CISA: Cyber Resilience Review (CCR): NIST Cybersecurity Framework Crosswalks. https://www.cisa.gov/sites/default/files/publications/4_CRR_4.0_Self_Assessment-NIST_ CSF_v1.1_Crosswalk-April_2020.pdf. Accessed 29 July 2020
14. EU Agency for Cybersecurity: Cybersecurity Procurement Guide for Hospitals. https://www. enisa.europa.eu/news/enisa-news/prevention-is-the-cyberdefence-for-hospitals. Accessed 29 July 2020
15. Health Industry Cybersecurity Practices: Managing Threats and Protecting Patients. https:// www.phe.gov/Preparedness/planning. Accessed 29 July 2020

# The Use of High-Tech Medical Care in Patients Older Than Working Age in the Russian Federation

Sofia Shlyafer[ORCID] and Irina Shikina[⊠][ORCID]

Federal Research Institute for Health Organization and Informatics of the Ministry of Health of the Russian Federation, Bld. 11, Dobrolyubova Street, Moscow 127254, Russia

**Abstract.** For many decades in Russia, the leading place in the structure of the causes of morbidity, disability and mortality is occupied by diseases of circulatory system. Therefore, the priority direction of the state policy in healthcare is the fight against cardiovascular diseases. The aim of our paper is to evaluate the results of surgical interventions on the heart with the use of high-tech medical care in patients older than working age in Russia. The following research methods are used in the research: analytical, statistical (calculations of extensive indicators, ratio indicators), content analysis. The data are taken from the federal statistical observation form No. 14 'Information on the activities of departments of medical organizations providing medical care in inpatient conditions' for patients older than working age (women aged 55 years and older, men – 60 years and older) for 2011–2019. The results of the federal project 'Fight against cardiovascular diseases' allowed to increase the number of heart surgeries performed by 3.1 times, including with the use of high medical technologies – by 3.6 times, as well as to reduce the frequency of postsurgical heart complications from 2.41 to 1.18%, with the use of HTMC - from 2.65 to 1.19%. The target indicator of the federal project 'the ratio of the number of X-ray endovascular interventions for therapeutic purposes to the total number of retired adult patients who suffered acute coronary syndrome, %' in 2019 was 51.9% and was higher than the target value approved for 2021 (50.0%). This indicator in discharged patients older than working age is 47.1%.

**Keywords:** Surgery · Heart · High-tech medical care · A patient older than working age · The frequency of postoperative complications · X-ray endovascular interventions for therapeutic purposes

## 1 Introduction

One of the most significant challenges to the national safety of the country in the field of public health protection is the aging of the population [1]. The growth of the population over the working age leads to the accumulation of the number of chronic non-communicable diseases, primarily diseases of the circulatory system [2–5], which occupy a leading place in the structure of the causes of morbidity, disability and mortality of the adult population [6–13].

T. Antipova (Ed.): DSIC 2021, LNNS 381, pp. 379–391, 2022.
https://doi.org/10.1007/978-3-030-93677-8_33

Currently, the Russia is implementing the national project 'Healthcare', approved by Decree of the President of the Russian Federation No. 204 dated 07.05.2018 'On the national goals and strategic objectives of the development of the Russian Federation for the period up to 2024'. One of the goals of the National Project in the field of healthcare is to reduce the mortality rates of the population from diseases of the circulatory system to 450 cases per 100 thousand population by 2024. To achieve this, it is essential to solve the develop and implement a program to combat cardiovascular diseases [14].

The main contribution to the structure of the incidence of circulatory system diseases in the Russian Federation is made by coronary heart disease [15, 16], which is still the main cause of disability and mortality in the population [17]. An effective way to increase the duration and improve the quality of life of patients with severe coronary heart disease is surgical treatment of heart and vascular diseases, including one with the use of high-tech medical care (HTMC) [18].

Article 34 of the Federal Law No. 323-FZ of November 21, 2011 'On the basics of public health protection in the Russian Federation' defines that 'high-tech medical care is part of specialized medical care and includes the use of new (complex) and (or) unique methods of treatment, as well as resource-intensive methods of treatment with scientifically proven effectiveness, including cellular technologies, robotic technology, information technologies and methods of genetic engineering developed on the basis of achievements of medical science and related branches of science and technology. The organization of high-tech medical care is carried out using the unified state information system in the field of healthcare' [19, 20].

At present, the leading methods of treatment are endovascular myocardial revascularization using coronary stents, coronary artery bypass graft and angioplasty of the coronary arteries [17].

**The aim of our paper** is to evaluate the results of surgical interventions on the heart with the use of high-tech medical care in patients older than working age in Russia.

## 2   Data and Methods

To study the number of surgical interventions, including ones with the use of HTMC, performed in a hospital for persons older than working age, we have analyzed the data from the federal statistical observation form No. 14 'Information on the activities of departments of medical organizations providing medical care in inpatient conditions' for 9 years (for the period from 2011 to 2019). Since 2011, the form contains information about the surgical work of the organization to provide medical care to patients over the working age (women aged 55 years and older, men—60 years and older), including with the use of HTMC.

One of the indicators specified in the passport of the federal project 'Fight against cardiovascular diseases' was 'the ratio of the number of X-ray-endovascular interventions for therapeutic purposes to the total number of retired adult patients who suffered acute coronary syndrome, %'. The indicator is calculated as the ratio of the number of X-ray endovascular interventions for therapeutic purposes (coronary artery angioplasty operations) to the total number of retired (discharged and deceased) adults (18 years and older) with acute coronary syndrome (with unstable stenocardia, acute myocardial infarction, repeated myocardial infarction, other forms of acute ischemic heart diseases) [21 (pp. 26–27), 22].

The following research methods are used in the research: analytical, statistical (calculations of extensive indicators, ratio indicators), content analysis.

# 3   Results

To highlight the significance of the number of heart operations, we studied the structure of surgical interventions performed in the hospital on other organs and systems in patients older than working age.

In the Russian Federation in 2019, 3.5 million surgical interventions were performed for patients older than working age, of which 663.2 thousand were performed using HTMC (18.9% of the total number of surgeries performed for people of retirement age). The largest number of surgeries in patients older than working age were performed on abdominal organs (19.3%), the visual organ (18.8%), the musculoskeletal system (13.4%), female genitalia (9.0%), skin and subcutaneous tissue (8.5%), and heart (8.0%). 77.0% of all cases of surgical care for the elderly were accounted for by the six types of surgical interventions mentioned above.

In 2011–2019, there was an increase in the number of surgeries performed in the hospital by 63.5% (from 21468 to 3500938), including on the heart - by 3.1 times (from 88936 to 279539), on the nervous system - by 2.6 times (from 24649 to 64574), on the kidneys and ureters - by 2.3 times (from 74154 to 169845), on the respiratory organs - by 2.0 times (from 39316 to 79900), on the musculoskeletal system-by 1.9 times (from 244999 to 468161), on vessels - by 1.9 times (from 84750 to 162474), etc. (Table 1, 2).

**Table 1.** The number of surgeries performed in hospital for patients older than working age in the Russian Federation in 2011 (abs. number, share in %) (according to the federal statistical observation form No. 14)

| Surgical procedure | 2011 | | | | |
|---|---|---|---|---|---|
| | Number of surgeries | | of which | | |
| | abs. number | in % of the total | using HTMC | | |
| | | | abs. number | in % of the total | in % of total surgeries |
| **Total of surgeries** | **2141468** | **100,0** | **255946** | **100,0** | **11,95** |
| Including on: nervous system | 24649 | 1,15 | 5545 | 2,2 | 22,5 |
| Endocrine system | 18765 | 0,9 | 1419 | 0,55 | 7,6 |
| Visual organ | 459996 | 21,5 | 86967 | 34,0 | 18,9 |
| Organs of the ear, throat, nose | 51996 | 2,4 | 1591 | 0,6 | 3,1 |
| Respiratory organs | 39316 | 1,8 | 2360 | 0,9 | 6,0 |
| Heart | 88936 | 4,15 | 69094 | 27,0 | 77,7 |

(*continued*)

**Table 1.**  (*continued*)

| Surgical procedure | 2011 | | | | |
|---|---|---|---|---|---|
| | Number of surgeries | | of which | | |
| | abs. number | in % of the total | using HTMC | | |
| | | | abs. number | in % of the total | in % of total surgeries |
| Vessels | 84750 | 4,0 | 15415 | 6,0 | 18,2 |
| Abdominal organs | 449257 | 21,0 | 15262 | 6,0 | 3,4 |
| Kidneys and ureters | 74154 | 3,5 | 6282 | 2,45 | 8,5 |
| Male genitalia | 69510 | 3,25 | 2248 | 0,9 | 3,2 |
| Male sterilization | 1783 | 0,1 | 67 | 0,03 | 3,8 |
| Female genitalia | 194752 | 9,1 | 5071 | 2,0 | 2,6 |
| obstetric | 3519 | 0,2 | 55 | 0,02 | 1,6 |
| Musculoskeletal system | 244999 | 11,4 | 37010 | 14,4 | 15,1 |
| Mammary gland | 46078 | 2,15 | 1988 | 0,8 | 4,3 |
| Skin and subcutaneous tissue | 216989 | 10,1 | 1840 | 0,7 | 0,8 |
| Mediastinum | - | - | - | - | - |
| Esophagus | - | - | - | - | - |
| Others | 72019 | 3,3 | 3732 | 1,45 | 5,2 |

**Table 2.**  The number of surgeries performed in hospital for patients older than working age in the Russian Federation in 2019 (abs. number, share in %) (according to the federal statistical observation form No. 14)

| Surgical procedure | 2019 | | | | |
|---|---|---|---|---|---|
| | Number of surgeries | | of which | | |
| | abs. number | in % of the total | using HTMC | | |
| | | | abs. number | in a % of the total | in % of total surgeries |
| **Total of surgeries** | **3500938** | **100,0** | **663232** | **100,0** | **18,9** |
| Including on: nervous system | 64574 | 1,8 | 28932 | 4,4 | 44,8 |
| Endocrine system | 27118 | 0,8 | 4726 | 0,7 | 17,4 |
| Visual organ | 658649 | 18,8 | 121480 | 18,3 | 18,4 |
| Organs of the ear, throat, nose | 81565 | 2,3 | 4415 | 0,7 | 5,4 |
| Respiratory organs | 79900 | 2,3 | 6666 | 1,0 | 8,3 |
| Heart | 279539 | 8,0 | 245828 | 37,1 | 87,9 |

(*continued*)

**Table 2.**  (*continued*)

| Surgical procedure | 2019 | | | | |
| --- | --- | --- | --- | --- | --- |
| | Number of surgeries | | of which | | |
| | abs. number | in % of the total | using HTMC | | |
| | | | abs. number | in a % of the total | in % of total surgeries |
| Vessels | 162474 | 4,6 | 30533 | 4,6 | 18,8 |
| Abdominal organs | 674270 | 19,3 | 30102 | 4,5 | 4,5 |
| Kidneys and ureters | 169845 | 4,85 | 13292 | 2,0 | 7,8 |
| Male genitalia | 99924 | 2,85 | 5743 | 0,9 | 5,7 |
| Male sterilization | 627 | 0,02 | - | - | - |
| Female genitalia | 314864 | 9,0 | 17836 | 2,7 | 5,7 |
| Obstetric | 49 | 0,001 | - | - | - |
| Musculoskeletal system | 468161 | 13,4 | 127086 | 19,1 | 27,15 |
| Mammary gland | 59753 | 1,7 | 9625 | 1,45 | 16,1 |
| Skin and subcutaneous tissue | 298750 | 8,5 | 10507 | 1,6 | 3,5 |
| Mediastinum | 2166 | 0,1 | 223 | 0,03 | 10,3 |
| Esophagus | 9943 | 0,3 | 1686 | 0,25 | 17,0 |
| Others | 48767 | 1,4 | 4552 | 0,7 | 9,3 |

In 2019, in the structure of surgeries performed in a hospital with the use of HTMC for people older than working age, more than 1/3 of operations were performed on the heart (37.1%), almost every fifth-on the musculoskeletal system (19.1%) and on the visual organ (18.3%), respectively, and others (Table 1, 2; Fig. 1).

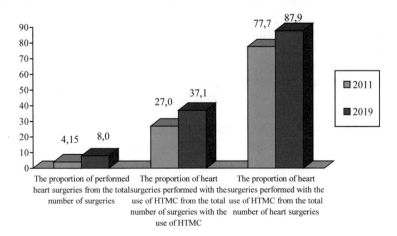

**Fig. 1.**  The proportion of heart surgeries performed, including those with the use of high-tech medical care, in patients older than working age in Russia in 2011 and 2019 (in %)

For the period from 2011 to 2019, the number of surgeries performed with the use of HTMC on people older than working age increased by 2.6 times (from 255946 to 663232), including surgeries on the skin and subcutaneous tissue - by 5.7 times (from 1840 to 10507), the nervous system - by 5.2 times (from 5545 to 28932), the mammary gland - by 4.8 times (from 1988 to 9625), the heart - by 3.6 times (from 69094 to 245828), female genitals - by 3.5 times (from 5071 to 17836), the musculoskeletal system - by 3.4 times (from 37010 to 127086), and the endocrine system - 3.3 times (from 1419 to 4726) (Table 1, 2; Fig. 1).

In 2019, the largest share of surgeries performed using HTMC out of all those performed on patients older than working age was on the heart (87.9%) and the nervous system (44.8%), every fourth - on the musculoskeletal system (27.15%), almost every fifth - on the vessels (18.8%) and on the visual organ (18.4%) (Table 2; Fig. 1).

As can be seen on the Table 3, during the analyzed period, patients over the working age had an increase in the number of heart surgeries for coronary artery angioplasty (X – ray cardiovascular interventions for therapeutic purposes) for coronary heart disease by 5.6 times (from 29573 to 166736), of which with stenting - by 6.8 times (from 23147 to 156764).

An annual increase in the proportion of coronary artery angioplasty performed with stenting was from 78.3% (in 2011) to 94.0% (in 2019). In 2019, the majority of heart operations in this category of individuals were performed using HTMC (87.9%) - coronary artery bypass graft (98.7%) and coronary artery angioplasty (88.9%).

An increase in the proportion of heart operations using HTMC in people older than working age was found by 10.2% (from 77.7 to 87.9%), of which in connection with coronary artery bypass grafting - by 9.4% (from 89.3 to 98.7%) and coronary artery angioplasty - by 1.0% (from 87.9 to 88.9%), of which with stenting - by 5.0% (from 87.5 to 92.5%) (Table 4).

The frequency of complications in the surgical treatment of heart diseases is still quite high. Common complications that affect the result of treatment are arterial thromboembolism and rhythm disturbances (atrial fibrillation), others [23, 24].

In the Russian Federation in 2019, the frequency of postsurgical heart complications was 1.18%, including 1.19% after operations using HTMC (0.64% and 0.97% of all operations, respectively). The highest frequency of postsurgical heart complications occurs during coronary artery bypass grafting 3.4%, with the use of HTMC-3.44%.

During the analyzed period (2011–2019), there was a decrease in the frequency of postsurgical heart complications from 2.41 to 1.18%, with the use of HTMC - from 2.65 to 1.19% (in all surgeries - from 1.06 to 0.64%, with the use of HTMC - from 1.74 to 0.97%). The greatest decrease in the indicator was determined after coronary artery bypass graft by 2.4% (from 5.8 to 3.4%), of which after the use of HTMC - by 2.59% (from 6.03 to 3.44%), after angioplasty of the coronary arteries - by 0.6% (from 1.35 to 0.75%), of which after the use of HTMC - by 0.72% (from 1.38 to 0.66%), after stenting - by 0.72% (from 1.41 to 0.69%), of which after the use of HTMC - by 0.85% (from 1.52 to 0.67%) (Table 5).

**Table 3.** The number of heart operations performed in hospital on patients older than working age in the Russian Federation in 2011, 2013, 2015, 2017 and 2019 (in abs. number, in % of total)

| Surgical procedure | 2011 | | 2013 | | 2015 | | 2017 | | 2019 | |
|---|---|---|---|---|---|---|---|---|---|---|
| | abs. number | in % of the total | abs. number | in % of the total | abs. number | in % of the total | abs. number | in % of the total | abs. number | in % of the total |
| **Total of surgeries** | **2141468** | **100,0** | **2459346** | **100,0** | **2865972** | **100,0** | **3144766** | **100,0** | **3500938** | **100,0** |
| Including on: | | | | | | | | | | |
| *Heart* | *88936* | *4,15* | *132439* | *5,4* | *183437* | *6,4* | *230938* | *7,34* | *279539* | *8,0* |
| From them: for coronary heart disease | 47169 | 2,2 | 76465 | 3,1 | 121929 | 4,25 | 157351 | 5,0 | 202946 | 5,8 |
| From them: coronary artery bypass graft | 13789 | 0,6 | 16747 | 0,7 | 19291 | 0,7 | 22299 | 0,7 | 25064 | 0,7 |
| Coronary artery angioplasty | 29573 | 1,4 | 53087 | 2,2 | 87431 | 3,05 | 124119 | 3,95 | 166736 | 4,8 |
| From them: with stenting | 23147 | 1,1 | 43101 | 1,75 | 89522 | 3,1 | 112692 | 3,6 | 156764 | 4,5 |

**Table 4.** The number of heart operations performed with the use of high medical technologies in a hospital for persons over the working age in the Russian Federation in 2011, 2013, 2015, 2017 and 2019 (in abs. number, in %)

| Surgical procedure | 2011 | | | 2013 | | | 2015 | | | 2017 | | | 2019 | | |
|---|---|---|---|---|---|---|---|---|---|---|---|---|---|---|---|
| | in abs. number | in % of the total | in % of total surgeries | in abs. number | in % of the total | in % of total surgeries | in abs. number | in % of the total | in % of total surgeries | in abs. number | in % of the total | in % of total surgeries | in abs. number | in % of the total | in % of total surgeries |
| Total of surgeries | 255946 | 100 | 11,95 | 306166 | 100 | 12,45 | 411866 | 100 | 14,4 | 530014 | 100 | 16,85 | 663232 | 100 | 18,9 |
| Including on: | | | | | | | | | | | | | | | |
| Heart | 69094 | 27,0 | 77,7 | 102283 | 33,4 | 77,2 | 150106 | 36,5 | 81,8 | 194005 | 36,6 | 84,0 | 245828 | 37,1 | 87,9 |
| From them: for coronary heart disease | 39830 | 15,6 | 84,4 | 58323 | 19,1 | 76,3 | 101139 | 24,6 | 82,9 | 131167 | 24,7 | 83,4 | 175991 | 26,5 | 86,7 |
| From them: coronary artery bypass graft | 12308 | 4,8 | 89,3 | 15693 | 5,1 | 93,7 | 18027 | 4,4 | 93,45 | 21455 | 4,05 | 96,2 | 24726 | 3,7 | 98,7 |
| Coronary artery angioplasty | 25983 | 10,15 | 87,9 | 37932 | 12,4 | 71,45 | 76636 | 18,6 | 87,65 | 105909 | 20,0 | 85,3 | 148296 | 22,4 | 88,9 |
| From them: with stenting | 20246 | 7,9 | 87,5 | 32663 | 10,7 | 75,8 | 64837 | 15,7 | 72,4 | 102115 | 19,3 | 90,6 | 144955 | 21,9 | 92,5 |

**Table 5.** The frequency of postsurgical heart complications in people over working age in the Russian Federation in 2011, 2013, 2015, 2017 and 2019 (in %)

| Surgical procedure | Частота послеоперационных осложнений (в %) | | | | | | | | | |
| --- | --- | --- | --- | --- | --- | --- | --- | --- | --- | --- |
| | 2011 | | 2013 | | 2015 | | 2017 | | 2019 | |
| | Total of surgeries | Of them with the use of HTMC | Total of surgeries | Of them with the use of HTMC | Total of surgeries | Of them with the use of HTMC | Total of surgeries | Of them with the use of HTMC | Total of surgeries | Of them with the use of HTMC |
| Total of surgeries | 1,06 | 1,74 | 0,99 | 1,53 | 0,81 | 1,23 | 0,73 | 1,12 | 0,64 | 0,97 |
| Including on: | | | | | | | | | | |
| *Heart* | 2,41 | 2,65 | 2,07 | 1,86 | 1,51 | 1,48 | 1,33 | 1,38 | 1,18 | 1,19 |
| From them: for coronary heart disease | 2,48 | 2,65 | 1,85 | 1,88 | 1,44 | 1,36 | 1,2 | 1,25 | 1,05 | 1,04 |
| From them: coronary artery bypass graft | 5,8 | 6,03 | 5,52 | 4,9 | 4,19 | 4,17 | 4,15 | 4,26 | 3,4 | 3,44 |
| Coronary artery angioplasty | 1,35 | 1,38 | 1,32 | 0,82 | 1,0 | 0,71 | 0,73 | 0,67 | 0,75 | 0,66 |
| From them: with stenting | 1,41 | 1,52 | 1,33 | 0,79 | 0,83 | 0,7 | 0,7 | 0,67 | 0,69 | 0,67 |

The goal of the project 'Fight against cardiovascular diseases' is to reduce mortality from diseases of the circulatory system.

The objectives of the project are to develop and implement regional programs to combat cardiovascular diseases, which include measures aimed at preventing the development of cardiovascular diseases, timely identification of factors for the development of complications of these diseases, increasing the availability and quality of specialized medical care for patients with cardiovascular diseases, increasing satisfaction with the medical care received [13, 25–29]. To monitor the implementation of measures and planned results, target indicators have been approved [21 (p. 23), 30].

One of the indicators specified in the passport of the federal project 'Fight against cardiovascular diseases' was 'the ratio of the number of X-ray-endovascular interventions for therapeutic purposes to the total number of retired adult patients who suffered acute coronary syndrome, %'. [21 (pp. 26–27), 22]. The target value of the indicator should reach 60.0% by 2024 (in 2019 - 43.0%, in 2020 - 46.5%, in 2021 - 50.0%; by 2022 - 53.5%; by 2023–57.0%). In 2019, this indicator was 51.9% and was higher than the target value (Table 5).

The indicator in discharged patients older than working age for 9 years of study increased from 6.7 to 47.1% (Table 6).

**Table 6.** The ratio of the number of X-ray endovascular interventions for therapeutic purposes to the total number of discharged and deceased patients who underwent acute coronary syndrome in the Russian Federation in 2011, 2013, 2015, 2017 and 2019 (in %)

| Age groups | Year | | | | |
|---|---|---|---|---|---|
| | 2011 | 2013 | 2015 | 2017 | 2019 |
| Total adult population | 9,8 | 15,8 | 24,6 | 36,1 | 51,9 |
| Working-age population | 16,6 | 24,0 | 37,5 | 47,8 | 64,8 |
| The population older than the working age | 6,7 | 12,2 | 19,4 | 31,4 | 47,1 |

# 4   Conclusion

Thus, in Russia, the state health policy aimed at reducing the main cause of morbidity and mortality of the population older than working age from diseases of the circulatory system in 2011–2019 led to an increase in the number of heart surgeries performed in a hospital by 3.1 times, including with the use of high medical technologies—by 2.6 times. There also was an increase in the proportion of heart surgeries with the use of HTMC in people older than working age from 77.7 to 87.9%, of them with coronary artery bypass grafting - from 89.3 to 98.7% and coronary artery angioplasty - from 87.9 to 88.9%, with stenting - from 87.5 to 92.5%. It is important that during the study period, there was a decrease in the frequency of postsurgical heart complications from 2.41 to 1.18%, with the use of HTMC - from 2.65 to 1.19%.

The main target indicator of the federal project 'Fight against cardiovascular diseases' was 'the ratio of the number of X-ray endovascular interventions for therapeutic purposes to the total number of retired adult patients who suffered acute coronary syndrome,%', which in 2019 was 51.9% and was higher than the target value approved for 2021 (50.0%). This indicator in discharged patients older than working age is 47.1%.

# References

1. Decree of the President of the Russian Federation of June 6, 2019 № 254 «On the Strategy for the development of health care in the Russian Federation for the period up to 2025» [Electronic resource]. https://www.garant.ru/products/ipo/prime/doc/72164534/. Accessed 25 Dec 2020. (In Russ.)
2. Stepchuk, M.A.: General and primary morbidity of the population older than the working age of the Belgorod region. Adv. Gerontol. **32**(4), 658–663 (2019). https://elibrary.ru/item. asp?id=39541896. (In Russ.)
3. Shlyafer, S.I.: Morbidity of the population older than the working age of the Russian Federation. Curr. Probl. Health Care Med. Stat. **1**, 16–27 (2014). https://cyberleninka.ru/ article/n/zabolevaemost-naseleniya-starshe-trudosposobnogo-vozrasta-rossiyskoy-federatsii. (In Russ.)
4. Shlyafer, S.I., Ivanova, M.A.: Hospitalized morbidity of the population older than the working age of the Russian Federation. Adv. Gerontol. **29**(5), 690–694 (2016). http://www. gersociety.ru/netcat_files/userfiles/10/AG_2016-29-05.pdf. (In Russ.)
5. Shikina, I.B.: The maintenance of security of the elderly and gerontic patients in the hospital conditions. Probl. Soc. Hyg. Health Care Hist. Med. **6**, 44–45 (2007). (In Russ.)
6. Oshchepkova, E.V.: Population mortality from cardiovascular diseases in the Russian Federation in 2001–2006 and ways to reduce it. Cardiology **2**, 62–72 (2009). https://elibrary. ru/item.asp?id=16606099
7. Oganov, R.G., Kontsevaya, A.V., Kalinina, A.M.: Economic damage from cardiovascular diseases in the Russian Federation. Cardiovasc. Therapy Prev. **4**, 4–9 (2011). https:// cyberleninka.ru/article/n/ekonomicheskiy-uscherb-ot-serdechno-sosudistyh-zabolevaniy-v-rossiyskoy-federatsii. (In Russ.)
8. Starodubov, V.I., Son, I.M., Leonov, S.A., Pogonin, A.V.: Assessment of the impact of healthcare modernization on the dynamics of morbidity of the adult population of the country. Manag. Health Care **5**, 6–17 (2013). https://cyberleninka.ru/article/n/otsenka-vliyaniya-modernizatsii-zdravoohraneniya-na-dinamiku-zabolevaemosti-vzroslogo-naseleniya-strany. (In Russ.)
9. Bunova, S.S., Usacheva, E.V., Zamakhina, O.V.: Dynamics of the incidence of myocardial infarction in the regions of the Russian Federation over an 11-year period (2002–2012). Soc. Aspects Pop. Health **6**(40), 3 (2014). http://vestnik.mednet.ru/content/view/624/30/lang,ru/. (In Russ.)
10. Semenov, V.Yu., Stupakov, I.N.: Efficiency of using the new medical and economic standard of coronary angiography in ICISC. Bulletin of the A.N. Bakulev National Research Center of the Russian Academy of Medical Sciences "Cardiovascular Diseases". Publishing House of the Bakulev Scientific Center of Cardiovascular Surgery **6**(15), 296 (2014). https:// elibrary.ru/item.asp?id=22894432. (In Russ.)
11. Boitsov, S.A., et al.: Comparative analysis of population mortality from acute forms of coronary heart disease over a fifteen-year period in the Russian Federation and the USA and factors influencing its formation. Ther. Arch. **9**(89), 53–59 (2017). https://doi.org/10.17116/ terarkh201789953-59. (In Russ.)

12. Botvinova, N.V.: The prevalence of diseases of the circulatory system in the emergency medical care of the city of Rudny. Knowledge **1–2**(53), 38–43 (2018). https://elibrary.ru/item.asp?id=32334524. (In Russ.)
13. Savina, A.A., Feiginova, S.I.: Dynamics of the incidence of diseases of the circulatory system of the adult population of the Russian Federation in 2007–2019. Soc. Aspects Pop. Health Care **2**(67), (2021). http://vestnik.mednet.ru/index2.php?option=com_content&task=view&id=1243&pop=1&page=0&Itemid=27. дата обращения 01.06.2021 г. https://doi.org/10.21045/2071-5021-2021-67-2-1. (In Russ.)
14. Decree of the President of the Russian Federation No. 204 of May 7, 2018 'On the national goals and strategic objectives of the development of the Russian Federation for the period up to 2024'. https://base.garant.ru/71937200. Accessed 21 May 2021. (In Russ.)
15. Boitsov, S.A., Samorodskaya, I.V.: Comparison of mortality rates from myocardial infarction in the regions of the Russian Federation in 2006 and 2015. Prev. Med. **3**(20), 11–16 (2017). https://doi.org/10.17116/profmed201720311-16. (In Russ.)
16. Goloshchapov-Aksenov, R.S., Semenov, V.Yu., Kicha, D.I., Ivanenko, A.V.: Dynamics of the incidence of diseases of the circulatory system of the adult population of the Moscow region. Public Health Life Environ. **7**(316), 4–8 (2019). https://doi.org/10.35627/2219-5238/2019-316-7-4-8. (In Russ.)
17. Chernyayev, M.V., Faybushevich, A.G., Muzganova, Yu.S.: Endovascular treatment with limus eluting stents in patients with acute coronary syndrome. Russ. Sklifosovsky J. 'Emerg. Med. Care' **8**(1), 45–52 (2019). https://doi.org/10.23934/2223-9022-2019-8-1-45-52. (In Russ.)
18. Filippovskaya, Zh.S., et al.: Oxidative stress and early post-operative complications in cardiac surgery. Messenger Anesthesiol. Resusc. **13**(6), 13–21 (2016). https://doi.org/10.21292/2078-5658-2016-13-6-13-21. (In Russ.)
19. Federal Law No. 323-FZ of November 21, 2011. On the basics of Public Health Protection in the Russian Federation. kremlin.ru›acts/bank/34333. (In Russ.)
20. Order of the Ministry of Health of the Russian Federation No. 746n dated December 2, 2014. On approval of the 'Regulations on the organization of specialized, including high-tech, medical care'. https://base.garant.ru/70859232/. (In Russ.)
21. Son, I.M., et al.: Methodological recommendations on algorithms for calculating indicators of the national project 'Healthcare'. CRIOIHC of the Ministry of Health of the Russian Federation, Moscow, 108 p. (2020). https://mednet.ru/images/materials/news/2020_-_mr_po_nacproektu_zdravoohranenie.pdf. (In Russ.)
22. Order of the Ministry of Health of the Russian Federation No. 179 dated March 29, 2019. On approval of methods for calculating additional indicators of the federal project 'Fight against Cardiovascular Diseases', included in the national project 'Healthcare'. base.garant.ru›72227462. (In Russ.)
23. Belov, Yu.V., Rosseikin, E.V.: The concept of 'adequate' myocardial revascularization—a new direction in the surgical treatment of coronary heart disease. Russ. J. Thoracic Cardiovasc. Surg. **1**, 50–53 (2001). https://elibrary.ru/item.asp?id=27455609. (In Russ.)
24. Boitsov, V.A., Podlesov, A.M.: Cardiac arrhythmias in chronic heart failure. Heart Fail. J. **5**(2), 224–227 (2001). https://cyberleninka.ru/article/n/narusheniya-ritma-serdtsa-u-patsientov-s-hronicheskoy-serdechnoy-nedostatochnostyu-i-sohranennoy-fraktsiey-vybrosa-obzor-literatury. (In Russ.)
25. Buzin, V.N., Mikhaylova, Yu.V., Buzina, T.S., Chuhrienko, I.Yu., Shikina, I.B., Mikhaylov, A.Yu.: Russian healthcare through the eyes of the population: dynamics of satisfaction over the past 14 years (2006–2019): review of sociological studies. Russ. J. Prev. Med. **23**(3), 42–47 (2020). https://doi.org/10.17116/profmed20202303142. (In Russ.)

26. Armashevskaya, O.V., Ivanova, M.A., Chuchalina, L.Yu.: Age features of pathology of women in the peri - and postmenopausal period. Adv. Gerontol. **30**(3), 363–367 (2017). https://elibrary.ru/item.asp?id=29823536. (In Russ.)
27. Son, I.M., Ivanova, M.A., Sokolovskaya, T.A., Liutsko, V.V., Dezhurny, L.I.: Activity and the density of rheumatologists in Russian Federation, 2013–2017. Cardiovasc. Therapy Prev. **18**(1), 134–142 (2019). https://doi.org/10.15829/1728-8800-2019-1-134-142
28. Liutsko, V.V., Ivanova, M.A., Son, I.M., Zimina, E.V., Perkhov, V.I.: The provision and full strength of general practitioners (family doctors) rendering outpatient primary health care to the population of the Russian Federation from 2007 to 2016. Russ. J. Prev. Med. **22**(1), 43–48 (2019). https://doi.org/10.17116/profmed20192201143. (In Russ.)
29. Passport of the federal project "Fight against cardiovascular diseases". https://minzdrav.gov.ru/poleznye-resursy/natsproektzdravoohranenie/bssz. (In Russ.)
30. Amlaev, K.R., Zafirova, V.B., Aibazov, R.U., Khubieva, A.A., Shikina, I.B., Tretyakov, A.A.: Medical and social aspects of lifestyle and literacy in the health of patients with cardiac surgery. Med. News North Caucasus **1**, 91–95 (2015)

# Improved Prosthetic-Orthopedic Products for Rehabilitation of Patients with Post-stroke Deformities of the Upper Extremity

Natalia Shchekolova🔟, Aleksandr Zinovev🔟, Irina Balandina$^{(\boxtimes)}$🔟,
Viacheslav Ladeishchikov🔟, and Aleksandr Tokarev🔟

E.A. Vagner Perm State Medical University, Petropavlovskaya Street 26,
614000 Perm, Russia

**Abstract.** Patients who have had a stroke acquire marked limitations in their life activities, primarily in mobility and physical independence, as well as in their daily activities. At the same time, the function of the upper extremity is severely impaired. Arthropathy and instability of the shoulder joint are formed against the background of spasticity. Fine motor skills are impaired; grip function of the hand is predominantly affected. Patients are unable to work and maintain themselves. We considered orthotics as one component of the overall complex of rehabilitation measures after a stroke. The task of orthotics for orthopedists is to ensure joint stability, control pain and inflammation, preserve hand function, and reduce the load on the affected joint, which can be the cause of its instability. The aim of the study was to improve prosthetic and orthopedic products for correction of pathological upper limb settings during rehabilitation of patients with post-stroke upper limb deformities. We evaluated the dynamics of pain syndrome, muscle tone, strength, and angulometric indices in 94 disabled patients when using prosthetic-orthopedic devices in the treatment of post-stroke upper limb deformities. Bandage with individual adjustment was used to form a physiological position of the limb. Pathological positioning of the hand and fingers was eliminated with orthoses with a three-stage change in their position. The work experience showed that the adequacy of the prescription of prosthetic-orthopedic devices is the key to the effectiveness of rehabilitation measures in the late recovery period of cerebral stroke.

**Keywords:** Post-stroke deformities · Upper extremity · Bandages · Orthoses · Correction

## 1 Introduction

The high level of disability after cerebral stroke is caused not only by cognitive, speech and psycho-emotional, but also by motor disorders. Patients who have had a stroke acquire marked limitations on their vital activities, primarily mobility and physical independence, and daily activities. At the same time, the function of the upper extremity is especially severely impaired [1]. Against the background of spasticity, arthropathy and instability of the shoulder joint are formed. Fine motor skills are impaired, mainly hand grip function suffers. Patients are unable to work and serve themselves. Restorative treatment of impaired function of the upper extremity requires the attention of specialists from various medical fields - rehabilitation therapists, neurologists, orthopedists [2–6]. However, prosthetic and orthopedic products for correction of post-stroke limb deformities are often used without taking into account the mechanisms of orthopedic pathology formation and the dynamics of biomechanical and stabilometric parameters during treatment [7].

Orthotics are not used as an independent method of therapy, since their effectiveness is significantly inferior to medication and surgery. We consider orthotics as one of the components of the general complex of rehabilitation measures after a stroke. The task of orthotics for orthopedists was to ensure joint stability, relieve pain and inflammation, preserve hand function, and reduce the load on the affected joint, which can cause instability [8–10].

## 2  Research Objective

Improvement of prosthetic-orthopedic products for correction of pathological upper limb settings in rehabilitation of patients with post-stroke upper limb deformities.

## 3  Methodology

A comprehensive examination of the health status of 94 disabled people with disorders of the musculoskeletal system, who were treated in the KSAI "Center for Comprehensive Rehabilitation of the Disabled" in Perm from 2015 to 2020 was carried out. The research was approved by the Ethical Committee of the Federal State Budgetary Educational Institution of E.A. Vagner Perm State Medical University and was conducted in compliance with the ethical standards set forth in the Declaration of Helsinki. All patients gave informed consent for inclusion in the study. We conducted a course of rehabilitation treatment in the conditions of the rehabilitation center for 21 days on the average 14 months after the cerebral stroke. In all patients, advanced prosthetic-orthopedic devices for correction of post-stroke pathological settings of the upper extremity were used. The examination was conducted on admission and on discharge from the rehabilitation center. Male patients dominated among the patients. The average age of the patients studied was 49.5 years. Most patients had ischemic stroke, one-third had hemorrhagic stroke, and mixed stroke was extremely rare. There were assessed the dynamics of pain syndrome, muscle strength and angulometric parameters [11, 12].

Statistical processing was performed on a personal computer using STATISTICA 10.0 and MICROSOFT EXCEL 2012 application software package. Quantitative signs with a normal distribution are presented as $M \pm \sigma$, where M is the mean value and $\sigma$ is the standard deviation.

## 4   Results and Its Discussion

The specifics of upper limb function restoration consisted of functionally-oriented sequential correction of the shoulder, forearm, and hand by using a shoulder brace and various orthoses. A modified universal bandage was used to correct orthopedic disorders after a stroke [13].

Figure 1 shows the general view of the bandage on the patient.

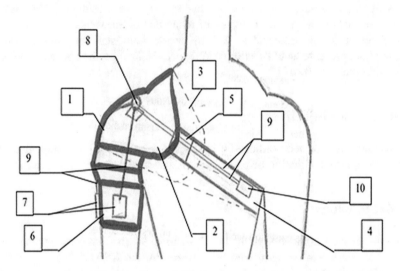

**Fig. 1.** Scheme of a modified bandage for correction of post-stroke pathology of the shoulder joint (1 - sleeve made of elastic material; 2 - anterior (ventral) lobe of the bandage sleeve; 3 - posterior (dorsal) lobe; 4 - textile strap holding the bandage; 5 - textile clasp "Contact"; 6 - distal part of the sleeve, covering the upper third of the shoulder; 7 - metal rings of the distal part of the sleeve; 8 - metal ring in the shoulder area; 9 - two non-stretchable textile cords; 10 - plastic lock for individual regulation of the degree of tension of the cords).

The bandage was used to form a physiological position of the limb, providing adequate correction of the shoulder joint condition in instability and distension of the joint capsule arising after stroke, including paresis or plegia of the hand. The bandage had individual adjustment of the degree of fixation, contributed to stabilization of the shoulder joint, reduction of pain syndrome, prevention of formation of post-stroke arthropathies and deformities. The bandage was a sleeve covering the shoulder joint with dorsal and anterior surfaces made of elastic material. From the dorsal lobe of the sleeve, the descending part of the cloth belt held the bandage. Then the strap went through the opposite axillary area, then its ascending part was attached to the anterior (ventral) lobe of the sleeve with a textile clasp. A distinctive feature of the device was that two metal rings were sewn to the anterior and posterior surfaces of the distal part of the sleeve covering the upper third of the upper arm. A third metal ring was sewn to the bandage at the shoulder area. Two non-stretchable textile cords were tied to the two lower rings. Both cords went up separately on the anterior and posterior surfaces of the upper third of the upper arm, then passed through the third ring in the shoulder area, and then together were attached on the anterior surface of the chest to the ascending part of the strap holding the bandage in place with a special plastic lock. The degree of tension of both cords was adjusted individually, which ensured a constant dosed stabilizing effect. Positive dynamics of orthopedic disorders in the patients during rehabilitation was noted. The range of motion in the right shoulder joint increased by more than one-third.

Pathological installation of the hand and fingers was eliminated by orthoses with a three-step change of their position [14]. In the course of treatment, the physiological position of the hand was formed, the tone and muscle strength in the hand were corrected, and fine motor skills, in particular grip function, were normalized. The orthosis was made of sheet thermoplastic material by vacuum molding in the form of a sleeve. The sleeve had palm and back surfaces and encompassed the hand and part of the forearm. The sleeve had distal and proximal ends and an opening to ensure the function of the first finger. Fixation of the orthosis on the hand was performed using any clasps, e.g., three "Contact" clasps (Russia). Three-stage controlled orthopedic correction of the increased muscle tone of the hand was performed to bring it to a physiological position. The main indications for the use of the developed utility model were the presence of increased muscle tone in the hand and forearm without contractures in the wrist joints. The use of the orthosis allowed stretching the muscles in which spasticity was detected, thus preventing contractures. The orthosis provided fixation, stabilization and correction of muscle tone of the distal parts of the upper extremity, was easily put on by the patient, and could be repeatedly treated with disinfectant solutions during repeated use. Correction of spastic hand alignment in patients using the unified orthosis is performed in three stages over 21 days. A general view of the orthosis is shown in Fig. 2 and 3.

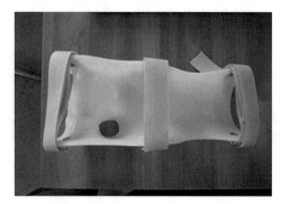

**Fig. 2.** Upper limb spasticity correction orthosis (top view).

Figure 4 shows the initial position of the hand during the first phase of rehabilitation (days 1–7).

The distal end of the orthosis is bent toward the palm surface at a 45° angle at the wrist joint location in relation to its proximal end. At a 60° angle at the location of the metacarpophalangeal joints of the fingers and at a 70° angle at the location of the proximal interphalangeal joints of the 2–5 fingers of the hand. At the second stage (day 7–14 of rehabilitation), we performed palmar flexion in the wrist joint at an angle of 25°, flexion in the metacarpophalangeal joints at 60°, and flexion in the proximal interphalangeal joints at 25° (Fig. 5).

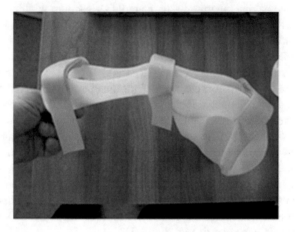

**Fig. 3.** Upper extremity spasticity correction orthosis (side view).

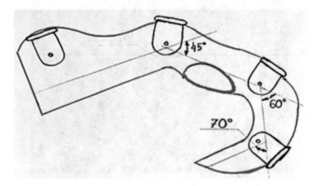

**Fig. 4.** Diagram of an upper limb orthosis for the first stage of rehabilitation.

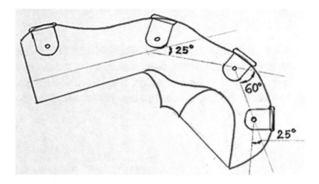

**Fig. 5.** Diagram of an upper limb orthosis for the second stage of rehabilitation.

At the third stage (15–21 days of rehabilitation), the hand was moved to the following position: dorsiflexion in the wrist joint of 25°; flexion in the metacarpophalangeal joints of 25°; and flexion in the proximal interphalangeal joints of 25° (Fig. 6).

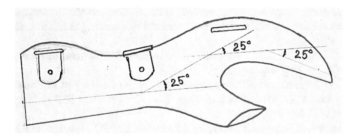

**Fig. 6.** Diagram of an upper limb orthosis for the third stage of rehabilitation.

The effectiveness of the presented prosthetic and orthopedic devices was proved with regard to objective clinical-functional and morphometric parameters. The effectiveness of the orthosis was evaluated by taking into account the dynamics of pain syndrome intensity in points according to the Visual Analogue Pain Scale, which decreased almost twofold. The intensity of the pain syndrome before treatment was $5.95 \pm 0.05$ points, but after the course of rehabilitation decreased to $2.5 \pm 0.01$ points. Extension in the right shoulder joint improved by more than a third. Flexion in the right shoulder joint before treatment was $30.1 \pm 0.5°$, after treatment, it was $45.5 \pm 0.5°$. In the left shoulder joint before treatment, the extension was $33.0 \pm 0.4°$, and after treatment it was $49.5 \pm 0.5°$.

Muscle strength in the right hand increased by 20% and in the left hand by 40% after the course of rehabilitation, which contributed to the ability to hold an object and accuracy of fine motor skills. We revealed positive dynamics of functional and morphometric motor parameters when using the developed prosthetic-orthopedic devices, which had practical significance for assessing the effectiveness of correction of post-stroke limb deformities in the complex rehabilitation of patients.

# 5    Conclusion

We recommend using modified prosthetic-orthopedic devices for normalization of upper limb function in complex rehabilitation of patients in the late recovery period of cerebral stroke. New technical solutions of prosthetic-orthopedic products improvement contribute to function-oriented restoration of impaired limb functions after cerebral stroke.

# References

1. Dos Santos, G.L., Souza, M.B., Desloovere, K., Russo, Th.L.: Elastic tape improved shoulder joint position sense in chronic hemiparetic subjects: a randomized sham-controlled crossover study. PLoS One **12**(1), e0170368 (2017). https://doi.org/10.1371/journal.pone.0170368
2. Bondarenko, F.V., Makarova, M.R., Turova, E.A.: Restoration of complex motor functions of the upper limb in patients after an ischemic stroke. Probl. Balneol. Physiother. Phys. Therapy **1**, 11–15 (2016). https://doi.org/10.17116/kurort2016111-15
3. Bondarenko, F.V., Makarova, M.R., Turova, E.A.: Basic approaches to the treatment of shoulder pain syndrome after a stroke **5**, 50–55 (2014)
4. Telenkov, A.A., Kadykov, A.S., Vuytsik, N.B., Kozlova, A.V., Krotenkova, I.A.: Post-stroke arthropathy: phenomenology and structural changes in joints. Alm. Clin. Med. **39**, 39–44 (2015)
5. Bernhardt, J., Cramer, S.C.: Giant steps for the science of stroke rehabilitation. Int. J. Stoke **8**(1), 1–2 (2013). https://doi.org/10.1111/ijs.12028
6. Pollock, A.: Physical rehabilitation approaches for the recovery of function and mobility following stroke. Cochrane Database Syst. Rev. **4**, CD001920 (2014). https://doi.org/10.1002/14651858.CD001920.pub3
7. Barrios-Muriel, J., Romero-Sánchez, F., Alonso-Sánchez, F.J., Salgado, D.R.: Advances in orthotic and prosthetic manufacturing: a technology review. Mater. (Basel) **13**(2), 295 (2020). https://doi.org/10.3390/ma13020295
8. Healy, A., Farmer, S., Pandyan, A., Chockalingam, N.: A systematic review of randomised controlled trials assessing effectiveness of prosthetic and orthotic interventions. PLoS One **13**(3), e0192094 (2018). https://doi.org/10.1371/journal.pone.0192094
9. McCaughan, D., et al.: Orthotic management of instability of the knee related to neuromuscular and central nervous system disorders: qualitative interview study of patient perspectives. BMJ Open **9**(10), e029313 (2019). https://doi.org/10.1136/bmjopen-2019-029313
10. Fromme, N.Ph., Camenzind, M., Riener, R., Rossi, R.M.: Need for mechanically and ergonomically enhanced tremor-suppression orthoses for the upper limb: a systematic review. J. Neuroeng. Rehabil. **16**, 93 (2019). https://doi.org/10.1186/s12984-019-0543-7
11. Ivanova, G.E., Skvortsov, D.V., Kaurkin, S.N.: Technique of objective registration of movements in the shoulder joint. Bull. Regen. Med. **3**(61), 8–13 (2014)
12. Shchekolova, N.B., Zinoviev, A.M.: Possibilities of conservative orthopedic correction of post-stroke spasticity of the upper extremity. Perm Med. J. **2**(34), 15–19 (2017)
13. Zinovev, A.M., Shchekolova, N.B., Bronnikov, V.A., Skliannaia, K.A., Saitov, S.A.: Shoulder bandage. Utility model patent RU 173102 U1, 08.11.2017. Application Rospatent No. 2017114479, 25 April 2017 (2017)
14. Zinovev, A.M., Shchekolova, N.B., Bronnikov, V.A., Skliannaia, K.A., Saitov, S.A.: Orthosis for correcting spastic hand positioning. Utility model patent RU 167069 U1, 20.12.2016. Application Rospatent No. 2016118370/14, 05 November 2016 (2016)

# Digital Media

# Brand Concept Rather Than Brand Image?: Consumer-Oriented Evaluation Items that Should Be Managed as Factors of Brand Loyalty

Takumi Kato[(⊠)] ⓘ

Saitama University, 255 Shimo-Okubo, Sakura-ku,
Saitama City 338-8570, Saitama, Japan
takumikato@mail.saitama-u.ac.jp

**Abstract.** The brand concept is the starting point of value creation and represents the essence of a product. However, research on loyalty factors has dealt with specific product features, such as performance and design, or brand image. Although the importance of the brand concept has long been debated in theory, it has been neglected in the practice of brand management. Therefore, this study tests the following hypothesis for Japanese personal computers and smartphones: "Consumers who recall the brand concept in terms of product attraction are more likely to have repurchase intention than consumers who recall the brand image." Based on the results of an online survey in which users of the target products were asked about repurchase intention and the attractiveness of the products, the property score was applied; as a result, the hypothesis was supported. There are two possible reasons for this finding. First, the brand image is not monopolized by one brand, as expressed by general adjectives. On the other hand, the brand concept is unique and irreplaceable for a strong brand. Second, the brand image is the perception of superficial results from the experience of the product, while the brand concept refers to the essence of the product. Thus, the psychological barrier of the brand concept is higher than that of the brand image. To win consumer loyalty, it is important to establish a clear brand concept and promote it consistently.

**Keywords:** Brand management · Loyalty · Brand concept · Brand image · Personal computer · Smartphone

## 1 Introduction

The brand concept provides a solution to a consumer's problem; that is, the brand concept meets a consumer's needs [1]. For example, Starbucks states its concept as the "third place" [2]—a place outside of the home and workplace where people can relax alone or gather and feel a sense of community. The brand image is cultivated within consumer perception through products and advertisements based on the concepts devised by marketers [3]. It is also said that consumers form an attitude and image of the brand through their experience of its products and advertising, which accumulates

T. Antipova (Ed.): DSIC 2021, LNNS 381, pp. 401–411, 2022.
https://doi.org/10.1007/978-3-030-93677-8_35

as brand knowledge [4, 5]. Therefore, the brand concept is the root of the product/service, and consumers develop the brand image and knowledge by experiencing the value generated from the concept.

Despite the large role of the brand concept, few studies have evaluated it as a loyalty factor. In brand management surveys and brand-related metrics, brand loyalty factors are commonly explained by product/service features such as performance [6, 7], quality [8–10], design [11, 12], and usability [13–15]. Alternatively, from the brand perspective, studies that have incorporated factor evaluation based on the brand image are common. As such, although much research has been conducted on brand loyalty factors, the brand concept has been overlooked.

In this study, the loyalty of consumers who are attracted to the brand concept is verified in comparison with the brand image. The target products are personal computers (PCs) and smartphones (SPs) in the Japanese market. Although the importance of the brand concept has been widely asserted in research on business administration and marketing, the effects specifically related to the brand concept have not been quantitatively evaluated. Thus, this study encourages the reaffirmation of the importance of the brand concept and its treatment as a loyalty factor rather than a brand image.

## 2   Literature Review and Hypothesis Derivation

Consumers form brand images using products [16] and marketing communications [17]. The brand image is recognized as an important precursor to loyalty [18]. To date, many studies have reported the impact of brand image on loyalty [19–26]. Alternatively, the brand image has been treated as a role in mediating the influence of measures such as promotion and satisfaction on loyalty [27, 28]. The viewpoints of the adverse effects are similar. For example, greenwashing, which refers to activities that make people believe that a company is doing more to protect the environment than in reality, damages the brand image and has a negative impact on loyalty [29]. Another example is religious animosity, which adversely affects the brand image and weakens loyalty [30]. Thus, the brand image plays a central role as a factor of loyalty.

However, the brand image is not unique to the brand. For example, it is impossible for one brand to monopolize specific brand images (e.g., as advanced or stylish). In short, consumers can easily switch brands when a brand with better images appears. Meanwhile, the brand concept is unique for a strong brand. A good brand concept is difficult for competitors to imitate and has no alternative. Therefore, consumers with high loyalty are considered to find value in the brand concept. It has been shown that consumers who recall the brand concept have higher loyalty compared to performance and design [31]. In short, the brand image is the perception of superficial results that stem from the experience of the product, while the brand concept refers to the essence of the product. Accordingly, the brand concept plays a stronger role in predicting loyalty than the brand image; thus, the following hypothesis is proposed:

**Hypothesis:** Consumers who recall the brand concept in terms of product attraction are more likely to have repurchase intention than consumers who recall the brand image.

## 3 Methodology

### 3.1 Survey

Generally, for consumer goods, the level of involvement is low, and unconscious continuous purchases are likely to occur. Alternatively, a brand switch can easily occur because of temporary factors such as discounts and novelty. Conversely, durable consumer goods are expensive and long-lasting, so the reason for continuous purchases/brand switches is clearer. This study focused on durable consumer goods and selected PCs and SPs, as they have many products with a clear brand concept. This study conducted an online survey in Japan from July 1–10, 2021 for PCs/SPs. The target subjects had two conditions: (a) age range of 20s–30s and (b) people who use PCs or SPs. The survey for PCs and SPs was different, and was controlled so that the same person could not participate in both. As shown in Table 1, the survey was distributed until the sample size was 1,000. Since the utilization rate of PCs was lower than that of SPs, the ratio of meeting the conditions was low.

The questions in the survey included the following 10 items: (1) sex, (2) age, (3) possession status of PCs/SPs, (4) brand of PCs/SPs, (5) purchase channel of PCs/SPs, (6) frequency of use of PCs/SPs, (7) elapsed time from the purchase of PCs/SPs, (8) loyalty indices (satisfaction, recommendation intention, and repurchase intention), (9) attractiveness of the product, and (10) clarity of the brand concept.

Conditions (a) and (b) were judged in questions (2) and (3), and the survey was completed at that point for those who did not meet the conditions. Table 2 shows that sex and age are collected almost evenly. Regarding the distribution of brands, Apple, which applies to both PCs/SPs, has the highest frequency. Question (8) was scored using a 5-point Likert scale (1 = very unsatisfied, 5 = very satisfied). As shown in Fig. 1, the high loyalty of Apple and Sony can be confirmed by looking at the average values of recommendation intention and repurchase intention by brand. The other brands have similar scores and are plotted close together.

Question (9) was important for testing the hypothesis of this study. Each respondent was randomly assigned to a brand concept group and brand image group. Then, for the question: "What is the most attractive point about your PCs/SPs?", the following options were presented to both groups: design, functionality, price, usability, and other (Table 3). As shown in Table 3, 11.5% of the brand image group selected the brand image, while 6.7% of the brand concept group selected the brand concept. In short, compared to the brand image, the brand concept was considered to have a higher psychological barrier in its selection.

The final question (10) was asked using a 5-point Likert scale to clarify the brand concept of the PCs/SPs owned for use in the propensity score, described later. When the brand concept was ambiguous, it was difficult to provide an answer in terms of its attractiveness. In addition, information on the devices (PCs or SPs) used by the respondents was automatically acquired using the function of the survey system.

**Table 1.** Survey delivery results.

| Target | No. of people who delivered the survey | No. of people who met the conditions | Percentage that met the conditions |
|---|---|---|---|
| PCs | 604 | 467 | 77.3% |
| SPs | 591 | 533 | 90.2% |

**Table 2.** Summary of respondent attributes.

| Category | Item | No. of respondents | Category | Item | No. of respondents |
|---|---|---|---|---|---|
| Sex | Male | 519 | Brand | Apple | 394 |
| | Female | 481 | | Dell | 32 |
| Age | Early 20s | 228 | | Fujitsu | 72 |
| | Late 20s | 289 | | Hp | 43 |
| | Early 30s | 234 | | Lenovo | 55 |
| | Late 30s | 249 | | Nec | 87 |
| Response device | PCs | 351 | | Samsung | 30 |
| | SPs | 649 | | Sharp | 62 |
| Target device | PCs | 467 | | Sony | 50 |
| | SPs | 533 | | Other | 175 |

**Table 3.** Results of answers to attractiveness of the product.

| Attractiveness | Brand concept group | | Brand image group | |
|---|---|---|---|---|
| | No. of people | Ratio | No. of people | Ratio |
| Brand concept/image | 34 | 6.7% | 57 | 11.5% |
| Design | 51 | 10.1% | 29 | 5.8% |
| Functionality | 145 | 28.8% | 118 | 23.8% |
| Price | 96 | 19.0% | 67 | 13.5% |
| Usability | 150 | 29.8% | 182 | 36.7% |
| Other | 28 | 5.6% | 43 | 8.7% |
| Total | 504 | 100.0% | 496 | 100.0% |

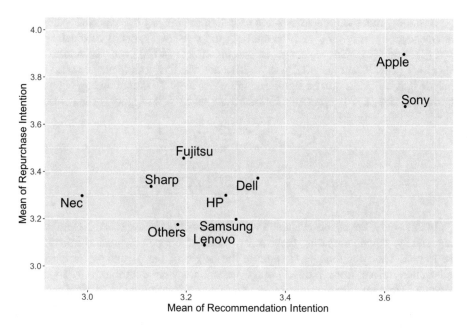

**Fig. 1.** Plot of mean loyalty indices for each brand.

## 3.2    Verification

The propensity score was used as a hypothesis verification method, where multiple covariates are aggregated into one variable called the propensity score and then used to correct the covariates in the treatment and control groups. In the brand concept group, the difference in loyalty was verified between those who chose the brand concept (equivalent to the treatment group) and those who did not (equivalent to the control group). Similarly, in the brand image group, the difference in loyalty was verified between those who chose the brand image (treatment group) and those who did not (control group). The hypothesis of this study was verified by comparing the differences between the two effect sizes.

Since the true value of the propensity score of each subject is unknown, estimation using a logistic regression model is common. When estimating the effect, one method involves matching people with similar propensity scores between both groups, but this reduces the sample size. In this study, only 34 out of the 504 people were in the treatment group in the brand concept group, making this method unsuitable. Accordingly, the effects were estimated by inverse probability weighting (IPW), which is the reciprocal weight of the propensity score [32]. As a logistic regression model for estimating the propensity score, the objective variable was recalling the brand concept/image for the attractiveness of the product (No. 4 in Table 4), and the explanatory variables were clarity of concept (No. 5), respondent attributes (Nos. 6–7), response device, and items from questions (4)–(7) (Nos. 8–24). As there were many variables, a stepwise method was adopted to select the variables.

Assuming that the propensity score of individual $i$ is $ps_i$, the expected value of the treatment group, result $Y_1$, and the control group, result $Y_0$, could be obtained using Eqs. (1) and (2). The average treatment effect (ATE) was calculated using $E(Y_1) - E(Y_0)$. When the ATE was positive and the 95% confidence interval did not include 0, it was judged to be significant at the 5% level. In the following, N represents the sample size, z = 1 represents the treatment, and z = 0 represents the control:

$$E(Y_1) = \sum_{i=1}^{N} \frac{z_i Y_1}{ps_i} / \sum_{i=1}^{N} \frac{z_i}{ps_i} \tag{1}$$

$$E(Y_0) = \sum_{i=1}^{N} \frac{(1 - z_i) Y_1}{1 - ps_i} / \sum_{i=1}^{N} \frac{1 - z_i}{1 - ps_i} \tag{2}$$

In this study, the effects evaluated were three loyalty indices (Nos. 1–3 in Table 4). Incidentally, when comparing these mean values, the recommendation intention was the lowest. This was thought to indicate the height of the psychological barriers, as satisfaction arises from product usage experience and contributes to repurchase intention. However, these are just personal feelings. Psychological barriers are higher for willingness to encourage others to make recommendations compared with the satisfaction or repurchase intention generated by individual emotions. Thus, being confident enough to recommend to others is a sign of commitment to the brand [36].

## 4   Results and Discussion

### 4.1   Results

Table 5 shows the results of the logistic regression model used to estimate the propensity score, and the suitability of these models is confirmed. Based on McFadden's pseudo R-squared, a value from 0.2–0.4 implies an excellent fit [33]. Further, the c-statistic (area under the curve) should be 0.7 or greater [34, 35]. In this study, all models meet both indices, confirming a certain degree of validity. Regarding the significant variables at the 5% level in the brand concept model group, brands such as Apple and Sony, as well as the variable ConceptClarity, show a positive effect. Thus, consumers who feel that the brand concept is clearer also tend to feel that it is attractive. However, in the brand image group, ConceptClarity is not selected as a model variable, so even the consumers who feel that the brand concept is not clear can easily answer the brand image.

Table 6 presents the estimation results of the ATE using the IPW estimator. In the brand concept group, the ATE of satisfaction is 0.307, 0.497 for recommendation intention, and 0.619 for repurchase intention. Each 95% confidence interval does not include 0, so the results are significant at the 5% level. However, in the brand image group, no significant effect is confirmed because each 95% CI contains 0. Accordingly, the hypothesis of this study is supported. In addition, it clarifies that repurchase intention is the largest as the ATE of the brand concept, followed by recommendation intention and satisfaction.

**Table 4.** Variable list.

| No | Variable | Description | Data type | Brand concept (504 respondents) | | Brand image (496 respondents) | |
|---|---|---|---|---|---|---|---|
| | | | | Mean | SE | Mean | SE |
| 1 | Satisfaction | Loyalty indices | 5-point Likert scale | 3.768 | 0.037 | 3.774 | 0.043 |
| 2 | Recommendation | | 5-point Likert scale | 3.377 | 0.042 | 3.385 | 0.043 |
| 3 | Repurchase | | 5-point Likert scale | 3.530 | 0.044 | 3.542 | 0.047 |
| 4 | Recalling_Brand | Recalling brand concept/image for attractiveness of the product | 0/1 | 0.067 | 0.011 | 0.115 | 0.014 |
| 5 | ConceptClarity | Clarity of concept | 5-point Likert scale | 3.381 | 0.054 | 3.456 | 0.054 |
| 6 | Female | Sex | 0: Male, 1: female | 0.482 | 0.022 | 0.480 | 0.022 |
| 7 | Age | Age | 1: Early 20s, 2: Late 20s, 3: Early 30s, 4: Late 20s | 2.498 | 0.049 | 2.510 | 0.049 |
| 8 | ResponseDevice_SP | Response device | 0: PCs, 1: SPs | 0.637 | 0.021 | 0.661 | 0.021 |
| 9 | TargetDevice_SP | Target device | 0: PCs, 1: SPs | 0.536 | 0.022 | 0.530 | 0.022 |
| 10 | Apple | Owned brand | 0/1 | 0.375 | 0.022 | 0.413 | 0.022 |
| 11 | Dell | | 0/1 | 0.022 | 0.007 | 0.042 | 0.009 |
| 12 | Fujitsu | | 0/1 | 0.077 | 0.012 | 0.067 | 0.011 |
| 13 | Hp | | 0/1 | 0.040 | 0.009 | 0.046 | 0.009 |
| 14 | Lenovo | | 0/1 | 0.054 | 0.010 | 0.056 | 0.010 |
| 15 | Nec | | 0/1 | 0.075 | 0.012 | 0.099 | 0.013 |
| 16 | Samsung | | 0/1 | 0.026 | 0.007 | 0.034 | 0.008 |
| 17 | Sharp | | 0/1 | 0.083 | 0.012 | 0.040 | 0.009 |
| 18 | Sony | | 0/1 | 0.052 | 0.010 | 0.048 | 0.010 |
| 19 | BrandOnlineStore | Purchase channel | 0/1 | 0.163 | 0.016 | 0.151 | 0.016 |
| 20 | BrandStore | | 0/1 | 0.111 | 0.014 | 0.141 | 0.016 |
| 21 | E-commerce | | 0/1 | 0.125 | 0.015 | 0.119 | 0.015 |
| 22 | RetailStore | | 0/1 | 0.530 | 0.022 | 0.482 | 0.022 |
| 23 | Frequency | Frequency of use | 1: Less than once a month, 2: Once a month, 3: Once every two weeks, 4: Once a week, 5: 2–3 times a week, 6: 4–6 times a week, 7: Every day | 5.915 | 0.085 | 5.875 | 0.089 |
| 24 | PurchaseTime | Elapsed time from purchase | 1: 1 year, 2: 2 years, …, 10: 10 years, 11: 10+ years | 3.159 | 0.109 | 2.958 | 0.106 |

Note: SE = standard error

**Table 5.** Logistic regression model estimation results.

| Variable | Brand concept | | | | Brand image | | | |
|---|---|---|---|---|---|---|---|---|
| | Estimate | SE | p-value | | Estimate | SE | p-value | |
| (Intercept) | −5.437 | 1.458 | 0.000 | *** | −1.906 | 0.551 | 0.001 | ** |
| ConceptClarity | 0.698 | 0.213 | 0.001 | ** | | | | |
| Female | −0.599 | 0.400 | 0.134 | | −0.817 | 0.318 | 0.010 | * |
| Age | −0.314 | 0.200 | 0.117 | | | | | |
| TargetDevice_SP | 1.485 | 0.765 | 0.052 | | 0.693 | 0.449 | 0.123 | |
| Apple | 2.556 | 1.063 | 0.016 | * | 1.013 | 0.407 | 0.046 | * |
| Fujitsu | 2.266 | 1.478 | 0.125 | | | | | |
| Nec | 2.015 | 1.217 | 0.001 | ** | | | | |
| Sony | 3.332 | 1.160 | 0.004 | ** | | | | |
| Frequency | −0.359 | 0.135 | 0.058 | | −0.220 | 0.073 | 0.053 | |
| McFadden | 0.260 | | | | 0.141 | | | |
| c-statistics | 0.872 | | | | 0.750 | | | |

Note: SE = standard error; *** p < 0.001; ** p < 0.01; * p < 0.05.

**Table 6.** ATE estimation results.

| Group | Variable | E(Y1) | E(Y0) | ATE | 95% CI | | |
|---|---|---|---|---|---|---|---|
| Brand concept | Satisfaction | 4.087 | 3.780 | 0.307* | 0.150 | − | 0.464 |
| | Recommendation | 3.878 | 3.381 | 0.497* | 0.326 | − | 0.667 |
| | Repurchase | 4.155 | 3.536 | 0.619* | 0.445 | − | 0.793 |
| Brand image | Satisfaction | 3.775 | 3.779 | −0.003 | −0.157 | − | 0.150 |
| | Recommendation | 3.419 | 3.376 | 0.043 | −0.130 | − | 0.215 |
| | Repurchase | 3.650 | 3.519 | 0.131 | −0.046 | − | 0.309 |

Note: CI = confidence interval; * p < 0.05.

## 4.2   Implications and Limitations

The implication of this study is that the results show the significance of evaluation from the viewpoint of brand management. As aforementioned, much research has been conducted on loyalty factors. However, the focus has been on the specific features of the product, such as performance and design [6–15]. From the brand's viewpoint, the brand image has been explored many times [19–30], while the brand concept has not. The results of this study show that consumers who are attracted to the brand image do not tend to have high loyalty, while the consumers who are attracted to the brand concept have significantly high loyalty. There are two possible reasons for this finding. First, the brand image is not monopolized by one brand, as expressed by general adjectives. However, the brand concept is unique and irreplaceable for a strong brand. Second, the brand image is the perception of superficial results from the experience of the product/service, while the brand concept refers to the essence of the product. Thus, the psychological barrier of the brand concept is higher than that of the brand image.

This study proposes three practical suggestions for this issue. First, marketers should not focus on acquiring the brand image while neglecting the brand concept. The brand image is easily affected not only by products but also by temporary advertisements, so it is easy to use for measuring marketing results. However, since many companies promote the same brand image, this is likely to have a temporary effect. Even if a brand image can be acquired, it is difficult to contribute it to consumer loyalty. Second, to win loyalty, a clear brand concept should be established before product development and advertising. As the results of this study show, if the brand concept is not clear, then consumers will not be attracted to it. Third, in brand management, consumers' recognition of the brand concept and recall of the brand concept in terms of attractiveness should be regularly observed. By defining a clear brand concept and consistently promoting it, it is possible to retain consumers as repeaters and spread favorable word of mouth as the "sales staff."

A limitation of this study is that the brand image was not classified. It has been clarified that brand images that belonging to emotional values (such as innovation and stylish) contribute to favorable attitudes toward the brand rather than the brand image belonging to functional values [37]. Hence, the results may fluctuate by dividing the brand image options by category. Additionally, it should also be noted that generalization is limited because the target industries were limited to PCs and SPs. These points can be explored in future research.

## 5   Conclusion

This study tested the following hypothesis for the Japanese PCs and SPs markets: "Consumers who recall the brand concept in terms of product attraction are more likely to have repurchase intention than consumers who recall the brand image." As a result of the analysis by propensity score based on the results of the online survey, a significant effect was confirmed; hence, the hypothesis is supported. In brand management, for loyalty factors, specific features of products such as performance and design or brand images have been dealt with, while the brand concept has been neglected. However, the origin of value creation concerns the brand concept, as loyal consumers understand the brand concept, empathize with it, and find value in it. If inconsistent marketing communications are repeated based on an ambiguous brand concept, then it will be difficult to establish a strong brand, even if a temporary brand image can be acquired. Therefore, this study encourages the reaffirmation of the importance of the brand concept.

## References

1. Park, C.W., Jaworski, B.J., MacInnis, D.J.: Strategic brand concept-image management. J. Mark. **50**(4), 135–145 (1986)
2. Schultz, H.: Pour your heart into it: how Starbucks built a company one cup at a time. Hyperion (1997)

3. Zenker, S.: Measuring place brand equity with the advanced Brand Concept Map (aBCM) method. Place Brand. Public Dipl. **10**(2), 158–166 (2014)
4. Campbell, M.C., Keller, K.L.: Brand familiarity and advertising repetition effects. J. Consum. Res. **30**(2), 292–304 (2003)
5. Hoeffler, S., Keller, K.L.: The marketing advantages of strong brands. J. Brand Manag. **10**(6), 421–445 (2003)
6. Kumar, V., Batista, L., Maull, R.: The impact of operations performance on customer loyalty. Serv. Sci. **3**(2), 158–171 (2011)
7. Yeh, C.H., Wang, Y.S., Yieh, K.: Predicting smartphone brand loyalty: consumer value and consumer-brand identification perspectives. Int. J. Inf. Manage. **36**(3), 245–257 (2016)
8. Devaraj, S., Matta, K.F., Conlon, E.: Product and service quality: the antecedents of customer loyalty in the automotive industry. Prod. Oper. Manag. **10**(4), 424–439 (2001)
9. Shen, C., Yahya, Y.: The impact of service quality and price on passengers' loyalty towards low-cost airlines: the Southeast Asia perspective. J. Air Transp. Manage. **91**, 101966 (2021)
10. Zehir, C., Şahin, A., Kitapçı, H., Özşahin, M.: The effects of brand communication and service quality in building brand loyalty through brand trust: the empirical research on global brands. Procedia Soc. Behav. Sci. **24**, 1218–1231 (2011)
11. Homburg, C., Schwemmle, M., Kuehnl, C.: New product design: concept, measurement, and consequences. J. Mark. **79**(3), 41–56 (2015)
12. Hsu, C.L., Chen, Y.C., Yang, T.N., Lin, W.K., Liu, Y.H.: Does product design matter? Exploring its influences in consumers' psychological responses and brand loyalty. Inf. Technol. People **31**(3), 886–907 (2018)
13. Chen, Y.Y.: Why do consumers go Internet shopping again? Understanding the antecedents of repurchase intention. J. Organ. Comput. Electron. Commer. **22**(1), 38–63 (2012)
14. Chiu, C.M., Chang, C.C., Cheng, H.L., Fang, Y.H.: Determinants of customer repurchase intention in online shopping. Online Inf. Rev. **33**(4), 761–784 (2009)
15. Lee, D., Moon, J., Kim, Y.J., Mun, Y.Y.: Antecedents and consequences of mobile phone usability: linking simplicity and interactivity to satisfaction, trust, and brand loyalty. Inf. Manage. **52**(3), 295–304 (2015)
16. Jin, N., Lee, S., Huffman, L.: Impact of restaurant experience on brand image and customer loyalty: moderating role of dining motivation. J. Travel Tour. Mark. **29**(6), 532–551 (2012)
17. Chinomona, R.: Brand communication, brand image and brand trust as antecedents of brand loyalty in Gauteng Province of South Africa. Afr. J. Econ. Manag. Stud. **7**(1), 124–139 (2016)
18. Bauer, H.H., Sauer, N.E., Exler, S.: The loyalty of German soccer fans: does a team's brand image matter? Int. J. Sports Mark. Spons. **7**(1), 8–16 (2005)
19. Anselmsson, J., Bondesson, N.V., Johansson, U.: Brand image and customers' willingness to pay a price premium for food brands. J. Prod. Brand Manag. **23**(2), 90–102 (2014)
20. Da Silva, R.V., Alwi, S.F.S.: Online corporate brand image, satisfaction, and loyalty. J. Brand Manag. **16**(3), 119–144 (2008)
21. Espinosa, J.A., Ortinau, D.J., Krey, N., Monahan, L.: I'll have the usual: how restaurant brand image, loyalty, and satisfaction keep customers coming back. J. Prod. Brand Manag. **27**(6), 599–614 (2018)
22. He, Y., Lai, K.K.: The effect of corporate social responsibility on brand loyalty: the mediating role of brand image. Total Qual. Manag. Bus. Excell. **25**(3–4), 249–263 (2014)
23. Lin, J., Lobo, A., Leckie, C.: Green brand benefits and their influence on brand loyalty. Mark. Intell. Plan. **35**(3), 425–440 (2017)
24. Ogba, I.E., Tan, Z.: Exploring the impact of brand image on customer loyalty and commitment in China. J. Technol. Manage. China **4**(2), 132–144 (2009)

25. Song, H., Wang, J., Han, H.: Effect of image, satisfaction, trust, love, and respect on loyalty formation for name-brand coffee shops. Int. J. Hosp. Manag. **79**, 50–59 (2019)

26. Unal, S., Aydın, H.: An investigation on the evaluation of the factors affecting brand love. Procedia Soc. Behav. Sci. **92**, 76–85 (2013)

27. Hsieh, A.T., Li, C.K.: The moderating effect of brand image on public relations perception and customer loyalty. Mark. Intell. Plan. **26**(1), 26–42 (2008)

28. Nyadzayo, M.W., Khajehzadeh, S.: The antecedents of customer loyalty: a moderated mediation model of customer relationship management quality and brand image. J. Retail. Consum. Serv. **30**, 262–270 (2016)

29. Chen, Y.S., Huang, A.F., Wang, T.Y., Chen, Y.R.: Greenwash and green purchase behaviour: the mediation of green brand image and green brand loyalty. Total Qual. Manag. Bus. Excell. **31**(1–2), 194–209 (2020)

30. Abosag, I., Farah, M.F.: The influence of religiously motivated consumer boycotts on brand image, loyalty, and product judgment. Eur. J. Mark. **48**(11/12), 2262–2283 (2014)

31. Kato, T.: Contribution of concept recall to brand loyalty: an empirical analysis of design and performance. J. Consum. Behav. 1–10 (2021). https://doi.org/10.1002/cb.1983

32. Rosenbaum, P.R.: Model-based direct adjustment. J. Am. Stat. Assoc. **82**(398), 387–394 (1987)

33. Hensher, D., Stopher, P.: Behavioural Travel Modeling. Croom Helm, London (1979)

34. Ferreira-González, I., et al.: Patterns of use and effectiveness of early invasive strategy in non-ST-segment elevation acute coronary syndromes: an assessment by propensity score. Am. Heart J. **156**(5), 946–953 (2008)

35. Yao, X., et al.: Effectiveness and safety of dabigatran, rivaroxaban, and apixaban versus warfarin in nonvalvular atrial fibrillation. J. Am. Heart Assoc. **5**(6), e003725 (2016)

36. Aaker, D.A.: Managing Brand Equity. The Free Press, New York (1991)

37. Kato, T.: Functional value vs emotional value: a comparative study of the values that contribute to a preference for a corporate brand. Int. J. Inf. Manag. Data Insights **1**(2), 100024 (2021)

# Creating and Using Synthetic Data for Neural Network Training, Using the Creation of a Neural Network Classifier of Online Social Network User Roles as an Example

A. N. Rabchevskiy[1,2(✉)] and L. N. Yasnitskiy[2,3]

[1] JSC "SEUSLAB", Perm, Russia
ran@psu.ru
[2] Perm State University, Perm, Russia
[3] National Research University "Higher School of Economics", Perm, Russia

**Abstract.** The use of synthetic data to train neural networks is becoming increasingly popular. Neural network classification of social network user profiles involves collecting large amounts of personal data, which is associated with high costs and the risk of leaks of confidential information. In this article we suggest using synthetic data to train the neural network classifier of social roles of users in online social networks that actively publish various kinds of materials (posts, reposts, comments) in social networks during the most active phase of the political protests, which can reduce the cost of data acquisition and maintain confidentiality of personal user data. Here is an example of dataset creation based on the algorithm that takes into account the ranges of neural network input parameters' values obtained from the analysis of real data and expert knowledge about correlations between values of different parameters for different user roles. Training and testing of the neural network has been carried out in several packages. Neural network classifier validation was done by comparing classification results of the synthetic neural network model with real data from several user samples. The validation results showed the adequacy of the synthetic neural network model to the real data. The effectiveness of dataset synthesis in cases where it is difficult or impossible to obtain the real data has been shown.

**Keywords:** Synthetic data · Dataset synthesis · Dataset creation · Social network · User roles · Neural networks · Classification

## 1 Introduction

The development of modern artificial intelligence technology is impossible to imagine without the use of synthetic data. As the name implies, this is data that is created artificially rather than from real events. They are often created by algorithms and are used for a wide range of activities. According to [1, 2], synthetic data is cheap to produce and can be useful for developing artificial intelligence models, deep learning and software testing. Data privacy provided by synthetic data is one of the most important advantages of synthetic data. User data often includes personal information

T. Antipova (Ed.): DSIC 2021, LNNS 381, pp. 412–421, 2022.
https://doi.org/10.1007/978-3-030-93677-8_36

and personal health information, and synthetic data allows companies to build software without disclosing user data to developers or software tools.

Most machine learning models require a lot of data to be more accurate. Synthetic data can be used to increase the size of training data for machine learning models.

Synthetic data generation creates labelled data instances that are ready to be used in training. This reduces the need for labour-intensive labelling efforts.

Over the last few years, generative models based on deep learning have gained increasing interest and offer some surprising improvements in this area [3]. Relying on huge amounts of data, well-designed network architectures and intelligent learning techniques, deep generative models have demonstrated an incredible ability to produce very realistic pieces of content of various types, such as images, texts and sounds.

Although synthetic data first began to be used in the 1990s, the abundance of computing power and storage space in the 2010s led to the wider use of synthetic data.

There is now a whole industry for the production of synthetic data. Basic use cases and software tools for synthetic data production are presented in [4]. It seems that some of these scenarios can be applied to the actual task of creating a neural network classifier for online social network users.

Online social networks are now a significant factor in people's daily lives and can serve both to communicate between users and to influence users to achieve marketing goals or to propagandize, agitate and mobilize people for certain protests and other illegal actions. To counteract such negative influences, social media has been extensively studied in academic circles [5–10], including in-depth analysis to identify structural and informational evidence of purposeful influence on the network. One element of this analysis is to identify the roles that users play in various social phenomena in online social networks. Various methods have been used to identify users' roles. Recently, neural networks have been increasingly used to classify users into roles. Qualitative classification requires a dataset corresponding to the subject area being modelled and an optimal neural network model. However, professionals often face the problem of lack of quality datasets for training, validation and testing of neural networks.

This paper presents an example of creating and using an artificially synthesised dataset to train, test and validate a neural network classifier for the roles of online social network users who actively publish various types of content (posts, reposts, comments) on social networks during the most active phase of political protest actions.

## 2   Materials

Social roles manifest themselves in different forms of user activity online. Various data can be used to categorise users into classes, including: number and type of publications, user behaviour patterns, etc. The aggregate of such data is interpreted as a conditional user profile. Neural network classification techniques can be used to classify such profiles into groups with similar parameter values.

In particular, [11] proposes a hybrid neural network to classify text in order to detect users' intentions. In [12], the use of deep neural networks to classify the sentiments of Twitter users is presented. Judging by the dates of these publications, classification of sets using neural networks is becoming increasingly popular.

In order to perform neural network classification of user roles, it is necessary to have high-quality training and validation sets. To solve this problem, the authors of [13] developed a special dataset of 1000 user profiles to be used to train a neural network identifying the social roles of Twitter users. Creating a dataset based on 740 thousand messages is presented in [14], while work [15] used a dataset consisting of more than 1.2 million text messages extracted from an online higher education community in Australia. To train their neural network, the authors [16] created a dataset based on content analysis of 350 million messages on Twitter. The authors of [17, 18] used ready-made datasets to analyze online social networks. The use of off-the-shelf datasets is convenient, but can be associated with both difficulty in obtaining them and incomplete correspondence of the off-the-shelf dataset to the subject area for which it is to be applied. In addition, neural networks trained on some networks may be unsuitable for other networks.

Because the distribution of roles among users in social networks is extremely heterogeneous, a high-quality dataset requires collecting data on several hundred thousand users, which is very expensive and implies a possible risk of leakage of sensitive data. Using an artificially synthesised dataset solves the problem of preserving user privacy and significantly reduces costs.

## 3  Method

The synthetic dataset was created based on a random data generation algorithm that takes into account the ranges of input parameter values of the neural network derived from the analysis of real social network users' data and expert knowledge about the correlation between the values of different neural network input parameters for different user roles.

### 3.1  Description of User Roles

Before creating a classifier of user roles in social media, the term 'social role' needs to be defined. The paper [19] suggests standardising the use of the term 'social role' in online communities as a set of social, psychological, structural and behavioural attributes, and proposes strategies for defining the social roles of users in some online communities. However, the authors do not propose a strict classification of users' social roles, as the set and definition of social roles depends on both the type of social community and the context in which user roles are considered.

In creating the expert neural network, we used the following classes of social media users:

1. A poster is an idea generator, a content creator, often an opinion leader, and with a lot of connections can unite many users around him or her.
2. Reposter - a distributor of ideas, rarely creates content, mostly reposts ready-made publications, aims to spread other people's publications as much as possible.

3. Commentator - does not create content, does not repost, but leaves lots of comments, participates in discussions and debates. Often he/she creates superfluous comments to increase the popularity of the topic of discussion.
4. Universal - a member who actively publishes posts, reposts and comments without a clear predominance of any one type of material.
5. Passive participant - a user who is not very active in the network in terms of creating content, reposts or comments, but regularly visits various pages of the social network. He or she is a recipient of all the information created by the Posters, Reposters and Commenters.

## 3.2 Neural Network Input and Output Parameters

The following parameters were used as input data to classify users:

X1 - Age of the account
X2 - Number of friends
X3 - Number of posts published
X4 - Number of published reposts
X5 - Number of published comments.

The outputs of the neural network model were:

D1 - takes the value 1 if the user is a Poster and 0 if not.
D2 - takes on a value of 1 if the user is a Reporter and 0 if not.
D3 - takes a value of 1 if user is a Commentator and 0 if not.
D4 - takes value 1 if user is a Universal and 0 if not.
D5 - takes value 1 if the user is a Passive Participant and 0 if not.

## 3.3 Dataset Synthesis

The main task in generating the dataset was to determine the ranges of each of the input parameters for each role and to introduce certain patterns into the dataset. The values needed to generate the dataset were the ranges obtained from the analysis of material published by users of the Vkontakte social network regarding a fake news blast about the existence of the so-called "Putin's Palace" (see Table 1).

In order for the neural network to learn well and to classify the input sets qualitatively, it was necessary to provide enough examples for each role. In our case, 400 examples were generated for each role.

## 3.4 Set Synthesis Algorithms for Each Role

For the synthesis of the sets for each role, the value ranges and expert parameter ratios corresponding to the subject area presented in Table 1 were used. The generation of the set was done using Microsoft Excel 2016. To generate the set, a random function was used to select a value from the range of values of the specified package. Let us denote this function as

**Table 1.** Value ranges for each role derived from the analysis of Vkontakte users' postings regarding the so-called "Putin's Palace"s.

| Parameter | Poster | Reposter | Commentator | Universal | Passive |
|---|---|---|---|---|---|
| X1 - Age of account (days) | 311–4382 | 0–4881 | 0–4183 | 545–4553 | 86–5170 |
| X2 - Number of friends | 0–32368259 | 0–31586803 | 0–28990781 | 0–34031059 | 0–36014728 |
| X3 - Number of posts | 2–94 | 0–94 | 0–6 | 0–17 | 0–1 |
| X4 - Number of reposts | 0–7 | 2–160 | 1–3 | 1–19 | 0–1 |
| X5 - Number of comments | 0–7 | 0–48 | 2–48 | 0–11 | 0–1 |

$$R(X_{min}; X_{max}) \tag{1}$$

Passive participants are defined as those who do not have a high activity rating, the ratio between the different types of material does not matter, as long as the values are within the maximum and minimum values. Table 2 shows the formulas used to generate the set for the Passive Participant role. The step of decreasing age of the account is denoted by $\Delta$.

**Table 2.** Formulas for generating a set for the Passive Participant role

| N | X1 | X2 | X3 | X4 | X5 |
|---|---|---|---|---|---|
| 1 | $X1_{max}$ | $X2_{max}$ | $X3_{max}$ | $X4_{max}$ | $X5_{max}$ |
| 2 | $X1_{max}-\Delta$ | $R(X2_{min}; X2_{max})$ | $R(X3_{min}; X3_{max})$ | $R(X4_{min}; X4_{max})$ | $R(X5_{min}; X5_{max})$ |
| 3 | $X1_{max}-2\Delta$ | $R(X2_{min}; X2_{max})$ | $R(X3_{min}; X3_{max})$ | $R(X4_{min}; X4_{max})$ | $R(X5_{min}; X5_{max})$ |
| .. | .. | .. | .. | .. | .. |
| N | $X1_{min}$ | $X2_{min}$ | $X3_{min}$ | $X4_{min}$ | $X5_{min}$ |

A poster is an active member whose main activity is creating posts. Let

$p_i$ - the number of posts published by user $i$,
$r_i$ - the number of reposts published by user $i$,
$k_i$ - the number of comments posted by user $i$,

then the total number of submissions by this user $m_i$ can be expressed as

$$m_i = p_i + r_i + k_i, \tag{2}$$

According to experts, a Poster is a user who mainly creates content, often reposts and occasionally comments on other users' posts. Thus, a Poster is a user, whose posts constitute at least 60% of all the materials published by the Poster, whose number of reposts does not exceed 37% and whose number of comments does not exceed 3%. That is, the Poster must meet the following conditions:

$$p_i \geq \alpha m_i, r_i \leq \beta m_i, k_i \leq \gamma m_i, \tag{3}$$

where $\alpha = 0.6$, $\beta = 0.37$, $\gamma = 0.03$. In this case the values of $r_i$ and $k_i$ can be expressed as

$$r_i \leq \delta p_i \text{ and } k_i \leq \varepsilon p_i, \text{ where } \delta = \frac{\beta}{\alpha} \text{ and } \varepsilon = \frac{\gamma}{\alpha}. \tag{4}$$

Using these relations and the values for a given role from Table 1, we present a set of formulas for generating the role Poster (see Table 3).

**Table 3.** Formulas for generating a set for the Poster role

| n | X1 | X2 | X3 | X4 | X5 |
|---|----|----|----|----|----|
| 1 | $X1_{max}$ | $X2_{max}$ | $X3_{max}$ | $R(0; \delta \, X3)$ | $R(0; \varepsilon \, X3)$ |
| 2 | $X1_{max}-\Delta$ | $R(X2_{min}; X2_{max})$ | $R(X3_{min}; X3_{max})$ | $R(0; \delta \, X3)$ | $R(0; \varepsilon \, X3)$ |
| 3 | $X1_{max}-2\Delta$ | $R(X2_{min}; X2_{max})$ | $R(X3_{min}; X3_{max})$ | $R(0; \delta \, X3)$ | $R(0; \varepsilon \, X3)$ |
| .. | .. | .. | .. | .. | .. |
| N | $X1_{min}$ | $X2_{min}$ | $X3_{min}$ | $X4_{min}$ | $X5_{min}$ |

According to experts, a Reposter is a user who mainly reposts content created by other users, often creates content themselves and occasionally comments on other users' posts. Thus, a Reposter is a user whose number of reposts is at least 60% of all the materials they have published, whose number of posts is no more than 37% and whose number of comments is no more than 3%. In other words, a user has to meet the following conditions:

$$r_i \geq \alpha m_i, p_i \leq \beta m_i, k_i \leq \gamma m_i, \tag{5}$$

where $\alpha = 0.6$, $\beta = 0.37$, $\gamma = 0.03$. In this case the values of $p_i$ and $k_i$ can be expressed as

$$p_i \leq \delta r_i \text{ and } k_i \leq \varepsilon r_i, \text{ where } \delta = \frac{\beta}{\alpha} \text{ and } \varepsilon = \frac{\gamma}{\alpha} \tag{6}$$

The set of formulas for generating the Reposter role is shown in Table 4.

A commenter is a user who mainly comments on other users' posts, does not often repost and rarely creates content himself. Thus, a Commentator is a user who has at least 60% of their comments, no more than 10% of their posts and no more than 30% of their reposts. That is, the user must meet the following conditions:

**Table 4.** Formulas for generating a set for the Reposter role

| n | X1 | X2 | X3 | X4 | X5 |
|---|----|----|----|----|----|
| 1 | $X1_{max}$ | $X2_{max}$ | $R(0; \delta\ X4)$ | $X4_{max}$ | $R(0; \varepsilon\ X4)$ |
| 2 | $X1_{max}-\Delta$ | $R(X2_{min}; X2_{max})$ | $R(0; \delta\ X4)$ | $R(X4_{min}; X4_{max})$ | $R(0; \varepsilon\ X4)$ |
| 3 | $X1_{max}-2\Delta$ | $R(X2_{min}; X2_{max})$ | $R(0; \delta\ X4)$ | $R(X4_{min}; X4_{max})$ | $R(0; \varepsilon\ X4)$ |
| .. | .. | .. | .. | .. | .. |
| N | $X1_{min}$ | $X2_{min}$ | $X3_{min}$ | $X4_{min}$ | $X5_{min}$ |

$$k_i \geq \alpha m_i, p_i \leq \beta m_i, r_i \leq \gamma m_i, \tag{7}$$

where $\alpha = 0.6$, $\beta = 0.1$, $\gamma = 0.3$. In this case the values of $p_i$ and $r_i$ can be expressed as

$$p_i \leq \delta k_i \text{ and } r_i \leq \varepsilon k_i, \text{ where } \delta = \frac{\beta}{\alpha} \text{ and } \varepsilon = \frac{\gamma}{\alpha}. \tag{8}$$

The set of formulas for generating the Commentator role is shown in Table 5.

**Table 5.** Formulas for generating a set for the Commentator role

| n | X1 | X2 | X3 | X4 | X5 |
|---|----|----|----|----|----|
| 1 | $X1_{max}$ | $X2_{max}$ | $R(0; \delta\ X5)$ | $R(0; \varepsilon\ X5)$ | $X5_{max}$ |
| 2 | $X1_{max}-\Delta$ | $R(X2_{min}; X2_{max})$ | $R(0; \delta\ X5)$ | $R(0; \varepsilon\ X5)$ | $R(X5_{min}; X5_{max})$ |
| 3 | $X1_{max}-2\Delta$ | $R(X2_{min}; X2_{max})$ | $R(0; \delta\ X5)$ | $R(0; \varepsilon\ X5)$ | $R(X5_{min}; X5_{max})$ |
| .. | .. | .. | .. | .. | .. |
| N | $X1_{min}$ | $X2_{min}$ | $X3_{min}$ | $X4_{min}$ | $X5_{min}$ |

A Universal is a user who does not meet conditions (3, 5, 7). In fact, a Universal is an active user who is not a Poster, Reporter or Commentator. Using this representation and the graphical values for this role from Table 1, we present a set of formulas for generating the set of Universal role (see Table 6).

**Table 6.** Formulas for generating a set for the Universal role

| n | X1 | X2 | X3 | X4 | X5 |
|---|----|----|----|----|----|
| 1 | $X1_{max}$ | $X2_{max}$ | $X3_{max}$ | $X4_{max}$ | $X5_{max}$ |
| 2 | $X1_{max}-\Delta$ | $R(X2_{min}; X2_{max})$ | $R(X3_{min}; X3_{max})$ | $R(X4_{min}; X4_{max})$ | $R(X5_{min}; X5_{max})$ |
| 3 | $X1_{max}-2\Delta$ | $R(X2_{min}; X2_{max})$ | $R(X3_{min}; X3_{max})$ | $R(X4_{min}; X4_{max})$ | $R(X5_{min}; X5_{max})$ |
| .. | .. | .. | .. | .. | .. |
| N | $X1_{min}$ | $X2_{min}$ | $X3_{min}$ | $X4_{min}$ | $X5_{min}$ |

The example sets for each role derived from the results of the algorithm have been combined, mixed and split into two parts:

- Teaching - 1700 examples,
- Test - 300 examples.

The prepared dataset was used for training and testing the neural network model on Neurosimulator 5.0 platform Nsim5sc [20] (access www.LbAi.ru). As a result of numerous iterations, the best result was obtained by a perceptron neural network with five input neurons, one hidden layer with seven neurons and five output neurons. The hyperbolic tangent was used as the activation functions of all neurons. The formula was used to estimate the error in the neural simulator:

$$
E = \frac{\sqrt{\dfrac{\sum\limits_{n=1}^{N}(d_n - y_n)^2}{N}}}{|\max(d_n) - \min(d_n)|}\ 100\%,
\tag{9}
$$

where $N$ is the number of sample elements, $d_n$ is the declared role of the $n$-th user, and $y_n$ is its role evaluated by the neural network. The testing error of the neural network model of user role classification is presented in Table 7.

**Table 7.** Test result of a neural network model of a user role classifier based on the Nsim5sc Neural Simulator

| No | Role name | Error % |
|----|-----------|---------|
| Y1 | Poster | 10,3% |
| Y2 | Reposter | 10,2% |
| Y3 | Commentator | 2,9% |
| Y4 | Universal | 16,5% |
| Y5 | Passive member | 6,3% |

In addition, a neural network model based on a synthetic dataset was trained and tested in other neural network packages (TensorFlow, Apple Create ML, Orange Data Mining) and in all cases the neural network with the same hyperparameters showed the best result.

The validation of the neural network model, based on the artificially synthesised dataset, was carried out by classifying real users and then analyzing the identified user roles. The validation was carried out on a sample of users actively posting different types of material (posts, reposts, comments) on social media during the most active phase (10–15 days) of protest political actions related to the fake news blast about the existence of the so-called "Putin's Palace". In addition, a validation was carried out on a sample related to protest actions around the 2020 presidential elections in Belarus. The results of the validation showed full coincidence of the classification results with the proposed algorithm and the results of analytical research of the data of real social network users, performed by expert-analysts.

## 4   Conclusion

The synthetic dataset generation algorithm used parameter value ranges derived from real data analysis and expert knowledge of the relationships between different parameter values for different roles. Using synthetic data to create a neural network classifier of user roles significantly reduced the cost of creating a dataset from real data and eliminated the risk of leakage of confidential data of social network users. At the same time, the neural network model showed low error rate and adequacy to real data of the target area.

Synthetic data for training and testing the neural network model has been registered as a computer database [21] and is available for use by completing the web request form on the website [22].

The use of synthetic datasets is a universal method and can be recommended for use when it is impossible or difficult to obtain real data for a dataset or when data confidentiality is required. In addition, it can be suggested to use this method to increase the number of examples in datasets when the number of real examples is not enough for quality training of a neural network.

## References

1. Dilmegani, G.: The Ultimate Guide to Synthetic Data in 2021. https://research.aimultiple. com/synthetic-data/
2. Dilmegani, G.: Synthetic Data Generation: Techniques, Best Practices & Tools. https:// research.aimultiple.com/synthetic-data-generation/
3. Lauterbach, A., Bonime-Blanc, A., Bremmer, I.: The Artificial Intelligence Imperative: A Practical Roadmap for Business. ABC-CLIO (2018)
4. Dilmegani, G.: Top 20 Synthetic Data Use Cases & Applications in 2021. https://research. aimultiple.com/synthetic-data-use-cases/
5. Castells, M.: Networks of Outrage and Hope. Social Movements in the Internet Age. Polity, Cambridge (2012)
6. Faris, D.M.: Dissent and Revolution in a Digital Age. I.B.Tauris (2013). https://doi.org/10. 5040/9780755607839
7. Gerbaudo, P.: Tweets and the Streets. Social Media and Contemporary Activism. Pluto Books, London (2012)
8. Tindall, D.B.: From metaphors to mechanisms: critical issues in networks and social movements research. Soc. Netw. **29**, 160–168 (2007). https://doi.org/10.1016/j.socnet.2006. 07.001
9. Bennett, W.L., Segerberg, A.: The logic of connective action. Inf. Commun. Soc. **15**, 739–768 (2012). https://doi.org/10.1080/1369118X.2012.670661
10. Juris, J.S.: Reflections on #Occupy Everywhere: social media, public space, and emerging logics of aggregation. Am. Ethnol. **39**, 259–279 (2012). https://doi.org/10.1111/j.1548-1425. 2012.01362.x
11. Liu, Y., Liu, H., Wong, L.-P., Lee, L.-K., Zhang, H., Hao, T.: A hybrid neural network RBERT-C based on pre-trained RoBERTa and CNN for user intent classification. In: Zhang, H., Zhang, Z., Wu, Z., Hao, T. (eds.) NCAA 2020. CCIS, vol. 1265, pp. 306–319. Springer, Singapore (2020). https://doi.org/10.1007/978-981-15-7670-6_26

12. Abdelhade, N., Soliman, T.H.A., Ibrahim, H.M.: Detecting twitter users' opinions of arabic comments during various time episodes via deep neural network. In: Hassanien, A.E., Shaalan, K., Gaber, T., Tolba, M.F. (eds.) AISI 2017. AISC, vol. 639, pp. 232–246. Springer, Cham (2018). https://doi.org/10.1007/978-3-319-64861-3_22
13. Sunghwan, M.K., Stephen, W., Cecile, P.: Detecting social roles in twitter. In: Proceedings of the Fourth International Workshop on Natural Language Processing for Social Media, Austin, TX, pp. 34–40 (2016)
14. Matsumoto, K., Yoshida, M., Kita, K.: Classification of emoji categories from tweet based on deep neural networks. In: Proceedings of the 2nd International Conference on Natural Language Processing and Information Retrieval - NLPIR 2018, New York, NY, USA, pp. 17–25. ACM Press (2018). https://doi.org/10.1145/3278293.3278306
15. Wijenayake, P., de Silva, D., Alahakoon, D., Kirigeeganage, S.: Automated detection of social roles in online communities using deep learning. In: Proceedings of the 3rd International Conference on Software Engineering and Information Management, New York, NY, USA, pp. 63–68. ACM (2020). https://doi.org/10.1145/3378936.3378973
16. Lin, H., et al.: User-level psychological stress detection from social media using deep neural network. In: Proceedings of the 22nd ACM International Conference on Multimedia, New York, NY, USA, pp. 507–516. ACM (2014). https://doi.org/10.1145/2647868.2654945
17. Jabłońska, M.R., Zajdel, R.: Artificial neural networks for predicting social comparison effects among female Instagram users. PLoS ONE **15** (2020). https://doi.org/10.1371/journal.pone.0229354
18. Segalin, C., et al.: What your facebook profile picture reveals about your personality. In: Proceedings of the 25th ACM International Conference on Multimedia, New York, NY, USA, pp. 460–468. ACM (2017). https://doi.org/10.1145/3123266.3123331
19. Gleave, E., Welser, H.T., Lento, T.M., Smith, M.A.: A conceptual and operational definition of "social role" in online community. In: 2009 42nd Hawaii International Conference on System Sciences. IEEE (2009). https://doi.org/10.1109/HICSS.2009.6
20. Cherepanov, F.M., Yasnitsky, L.N.: Neurosimulator 5.0: Rospatent Certificate of State Registration of Computer Programme No. 2014618208 dated 12.07.2014
21. Rabchevskiy, A.N., Zayakin, V.S.: A database for the classification of roles of social network users. State Registration Certificate for the Computer Database No. 2021621533 dated 15.07.2021
22. https://seuslab.ru/registered_db/2021621533?lang=en

# World Heritage on Institutional Pages, Personal Pages and Blogs

Artur Filipe dos Santos[1(✉)] and Marta Loureiro dos Santos[2]

[1] ISLA-IPGT, Instituto Politécnico de Gestão e Tecnologia, Oporto, Portugal
`artur.santos@islagaia.pt`
[2] R. Diogo Macedo, 4430-999 Vila Nova de Gaia, Portugal

**Abstract.** At the current juncture in which the World Heritage, its classification, its study, preservation and dissemination depends to much on its ability to
generate economic benefits for the cultural tourism industry, it is essential to
promote heritage in the largest communication platform of the moment, the
Internet, the largest standard of new information and communication technologies. Fundamental tools of dissemination in the context of communication
and public image in the current business vision, it is important to reflect whether
public bodies and private companies that campaign around the promotion of
cultural heritage bet on these platforms to reach their target audiences. In these
studies we focus on the Iberian reality, presenting a theoretical introduction that
focuses on communication in the sphere of the dissemination of cultural heritage, with emphasis on sites classified as World Heritage within UNESCO,
emphasizing with examples of institutional pages, personal pages and Portuguese and Spanish blogs.

**Keywords:** Social media · Communication · World Heritage · Sites · Blogs

## 1 Introduction to the Concept of Cultural Heritage Focusing on UNESCO Vision

It is not now that humanity has felt at the heart the will and importance of preserving its
legacy, its work done in the present and in the past and defending it, so that one day the
generations of the future could witness to history, which touches, and thus know,
through monuments, the heritage erected, much of what the ancestors were and created:
"Our ancestors knew perhaps that the gardens of Kahore, the mosques of Cairo, the
Cathedral of Amiens and the Hypogeus of Malta were sumptuous, rare, strange
monuments. At times they were sensitive to the splendor of a mountain, a large river
and even a jungle populated with wild animals and came to admit that these elements
could make the pride of a people and witness the nobility of their history or that these
geographical accidents could symbolize a nation, its adventures and its misadventures.
But they would not have been told that this had a 'universal value'" [1].

It was in 1972 that the United Nations Educational, Scientific and Cultural Organization (UNESCO) created the Convention concerning the Protection of the World
Cultural and Natural Heritage, designed to, citing UNESCO "to ensure, as far as
possible, the proper identification, protection, conservation and presentation of the

T. Antipova (Ed.): DSIC 2021, LNNS 381, pp. 422–432, 2022.
https://doi.org/10.1007/978-3-030-93677-8_37

world's irreplaceable heritage" [2]. A colossal goal in an attempt to preserve a heritage that is, UNESCO states "among the most precious and irreplaceable possessions, not only of each nation, but also of humanity as a whole", as it declares this United Nations organization "the loss, through deterioration or disappearance, of any of these most prized assets constitutes an impoverishment of the heritage of all the peoples of the world" [3].

## 2 The Importance of Communication for the Dissemination of Cultural Heritage

All efforts to preserve the cultural heritage of humanity would never have the expression obtained if it had not been made known at an international level, if the various measures taken by the international institutions in favour of a common goal, which is to preserve the tangible witness of our history, were not disclosed.

Thus, with a fixed effort by the various bodies to develop a concerted institutional communication, it has been achieved today with bibliographic references and documents of great importance to understand not only the concrete objectives of preserving heritage, but the fact that these same documents exist are the physical proof that the various institutions involved have actually practiced a communication policy.

César Carreras Monfort advocates that "in recent years, the world of heritage institutions has undergone significant changes both in its social dimension and in its priorities" [4], pointing to new information and communication technologies as the engine of this transformation, driven mainly by the need for evaluation and dissemination of cultural and educational offer:

> "It is through this constant evaluation of the educational offer of the institutions of memory, of the languages of communication and the elaboration of the contents that the center will be able to satisfy little by little the expectations of its potential audience. To all this already complex problem, information and communication technologies (ICT) have been added in recent years. These technologies make it possible to incorporate new media for the communication of heritage content to the general public and, therefore, the analysis of their applications and the results obtained by them can be of great help to educators" [4].

And institutions that deal directly with classified goods and that do not necessarily have a profit- and profit-taking, such as state bodies, supra-national organizations or academic institutes know the importance of the Internet as a privileged platform to make themselves known as well as disseminate initiatives in favor of the defense and dissemination of universal heritage. Proving this statement are the various pages and sites of the most varied bodies linked to the heritage that it maintains, increasingly strengthening its presence in the global network for philanthropic, academic and also political-economic purposes.

And in these institutional sites there is in the vast majority of examples a page exclusively dedicated to communication, that is, in addition to the site itself is already built to make known the dynamics of institutional communication of the organization, includes a page dedicated to contact with the *media,* with *neswsletters* dedicated to the media but also to the researcher of communication issues and the ordinary citizen who wishes to be aware of the initiatives developed by this or that body.

In any case, all information relating to the effort made by UNESCO and the cities classified in order to preserve the valuable historical and universal natural estate is present in the eyes of all who wish to consult it, at the disposal of public scrutiny and is evident in various media, since the book, the website, with resources for videos, sounds and manifestos, not only for ordinary people but especially for researchers, public and private bodies and the school population, are the example of the communication policy developed by UNESCO, which nowadays becomes essential with regard to the dissemination of organized activities, but above all in building a spirit of awareness about these matters, contributing to greater credibility and appreciation in the eyes of public opinion, but also as a good "advertising brochure" for those who wish to discover new destinations. Carlos Sotelo Enríquez bets on the proposition that:

> "The policy of institutional communication, according to an integrative philosophy, takes over internal and external areas. It is a simultaneous process, and therefore does not happen first within the organization, and then to spread abroad, as some doctrines intended. Recognizing the participation of other individuals and legal entities, in addition to the members of the institution, in the development of identity, supposes opening the organization to other informative related that occur at the same time" [5].

The policy of institutional communication, according to an integrative philosophy, takes over internal and external areas. It is a simultaneous process, and therefore does not happen first within the organization, and then to spread abroad, as some doctrines intended. Recognizing the participation of other individuals and legal entities, in addition to the members of the institution, in the development of identity, supposes opening the organization to other informative related that occur at the same time [6].

And because the application of a heritage site often involves at the same time weighted exercises of citizenship, the internet also demonstrates a relevant role, due to the attention that public opinion pays to the heritage and the tourism industry that largely streamlines the promotion and continuous dissemination of heritage: "As with much of the internet, diplomacy is increasingly taking place within public view. One benefit of this stage of evolution, is the ability for observers to exercise greater scrutiny of "the history, techniques, and effects of diplomatic engagements with foreign audiences" [7].

Finally, and focusing on heritage in a nature of tourist dissemination, it is worth noting the online page of the World Tourism Organization (https://www.unwto.org/), as advocated By Auvo Kostianien:

"Anybody interested in exotic travelling may have a look at the silk road pages. The Silk Road travel is a multi-state joint venture started in 1994 under the auspices of the World Tourism Organization (...). This venture is a goo example of cooperative effort of many states and cultures. Silk road web sites ar an example of the pages referring to historical roads and cultural connections capitalized by the travel industry" [8].

## 3 The Context of Online Promotion of Cultural Heritage for National and Local Public Organizations in Portugal

In Portugal, one of the institutions to invest in the creation of an institutional page to disseminate information on the activities of the organization and the rest of the institutional communication was exactly the Portuguese Institute of Architectural Heritage (IPPAR), a body created in 1992 and asperated by the Ministry of Culture and which protects Portuguese assets classified by UNESCO.

IPPAR, which changed its name to *Direção-Geral do Património Cultural* (DGPC) has a complete website where it is possible for the internet user to know the Portuguese goods recognized with the Classification of World Heritage and where it is pointed out on the message page of the Board of this body the importance of communication with the public in various aspects:

> "The purpose of this website is to provide access to a wide range of information on the different areas of action of the direct responsibility of the Directorate-General for Cultural Heritage and also on contents related to the universe of work in the field of Cultural Heritage, namely urban, architectural and archaeological built heritage, mobile heritage, intangible heritage and museums and monuments" [9].

The DGPC also stresses the importance of constantly improving "the communication between the public and the Public Service that we wish to be, with excellence" [9].

This site presents information on the most diverse aspects of heritage, and can even make a search for the entire estate of Portuguese monuments, their history, the interventions developed by the DGPC in the conservation of heritage, in order to facilitate access to information to students, researchers, municipal agencies and also professionals in the sector as well as an area also focused exclusively on the citizen:

> "In order to facilitate the use of services by the public, DGPC seeks to provide online the largest number of features; you can already access, among others, information on the conditions of use and transfer of monuments under the management of the Institute, to heritage inventory KIT's (general architectural heritage, single-family housing of the 20th century and industrial heritage), to libraries and archives of the Institute and download various forms" [9].

Also in relation to Portuguese institutional pages, there are examples of importance especially in the sites of the municipal councils whose historical center is considered a World Heritage Site, as is the case of the cities of Angra do Heroísmo (classified in 1983), Évora (1986), Porto (1996) and also the city of Guimarães (2001), among others.

With its own website or with a page added to the site of the building, the information on classified heritage is well documented, available to other institutions, professionals, the public and the media. One of the best examples is the site of the Municipality of Porto, more specifically The Tourism of Porto, institution responsible for the tourist promotion of Porto, within the scope of the port building and that includes a page exclusively dedicated to all kinds of information related to goods classified by UNESCO:

"The unique characteristics of the historic center of Porto made UNESCO classify it as "Cultural Heritage of Humanity" in December 1996. To know the stages of classification, the classified and protected area and discover the list of world and artistic heritage" [10].

This page presents in detail all the contours of the application that led the historic center of Porto to be considered world heritage, the classified area, as well as the area of protection and also the efforts that have been developed by the Urban Rehabilitation Society of the Municipality of Porto and public and private bodies, with updated information dedicated to citizens, to the public but also with press releases with a view to disseminating the initiatives that the Chamber carries out around the nominated heritage.

Spain also presents examples of online sites of the responsibility of buildings whose "historical hull" is pointed by UNESCO.

The website of the *Ayuntamiento* de Santiago de Compostela presents an exclusive page for the World Heritage, presenting a chronological description of all the facts, episodes and events that led to the historical center of the "apostle's city" first considered a historical set of national interest in 1941 and finally to be crowned with the title of World Heritage [11], this option being an innovation in relation to other pages founded.

It should be noted that also the medieval path to Santiago (Way of Saint James), called the Primitive Way and the French Way were also awarded the title of World Heritage, in 1993, routes that crosses several Spanish cities.

With the classification of the historic center of the city of Salamanca, dated 1988, there are two relevant online sites, thus witnessing the lively effort of the salmantin authorities to disseminate in an exemplary way the classified heritage of this city of the province of Castilla y Leon. The most complete site about the World Heritage "Charro" is the portal of Tourism of Salamanca [12], the responsibility of the Ayuntamiento de Salamanca, in which it presents a detailed record on the heritage of this city but that despite the vast information does not contain on this site a page dedicated to the World Heritage exclusively, having only in the other pages references to the title granted by UNESCO. This is because one of the sites developed by the Salamanca authorities was developed in particular to specifically disseminate classified heritage. This portal, entitled "Salamanca, World Heritage Site", addresses the heritage classified in the most varied cultural and historical aspects, with a detailed list of protected heritage, even containing a page where you can find the ICOMOS evaluation dossier, when submitting the application:

"The International Council of Monuments and Sites (ICOMOS), as unesco's advisory body for the inscription of monuments and sites on the World Heritage List, carried out the technical evaluation of the Old City of Salamanca. All the original documentation on Salamanca, sent by Spain to UNESCO, is available at the ICOMOS Documentation Centre in Paris" [13].

Other examples worthy of emphasis on the bet of local authorities on the Internet are the cases of Ávila [14], whose estate of its historic center was classified in 1985, (presenting this city a project for dissemination of classified heritage dedicated to the younger *public, entitled Patrimonitos de Ávila,* also using the Facebook platform, in an initiative recognized internationally by UNESCO).

As far as institutional sites are concerned, there are several Spanish national organisations promoting Spanish heritage on the internet.

The website of the Spanish Ministry of Culture has a page dedicated exclusively to the World Heritage of this country:

"World Heritage sites of exceptional universal value to humanity that, as such, have been inscribed on the UNESCO World Heritage List in order to ensure their protection and conservation for future generations.

World Heritage-UNESCO: The contents of these pages inform about World Heritage: what it is, what is the process of inscription of goods in the World Heritage List, what organisms intervene in its management and how its conservation is guaranteed.

World Heritage in Spain: Spanish assets declared World Heritage: what are they, upcoming candidacies and conservation and protection actions on these assets" [15].

In Spain there is even an organization that brings together all the cities whose historic center is classified as world heritage, they are the cities of: Alcalá de Henares, Ávila, Cáceres, Córdoba, Cuenca, Ibiza/Eivissa, Mérida, Salamanca, San Cristobál de La Laguna, Santiago de Compostela, Segovia, Tarragona, Toledo [16].

This institution, which enjoys a prestigious international status, even with the institutional support of UNESCO, has a very complete space on the internet, presenting, like the website of the Spanish Ministry of Culture, information related to unesco classification application statutes, international letters of cooperation, press information, documents for researchers and academics, information dedicated to tourism agents and the general public, with the aim of disseminating on a large scale the rich pointed heritage:

"The promotion, which is so important to achieve the objective of profitification proposed by UNESCO, is a great achievement of the Association, which has saved efforts and budget through joint work, apart from the work that, individually, is carried out from each of the municipalities. From the first moment it has been seen the possibilities that were obtained by offering potential tourists, not a competition, but a complement to the offer of Sun and Beach tourism, proposing this other alternative of Cultural Tourism and historical and artistic quality of our cities" [16].

# 4 World Heritage on Personal Pages and Blogs

But it is not only in the institutional pages of a local, national or even international nature that information on the World Heritage of the historic centers of the cities is found. There are hundreds of personal pages with lots and varied information on this specific topic, not forgetting the universe of blogs.

The following personal page is one of several examples found:

"Fortifications of Elvas to World Heritage

Because it is of interest to my readers, some of whom are from the city of Elvas, here I leave you the news of the writing of the prestigious periodical "Linhas de Elvas".

The raiano municipality of Elvas has already handed over the application dossier of the military fortifications of the city to World Heritage to the National Commission of UNESCO, revealed the councilor of the municipality Elsa Grilo" [17].

Another personal page shows a cybernaut committed to make known the objectives and purpose of the UNESCO classification, presenting examples of spoils already awarded with the title of World Heritage:

"World Heritage Site (I)
Bringing the multiple facets of a complex world closer together; contact with a common heritage of humanity; to know new cultures; learn in a playful way; are some of the aspects that travel can provide us. Since 1978, aiming at the protection of what humanity has most precious, UNESCO (United Nations Educational, Scientific and Cultural Organization) has already classified as "humanity, for its exceptional value, about 790 different sites in 134 countries – in the points of natural interest (landscapes – 154), of cultural point of interest (monuments – 611) or mixed (23) –, of which 13 in Portugal. The awareness of the existence of a world heritage to be preserved dates back to the late 1950s when, following the construction of the Assuan dam (Egypt), it became necessary to move the nubian temples of Abu Simbel to save them from the waters of the Nile, a $40 million project" [18]

Due to the historical and economic importance and especially the regionalist character of its people, there are many personal pages that address the city of Porto and its heritage classified, since 1986, by UNESCO. One of the most important amateur photographers in the city of Porto, António Amen, created a vast personal page, in which, in addition to exhibiting his vast photographic estate about this Portuguese city, he included a page only dedicated to unesco classification, presenting credible and complete texts about the classification, as well as a set of photographs illustrating the heritage granted:

"On December 5, 1996, at a UNESCO meeting held in Mexico City, the Historic Center of The City of Porto was classified as a World Heritage Site, opening new perspectives to the Invicta City, integrating it in the route of the great values of humanity.
The application process took four years, but was brought to a good conclusion after difficult struggles to make prevail the arguments of its more than just belong" [19].

With regard to examples of Spanish personal pages are many, varied, most of good quality, others not so, and many already inactive or without updating.

The first example refers to Ávila, with the title of the page referring to the same nomenclature "Avila Patrimonio de la Humanidad" [20], a site that presents mainly *links related* to classified heritage, with links to museums, archeosites, albums of old photos of Ávila and of course to the institutional sites that govern the heritage.

An example, also related to the classified heritage of Ávila is the page of a group of students of the 2nd Year of IES Universidad Laboral de Toledo that presents a site very much with chapters on historical sources, legends related to Ávila, heritage, historical routes, as well as other aspects related to the UNESCO classification of the historic center of this city, with very concrete objectives:

"Hello! We are a group of students of 2º ESO of the IES Labor University of Toledo. We are participating in the Youth for Heritage project and this year the heritage city is Ávila. On this website we will collect all the information, images, videos and games made by us. Hopefully you like it!
Hello! We are a group of students of 2º ESO of the IES Labor University of Toledo. We are participating in the Youth for Heritage project and this year the heritage city is Ávila. On this website we will collect all the information, images, videos and games made by us. Hopefully you like it" [21].

The universe of blogs, a platform that emerged in 1997, is also attentive to the theme of World Heritage from the first hour to the present day.

There are many examples of blogs that address this subject directly or indirectly, to the point that when comworking a simple search in Google search engines or more specifically in one of the engines specialized in blog search, Technorati, using "Tags", that is, search expressions such as "World Heritage" or "World Heritage" it is verified that only on Google (choosing the advanced option of search for blogs) there are 810 references to blogs that present the "World Heritage" tag, thus finding blogs using the Portuguese language (these being mostly Brazilian) and others the Castilian language (this time with greater predominance are the Spanish blogs). Using the "tag" "Património da Humanidade" then we found, using again google blog search, about 870 blog references in the language of Camões. But if we translate this last expression into Castilian, to "Patrimonio de la Humanidad" we found about 780 blog references written in Spanish.

Using the Technorati search engine the values that arise to us are quite similar given the increasing convergence between these two search platforms.

Given such a wide number of blogs that feature "posts" on this topic, let's look at some examples Portuguese and Castilian.

One of the good examples is the blog "Macau Antigo", developed by João Botas, which covers the history and heritage of this former Portuguese colony, whose historic center was inscribed in 2005 on the World Heritage List:

"World Heritage Signage
There are 32 new signboards that will be erected throughout the city to accompany the buildings marked as world heritage sites in Macao. In addition to overshadowing UNESCO symbols, these brands were designed with minimalist shapes complementing their surroundings and integrating in line with the most visited places in the territory. Prepared by the Cultural Institute of Macao, the design, in a study begun three years ago, was in charge of the Macanese architect Carlos Marreiros" [22].

With regard to Castilian-language blogs, one of the best examples found is the blog of the photographic estate of the *influencer* Francisco Curbelo, which, in addition to covering photographs, also contains texts that frame the images in concrete themes, being varied the "posts" that this photographer dedicates to the classified heritage, not always missing a link to the List of World Heritage classified, made available by the online site of the World Heritage Center:

"San Cristobal de La Laguna Património de la Humanidad.
San Cristobal de La Laguna Património de la Humanidad.
The municipality of San Cristóbal de La Laguna, is not only known for the richness of its natural landscape of part of the Anaga massif that it shares with Santa Cruz, forming the Rural Park of Anaga, but also, and perhaps this is what it is best known for internationally: World Heritage City.
All this brings us to a December 2, 1999, where the UNESCO World Heritage Committee meeting in Marrakech (Morocco) made public its approval of the title of World Heritage Site for the city of San Cristóbal de La Laguna, Tenerife" [23].

The blog "Historia del Arte", taught by a teacher of this discipline, Amparo Santos, important cultural influencer in Spain, also presents "posts" that make extensive allusion to the Spanish classified heritage, among others, using videos and didactic material. In the following "post" the author addresses the classified historical center of Cuenca:

Cuenca, "The Enchanted City"... this will be our route today.

The city of Cuenca is one of the main monumental complexes of Spain, with a great international projection that made it declared a World Heritage Site by UNESCO.

Las Casas Colgadas, also known as Casas Voladas, Casas del Rey and, erroneously, Casas Colgantes, is a set of civil buildings located on the edge of a rock that represent the city of Cuenca worldwide. In the past this architectural element was frequent on the eastern edge of the old city, located in front of the sickle of the Huécar River, although today only a small part of them survive. Of all, the best known are a set of three of these structures with wooden balconies [24].

## 5   Conclusion

Today's digital communication platforms are the key vehicles for promoting the tourism industry, an essential foundation for the effort to preserve and promote the built cultural heritage, as well as the natural heritage classified by UNESCO as a World Heritage Site.

Although much of the dissemination goes largely through the universe of social networks, with special emphasis on *Instragram* and *Facebook*, site of promotion of image content such as *Pinteret, Flickr* or *youtube*, the institutions pages of governments, municipalities and cultural organizations that campaign around the classified heritage still today make an undeniable contribution to the dissemination of cultural and universal heritage, assisting travelers in the transmission of all knowledge that concerns not only the nature of the heritage and its characteristics, but also as a focus of promotion of the city itself and the region, acting as a factor in promoting other activities that orbit the sphere of heritage, such as hospitality, handicrafts, food restoration, and tourism.

With the democratization of the internet and the web 2.0 revolution we have witnessed, among others, the proliferation of personal pages and blogs, which, although they have now lost relevance to social networks, are still credible sites for internet users for more information about heritage, without forgetting the opinion that the creators of these same homepages and blogs have of a particular place or heritage classified.

In this last context, the role of so-called *influencers* plays a fundamental role in the dissemination of heritage not only from a tourist perspective but above all from a cultural, artistic root and where the history and importance of a particular place or heritage has held in the course of relevant historical events.

## References

1. Comissão Nacional da UNESCO: A Protecção do Património Mundial, Cultural e Natural. UNESCO Portugal, Lisboa (1992)
2. UNESCO: Convention Concerning the Protection of the World Cultural and Natural Heritage. World Heritage Center, 20 de Junho de 2020. https://whc.unesco.org/en/conventiontext/

3. UNESCO: General Policies Regarding the World Heritage Convention. World Heritage Policy Compendium (2019). https://whc.unesco.org/en/compendium/50. acedido em Junho de 2021
4. Monfort, C.C.: Comunicación y educación no formal en los centros patrimoniales ante el reto del mundo digital. La comunicación global del Patrimonio Cultural (Ediciones Trea) 287 (2008)
5. Sotelo, C.E.: Introducción a la Comunicación Institucional. Ariel, Barcelona (2001)
6. Subires, M.P.: Internet como medio para la salvaguardia del patrimonio cultural inmaterial. Fundación Telefónica (2018). https://telos.fundaciontelefonica.com/archivo/numero091/internet-como-medio-para-la-salvaguardia-del-patrimonio-cultural-inmaterial/. acedido em Julho de 2021
7. Niglio, O., Lee, E., Ramachanderan, R.: Diplomacy in the Age of the Internet: Challenges for the World Heritage Convention. Eda Esempi di Architettura 10 (2021)
8. Kostiainen, A.: Internet web pages related to 'history, heritage and tourism.' Int. J. Tour. Res. (Wiley) **3**, 65 (2001)
9. DGPC: Bem-vindo ao website da Direção-Geral do Património Cultural (2020). http://www.patrimoniocultural.gov.pt/pt/quem-somos/. acedido em Julho de 2021
10. CMP: Património Mundial. s.d. https://cultura.cm-porto.pt/patrimonio-cultural/patrimonio-mundial
11. Ayuntamiento de Santiago de Compostela: Patrimonio de la Humanidad (2021). http://www.santiagodecompostela.org/turismo/interior.php?txt=t_patrimonio&lg=cas. acedido em Junho de 2021
12. Ayuntamiento de Salamanca: Salamanca cidade, Patrimonio da Humanidade (2021). http://www.salamancaemocion.es/pt-pt/que-hacer/cultural/salamanca-cidade-patrimonio-da-humanidade. acedido em Junho de 2021
13. Ayuntamiento de Salamanca: Salamanca Patrimonio Mundial Evaluación del ICOMOS (2018). http://www.salamancapatrimonio.com/icomos.htm. acedido em Junho de 2021
14. Ayuntamiento de Ávila: Patrimonio Mundial en Ávila (2020). http://www.avila.es/opencms/opencms/AVIL/paginas/MENU/ORGA/MAUPH/PATR/PATR.html. acedido em Junho de 2021
15. Património Histórico: Património Mundial (2018). http://www.mcu.es/patrimonio/CE/PatrMundial/Introduccion.html. acedido em Junho de 2021
16. Organización de las Ciudades del Patrimonio Mundial de España. Las Ciudades (2021). http://www.ciudadespatrimonio.org/default.aspx. acedido em Junho de 2021
17. Pardal, J.C.: Fortificação de Elvas a Património Mundial (2010). http://www.josepardal.com/2010/04/07/fortificacoes-de-elvas-a-patrimonio-mundial/. acedido em Junho de 2021
18. Vicente, L.: Património Mundial da Humanidade (2005). http://memoriavirtual.net/2005/06/sociedade/patrimonio-mundial-da-humanidade-i/. acedido em Junho de 2021
19. Amen, A.: Amen Fotografia (2018). http://amen.no.sapo.pt/Porto%20Patrimonio%20Mundial.htm. acedido em Junho de 2021
20. Venueza, M.: Avila Patrimonio de la Humanidad. S/D. http://www.terra.es/personal/jmh00005/index.html. acedido em Junho de 2021
21. 2° IES Toledo. Ávila Patrimonio de la Humanidad (2010). http://www.avilapatrimoniodelahumanidad.com. acedido em Junho de 2021
22. Botas, J.: Macau Antigo (2010). http://macauantigo.blogspot.com/2010/07/sinaletica-do-patrimonio-mundial.html. acedido em Junho de 2021
23. Curbelo, F.: San Cristóbal de la Laguna Património de la Humanidad (2010). http://fotoscurbelo.blogspot.com/2010/07/san-cristobal-de-la-laguna-patrimonio.html. acedido em Junho de 2021

24. Santos, A.: Cuenca "La Ciudad Encantada"… Ésta será Nuestra Ruta de Hoy (2010). http://historiadelarte-amparosantos.blogspot.com/search?q=patrimonio+de+la+humanidad.    acedido em Junho de 2021
25. Morkes, J., Nielsen, J.: Nielsen Norman Group - Papers and Esseys. Nielsen Norman Group (1997). http://district4.extension.ifas.ufl.edu/Tech/TechPubs/HowToWrMritefortheWeb.pdf. acedido em Dezembro de 2020
26. Organización de las Ciuidades Patrimonio de la Humanidad de España: La organización (2018). http://www.ciudadespatrimonio.org/DesktopDefault.aspx?tabID=8421

# Eradicating Terrorist Networks on Social Media: Case Studies of Indonesia

Paryanto[1,2] , Achmad Nurmandi[2(✉)] , Zuly Qodir[2] ,
and Danang Kurniawan[3] 

[1] Political Islam, Universitas Muhammadiyah Yogyakarta,
Yogyakarta, Indonesia
[2] Political Islam, Jusuf Kalla School of Government,
Universitas Muhammadiyah Yogyakarta, Yogyakarta, Indonesia
{nurmandi_achmad, zuliqodir}@umy.ac.id
[3] Government Affairs and Administration, Jusuf Kalla School of Government,
Universitas Muhammadiyah Yogyakarta, Yogyakarta, Indonesia

**Abstract.** This study aims to analyze the countermeasures of terrorism in Indonesia through the social media function. Social media is a means of terrorism network to spread radical views and recruit new members. Through the BNPT (National Counterterrorism Agency), the Indonesian Government carries out government tasks in counterterrorism. This research will look at the social media function of BNPT in countering terrorism. This study uses the Q-DAS (Qualitative Data Analysis Software) method approach. The results showed that Twitter's social media role could be a strategic tool to counter terrorists and destroy the narrative of radicalism in social media. The Indonesian Government, through the BNPT, has taken an information dissemination approach related to understanding radicalism and preventing the spread of radicalism on social media. Prevention is carried out by disseminating narrative information to strengthen unity through hashtags and content, inviting the public to work together with the Government to tackle terrorism. Content information delivered on BNPT social media has shifted the issue of terrorism content to Covid19; in substance, the BNPT account still functions as disseminating information on the prevention and countermeasures of terrorism. Also, BNPT's strategy in optimizing social media Twitter is collaborating with social media accounts of government agencies, state security tools (TNI, Police of the Republic of Indonesia), and mass media as partners in publishing information to the public.

**Keywords:** National Agency for Contra Terrorism (BNPT) · Terrorism · Social media · Communication

## 1 Introduction

This study aims to see the countermeasures for terrorists in Indonesia through mass communication on social media Twitter. In the last two decades, Indonesia has been shocked by bomb attacks such as Bali, the Plaza Atrium, the JW Marriot Hotel, The Ritz Carlton Hotel, the Australian and Philippine Embassies, the Myanmar Embassy, Thamrin-Jakarta, Surabaya, and many other small-scale attacks. terror [1, 2]. The last

T. Antipova (Ed.): DSIC 2021, LNNS 381, pp. 433–442, 2022.
https://doi.org/10.1007/978-3-030-93677-8_38

case of terrorist attack in Surabaya on 13 and 14 May 2018 targeted three churches and one police station, and the attack managed to get considerable media coverage from the media, prompting policymakers to follow up on the revised state anti-terrorism law [3–5]. So government needs to take various approaches to countermeasures strategies in breaking the terrorist network in Indonesia (Fig. 1).

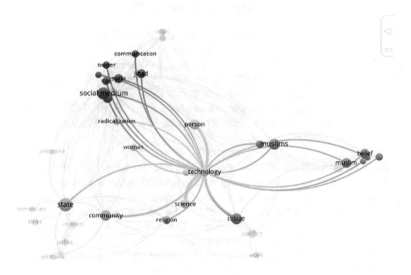

**Fig. 1.** Terrorism network in Indonesia

Results of analysis a collection of research related to terrorists in Indonesia in the last ten years using Vosviewer, show the visualization of the relationship between the use of technology as an important part in building a network of radical under-standing. Terrorism networks in Indonesia utilize Twitter social media technology to spread ideology, recruit new members and guide suicide bombers as a recruitment process for Indonesian Muslims to carry out jihad (holy war or struggle) in Iraq and Syria [4, 6, 7]. Several governments have emphasized the importance of online programs aimed at undermining terrorist recruitment, including the use of state-run accounts on various social media platforms to respond directly to terrorist messages [8]. Through the Counter-Terrorism Agency (BNPT), the Indonesian Government has implemented many strategies in the field of prevention, one of which is tracking a large number of followers on social media platforms, especially Facebook and Instagram [1, 9]. BNPT is a non-ministerial government agency (LPNK) that carries out government tasks in counterterrorism.

Indonesian government through the BNPT is active in preventing and breaking terrorist networks that develop through social media. The current counterterrorism shift has entered into Internet intermediaries that provide a platform for and curate third-party content to monitor and supervise on behalf of the State [10]. The Government is required to be innovative and active in counterterrorism through the use of technology. This research focuses on utilizing the social media function of Twitter @BNPT_RI

Indonesia's National Counterterrorism Agency (BNPT) in cutting off terrorism net-works on social media. This research is expected to explain how the Government disseminates narrative information to the public in cutting off the terrorism network in Indonesia.

## 2 Literature Review

In the era of ever-connected social networks, terrorists exploit social media platforms through sophisticated approaches [11]. Twitter has emerged as a central platform where violent Islamist extremist groups spread radical propaganda to recruit civilians, espe-cially young people vulnerable to radicalization [12]. Recognizing social media's benefits as a networking platform, extremist organizations use social media to spread ideology, recruit new members, and guide suicide bombers [6].

Various extremist and terrorist organizations have started to actively use the pos-sibilities of modern communication technology to influence the consciousness of various social groups through destructive information campaigns [13]. Identifying the main actors related to terrorism on social media is very important for law enforcement agencies and social media organizations in their efforts to fight terrorism-related online activities [14]. Police organizations increasingly use social media and other forms of new communication technology to communicate with the public in diverse and inno-vative ways [15].

Strategic communication and content moderation are two broad responses that need to be considered in policy development to challenge extremist exploitation of social media [16]. Best countermeasures against terrorism Understanding, Assessing and Responding to Terrorism: Protecting Critical Infrastructure and Personnel provides a detailed description on how to implement preventive and protective measures to ensure the safety of personnel and facilities, including vulnerability analyzes, security pro-cedures, emergency response procedures, and training programs [17]. Building a system of coordination of cooperation and collective preventive measures against terrorism is needed to overcome more complex forms of terrorist attacks such as simultaneous attacks and cyber-attacks [18].

## 3 Method

This study conducted a study on how the social media role of the National Coun-terterrorism Agency Twitter @BNPT_RI in tackling terrorism through social media. This study uses a qualitative approach with descriptive application to convey the results and values of social media content in countering terrorism. This study also uses Q-DAS (Qualitative Data Analysis Software) to analyze social media data and Nvivo 12Plus to analyze social media content and data. Twitter account activity data was taken from 2016–2021 based on hashtags, mentions, and narrative messages (Fig. 2).

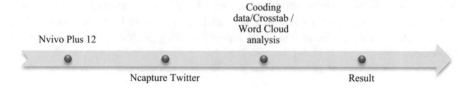

**Fig. 2.** Data collection and analysis

Data collection is carried out through BNPT's Twitter social media account using Ncapture which can record social media activities. The Ncapture data is downloaded in the form of a RIS file and then processed using manual coding NVivo Plus 12. In analyzing the research results, the data is presented in the form of word frequency, word cloud and Cluster Analysis. The presentation of the data can explore and visualize BNPT's Twitter social media activities in counterterrorism activities in Indonesia through Twitter social media.

## 4    Result and Discussion

Combating terrorism through social media Twitter can be seen from the activity of user accounts. The analysis results show that the BNPT social media account has had dynamic activities in the last four years. The highest @BNPT account activity was in 2020, with a total of 1023 content activities. When viewed from 2017–2019, it shows a decrease in @BNPT activity in social media communication, and then in 2020, account activity has increased (Fig. 3).

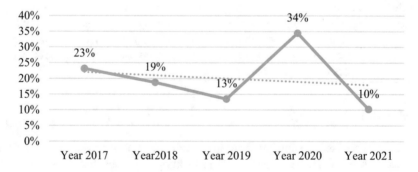

**Fig. 3.** The activity of BNPT's Twitter social media accounts for the last five years

Counterterrorism can be carried out through BNPT's social media activities, massive in providing information related to eradicating terrorists so that the function of the BNPT Twitter social media can become a tool for global information warfare that must be done to fight terrorists and protect citizens in terrorist networks [19, 20]. The government needs to use social media as a strategy to counter radicalism and break terrorist networks. The role of social media is very strategic, where in Indonesia most people have a very high relationship with social media. So that every form of information submitted by the government can be received and understood quickly by the public.

## 4.1  BNPT RI Institutional Cooperation in Countering Terrorism

Social media can coordinate cooperation and collective preventive measures against terrorism to overcome more complex forms of terrorist attacks such as simultaneous attacks and cyber-attacks [21, 22]. The coordination carried out by BNPT in social media Twitter shows that there is collaboration in communication between stakeholders. As for those who are BNPT partners in counterterrorism in Indonesia in collaboration with three cluster groups. The cluster groups are differentiated based on the duties and functions of the authorities, which are the everyday tasks of counterterrorism. In addition, the relationship in each of these actors has a different configuration magnitude from one another.

The results of the collaboration analysis between each actor who collaborates with BNPT RI can be seen from the communication built on each actor's social media accounts. Currently, social media accounts are one of the essential delivery alternatives in public and mass communication, while the results of the analysis of the collaboration built by BNPT RI with internal and external actors are as follows (Fig. 4).

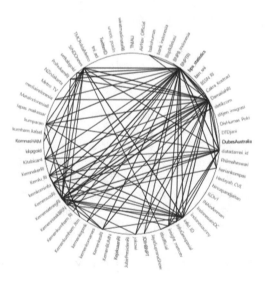

**Fig. 4.**  Social media communication activities @BNPT for the last four years

BNPT RI builds collaborations in countering terrorism on social media involving 85 actors, consisting of ministry agencies, public officials, and national media. The built cooperation can become a decision support system that will take over to coordinate, lead and control agencies in carrying out counter-terrorists [23, 24]. The collaboration carried out by BNTP of 85 actors with high intensity, and only 17 actors are active in collaborating. Collaboration across institutions has made the countermeasures to prevent the understanding of terrorism a shared task for every stakeholder, starting from High Ministry Institutions, Police, TNI, and Media to Social Institutions (Table 1).

**Table 1.** BNPT partner actors in countering terrorism in Indonesia

| Government | National media | Institutions security apparatus |
|---|---|---|
| President RI Jokowi | SINDOnews | Kostrad TNI AD |
| Ministry of Health | Media Indonesia | Police Republic of Indonesia |
| Ministry of Foreign Affairs | Kumparan | |
| Ministry of Communication and Information | Insight Metro Tv | |
| Ministry of State Secretariat | Harian Kompas | |
| Ministry of Coordinating and Economic Affairs | Detik.com | |
| Directorate General of Immigration | | |
| Indonesian National Agency for Disaster Management | | |

The highest government partner who collaborates with BNPT RI in countering terrorism. In addition to government institutions, the Indonesian National Narcotics Agency (BNPT) coordinated intensely with the President of the Republic of Indonesia, Jokowi, on social media and the military's state security tool Kostrad TNI AD, and the Police Republic of Indonesia. The coordination built by BNPT between ministries strengthens each other in countering terrorism in Indonesia. Also, BNPT provides data information to the public through cooperation with national media, including SINDOnews, Media Indonesia, Kumparan, Insight Metro Tv, Kompas Tv, and Detik.com.

## 4.2   BNPT's Information Strategy in Eradicating Terrorist Networks on Social Media

BNPT produces information on counterterrorism through social media activities related to information content. An Information Strategy on the National Incident Management System (NIMS) enables all public, private and non-governmental organizations to work together effectively to prepare for, prevent, respond to terrorism networks [25]. The analysis of social media content from BNPT, produced by hashtags, shows two sharing of information-shifting information provided to the public (Table 2).

**Table 2.** Cluster integrate communication

| Consistent issues of terrorism | | Shifting content issues | |
| --- | --- | --- | --- |
| Peaceful friends | 6% | New normal | 3% |
| Protect yourself protect the country | 5% | Adaptation to new habits | 9% |
| Work together to prevent terrorism prevent terrorism | 13% | National preparedness | 13% |
| Peace-loving religion | 1% | Covid 19 | 25% |
| Peaceful Indonesia | 2% | Just at home | 1% |
| Ready to | 3% | Prevent covid 19 | 1% |
| Peace greetings | 2% | Take care of health | 1% |
| Moderate media against radicalism | 1% | Wear a mask | 1% |
| Synergy | 3% | Keep a distance | 1% |
| Together with prevent terrorism | 1% | Covid 19 | 1% |
| Peace indonesia | 3% | Versus covid | 1% |
| Deradicalization | 2% | New normal | 1% |

BNPT RI social media information can be seen from hashtags as a symbol of information to be conveyed to the public. Based on the analysis results, there is a shift in information from the issue of counterterrorism to the handling of Covid 19. Comparison of the content of issues produced by BNPT RI related to terrorism is (42%), and the issue of Covid19 (58%). The production of Covid19 information content was high in 2020, which showed the highest level of BNPT account activity in the last five years.

Social media can promote the idea that all information is accessible to everyone (including the most confidential and proprietary information of organizations), has become the latest ideology to gain a wide following [26]. Overall, the information conveyed by BNPT RI can be seen from the word frequency analysis, which shows that the narrative that often appears is an essential issue for BNPT RI. Narrative information carried out by BNPT RI in overcoming the use of preventive information dissemination against the understanding of terrorism plays a vital role for the community (Fig. 5).

**Fig. 5.** Word Frequency of BNPT RI

The BNPT RI Twitter social media's role in tackling terrorism is consistent in producing information on terrorist issues. The use of the terms radical, preventive, countermeasures, and religious dominates the information conveyed. However, there is a shift in the issue where the words covid, adapt, information, and health have a significant frequency. Shifting the issue does not change the substance of the social media function of BNPT RI in counterterrorism; this condition can be seen from the cooperation of institutional partners built by BNPT RI, which has the same organizational vision related to terrorism cases.

## 5   Conclusion

Terrorist networks in Indonesia use social media technology to spread radicalism and recruit new members. This condition is carried out by terrorism groups considering the very high number of social media users in Indonesia, besides that, the nature of social media can attract the public's attention at large, which is one of the virtues of social media as an alternative. In addition, the attention of the Government of Indonesia regarding terrorism cases is a grave concern. The Government can see this condition of Indonesia to form a particular agency to tackle the understanding of terrorism through the Indonesian National Narcotics Agency. The nature of the BNPT RI is a non-ministerial institution that functions to combat terrorism, and currently, the Indonesian National Narcotics Agency has implemented a strategy to counter terrorism in Indonesia through the social media Twitter account @BNPT_RI. This strategy is undoubtedly a form of follow-up to the activities of terrorist groups in Indonesia which use social media as access to recruitment to become terrorists. BNPT RI utilizes social media as a forum for massive information dissemination related to the prevention of radicalism spread by terrorists, BNPT carries out a prevention information strategy by providing education about nationality. Information provided by BNPT through social media has experienced a shift in terrorism to covid19, but this shift did not last long and did not change the substance of BNPT's role in countering terrorism. In addition, BNPT RI collaborated with actors from the Central Ministry of Higher Education, Media, and Community Social Institutions, to tackle terrorism in Indonesia.

# References

1. Erikha, F., Rufaedah, A.: Dealing with terrorism in Indonesia: an attempt to deradicalize, disengage and reintegrate terror inmates with a social psychology approach. In: Terrorist Rehabilitation and Community Engagement in Malaysia and Southeast Asia, pp. 131–138. Taylor and Francis, Linguistics Department, Universitas Indonesia (UI), Indonesia (2019)
2. Milla, M.N., Hudiyana, J., Cahyono, W., Muluk, H.: Is the role of ideologists central in terrorist networks? A social network analysis of Indonesian terrorist groups. Front. Psychol. **11**, 333 (2020). https://doi.org/10.3389/fpsyg.2020.00333
3. Fenton, A., Price, D.: ISIS, Jihad and Indonesian law: legal impacts of the January 2016 Jakarta terrorist attacks. Issues Leg. Scholarsh. **14**(1), 1–26 (2016). https://doi.org/10.1515/ils-2016-0255
4. Amin, S.J.: Movement of islamic organization and its impact on radical perception with supply chain management in the digital era. Int. J. Supply Chain Manag. **9**(2), 864–871 (2020). https://www.scopus.com/inward/record.uri?eid=2-s2.0-85085384362&partnerID=40&md5=b3da098652dc275b3ea2854ecb88b841
5. Ikhwanuddin, K.D., Ellyani, E., Budiono, A., Naseh, M.: An analysis of psychological health of suicide bombers' family in Surabaya. Indian J. Public Health Res. Dev. **10**(8), 1259–1264 (2019). https://doi.org/10.5958/0976-5506.2019.02068.0
6. Aryuni, M., Miranda, E., Fernando, Y., Kibtiah, T.M.: An early warning detection system of terrorism in Indonesia from twitter contents using naïve Bayes algorithm. In: 5th International Conference on Information Management and Technology, ICIMTech 2020, pp. 555–559 (2020). https://doi.org/10.1109/ICIMTech50083.2020.9211261
7. Syafik, M., Seniwati: Promoting peace, countering terrorism in Indonesia: the role of Indonesian media. J. Eng. Appl. Sci. **12**(16), 4053–4056 (2017). https://doi.org/10.3923/jeasci.2017.4053.4056
8. Aistrope, T.: Social media and counterterrorism strategy. Aust. J. Int. Aff. **70**(2), 121–138 (2016). https://doi.org/10.1080/10357718.2015.1113230
9. Nurhasan Affandi, R.M.T., Dermawan, W., Alam, G.N.: In refer to societal security dealing with the radicalisation in west java, Indonesia. Cent. Eur. J. Int. Secur. Stud. **13**(4), 275–300 (2019). https://www.scopus.com/inward/record.uri?eid=2-s2.0-85083781856&partnerID=40&md5=dfaa4822c079b2cad466ada84351fd20
10. Huszti-Orban, K.: Internet intermediaries and counter-terrorism: between self-regulation and outsourcing law enforcement. In: 10th International Conference on Cyber Conflict: CyCon X: Maximising Effects, CyCon 2018, vol. 2018-May, pp. 227–243 (2018). https://doi.org/10.23919/CYCON.2018.8405019

11. Almoqbel, M., Xu, S.: Computational mining of social media to curb terrorism. ACM Comput. Surv. **52**(5), 1–25 (2019). https://doi.org/10.1145/3342101

12. Arpinar, I.B., Kursuncu, U., Achilov, D.: Social media analytics to identify and counter islamist extremism: systematic detection, evaluation, and challenging of extremist narratives online. In: 2016 International Conference on Collaboration Technologies and Systems, CTS 2016, pp. 611–612 (2016). https://doi.org/10.1109/CTS.2016.113

13. Azhmuhamedov, I.M., Machueva, D.A.: Development of measures to counter information extremism. In: 2018 Multidisciplinary Symposium on Computer Science and ICT, REMS 2018, vol. 2254, pp. 318–325 (2018)

14. Mitzias, P., et al.: Deploying semantic web technologies for information fusion of terrorism-related content and threat detection on the web: the case of the TENSOR EU funded project. In: 19th IEEE/WIC/ACM International Conference on Web Intelligence Workshop, WI 2019, pp. 193–199 (2019). https://doi.org/10.1145/3358695.3360896

15. Crittenden, S., et al.: Prostaglandin E2 promotes intestinal inflammation via inhibiting microbiota-dependent regulatory T cells. Sci. Adv. **7**(7), 1–16 (2021). https://doi.org/10.1126/sciadv.abd7954

16. Ganesh, B., Bright, J.: Countering extremists on social media: challenges for strategic communication and content moderation. Policy Internet **12**(1), 6–19 (2020). https://doi.org/10.1002/poi3.236

17. Laqueur, W.: A History of Terrorism. Taylor and Francis, Milton Park (2017)

18. Cuiqing, J., Xiangxiang, W., Zhao, W.: Forecasting car sales based on consumer attention. Data Anal. Knowl. Discov. **5**(1), 128–139 (2021). https://doi.org/10.11925/infotech.2096-3467.2020.0418

19. O'Donnell, E.C., Netusil, N.R., Chan, F.K.S., Dolman, N.J., Gosling, S.N.: International perceptions of urban blue-green infrastructure: a comparison across four cities. Water (Switz.) **13**(4), 544 (2021). https://doi.org/10.3390/w13040544

20. Jones, A., Kovacich, G.L.: Global Information Warfare: The New Digital Battlefield, 2nd edn. Taylor and Francis, Milton Park (2015)

21. Hwang, J.C., Schulze, K.E.: Why they join: pathways into Indonesian jihadist organizations. Terror. Polit. Violence **30**(6), 911–932 (2018). https://doi.org/10.1080/09546553.2018.1481309

22. Spaniel, W.: Rational overreaction to terrorism. J. Conflict Resolut. **63**(3), 786–810 (2019). https://doi.org/10.1177/0022002718756489

23. Kardaras, K., Lambrou, G.I., Koutsouris, D.: Security efficient command control communication and integration with digital television. Int. J. Electron. Secur. Digit. Forensics **12**(3), 293–301 (2020). https://doi.org/10.1504/IJESDF.2020.108298

24. Weinhardt, L.S., et al.: The role of family, friend, and significant other support in well-being among transgender and non-binary youth. J. GLBT Fam. Stud. **15**(4), 311–325 (2019). https://doi.org/10.1080/1550428X.2018.1522606

25. Bennett, B.T.: Understanding, Assessing, and Responding to Terrorism: Protecting Critical Infrastructure and Personnel. Wiley, New York (2017)

26. Sinai, J.: The nonviolent lone actor: the insider threat in information security. In: Understanding Lone Actor Terrorism: Past Experience, Future Outlook, and Response Strategies. Analytics & Business Intelligence, CRA, pp. 280–294. Taylor and Francis, Vienna (2016)

# Digital Medicine, Pharma and Public Health

# Problems and Negative Consequences of the Digitalization of Medicine

Irina Mirskikh[1]([✉]) [iD], Zhanna Mingaleva[2] [iD], Vladimir Kuranov[3] [iD], and Olga Dobrovlyanina[1] [iD]

[1] Perm State National Research University, Perm 614990, Russian Federation
[2] Perm National Research Polytechnic University, Perm 614990, Russian Federation
[3] E.A. Vagner Perm State Medical University, Perm 614990, Russian Federation

**Abstract.** Digital technologies are widespread and used in various fields and areas of healthcare. The undoubted advantages associated with the possibility of remote access to many medical services have led to the rapid spread of digital medicine. However, the diverse use of digital technologies in medicine is accompanied by a number of negative factors. Some of them can cause serious harm to the life and health of people, as well as cause significant damage for the society. The paper is devoted to the identification of problems and shortcomings in the provision of medical services using digital technologies. Based on the methods of bibliographic, content, formal and legal analysis, the main negative consequences of the use of digital technologies in the field of medicine are revealed and the solutions are suggested. The paper identifies potential and real opportunities for committing crimes in the field of medicine with the help of digital technologies, and gives their classification. Special emphasis is made on the contradictory consequences of the widespread use of telemedicine. It is concluded that the use of telemedicine is associated with many negative and dangerous social consequences. Recommendations for improving the Russian legislation in the field of telemedicine regulation and digitalization of the healthcare system as a whole are formulated.

**Keywords:** Digitization of medicine · Iatrogenic crimes · Protection of confidential information

## 1 Introduction

Digital technologies are the basis for the formation of an innovative society and are spreading in all areas of human life and sectors of the economy [1–3]. Digital technologies are also widespread in the healthcare sector [4, 5]. People are ready to make an electronic appointment at a polyclinic or for a consultation with a specialist, to maintain electronic patient records, to receive the results of tests and medical examinations by e-mail. The issuance of digital disability sheets and certificates of vaccination or a previous illness has become widespread. The practice of remote admission of patients using telecommunications technologies and telemedicine consultations is increasingly growing.

Digital technologies and telemedicine are widely used in the supervision of patients with chronic diseases that require constant monitoring (diabetes mellitus, heart diseases, etc.). A special field of application of telemedicine is surgery, which must be carried out under the supervision of doctors and specialties, as well as consultations of patients requiring special treatment in cases when there is no opportunity to provide assistance in person. Providing telemedicine consultations becomes the most common direction of digitalization of medical services in Russia nowadays.

Researchers and specialists engaged in the promotion of digital technologies in medicine are constantly expanding the list of benefits received by individuals and society as a whole from the development of digital medicine. The main advantages of remote (distant) healthcare that are traditionally called are significant expansion of access to medical services, the possibility of using individual solutions in the field of healthcare [6], which contributes to faster delivery of medical services, provides lower costs and provides wide access to healthcare services with higher quality. In general, the complex of various advantages and positive aspects associated with the development of digital medicine can be divided into several groups (see Fig. 1).

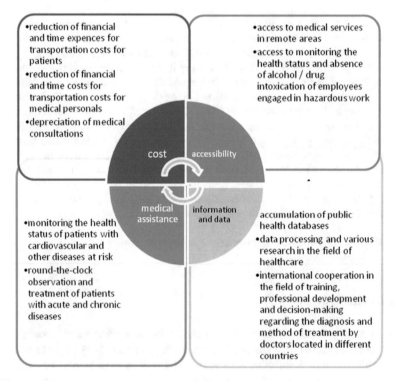

**Fig. 1.** The main advantages of using digital technologies in medicine and healthcare. Source: compiled by the authors

Figure 1 demonstrated the opportunity to organize mass health monitoring of employees engaged in hazardous work, in particular permanent monitoring the health status and absence of alcohol/drug intoxication for drivers, pilots, train operators, etc.

In addition, the use of digital technologies in healthcare in general made the access to various medical services less complicated, especially in the conditions of strict quarantine and isolation caused by the Covid-19 pandemic [7–9]. Digital technologies have significantly improved the monitoring of the health status of patients with various diseases at risk (for example, cancer, diabetes, cardiovascular diseases, etc.), allowing timely adjustments to treatment programs based on the prompt receipt of indicators of the effectiveness of treatment [10].

As a result, the level of digitalization of medicine in 55 regions of Russia reached a value from 65% to 100% by the end of 2020. And only in 29 regions it was less than 65% [11].

However, with the development of digital medicine, the number of issues and unresolved problems related to both the procedure for providing telemedicine services and the public effect of the use of digital technologies in this area also increases. Public opinion polls show contradictory trends in the development of digital medicine in Russia [12].

## 2    Theory and Methodology of Research

The study is based on legal and socio-humanitarian approaches to the definition of medical care provided with the use of telemedicine technologies.

In regulatory documents, telemedicine technologies are defined as information technologies that provide remote interaction of medical workers with each other, with patients and (or) their legal representatives, identification of these persons, documentation of their actions during consultations, remote medical monitoring of the patient's health [13]. Telemedicine services include a digital medical record, a telephone conversation, video conferences, the issuance of a prescription for medicines (also in digital form), the issuance of a medical report and other forms of assistance.

Within the broader framework of socio-humanitarian and technical approaches, telemedicine consultations can be defined as "organizing of the collection, processing and transmission of medical information using information and communication technologies and data transmission means" [14, p. 340]. "Telemedicine is a symbiosis of high-tech medicine and the latest achievements in the field of communication technologies" [15, p. 1039]. Telemedicine is the use of modern means of communication for the distant (remote) provision of medical and consulting services.

Currently, two methods of organizing telemedicine consultations are used – in real time and in the form of transmitting materials prepared during the consultation with a written prescription of a doctor (specialist) by e-mail. At the same time, both direct communication between the doctor and the patient (individual consultation = medical appointment) and collective consultations with the simultaneous presence of several doctors can be carried out in real time – the so-called medical consultations.

The main factor affecting the degree and spread of each of the methods of implementing telemedicine consultations is "the level of access to technical facilities and communication technologies of medical institutions and patients themselves, in particular, a computer, video camera, Skype and other special programs. Such programs allow visual interaction between the patient and the doctor and between doctors of different hospitals and regions" [14, p. 341]. In turn, this depends on the level of innovative receptivity of different territories (for example, underdeveloped, remote, provincial) to digitalization, on the institutional conditions for the development of innovative factors that ensure the successful implementation of information, communication and digital technologies [16, 17].

However, the development of telemedicine and the widespread use of other digital technologies in the healthcare system as a whole is characterized by a number of significant contradictions and increasingly growing problems of a public nature. Such complex problems relate to both conflicts in the field of legislative regulation of various aspects of the provision of medical services in remote and telecommunication forms, and the expansion of opportunities for committing crimes in the field of medicine based on the use of digital technologies, causing an increase in iatrogenic crimes[1].

The article provides the analysis of the complex of problems and contradictions that appeared in the telemedicine segment nowadays. The potential of telecommunications medicine development in Russia is analyzed. The main research methods are bibliographic, content and formal legal methods.

The materials from Internet sites in search engines Google and Yandex, a database of scientific publications on the development of telemedicine and the information from web sites of major medical organizations and associations are used as the information base of the research.

The analysis of the legal aspects of the use of digital technologies in public health and telemedicine was carried out on the basis of the regulatory documents of the Russian Federation. They are: Federal Law No. 242-FL "On Amendments to Certain Legislative Acts of the Russian Federation on the use of information technologies in the field of health protection" dated 01.01.2018 [18]; Federal Law No. 323-FL "On the basics of public health protection in the Russian Federation" dated 21.11.2011 (article 36.2. Features of medical care provided using telemedicine technologies) [13]; Federal Law No. 258-FL dated 31.07.2020 "On experimental legal regimes in the field of digital innovations in the Russian Federation)" [19]; Federal Law No. 149-FL dated 27.07.2006 "On Information, Information Technologies and Information Protection" [20]; Order of the Ministry of health of the Russian Federation from 30.10.2020 No. 1184н "On amendments to the order of the Ministry of health of the Russian Federation of 19 March 2020 No. 198н "On temporary order of organization of healthcare organizations to implement measures to prevent and mitigate the spread of the new coronavirus infection COVID-19" (Registered 12.11.2020 No. 60860) [21]; Federal Law No. 174-FZ dated 18.12.2001 "The Criminal Procedure Code of the Russian Federation" [22].

The study of regulatory documents in the field of the organization of telemedicine and the use of digital technologies in health care in general made it possible to identify the main problems and negative consequences in this area.

---

[1] Iatrogeny is a medical-legal term that means a deliberate or negligent act of a medical professional, resulting in the death or harm to the patient's health, in the event of violation of the medical care standards.

# 3   Results of the Study

Russian researchers pay attention to the danger of the growth of iatrogenic crimes and offenses connected with the development of telecommunications services and the complexity of their disclosure and proof [23]. Providing medical services without a license is the most widespread [24–26]. The possibility of providing services using telecommunications devices significantly expands the opportunities of illegal doctors and medical centers to offer and implement various types of consultations without official license and state control. The public (social) and personal danger of such actions increases if "a doctor who does not have the right to perform such medical actions provides assistance remotely" [26, p. 64].

The second serious problem in the field of digital medicine is the misuse or loss of information from patients' medical documents [14]. For example, based on the study of international experience in the use of personal data of patients in telemedicine consultations, O. V. Kalachev and I. S. Sushnikov come to the conclusion that there are currently significant problems in the field of legal regulation of personal data and medical secrecy regimes [27]. With regard to interaction with the patient himself, the increased difficulty of obtaining the patient's written consent to the collection, processing and transfer of personal data in the implementation of telemedicine services is revealed as well as the lack of clear boundaries for the application of exceptions to the rule on the patient's written consent to the processing of personal data. This can cause violations of the rights and legitimate interests of patients. Regarding the ethics of the personal data operator's behavior, it is possible to conclude that a reduced responsibility of the personal data operator, gives rise to the problem that the medical data of patients come into public domain or becomes known to third parties [28, 29].

With regard to the level of development of legal regulation as a whole (the third set of problems), there is a continuing uncertainty of the relationship between the legal regimes of medical secrecy and personal data [30].

Previous studies of the development of digitalization of medical services in Russia made it possible to reveal the presence of a number of factors and conditions that positively affect the spread of telemedicine in world and in Russia [31]. These are such factors as: the development of new digital medical technologies; wide coverage of remote territories by the Internet; expanding the possibilities of using the Internet on various mobile devices; increasing the age of Internet users; reducing the geographical differentiation of Internet users [14, p. 340]. At the same time the creation and rapid introduction of new digital medical technologies into practice refers to the expansion of the possibilities of using the most modern digital technologies by medical institutions, while the action of the remaining factors is more aimed at expanding the number of people who are ready and able to receive medical services remotely and relies on the innovative readiness and ability of specific territories (regions) to implement digital technologies [32].

However, the factors that contribute to the development of digital medicine are associated with a deep socio-economic paradox and a high inconsistency of the results of their influence. The complexity of the situation can be explained by the fact that the positive factors listed above are based on the potential for extremely high public danger of crimes that can be committed with the help of digital technologies in medicine, as a result of which significant harm to human health or death can be caused.

In particular, the specifics of the Internet, the ease and breadth of access to the Internet allows persons engaged in medical activities of inadequate quality, as well as illegal activities, to expand the geography of their activities, simultaneously covering thousands of people with remote consultations. As a result of such actions, harm can be caused to a much larger number of people compared to the activities of doctors who conduct face-to-face appointments. This leads to an increase in the number of iatrogenic crimes, which are defined as socially dangerous actions of doctors and medical personnel who violate the legal principles and conditions for providing medical care. These are crimes in the field of healthcare, and refer to a type of crimes against human life and health, which can be characterized by a high degree of public (social) danger. So, according to the Ministry of Health of Russia, medical errors annually cause about 70 thousand complications in patients [33].

Criminal cases on iatrogenic crimes are usually initiated by citizens against doctors and medical personnel guilty in improper provision of medical care that led to harm to the patient's health or death. They can be initiated also by the state and are based on the materials of prosecutor's investigation conducted on complaints of citizens about improper provision of medical care or at the request of a deputy, information in the mass media, including electronic form of such information. According to the Investigative Committee of Russia, about 7 thousand reports about crimes related to medical mistakes and inappropriate provision of medical care are received over the year, but the number of criminal cases initiated is much less. So in 2019, the Investigative Committee of Russia received 6.5 thousand messages from citizens, 2.1 thousand criminal cases were initiated, and only 332 cases were brought to the court [34].

On the other hand, the search for such remote "doctors" is significantly difficult, and bringing them to justice often becomes impossible both because of the imperfection of existing legislation (including Penal Code), and because of the complexity of the deportation of such criminals from abroad.

The second direction of the introduction of digital technologies in medicine, which causes problems and make society and individuals worry, is the process of digitizing personal data of patients related to their health, as well as procedures for storing and transmitting data [35–37]. And, recently, "cloud" services have become such places for storing medical confidential information. It is believed that "patient health data stored in the cloud is strictly confidential information, securely hidden to avoid unauthorized access to protect patient information" [38, p. 21165].

In order to increase the reliability of storing and transmitting such data, computer program developers are offering new technical ways of information protection, including new digital technologies.

First of all, Blockchain technology ought to be taken into account, which, according to experts, is able to provide the secure integration of medical records into a blockchain-based platform, high reliability of data storage and transmission [39, 40]. It is assumed that the Blockchain allows solving the problem of personal data protection, including providing detailed control of access to personal data and confidential information, determining what should be available, when and for whom, while complying with collective regulatory requirements. Modern researches in this area suggest combining technologies such as blockchain and scalable data lakes with the development and implementation of appropriate system procedures that guarantee secure access to confidential data. However, in practice, large amounts of personal data are increasingly becoming the property of third parties or come to the public domain.

As the cloud transfer of medical data becomes more and more widespread, it attracts more and more attention not only of researchers and scientists, but also of various kinds of scammers. Such confidential information is easy to steal and sell. Medical data transmitted through unreliable networks can be manipulated, and people can be compromised. Some researchers believe that the cost of medical confidential information related to the patient's health may be much higher than the cost of trade secrets and other types of confidential information [14]. The misuse of personal data and information from patients' medical documents is becoming one of the most serious problems in the field of digital medicine. However, the application of legal measures of responsibility for such illegal actions faces a number of problems.

## 4 Discussion

The further development of digital technologies in the field of healthcare and medicine is accompanied by the number of problems the most serious of which is the problem of iatrogenic crimes, the possibilities and spread of which are increasing with the growth of digitalization of the medical sphere. Providing medical services via the Internet can lead to medical mistakes, making wrong diagnosis (consultation), which can cause serious harm to health or even death of a person. Another area of concern for both patients and specialists in the field of medicine is the problem of compliance with the ethical principles of consulting by doctors as digital technologies develop [41]. Finally, serious concerns are connected with the increase of iatrogenic crimes related to illegal medical activities. At the same time, the procedures of investigation of such crimes, as well as holding persons criminally liable for committing iatrogenic crimes, is rather complicated.

In investigating such crimes, investigators rely primarily on the conclusions of judicial and medical experts, although, according to the article 17 of the Criminal Procedure Code of the Russian Federation, none of the evidence has a priority. However, investigators do not have special knowledge in the field of medicine, and due to the lack of a clear regulatory and professional-methodological base in the field of medical activity, practically do not understand these issues without help of experts and specialists. The volume of specific aspects studied, medical documentation is very significant and requires professional medical knowledge. In addition, life situations are different and can be atypical.

At the same time, experts often indicate in their conclusions that there are shortcomings in treatment. They practically and do not take into account the interrelation between the action (inaction) of the doctor and the consequences that have occurred, that is fundamentally important. At the same time, expert conclusions often contain a large number of special terms that are difficult for understanding. The results of the investigation are also negatively affected by the duration of conducting forensic medical examinations, in some cases exceeding the period of criminal prosecution (if the crime is of minor gravity). The complexity of proving violating professional duties in the actions of medical personnel is also caused by corporate solidarity, as well as the lack of uniform standards and methods of providing medical care.

It is possible to come to the conclusion that it is important to provide special training of investigators for the investigation of iatrogenic crimes, the State need to develop special methods of conducting investigations in this area [42].

Significant difficulties caused by both technical features and gaps in legal regulation of a number of issues are also accompanied by the process of introducing new digital technologies for providing medical services into the practical activities of medical institutions [43, 44].

Thus, the most important condition for the successful development of digitalization of the healthcare sector in Russia, in our opinion, is the creation of progressive legal regulation in the field of digital medicine, taking into account the need to protect personal data and confidential information from patients' medical documents. Unfair access to such personal confidential information can be potentially dangerous and lead to serious mistakes in treatment.

## 5   Conclusions

As foreign and domestic studies of the main results of the introduction of digital technologies in the provision of medical services have shown, this process is accompanied by both positive and negative consequences of its development. The study of gaps in the regulation of issues in the field of digital healthcare in Russia allowed to come to the following conclusions and formulate recommendations.

The use of telemedicine is associated with many negative and dangerous social consequences, which are the results of intentional and unintentional actions.

The misuse of personal data and information from patients' medical documents is becoming one of the most serious problems in the field of digital medicine.

The correlation between the legal regimes of medical secrecy and personal data remains uncertain, and requires examination and solution.

The creation of cloud computing environments in the field of healthcare, the protection and confidentiality of medical records become the priority task of applications for medical services.

In order to improve the security of integration and exchange of medical records in telecommunications, video consultations, sending research results by e-mail, etc., it is necessary to develop legal and technical solutions that provide secure data exchange. These solutions should ensure a high level of confidentiality of the exchanged data and include a strictly regulated procedure for accessing personal data, including a list of those who can have access to specific data and a list of open information.

For the entire communication network, it is necessary to create reliable access control to personal data, clearly record and determination what information can be available, who and when can have access to data. This is especially important when complying with collective regulatory requirements.

**Acknowledgment.** The work is carried out based on the task on fulfilment of government contractual work in the field of scientific activities as a part of base portion of the state task of the Ministry of Education and Science of the Russian Federation to Perm National Research Polytechnic University (topic # *FSNM-2020-0026*).

# References

1. Chirkunova, E., Anisimova, V.Y., Tukavkin, N.M.: Innovative digital economy of regions: convergence of knowledge and information. In: Ashmarina, S.I., Mantulenko, V.V. (eds.) Current Achievements, Challenges and Digital Chances of Knowledge Based Economy. LNNS, vol. 133, pp. 123–130. Springer, Cham (2021). https://doi.org/10.1007/978-3-030-47458-4_15

2. Mingaleva, Z., Mirskikh, I.: On innovation and knowledge economy in Russia. World Acad. Sci. Eng. Technol. **42**, 1018–1027 (2010)

3. Kudryavtseva, V.A., Vasileva, N.V.: Digitalization as the basis for the construction industry development. IOP Conf. Ser. Earth Environ. Sci. **751**(1), paper # 012100 (2021)

4. The digital revolution in healthcare: achievements and challenges, TASS, SPIEF-2017, May 29. https://tass.ru/pmef-2017/articles/4278264. Accessed 07 July 2021

5. Greenspun, H., Coughlin, S.: mHealth in a mWorld. How mobile technology is transforming health care, Deloitte Center for Health Solutions. https://www2.deloitte.com/content/dam/Deloitte/us/Documents/life-sciences-health-care/us-lhscmhealth-in-an-mworld-103014.pdf. Accessed 07 July 2021

6. Ventola, C.L.: Mobile devices and apps for health care professionals: uses and benefits. Pharm. Therap. **39**(5), 356–364 (2014)

7. Antipova, T.: Digital view on COVID-19 impact. In: Antipova, T. (ed.) ICCS. LNNS, vol. 186, pp. 155–164. Springer, Cham (2021). https://doi.org/10.1007/978-3-030-66093-2_15

8. Shakhabov, I.V., Melnikov, Y., Smyshlyaev, A.V.: Features of the development of digital technologies in healthcare in the context of the COVID-19 pandemic. Sci. Rev. Med. Sci. **6**, 66–71 (2020)

9. Platonova, N.I., Shakhabov, I.V., Smyshlyaev, A.V., Kuznetsov, D.V.: Telemedicine technologies in the context of the COVID-19 pandemic. Med. Law **1**, 21–28 (2021)

10. Antipova, T., Shikina, I.: Informatic indicators of efficacy cancer treatment. In: 12th Iberian Conference on Information Systems and Technologies (CISTI), pp. 1–5 (2017). https://doi.org/10.23919/CISTI.2017.7976049

11. Anfinogenov, I.: Digitalization of medicine and implementation of MIS 2021: latest news. https://archimed.pro/blog/tsifrovizatsiya-meditsiny-i-vnedrenie-mis-2021-svezhie-novosti. Accessed 20 July 2021

12. Buzin, V.N., Mikhaylova, Y., Buzina, T.S., Chuhrienko, I., Shikina, I.B., Mikhaylov, A.: Russian healthcare through the eyes of the population: dynamics of satisfaction over the past 14 years (2006–2019): review of sociological studies. Russ. J. Prevent. Med. **23**(3), 42–47 (2020). https://doi.org/10.17116/profmed20202303142

13. Federal Law No. 323-FZ "On the basics of public health protection in the Russian Federation" dated 21.11.2011 (article 36.2. Features of medical care provided using telemedicine technologies). http://www.consultant.ru/document/cons_doc_LAW_121895/. Accessed 18 July 2021

14. Mirskikh, I., Mingaleva, Z., Kuranov, V., Matseeva, S.: Digitization of medicine in Russia: mainstream development and potential. In: Antipova, T. (ed.) ICIS. LNNS, vol. 136, pp. 337–345. Springer, Cham (2021). https://doi.org/10.1007/978-3-030-49264-9_30

15. Demina, A.S., Rostovsky, N.S.: Features of the use of telemedicine technologies in domestic practice. Sci. Alley **4**(6/22), 1037–1040 (2018)
16. Belokur, O.S., Tsvetkova, G.S., Maslikhina, V.Y.: Digitalization of the economy in the provincial region: main trends and problems. IOP Conf. Ser. Earth Environ. Sci. **650**(1), paper # 012015 (2021)
17. Mingaleva, Z., Gayfutdinova, O., Podgornova, E.: Forming of institutional mechanism of region's innovative development. World Acad. Sci. Eng. Technol. **58**, 1041–1051 (2009)
18. Federal Law №242 "On amendments to the current legal acts of the Russian Federation on the issues of using information technologies and delivering medical telecommunication services in the field of health care" of 01.01.2018. http://www.consultant.ru/document/cons_doc_LAW_221184/. Accessed 13 July 2021
19. Federal Law No. 258-FZ dated 31.07.2020 "On experimental legal regimes in the field of digital innovations in the Russian Federation)". www.kremlin.ru/acts/bank/45796. Accessed 13 July 2021
20. Federal Law No. 149-FZ dated 27.07.2006 "On Information, Information Technologies and Information Protection". www.consultant.ru/document/cons_doc_LAW_61798/. Accessed 13 July 2021
21. Order of the Ministry of health of the Russian Federation from 30.10.2020 No. 1184н "On amendments to the order of the Ministry of health of the Russian Federation of 19 March 2020 No. 198н". "On temporary order of organization of healthcare organizations to implement measures to prevent and mitigate the spread of the new coronavirus infection COVID-19" (Registered 12.11.2020 No. 60860). Accessed 13 July 2021
22. Federal Law No. 174-FZ dated 18.12.2001 "The Criminal Procedure Code of the Russian Federation" (as amended on 07/01/2021). www.consultant.ru/document/cons_doc_LAW_34481. Accessed 13 July 2021
23. Rotkov, A.I., Chuprova, A.: Some questions of the application of criminal legislation for violations in the field of telemedicine. Legal Sci. Pract. Bull. Nizhny Novgorod Acad. Ministry Internal Affairs Russ. **4**(28), 176–178 (2014)
24. Petrova, R.E., et al.: The current state of the development of telemedicine in Russia: legal and legislative regulation. Prev. Med. **22**(2), 5–9 (2019)
25. Kozlova, A.S., Taraskin, D.S.: Telemedicine development trends and its impact on the Russian insurance market. Bull. Saratov State Socio-Econ. Univ. **2**(71), 144–148 (2018)
26. Nekrasov, V.N.: Features of criminal responsibility for crimes in the field of telemedicine. Inst. Bull. Crime Punishm. Correct. **13**(1), 63–67 (2019). https://doi.org/10.46741/2076-4162-2019-13-1-63-67
27. Kalachev, O.V., Sushilnikov, I.S.: International experience of using personal data of patients in telemedicine consultations. Mil. Med. J. **341**(7), 54–56 (2020)
28. Semeshko, A.I., Kuranov, V.G.: Regulation of the relationships in healthcare by the norms of professional medical organizations on the international regional level. Perm Univ. Herald Ser. Yuridical Sci. 3(25), 196–206 (2014)
29. Declaration of doctors of Russia. http://www.rmass.ru/publ/info/deklar.vrachey. Accessed 16 June 2021
30. Allamyarova, N.V., Sanakoeva, E.G.: Legal regulation of telemedicine in the health care system of the Russian Federation: state and prospects for improvement. Mil. Med. J. **340**(3), 15–22 (2019)
31. Lee, E., Han, S.: Determinants of adoption of mobile health services. Online Inf. Rev. **39**(4), 556–573 (2015)
32. Mingaleva, Z., Balkova, K.: Problems of innovative economy: forming of «innovative society» and innovative receptivity. World Acad. Sci. Eng. Technol. **59**, 838–843 (2009)

33. Kryazhev, A.: The Ministry of Health reported 70 thousand complications per year due to doctors' mistakes. https://ria.ru/20200208/1564435275.html?in=t. Accessed 21 June 2021
34. The Investigative Committee brought 332 criminal cases of medical errors to court in 2019. https://ria.ru/20200303/1567774151.html. Accessed 19 June 2021
35. Jutel, A., Lupton, D.: Digitizing diagnosis: a review of mobile applications in the diagnostic process. Diagnosis 2(2), 89–96 (2015)
36. Voskanyan, Y., Shikina, I., Kidalov, F., Musaeva, S., Davidov, D.: Latent failures of the individual human behavior as a root cause of medical errors. In: Antipova, T. (ed.) ICADS. AISC, vol. 1352, pp. 222–234. Springer, Cham (2021). https://doi.org/10.1007/978-3-030-71782-7_20
37. Khuntia, J., Yimb, D., Tanniru, M., Lim, S.: Patient empowerment and engagement with a health infomediary. Health Policy Technol. 6(1), 40–50 (2017)
38. Denis, R., Madhubala, P.: Hybrid data encryption model integrating multi-objective adaptive genetic algorithm for secure medical data communication over cloud-based healthcare systems. Multimed. Tools Appl. 80(14), 21165–21202 (2021). https://doi.org/10.1007/s11042-021-10723-4
39. Bowles, J., Webber, T., Blackledge, E., Vermeulen, A.: A blockchain-based healthcare platform for secure personalised data sharing. Stud. Health Technol. Inform. 281, 208–212 (2021). https://doi.org/10.3233/SHTI210150
40. Kadirov, A.O., Smykalo, N.V.: Digitization of Russian medicine using blockchain technology: retrospective analysis and development prospects. Innov. Invest. 12, 246–250 (2019)
41. Kuranov, V.G., Semeshko, A.I.: The legal significance of medical ethics. Med. Law Theory Pract. 1(1), 83–88 (2015)
42. Dobrovlyanina, O.V.: The introduction of new electronic technologies in criminal proceedings. Ex jure 2, 104–117 (2019). https://doi.org/10.17072/2619-0648-2019-2-104-117
43. Morozova, Y.: Digitization as a global, country and industry process in improving the effectiveness and efficiency of healthcare and medicine. Intellect. Innov. Invest. 4, 44–53 (2019)
44. Carman, K.L., Workman, T.A.: Engaging patients and consumers in research evidence: applying the conceptual model of patient and family engagement. Patient Educ. Couns. 100(1), 25–29 (2017)

# Management of the Flow of Patients in a Modern Medical Organization

Yuriy Voskanyan[1] , Irina Shikina[2(✉)] , Fedor Kidalov[3] ,
and Sergey Kurdyukov[1]

[1] Russian Medical Academy of Continuing Professional Education
of the Ministry of Health of Russia, Bld. 2/1, Barrikadnaya st.,
Moscow 125993, Russia
[2] Federal Research Institute for Health Organization and Informatics
of the Ministry of Health of the Russian Federation,
Bld. 11, Dobrolyubova st., Moscow 127254, Russia
[3] State Institution of the City of Moscow 'Information and Analytical Center
in the Field of Healthcare', Bld. 10, Novaya Basmannaya st.,
Moscow 107078, Russia

**Abstract.** With the use of agent-based model of human behavior 3 phases and
11 stages of patient's consumer cycle were identified: the phase of making a
decision about buying/choosing a medical organization, the phase of short-term
interaction (receiving medical services), the phase of forming long-term relations. Using observational and analytical research methods, the authors analyzed
the influence of relevant psychological factors on various stages of the consumer
cycle. Using the concept of 7p marketing, the authors analyzed the significance
of various elements of the value proposition of a medical organization for
various stages of the consumer cycle. In the logic of the concept of the role
model of consumption, the authors identified the main directions of promotional
communication with a potential patient in order to position the value proposition
and form a continuous flow of patients.

**Keywords:** Medical care · Patient · Consumer cycle · Role model of
consumption · Promotional communication

## 1 Introduction

The widespread introduction of the principles of insurance medicine in the Russian
Federation has divided the actors of the healthcare system into medical service providers
(medical organizations), consumers of medical services (patients) and intermediaries
(insurance funds and insurance medical organizations). Regardless of the form of
ownership, the formation of patient flow to any medical organization has become less
dependent on the executive authorities in the field of healthcare, and to a greater extent –
on the quality of medical care provided, by which we mean productiveness, safety,
efficiency, accessibility and patient orientation [1–5]. In this situation, understanding the

patient's consumer cycle in the holistic aspect, as well as the mechanisms of decision-making by the patient, and the basics of effective promotional communication with a potential consumer, in order to align communication gaps in the positioning of a value proposition, are of particular scientific and practical interest from the point of view of forming a continuous flow of patients.

## 2    Data and Methods

The paper is and analytical and observational study that aims to find an optimal strategy of consumer cycle management, promotional communication with the patient, and forming the patient's involvement in the process of providing medical care. The search for information was carried out by three researchers independently of each other for the period 1980–2020 with the use of scientific, technical and medical databases: Google Scholar, OSF, CORE, MEDLINE, Cochrane Collaboration; EMBASE, SCOPUS, ISI Web of Science. Analytical reviews, retrospective and prospective observational studies of high methodological quality were the subject of the analysis. The observational study of the consumer cycle was carried out through selective observation with the use of validated questionnaires. The assessment of the contribution of various factors to the decision-making process by a potential consumer was carried out by a team of 7 marketers using the method of independent expert evaluation with a range of estimates from 1 to 5 points.

## 3    Results

Using the behavioral approach, the analysis of the consumer cycle of a patient of a medical organization was conducted. The behavior of the patient, as a consumer of medical services, is considered by the authors as one of the expressions of individual human behavior, which is a cycle consisting of perception (with the formation of a meaningful image), information processing (with the formation of a mental model of the situation), decision-making, planning and behavioral act. In this cycle, the last observable stage (the physiological component - the behavioral act) is a consequence of the previous four psychological processes that do not strictly follow the laws of logic but are derived of a complex interaction of thinking, emotions, attitudes and existing motives [6–11]. Two types of thinking participate in decision-making [10]: logical (slow) and associative-intuitive (fast). Table 1 presents the main psychological factors that regulate individual human behavior.

**Table 1.** Psychological factors that regulate individual human behavior

| № | Factor | Definition | Characteristic |
|---|--------|-----------|----------------|
| 1 | **Logical thinking (System 2, according to D.Kahneman)** | A conscious subjective assessment of the surrounding world based on a system of axiomatic provisions (paradigm), which includes the concept (meaning), judgment (causal relations) and inference (conclusion) | Logical evaluation of the content and price of a value proposition |
| 2 | **Individual motives** | The desire to satisfy a certain need (be healthy, etc.) | - Meaningful (the desire to recover) <br> - Accompanying (desire for respect, recognition, service, comfort) |
| 3 | **Emotions** | Subconscious subjective assessment of the surrounding world | - Positive emotions (relief-joy, satisfaction, etc.) <br> - Negative emotions (excitement (fear, anger, shame, dissatisfaction, frustration, etc.) <br> - Ambivalent emotions |
| 4 | **Mental attitudes** | Predisposition to a certain activity in a certain situation (behavioral intentions). Mental attitudes form an appropriate evaluation hypothesis at the subconscious level in terms of perception and processing of information | The basis of mental attitudes are: <br> - Expectations *(which are based on the previous interaction experience);* <br> - Beliefs *(based on the available information);* <br> - Experienced emotions; |
| 5 | **Associative-intuitive or heuristics (System 1, according to D.Kahneman)** | Intuitive decision-making in the subconscious that later is processed through consciousness. Decisions are formed under the influence of emotions, motives and mental attitudes. | Evaluation of the content and price of a value proposition |

For further analysis, the consumer cycle of a potential patient was divided into 3 stages: the decision-making stage, the stage of short-term interaction (receiving services) and the stage of long-term relationship. The content of the described stages in the form of actions and events is presented in Table 2.

**Table 2.** Consumer cycle stages of a potential patient

| Consumer cycle stages | | | | | |
|---|---|---|---|---|---|
| The stage of making a purchase decision (choosing a medical organization) | | The stage of short-term interaction (receiving a service) | | The stage of forming a long-term relationship | |
| Action | Event | Action | Event | Action | Event |
| 1.Search for information | Information about the center is received by the patient | 5.Receiving the service and evaluating the process | The service is received, the conclusion about the quality of the process | 8.Compliance with medical prescriptions | Compliance (commitment to the quality of implementation of recommendation) |
| 2.Interest ('cold client') | Attention is fixed (concentration on the center > 7 seconds; click, view) | 6.Evaluation of the result | Conclusion about the quality and utility of the result | 9.Feedback with the center | The exchange of target content occurred |
| 3.Probing ('hot client') | Primary acquaintance with the center (a question via the website, a call, an appointment) | 7.Payment for the service | Transaction confirmation (payment receipt);Certificate for tax deduction | 10.Making a decision on re-receiving services in the center | The patient has made a new appointment |
| 4.Making a purchase decision | The patient has made an appointment | | | 11.Making a decision on the recommendation of the center to relatives and friends | Relatives or friends of the patient had made an appointment |

Table 3 shows the influence of the psychological factors described in Table 1 on various stages of the consumer cycle. At the stage of information search, the main role is played by the dominant motive (the desire to recover) and attitudes towards a fair price from the patient's point of view. At the stage of interest, emotions (fears) and attitudes about the clinic's capabilities and hypotheses about the center's ability to meet similar needs begin to play a major role. The stage of primary acquaintance is marked by the dominance of various types of emotions, on the basis of which new attitudes are formed (the ability or inability to meet the needs of a particular patient). Making a purchase decision (choosing a medical organization) depends, first of all, on hypotheses regarding the center's ability to meet the patient's needs and attitudes regarding a fair price in this case. Logical thinking is activated at this stage, mainly in the presence of any contradictions in the mental model, or when contradictory (ambivalent) emotions appear. The stage of receiving the service is the basis for the formation of new emotions in response to the assessment of the compliance of the process of providing medical care with the expectations of the patient. The evaluation of the result also takes

**Table 3.** Psychological factors of individual behavior and the consumer cycle

| № | Consumer cycle stage | Factors of individual human behavior of the consumer | | | | |
|---|---|---|---|---|---|---|
| | | Logical thinking | Perso-nal motives | Emotions and feelings | Attitudes | Associative-intuitive thinking |
| 1 | Search for information | | The desire to re-cover | | Fair price | |
| 2 | Interest ('cold client') | | | Fears (risk of damage and uncertainty) | Ability or inability of meeting similar needs | Ability or inability to satisfy the need to be healthy |
| 3 | Primary acquaintance with the center ('hot client') | | | Positive (compliance with the attitudes), Negative (non-compliance with the attitudes), ambivalent | Ability or inability of the center — the new attitude | |
| 4 | Making a purchase decision | It is acti-vated only when there are contradictions in the mental model, or there are ambivalent emotions | | | A fair price, taking into account the expected capabilities | Ability or inability to satisfy the need, based on new attitudes |
| 5 | Receiving the service and evaluating the process | | | Positive (compliance with the attitudes), Negative (non-compliance with the attitudes), ambivalent | Expectations (respect, attention, service and professionalism of performers) | |
| 6 | Evaluation of the result | Analysis of the rea-sons for non-complian-ce with ex-pectations | | Positive (compliance with the attitudes), Negative (non-compliance with the attitudes), ambivalent | Expected utility | Compliance/n on-compliance with expectations |
| 7 | Payment for the service | | | Derivatives of the result evaluation | A fair price, taking into account the evaluation of the result obtained | |
| 8 | Compliance with medical prescriptions | | | | Expected utility in the future – a new attitude | |
| 9 | Feedback with the center | | | Trust and a sense of gratitude | | |
| 10 | Making a deci-sion on re recei-ving the center's services | It is ena-bled if the result does not meet the expectations | | Trust | Expected utility | Ability or inability to satisfy the need, based on new attitudes |
| 11 | Recommendations of the center to relatives and friends | | The desire to help others | Trust and a sense of gratitude | Expected utility for others | Ability or inability to satisfy needs of others |

place on the basis of hypotheses that determine the correspondence (discrepancy) of the result obtained to the result expected by the patient, as well as emotions that arise in response to the outcome of such an assessment. At the stage of payment for services, the management of the patient's behavior is based on the attitudes towards a fair price and the emotional component, which is a derivative of the evaluation of the result obtained. The formation of the patient's compliance with medical prescriptions is based on the attitude associated with their expected utility in the future. Maintaining feedback with the center (the patient reflects his attitude to the clinic in the form of reviews, calls, letters, etc.) is a direct consequence of the feeling of gratitude to the clinic that arose during the treatment. The decision to re-visit is based on the sense of trust in the center that has arisen, new attitudes that have been formed (as a result of previous contact with the center) and hypotheses extracted from memory under their influence. The transcendental motive (the desire to help others), a sense of trust and gratitude to the center, together with new attitudes (expectation of future utility) and hypotheses (confidence in the center's ability to benefit other people with similar problems), encourage the patient to recommend the medical organization to his relatives and friends [12].

Using the concept of 7p marketing [12], the authors analyzed the significance of the necessary elements of a conditional value proposition, as well as equivalent elements of a consumer's value request (in terms of motives, emotions, attitudes, hypotheses and logic) for various stages of the consumer cycle (Table 4).

At the stage of **information search**, indirect communication technologies play a key role for the consumer, namely, their content and context (what is offered and how), to a lesser extent—the price of the service. At the stage of **interest ('cold client')**, the significance of the price of the service (package of services) increases somewhat and becomes equal to the communication content. At the same time, the vector of indirect communication begins to change-from the positioning of the service (package of services) to its evaluation by other consumers (reviews, social networks, etc.). At the stage of **primary acquaintance with the clinic ('hot client')**, the most important role for the patient is played by the quality of direct communication – the communication skills of the first contact person (congruence of elements of the psychological square (content, relationships, self-expression, appeal), verbal, paraverbal and non-verbal components, the relevance of content to the consumer), artifacts of corporate culture (behavior and appearance of staff), the visible state of the production environment. Along with the described elements, re-evaluation of the price becomes key in **making a purchase decision** (choosing a medical organization). At the stage of **receiving the service** and evaluating the patient's process, the content, comfort and apparent reliability (the difference between cumulative utility and cumulative risk) of the processes, the attitude, behavior and appearance of the staff begin to play a dominant role in the value proposition. The place and time of receiving the service, and the state of the environment have a slightly smaller impact. The stage of **evaluating the result** is one of the key ones from the point of view of building subsequent long-term relationships with the patient. The degree of satisfaction of the need in terms of the quality and safety of the medical care provided and, and, to a lesser extent, the joint assessment of the result with the doctor at this stage are the dominant factors of the value proposition. At the stage of **payment for services**, the dominant elements of the value proposition are the convenience of the place and time of the transaction, as well as the attitude of the staff

**Table 4.** The significance of the elements of the value proposition for different stages of the consumer cycle

| Stage of the consumer cycle | Service (package of services): (content and quality) / Need | Price / Costs | Place and time of service provision / Convenience of place and time | Promotion / Communication | Processes (operational efficiency) / Processes (reliability, comfort, patient orientation) | Staff (competency) / Staff (attitude, behavior, appearance) | Environment: (infrastructure, tools, objects of labor) / Environment (safety and comfort) |
|---|---|---|---|---|---|---|---|
| | **The concept of 7p marketing** — Value proposition / Consumer's value demand | | | | | | |
| Search for information | 1 | 4 | 1 | 5 | 1 | 1 | 1 |
| Interest | 1 | 5 | 1 | 5 | 1 | 1 | 1 |
| Primary acquaintance with the center | 1 | 1 | 1 | 5 | 1 | 5 | 5 |
| Making a purchase decision | 1 | 5 | 1 | 5 | 1 | 5 | 5 |
| Receiving the service and evaluating the process | 1 | 1 | 4 | 1 | 5 | 5 | 4 |
| Evaluation of the result | 5 | 1 | 1 | 3 | 1 | 1 | 1 |
| Payment for the service | 1 | 3 | 5 | 1 | 1 | 5 | 1 |
| Compliance with medical prescriptions | 5 | 1 | 1 | 5 | 1 | 1 | 1 |
| Feedback with the center | 5 | 3 | 3 | 1 | 4 | 4 | 3 |
| Making a decision on rereceiving the center's services | 5 | 3 | 3 | 1 | 4 | 4 | 3 |
| Recommendations of the center to relatives and friends | 5 | 5 | 5 | 5 | 5 | 5 | 5 |

Note:
- *Score 1 – minimum significance for the consumer;*
- *Score 5 – maximum significance for the consumer;*

**Table 5.** Role model of consumption and positioning of key success factors in a medical organization (MO)

| Role | Role content | Possible performers of the role | Communication with the role performer | |
|------|-------------|-------------------------------|-------------|----------|
| | | | **Direct** | **Indirect** |
| **Initiator** | Initiates the process of purchasing services in this MO | Doctors of the MO | Indoctrination(training in organizational norms); Training in technical skills | Through the internal motivation |
| | | Patient's family members | - Call-center - Reception and non-medical staff - Medical staff | Recommendations of relatives and friends; Internet (MO website, advertisements, bloggers, social networks); Recommendations of doctors; Advertising (except for advertising on the Internet) |
| | | Patient herself | | |
| **Influencer** | Recommends the purchase of ser-vices in this MO | Other patients of the MO | Doctors of the MO | Internet (website, advertising, bloggers, social networks) |
| | | Doctors of the MO | Indoctrination(training in organizational norms); Training in technical skills | Through the internal motivation |
| | | Doctors of other clinics | - Management of the center - Doctors of the MO | Internet (website of the MO); Scientific forums; Special magazines |
| | | Relatives and friends of the patient | - Call-center - Reception and non-medical staff - Medical staff | Recommendations of relatives and friends; Internet (the center's website, advertising, bloggers, social networks);Recommendations of doctors; Advertising (except for advertising on the Internet) |
| | | Patient's family members | | |
| | | Social networks | | |
| | | Bloggers | | |
| | | Advertising | Marketing and Advertising Department | Doctors of the MO; Patients of the MO; Family members of MO patients |
| | | Website of the MO | Marketing and Advertising Department; Doctors of the MO; Patients of the MO | |
| | | Patient | Reception of the MO; Legal department of the center; Sales depa-rtment; Call-center | |
| | | Relatives or family members of the patient | | |

(*continued*)

**Table 5.** (*continued*)

| Role | Role content | Possible performers of the role | Communication with the role performer | |
|---|---|---|---|---|
| | | | **Direct** | **Indirect** |
| Payer | Pays for MO's services | Insurance companies (VMI) | - Management of the MO<br>- Legal department of the MO | Internet (website of the MO MO) |
| | | The organizationemployer of the patient | | |
| | | Insurance companies (CHI) | | |
| | | Ministry of Health of the subject of Federation (state order) | | |
| | | Ministry of Health of Russia (state order) | | |
| User | The person to whom the MO's services are provided | Patient | Doctors of the MO; Reception of the MO; Secondary medical personnel; Management of the MO; Non-medical staff of the center | - Educational literature for the patient<br>- Internet (patient's personal account) |
| Appraiser | A person who evaluates the quality and content of MO's services and compares them with other MOs | Patient | - Doctors of the MO<br>- Management of the MO | Internet (patient's personal account) |
| | | Patient's family members | | Website (the results of the evaluations are publicly available) |
| | | Insurance companies (VMI) - expertise | - Management of the MO<br>- Doctors of the center<br>- Lawyers of the MO<br>- Economists of the MO | |
| | | Insurance companies (MHI) - expertise | | |
| | | The Ministry of Health of Russia and the Ministry of Health of the subject of Federation - expertise | | |
| | | Federal Service for Surveillance in Healthcare | | |
| | | Doctors of other clinics | Doctors of the MO | |

accepting the payment. The price plays a smaller role there (in terms of the possibility of incurring costs), mainly when the cost changes during the treatment process. **Compliance** with medical prescriptions is formed exclusively in patients who are satisfied with the result, and in patients who receive feedback from the clinic. This determines the dominant role of such elements of the value proposition as the quality of the service provided (especially its effectiveness) and communication with the patient in the form of feedback. **Receiving feedback** from the patient is a consequence of many factors of the value proposition: first of all, the result, secondly, memories of the processes, attitude and behavior of the staff, to a lesser extent-the price, place and time of receiving medical care and the state of the environment. These same factors have the greatest influence on the patient's **decision to re-visit the center** (if necessary). **The recommendation of the medical center to friends and relatives** is not only the result of the patient's satisfaction, but also the result of a feeling of gratitude to the clinic, a desire to help loved ones and a sense of responsibility for their words. That is why at this stage all the elements of the value proposition will play a dominant role (there are no small things in the patient's recommendations, and there cannot be such) [13–17].

Using the concept of a role model of consumption [18], Table 5 presents the main directions of promotional communication with the positioning of the value proposition by specific employees and services of a specific medical organization for various roles in the process of medical services consuming.

## 4   Conclusion

The patient's free choice of a medical service provider in the conditions of insurance medicine dictates the need for each medical organization to form its own strategy for managing the consumer cycle and a strategy for positioning a value proposition for a potential patient. Taking into account the psycholinguistic foundations of the regulation of individual human behavior, it is possible to understand the basic mechanisms of perception, information processing and decision-making of patients both at the stage of choosing a medical organization and subsequent interaction with it. Taking into consideration these mechanisms, an effective strategy for forming a patient's commitment to the clinic at the level of his involvement in the process of providing medical care can be built. An important addition to the consumer cycle management strategy is the strategy of effective promotional communication with a potential consumer of medical services, based on a role model of consumption, involving communication with the initiator, influencer, payer, user and appraiser. An integrated approach using the two described strategies will allow to form a continuous flow of patients and significantly increase the long-term reputational and financial stability of the clinic.

## References

1. Busse, R., Klazinga, N., Panteli, D., Quentin, W. (eds.): Improving healthcare quality in Europe Characteristics, effectiveness and implementation of different strategies. World Health Organization (acting as the host organization for, and secretariat of, the European Observatory on Health Systems and Policies) and OECD, 419 p. (2019)

2. Tello, J.E., Barbazza, E., Wadell, K.: Review of 128 quality of care mechanisms: a framework and mapping for health system stewards. Health Policy **124**, 12–24 (2020)

3. Shikina, I.B.: The maintenance of security of the elderly and gerontic patients in the hospital conditions. Prob. Soc. Hyg. Health Care Hist. Med. **6**, 44–45 (2007)

4. Voskanyan, Y., Shikina, I., Kidalov, F., Davidov, D., Abrosimova, T.: Risk management in the healthcare safety management system. J. Digit. Sci. **3**(1) (2021). https://doi.org/10.33847/2686-8296.3.1_4

5. Voskanyan, Y., Shikina, I., Kidalov, F., Musaeva, S., Davidov, D.: Latent failures of the individual human behavior as a root cause of medical errors. In: Antipova, T. (ed.) ICADS 2021. AISC, vol. 1352, pp. 222–234. Springer, Cham (2021). https://doi.org/10.1007/978-3-030-71782-7_20

6. Voskanyan, Y., Kidalov, F., Shikina, I., Kurdyukov, S., Andreeva, O.: Model of individual human behavior in health care safety management system. In: Antipova, T. (ed.) ICCS 2020. LNNS, vol. 186, pp. 413–423. Springer, Cham (2021). https://doi.org/10.1007/978-3-030-66093-2_40

7. St. Pierre, M., Hofinger, G., Buerschaper, C.: Crisis Management in Acute Care Settings Human Factors and Team Psychology in a High Stakes Environment. Springer, Heidelberg (2008)

8. Mattsson, S., Fast-Berglund, A., Stahre, J.: Managing production complexity by supporting cognitive process in final assembly. Procedia CIRP **17**, 212–217 (2014)

9. Akintunde, E.A.: Theories and consep for human behaivior in environmental preservation. J. Environ. Public Health. **1**(2), 120–133 (2017)

10. Kahneman, D., Slovic, P., Tversky, A.: Judgement Under Uncertainty: Heuristics and Biases. Cambridge University Press, Cambridge (1982)

11. Rasmussen, J.: Skills, rules, knowledge: signals, signs and symbols and other distinctions in human performance models. IEEE Trans. Syst. Man Cybern. **13**(3), 257–267 (1983)

12. Yaghoubi, M., et al.: A systematic review of factors influencing healthcare services marketing in Iran. Bali Med. J. **6**(2), 268–278 (2017)

13. Riurean, S., Antipova, T., Rocha, Á., Leba, M., Ionica, A.: VLC, OCC, IR and LiFi reliable optical wireless technologies to be embedded in medical facilities and medical devices. J. Med. Syst. **43**(10), 1 (2019). https://doi.org/10.1007/s10916-019-1434-y

14. Butovskaya, M.L., Chargaziya, L.D.: The involvement in competitive gaming and aggressive behavior of young men. Voprosy Psychologii **2**, 91–103 (2020). https://www.scopus.com/inward/record.uri?eid=2-s2.0-85096096365&partnerID=40&md5=7a15a6b6cbc18c39a57ad65757d9773c

15. Liutsko, V.V., Son, I.M., Ivanova, M.A., et al.: Working time costs of doctors-therapists of divisionals on a patient. Ther. Arch. **91**(1), 19–23 (2019). https://doi.org/10.26442/00403660.2019.01.000023

16. Mirskikh, I., Mingaleva, Z., Kuranov, V., Matseeva, S.: Digitization of medicine in Russia: mainstream development and potential. In: Antipova, T. (ed.) ICIS 2020. LNNS, vol. 136, pp. 337–345. Springer, Cham (2021). https://doi.org/10.1007/978-3-030-49264-9_30

17. Antipova, T.: Digital view on COVID-19 Impact. In: Antipova, T. (ed.) ICCS 2020. LNNS, vol. 186, pp. 155–164. Springer, Cham (2021). https://doi.org/10.1007/978-3-030-66093-2_15

18. Pistikou, A.M., Zyga, S., Sachlas, A., Katsa, M.E., Daratsianou, M., Rojas Gil, A.P.: Determinative factors of being an effective health-care role mode. Int. J. Occup. Health Public Health Nurs. **1**(3), 3–14 (2014)

# Impact of the Income Level of the Working-Age Population on Certain Mortality Indicators in Russian Federation

Nikita Golubev(ID), Aleksandr Polikarpov(ID), Irina Shikina$^{(\boxtimes)}$(ID), and Ekaterina Shelepova(ID)

Federal Research Institute for Health Organization and Informatics of the Ministry of Health of the Russian Federation, Bld. 11, Dobrolyubova st., Moscow 127254, Russia

**Abstract.** The article analyzes the following indicators in the Russian Federation for the 25-year period (1995–2019): the number of people with incomes below the subsistence minimum as a percentage of the total population, the dynamics of mortality in working age (per 1000 people of relevant age) and mortality from selected alcohol-related causes (per 100 000 people). The aim of our study is to examine the possible relationship between the level of income of the working-age population and the mortality rate of the working-age population, as well as mortality from selected alcohol-related causes. Analysis of the indicators shows a 12.5% decrease in the number of people with incomes below the subsistence level; mortality of the working-age population decreased by 42.0%, and mortality from certain alcohol-related causes decreased by 17.6%, which is important for the medical and demographic situation in the state. We also present a graph of the dynamic changes in the two studied indicators. From 1995 to 1998, against the background of the increase in the monetary income of the population before the financial and economic crisis, there was a steady downward trend in mortality among the working-age population, as well as mortality from selected alcohol-related causes. In 2005, against the background of a long gradual decline in the number of people with monetary incomes below the subsistence level, the lines of mortality in the working-age population and mortality from selected alcohol-related causes had a unidirectional peak.

**Keywords:** Income level of the population · Mortality · Working-age population · Alcohol · Selected alcohol-related causes

## 1 Introduction

World Health Organization reports have repeatedly noted the link between population health and socioeconomic factors such as income, access to health care, social protection, and employment. The most important socio-economic factor that determines differences between population groups and influences the disparity of health characteristics is the level of income [1]. Therefore, an increase in the birth rate, reduction of

T. Antipova (Ed.): DSIC 2021, LNNS 381, pp. 467–476, 2022.
https://doi.org/10.1007/978-3-030-93677-8_41

mortality, preservation of health, reduction in the incidence of chronic non-infectious diseases and factors causing them, an increase in population income, reduction in the number of population with monetary income below the subsistence minimum, increase in life expectancy while maintaining the labor potential of the population are the main goals of Russia's national security [2].

According to Izmerov et al., 409,900 (1990) to 739,900 (2005) people of working age died annually in Russia between 1990 and 2012. About 80% of them were male. In 2012, the number of female and male deaths fell to 496,300, 32.9% less than in 2005, but 21.1% more than in 1990. Consequently, supermortality continued to be observed in the working age [3, 4].

The supermortality of the working-age population began in the Soviet Union in the mid-1980s, when, after nearly two decades of growth in life expectancy, Gorbachev's short (1985–1987) but extensive anti-alcohol campaign stimulated the start of decrease. After the collapse of the USSR in 1991, due to the severe economic crises in most of the successor states, life expectancy drastically decreased and mortality among the population increased [5]. The minimum life expectancy in the Russian Federation was reached in 1994. It was followed by a rapid but short recovery [6], which was soon replaced by a sharp drop due to the financial and economic crisis of August 1998 and the decline in income levels, including below the subsistence level [5, 7, 8].

Mortality from alcohol-related causes is the most important and changeable component of mortality, reflecting the changes taking place in the country related to the living conditions of the population. Under the conditions of poorly organized political, economic and social transformations, a sharp weakening of the social protection system, a decrease in the income level of the population and an increase in unemployment and high psychosocial stress [9], hazardous alcohol consumption played an important role in fluctuations of mortality rates among the population, especially among people of working age [6, 7, 11–15].

## 2   Data and Methods

The study period (from 1995 to 2019) was chosen due to the availability of registered data from the Federal State Statistics Service (ROSSTAT) and federal statistical observation forms.

According to the ROSSTAT methodology, information about the number of people with incomes below the subsistence minimum was based on the data on the distribution of the population by per capita average income and was the result of its comparison with the subsistence minimum. Since 2000, the methodology for calculating the minimum subsistence level has been changed. Since 2005, the composition of the consumer goods basket for determining the minimum subsistence level has been changed, and since 2013, the whole procedure for calculating the minimum subsistence level has also been changed.

In accordance with ROSSTAT data, the mortality rate from selected alcohol-related causes includes the number of deaths from accidental alcohol poisoning, chronic alcoholism, alcoholic psychosis, alcoholic liver disease, alcoholic cardiomyopathy, alcohol-induced nervous system degeneration, alcoholic etymology chronic pancreatitis. Prior to 2005, only accidental alcohol poisoning, chronic alcoholism, alcoholic psychoses, and alcoholic liver disease were taken into account.

To estimate the incidence of alcoholism and alcoholic psychosis we used the data from the reporting form of federal statistical observation № 11 'Information about the diseases of narcological disorders', approved by ROSSTAT Order № 410 of 16.10.2013 [16].

The article indicates the dynamics of the above-mentioned indicators from the nineties of the last century to the present.

Analytical and statistical methods, as well as the method of content analysis, were used in the research.

## 3 Results

We've analyzed the dynamics of the following indicators: the number of people with income below the subsistence minimum as a percentage of the total population, deaths in working age per 1,000 people of relevant age, and mortality from selected alcohol-related causes (per 100,000 population) for the period from 1995 to 2019 in the Russian Federation (Table 1).

**Table 1.** Dynamics of changes in the number of people with incomes below the living wage as a percentage of the total population, deaths at working age (per 1,000 people of the corresponding age), and deaths from selected alcohol-related causes (per 100,000 people) for 1995–2019 in the Russian Federation

| Year | The number of people with incomes below the living wage as a percentage of the total population, % | Deaths at working age (per 1,000 people of the corresponding age) | Deaths from selected alcohol-related causes (per 100,000 people) |
|---|---|---|---|
| 1995 | 24,8 | 8,1 | 39,2 |
| 1996 | 22,1 | 7,2 | 30,2 |
| 1997 | 20,8 | 6,4 | 23,5 |
| 1998 | 23,4 | 6,1 | 21,6 |
| 1999 | 28,4 | 6,8 | 25,5 |
| 2000 | 29,0 | 7,3 | 32,8 |
| 2001 | 27,5 | 7,5 | 37,4 |
| 2002 | 24,6 | 7,8 | 42,3 |
| 2003 | 20,3 | 8,1 | 44,0 |
| 2004 | 17,6 | 8,1 | 43,2 |
| 2005 | 17,8 | 8,3 | 73,1 |
| 2006 | 15,2 | 7,5 | 63,1 |
| 2007 | 13,3 | 7,0 | 52,9 |
| 2008 | 13,4 | 6,9 | 53,7 |
| 2009 | 13,0 | 6,4 | 48,4 |
| 2010 | 12,5 | 6,3 | 47,5 |
| 2011 | 12,7 | 6,0 | 42,7 |
| 2012 | 10,7 | 5,8 | 39,1 |
| 2013 | 10,8 | 5,6 | 36,8 |
| 2014 | 11,3 | 5,7 | 39,6 |
| 2015 | 13,4 | 5,5 | 40,1 |
| 2016 | 13,2 | 5,3 | 38,4 |
| 2017 | 12,9 | 4,8 | 33,5 |
| 2018 | 12,6 | 4,8 | 33,2 |
| 2019 | 12,3 | 4,7 | 32,3 |

From 1995 to 1999 in Russia there was a decrease by 3.6% points in the number of the population with monetary incomes below the subsistence level from 24.8% in 1995 to 28.4% in 1999. The greatest rate (4.0% points) of decline in this indicator was from 1995 to 1997: from 24.8% to the minimum recorded of 20.8%.

The mortality rate in the working age population (per 1,000 of the corresponding age population) in the Russian Federation began to be applied in 1995. Analyzing it, we noted a decrease by almost 16.0%: from 8.1 in 1995 to 6.8 in 1999, with a minimum value of the index (6.1) in 1998.

During the same time period, a significant decrease in the death rate from selected alcohol-related causes (per 100,000 population) was recorded—by 34.9%: from 39.2 in 1995 to 25.5 in 1999. The minimum of the indicator (21.6) was registered in 1998.

After 1998, due to the economic crisis, all the considered positive trends of changes in the studied indicators began to change and fluctuate. The standards of living and incomes of a large part of the population declined sharply, working conditions worsened with the growth of unemployment, which increased the incidence of bad habits and unhealthy lifestyles.

The decrease in the availability and quality of medical care, the deterioration of drug provision (especially for working citizens, due to the liquidation of the system of medical and preventive care for the working population), and poor equipment of medical organizations have largely determined the increase in mortality from preventable causes [3, 4, 17].

As in the nineties of the twentieth century, in the period from 2000 to 2009 in Russia, the state continues to actively develop a medical and demographic policy. In 2001, the Concept of Demographic Development of the Russian Federation for the period until 2015 was adopted [18]. Nevertheless, even against the background of a general decline in the population's incomes, there has been a steady downward trend in the number of people with incomes below the subsistence level by 16% points, from 29.0% in 2000 to 13.0% in 2009. Yet, it should be taken into account that since 2005 the composition of the consumer goods basket for determining the value of the subsistence minimum has been changed (Table 1).

During the same period (2000–2009), state policy was particularly focused on the health care industry and supporting it with large financial flows. On January 1, 2006, the state began to implement the National Priority Project 'Health', which included measures for the revival of preventive medicine, and the development of primary medical care and high-tech medical care, which was reflected in the dynamic reduction of mortality in the working-age population [3, 19, 20]. The analysis reveals a continuing smooth, not strongly fluctuating trend of a 12.3% decline in the working age mortality rate (per 1,000 population of the relevant age): from 7.3% in 2000 to 6.4% in 2009.

However, in this decade, there has been a significant increase in deaths from selected alcohol-related causes (per 100,000 population) among the working-age population (by 47.6%): from 32.8 in 2000 to 48.4 in 2009. The problems of alcohol-associated mortality were associated with a sharp increase in the availability of alcohol and an equally sharp deterioration in its quality [4]. Seven years after the 1998 crisis, this indicator rose from 32.8 in 2000 to a peak of 73.1 in 2005, after what began to gradually decline, although it did not reach the 2000 level in the decade under review.

In the period from 2010–2019, as well as since the nineties, the analysis shows a continuing downward trend in the number of people with incomes below the subsistence level (by 0.2% points): from 12.5% in 2010 to 12.3% in 2019. Yet, we must take into account the fact that in 2013 the procedure for calculating the subsistence level was changed.

Analysis of the same period reveals a steady, yet smooth, decline in the working-age mortality rate (per 1,000 population of appropriate age) by 25.4%: from 6.3 in 2010 to 4.7 in 2019. In accordance with the two aforementioned rates, mortality from selected alcohol-related causes (per 100,000 population) decreased by 32.0%: from 47.5 in 2010 to 32.3 in 2019. It took nearly twenty years for the state's efforts to bring the rate of deaths from selected alcohol-related causes down to the 2000 level (32.8).

The reduction of mortality in the working age is associated with the systematically implemented policies in the field of health care, which led to an increase in the availability and quality of medical care, actively conducted preventive examinations and clinical examination. These measures allowed to maintain public health and prevent the development of chronic non-communicable diseases and their complications [21–25].

Further, we conducted a graphical comparison of the lines of the analyzed indicators (the number of the population with monetary income below the subsistence level; deaths in the working age; mortality from selected alcohol-related causes) (Fig. 1).

In 2019 in the Russian Federation, the share of the population with monetary income below the subsistence minimum was 12.3%. In the dynamics over the analyzed period (1995–2019), this indicator was subject to fluctuations but generally decreased twofold: from 24.8% in 1995 to 12.3% in 2019. From 1995 to 1997, the share of the population with monetary incomes below the subsistence minimum decreased from 24.8% to 20.8%, respectively, with an upward trend thereafter. From 2000 to 2012, there was a decline from 29.0% to 10.7%, and then the rate increased to 13.4% in 2015. After that the indicator decreases again to 12.3% in 2019 (Fig. 1 - blue line).

In general, over the analyzed period the mortality rate of the working-age population (per 1,000 people of the corresponding age) in the Russian Federation decreased from 8.1% in 1995 to 4.7% in 2019 (by 1.7 times). In our opinion, coinciding with the data of Shkolnikov V.M. et al. (2014) and Shurgaya M.A. et al. (2017), Russia has the longest decline in mortality of the working-age population since 1965 [9, 26].

The indicator fluctuated both as a smooth decline (by 24.7%: from 8.1% in 1995 to 6.1% in 1998) and as a 1.4-fold increase (from 6.1% in 1998 to 8.3% in 2005). From 2005 to 2019, there has been a 43.4% decline in the nationwide working-age mortality rate: from 8.3% in 2005 to 4.7% in 2019 (Fig. 1 - green line).

In the Russian Federation, the mortality rate from selected alcohol-related causes (per 100,000 population) was subject to significant fluctuations over the analyzed period, but overall decreased by 17.6%: from 39.2 in 1995 to 32.3 in 2019. Fluctuations were recorded in the form of a downward trend of 1.8 times, from 39.2 in 1995 to 21.6 per 100,000 population in 1997. The mortality rate increased 3.4 times over the period 1998–2005, from 21.6 in 1998 to 73.1 in 2005. Of note is the dramatic 1.7-fold increase in mortality from selected alcohol-related causes: from 43.2 in 2004 to 73.1 in 2005 per 100,000 people. Further, from 2005 to 2013, there was a gradual decrease in the mortality rate almost twofold: from 73.1 to 36.8 in 2013. The subsequent years 2013–2015 saw a wave of slight growth of 9.0%: from 36.8 in 2013 to 40.1 in 2015. Over the most recent 2015–2019 period, there has been a 19.5% decline in the rate of deaths from selected alcohol-related causes, from 40.1 in 2015 to 32.3 in 2019 (Fig. 1 - orange line).

Figure 1 highlights two dynamic changes in the studied indicators. From 1995 to 1998, against the background of the increase in the monetary income of the population before the financial and economic crisis, there was a steady downward trend in

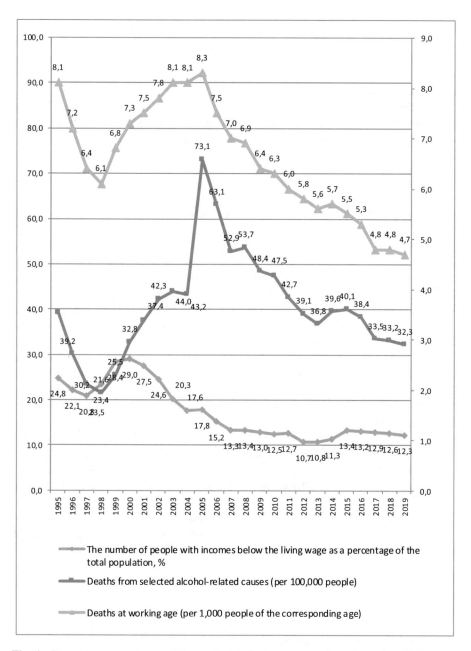

**Fig. 1.** Dynamic representation of the studied indicators: the number of people with incomes below the living wage as a percentage of the total population; deaths in working age per 1,000 people of the relevant age and mortality from selected alcohol-related causes (per 100,000 population), from 1995 to 2019 in the Russian Federation

mortality among the working-age population, as well as mortality from selected alcohol-related causes. In 2005, against the background of a long gradual decline in the number of people with monetary incomes below the subsistence level, the lines of mortality in the working-age population and mortality from selected alcohol-related causes had a unidirectional peak.

In order to confirm the relationship between the level of income of the working-age population and its mortality rate, as well as mortality from selected alcohol-related causes, a quantitative indicator is planned to be added in the future.

## 4   Conclusion

The analysis of 25 year-period (from 1995 to 2019) allows us to note that the population of the Russian Federation with monetary incomes below the subsistence minimum has changed over the analyzed period from 24.8% in 1995 to 12.3% in 2019. The lowest percentage of the population with monetary incomes below the subsistence minimum was reached in 2012 - 10.7. Yet, it is necessary to take into account that since 2000 the methodology for calculating the subsistence minimum has been changed. Since 2005 the composition of the consumer basket for determining the subsistence minimum has been changed, and in 2013 the procedure for calculating the subsistence minimum was changed as well.

There was a 42.0% decrease in the working-age population mortality rate in the Russian Federation over 1995–2019: 8.1 in 1995 and 4.7 in 2019 (respectively per 1,000 population of the relevant age). The lowest mortality rate in the working-age population was recorded in 2019 - 4.7, the highest one in 2005 - 8.3 per 1,000 people of the corresponding age.

Over the analyzed period from 1995 to 2019 there has been a reduction of the mortality rate from selected alcohol-related causes in the Russian Federation by 17.6%: from 39.2 per 100 thousand population in 1995 to 32.3 in 2019. The lowest mortality rate was achieved in 1998 - 21.6 cases per 100 thousand population, the highest rate in 2005 - 73.1.

It is noted that on the graph the lines of indicators of mortality in the working age population and mortality from selected alcohol-related causes, have a peak in 2005 against the background of a long decline in incomes below the subsistence level.

## References

1. Prokhorov, B.B., Ivanova, E.I., Shmakov, D.I., Shcherbakova, E.M.: Medico-Demographic Forecasting: Textbook, p. 360. MAKS Press, Moscow (2011)
2. Decree of the President of the Russian Federation of June 6, 2019 № 254 «On the Strategy for the development of health care in the Russian Federation for the period up to 2025» [Electronic resource]. https://www.garant.ru/products/ipo/prime/doc/72164534/. Accessed 25 Dec 2020. (in Russian)

3. Izmerov, N.F., Tikhonova, G.I., Gorchakova, T.: Mortality of the working-age population in Russia and developed countries of Europe: trends of the last twenty years. Ann. Russ. Acad. Med. Sci. **7–8**, 121–126 (2014). (in Russian)
4. Andreyev, E.M., Zhdanov, D.A., Shkolnikov, V.M.: Mortality in Russia after the collapse of the USSR: facts and explanations. SPERO **6**, 115–142 (2007)
5. Leon, D.A., et al.: Huge variation in Russian mortality rates 1984–94: artefact, alcohol, or what? Lancet **350**(9075), 383–388 (1997). https://doi.org/10.1016/S0140-6736(97)03360-6
6. Shkolnikov, V., McKee, M., Leon, D.A.: Changes in life expectancy in Russia in the mid-1990s. Lancet **357**(9260), 917–921 (2001). https://doi.org/10.1016/S0140-6736(00)04212-4
7. Shkolnikov, V.M., Andreev, E.M., Leon, D.A., McKee, M., Meslé, F., Vallin, J.: Mortality reversal in Russia: the story so far. Hygiea Internationalis Interdisc. J. Hist. Public Health **4** (1), 29–80 (2004). https://doi.org/10.3384/hygiea.1403-8668.044129
8. Shkolnikov, V.M., Andreev, E.M., Zhdanov, D.A.: Mortality of the working-age population, alcohol and life expectancy in Russia. In: Khalturina, D.A., Korotaeva, A.V. (eds.) Alcohol Catastrophe. The Possibilities of State Policy in Overcoming Alcohol Mortality in Russia, pp. 85–104. Lenand (2008). (in Russian)
9. Shkolnikov, V.M., Andreev, E.M., McKee, M., Leon, D.A.: Rising life expectancy in Russia of the 2000s. Demographic Rev. **2**(1), 5–37 (2014). (in Russian)
10. Walberg, P., McKee, M., Shkolnikov, V., Chenet, L., Leon, D.A.: Economic change, crime, and mortality crisis in Russia: regional analysis. BMJ **317**(7154), 312–318 (1998). https://doi.org/10.1136/bmj.317.7154.312
11. Chenet, L., McKee, M., Leon, D., Shkolnikov, V., Vassin, S.: Alcohol and cardiovascular mortality in Moscow: new evidence of a causal association. J. Epidemiol. Commun. Health **52**(12), 772–774 (1998). https://doi.org/10.1136/jech.52.12.772
12. Khalturina, D.A., Korotaev, A.V.: Introduction. Alcoholic catastrophe: how to stop the extinction of Russia. In: Khalturina, D.A., Korotaeva, A.V. (eds.) Alcoholic Catastrophe. The Possibilities of State Policy in Overcoming Alcohol Mortality in Russia, pp. 5–58. Lenand (2008). http://net.knigi-x.ru/24raznoe/535545-1-vvedenie-alkogolnaya-katastrofa-kak-ostan ovit-vimiranie-rossii1-halturina-korotaev-rossiyskiy-demografichesk.php. (in Russian)
13. Nemtsov, A.V.: Again about alcohol. Demoscope Weekly, pp. 567–568 (2013). http://demoscope.ru/weekly/2013/0567/index.php. (in Russian)
14. Levintova, M.: Russian alcohol policy in the making. Alcohol Alcohol. **42**(5), 500–505 (2007). https://doi.org/10.1093/alcalc/agm040
15. Leon, D.A., et al.: Hazardous alcohol drinking and premature mortality in Russia: a population based case-control study. Lancet **369**(9578), 2001–2009 (2007). https://doi.org/10.1016/S0140-6736(07)60941-6
16. ROSSTAT Order No. 410 of 16.10.2013 "On the approval of statistical tools for the organization by the Ministry of Health of the Russian Federation of federal statistical monitoring of the incidence of drug-related disorders in the population". http://www.consultant.ru/document/cons_doc_LAW_153560/. (in Russian)
17. Lindenbraten, A.L., Golubev, N.A., Afonina, M.A.: The evaluation of direct economic loss related to tobacco smoking in 2009 and 2016. Prob. Soc. Hygiene Public Health Hist. Med. **27** (4), 363–368 (2019). https://doi.org/10.32687/0869-866X-2019-27-4-363-368. (in Russian)
18. Decree of the Government of the Russian Federation of September 24, 2001 N 1270-r "The Concept of Demographic Development of the Russian Federation for the period until 2015". http://www.consultant.ru/document/cons_doc_LAW_98526/4094ed4b52a84a8d9cd2fefb5f 6799fc0c1d6e6e/
19. Federal State Statistics Service (ROSSTAT) [electronic resource]. http://www.gks.ru/free_ doc/new_site/population/demo/demo26.xls. (in Russian)

20. Ivanova, A.E., Sabgayda, T.P., Semenova, V.G., Zaporozhchenko, V.G., Zemlyanova, E.V., Nikitina, S.Yu.: Factors distorting structure of death causes in working population in Russia. Soc. Aspects Public Health [Electron. Sci. J.] **32**(4) (2013). http://vestnik.mednet.ru/content/view/491/27/lang,ru/. (in Russian)

21. Buzin, V.N., Mikhaylova, Yu.V., Buzina, T.S., Chuhrienko, I.Yu., Shikina, I.B., Mikhaylov, A.Yu.: Russian healthcare through the eyes of the population: dynamics of satisfaction over the past 14 years (2006—2019): review of sociological studies. Russ. J. Prev. Med. **23**(3), 42–47 (2020). https://doi.org/10.17116/profmed20202303142. (in Russian)

22. Voskanyan, Y., Shikina, I., Kidalov, F., Davidov, D.: Medical care safety - problems and perspectives. In: Antipova, T. (ed.) ICIS 2019. LNNS, vol. 78, pp. 291–304. Springer, Cham (2020). https://doi.org/10.1007/978-3-030-22493-6_26

23. Kalininskaia, A.A., Son, I.M., Shliafer, S.I.: The problems and perspectives of development of rural health care. Prob. Soc. Hygiene Public Health Hist. Med. **27**(2), 152–157 (2019). https://doi.org/10.32687/0869-866X-2019-27-2-152-157. (in Russian)

24. Topilin, M.A.: On the problems of demographic policy. Transcript conference on the implementation of the Demographic politicians». Government House, Moscow, 10 June 2015. (in Russian)

25. Voskanyan, Y., Shikina, I., Kidalov, F., Andreeva, O., Makhovskaya, T.: Impact of macro factors on effectiveness of implementation of medical care safety management system. In: Antipova, T. (ed.) ICIS 2020. LNNS, vol. 136, pp. 346–355. Springer, Cham (2021). https://doi.org/10.1007/978-3-030-49264-9_31

26. Shurgaya, M.A., Memetov, S.S., Ivanova, T.A., Karaeva, A.F., Lyalina, I.V.: Demographic situation in Russia: medical and social aspects. Med. Soc. Expert Eval. Rehabil. **20**(4), 214–220 (2017). https://doi.org/10.18821/1560-9537-2017-20-4-214-220. (in Russian)

# Digital Public Administration

# Analysis Website Quality Official Government Tweet Accounts to Campaign for Tourism Sites in the Lampung Area

Ningsih Wirandari[1] , Achmad Nurmandi[1]([⊠]) ,
Isnaini Muallidin[1] , Danang Kurniawan[1] , and Salahudin[2]

[1] Department of Government Affairs and Administration, Jusuf Kalla School of Government, Universitas Muhammadiyah Yogyakarta, Yogyakarta, Indonesia
[2] Department of Government Science, Universitas Muhammadiyah Malang, Indonesia, Tlogomas Street, Tegalgondo, Karang Ploso Malang, Indonesia
salahudinmsi@umm.ac.id

**Abstract.** This study aims to determine the quality of government websites in applying the principles of good governance and to analyze the official social media Twitter accounts of the Regional Government in conveying information to the public. The research object is the tourism and cultural official website and Twitter account of the Provincial Governments of Lampung the which one site way kambas. Sources of research data through surveys and observations. Website quality assessment is done by giving a score on indicators of information quality, service quality, perceived website quality.

In the research use method deskriptif qualitative. Twitter content is analyzed using the tools in Nvivo 12 Plus with the NCapture feature to get coding to document and analyze data systematically. The results showed that the overall website of the @waykambas has (High Quality) with a score above (70%). The @pemprov_Lampung website has the lowest score, with 65% (Low Quality). Twitter social media for the @waykambas and @pempro_Lampung has a high level of information delivery activity. The issues surrounding event tourism and culture were discussed on two local government Twitter accounts. Problems that often develop in these three regional governments have also resulted in increased Twitter account activity.

**Keywords:** E-government · Official Website · Tourism and cultural

## 1 Introduction

### 1.1 Background

Website government is an important promotional tool for a tourist destination as it is an information channel to introduce all potential tourists as the basis for the decision and selection of travel destinations [1]. According to [2], promotion using the website is a necessity for developing countries to help boost tourism. Indonesia as a developing country, in the last period, has been increasingly promoting tourism. [3] Lampung Province, East Lampung, for example, is a favorite tourism destination for both

T. Antipova (Ed.): DSIC 2021, LNNS 381, pp. 479–489, 2022.
https://doi.org/10.1007/978-3-030-93677-8_42

domestic and foreign tourists besides Bali. However, East Lampung as one of the regencies in Lampung Province also has tourism potential that is no less interesting. The most famous tourism potential in East Lampung Regency is WayKambas. Based on visit data for five holidays (16–20/6/2018), the number of visitors to TNWK is 22,987 people, with the number of four-wheeled vehicles (2,789 units) and two-wheeled vehicles (2,446 units). This figure has increased annually, this indicates that the @wayKambas has increased the level of visits per year. The promotions that have been carried out by the Department of Culture and Tourism of East Lampung Regency as the leading sector include books, leaflets, booklets, travel dialogue, and websites. The Lampung province tourism website can be accessed at the address https://dinaspariwisata.lampungprov.go.id//. However, promotion using the website is less than optimal and does not gain user trust, as can be seen from the level of user visits to the website of @Disbudpar, East Lampung Regency, which does not always increase per month and is only visited 872 times per month [3]. This is where the role of the government as the leading sector, should be able to act as a forum, spokesperson and marketing to promote existing tourism.

The government must be able to capture the development of information and communication technology (ICT) to disseminate this information to the public, one of which is through a Twitter account. As stated by [4], if the user believes that the Twitter account website is trusted, it will affect the user's intention to adopt the information and engage voluntarily in sharing the information with their relatives. This is the working principle of viral marketing. Viral marketing is one of the effective marketing efforts using the internet network where customers try to get someone to voluntarily convey marketing messages to others [5]. According to [6], there are seven main ways of communication in promoting a product called the promotional mix. Each has advantages and disadvantages. First, advertising. Second, sales. Third, public relations. Fourth, direct marketing. Fifth, personal selling. Sixth, viral marketing. Seventh, event marketing [7] of the several promotional tools above, viral marketing can be used as a reference for promotion in tourism. Besides being cheap, the information that is disseminated can be spread quickly because it uses social networks in disseminating information. According to data from the Ministry of Communication and Information tourism government, Indonesia's internet users amounted to 73.7%, up from 64.8% from 2018 in Indonesia, 2019 there were 266,911,900 million so that Indonesian internet users are estimated at 196.7 million. user. This number is up from 171 million with a penetration of 73.7% or an increase of about 8.9% or about 25.5 million users. Indonesia is ranked as the 5th largest Twitter user in the world. The use of viral marketing by utilizing social networks is indeed reliable, such as the results of research [8, 9]. With the internet media, tourism will have a greater opportunity to reach people around the world at no additional cost. To further optimize promotion by utilizing ICT, it is necessary to build a promotion model using a viral marketing approach. This study was conducted to determine the factors that influence the level of trust of users of the tourism Twitter account website in East Lampung Regency. By knowing these factors, it can be used as a reference in building a government website as a promotional medium for increasing tourism. This is because the level of trust is considered to be able to encourage the realization of viral marketing.

## 2   Literature Review

### 2.1   Website Used in Government

Many studies show viral marketing is proven to be an effective marketing tool on social networks. However, few researchers have investigated from the point of view of government websites or portals [10]. According to [11], the important thing that must be considered in growing tourism in developing countries is to focus on assessing the interactivity, trustworthiness, value aspects, and information aspects of the tourism website itself. Technology in government is known as E-Government, where technology plays a role in facilitating the management of information and services [12]. E-government emerged due to Information and Communication Technology (ICT) development to provide government services to stakeholders. Through today's rapidly evolving world, ICTs enable governments to deliver services more efficiently by facilitating sharing information across the globe. ICT provides access to information such as the internet, digital or wireless networks, and other systems [13]. There are three components of e-government—first, ICT, which includes the internet, web-based applications, devices, and mobile; second, stakeholders and consumers of electronic services [14]. The third is the yield component. These three components can increase the interaction between stakeholders and empower the community through transparent access to information to increase the efficiency and effectiveness of government administration [15]. With the growth of the internet, the government's public service model is gradually changing from the traditional manual method to an online e-government website [16]. E-Government services are usually provided through Web portals developed and managed by the government [17]. Government use of websites can increase transparency, foster trust in government, reduce corruption in the public sector, and strengthen accountability and transparency [18]. Government web portals have different functions. Some of them focus on access to information. Access to integrated and open government and citizen electronic participation transactions is also provided through governmentwide access to data, electronic petitions, electronic consultations, etc. [19]. Through the website, the government can provide services according to the needs of the community. Sometimes people use it to get information and exchange information. Some use it for transactions. Therefore, the government's completeness of service information needs to be considered by government [20] Information Quality can be used to predict information adoption [21] and is a determinant of user trust in commercial sites [22]. In the context of this study, the quality of information has a positive effect on the trustworthiness of tourism websites. Service Quality Service quality is a person's assessment or perception of the quality of services provided on the website. According to Iliachenko [23], service quality in tourism web promotion is seen from interactivity, design, information, and technical features. Service quality will have a positive effect on the trustworthiness of tourism websites. Perceived Website Quality According to [24], website quality refers to consumer perceptions of website performance in searching and sending information. Perceived website quality has a positive effect on tourism website trust. User Satisfaction with Previous Experience [25] shows that user satisfaction is measured by user satisfaction in general based on all cumulative customer experiences with companies, products, or services. Based on this, the researcher assumes that consumers who are satisfied with their previous experience in using the website will trust the website more. User Experience and Ability According to [26], the level of user experience using the internet can be taken into account as an antecedent of trust. More knowledge and experience of using the internet can spur greater confidence in using it, which will increase online trust [27]. Trust in the Website value of user trust is seen from the value and amount of information generated from the website [28]. In this study, it is expected that the user's trust in the website can influence the user's intention to adopt the information received and be involved in sharing the information with others. Information adoption can be defined as a process by which people are intentionally involved in using information [29]. Filieri's research [30] shows that trust in websites can be used to predict user intentions to adopt information and word of mouth.

Viral Marketing Viral means a virus. In viral marketing, the message conveyed develops through the internet network, like a computer virus that duplicates itself more and more. Someone buys something not necessarily because of the influence of that marketing activity but because of hearing 'positive news' from other independent sources, this is where viral marketing works [31].

**Tourism Website of Lampung Province**

Department of Culture and Tourism of East Lampung Regency is an agency whose main tasks and functions are to explore, preserve, and develop a regional culture in the Lampung region. One of the promotions that have been carried out is through a website that can be accessed at the address https://dinaspariwisata.lampungprov.go.id/. From statistical data, the number of news displayed on the website in 2018, on average only displays 9 news per month (*Lampung Tourism*, nd). This shows that the department of culture and tourism of east Lampung regency is lacking in displaying and disseminating information.

## 2.2    Government Social Media (Twitter Account)

Social media is seen as a means of interaction and provides more opportunities for the public to interact with government functions. Social media today offers significant benefits in shaping society and relationships with the government. Users can easily participate, share, interact, discuss, collaborate, and create content with various platforms such as YouTube, Facebook, WhatsApp, Instagram, Twitter, and others [32]. It is limited to the relationship between policymakers and policymakers and new interactions that can be mutually beneficial through social media [33]. Therefore, in theory, the current use of ICT has given birth to new constructions in government and society [34]. The use of social media in government tends to have a significant impact on improving public services. This increase can encourage public participation. Information on government programs can be conveyed through social media [35]. As one of the social media, the government has used Twitter to convey information quickly, such as disaster preparedness, and get a public response [23, 36]. Through microblogging, a social network, Twitter users can quickly read, respond, and send messages called tweets [37]. Twitter's existence has been widely accepted for providing direct and broad access to many of the latest news operations while still helping to spread incident reporting [38].

## 3   Method

This research uses qualitative research methods in [39], stating that this approach helps researchers investigate a phenomenon so that it is better understood and used in a limited sense to analyze an event. The focus of this research is the analysis of the city government's website and Twitter account. The city's official Twitter accounts were analyzed using NVivo 12 Plus to coding data with the NCapture to get count coding data function to record and analyze information systematically [40]. To be studied and researched carefully and thoroughly. By visiting the City Government website, data were obtained. Website output calculation using an instrument provided by Google, Page Speed Insights. On a percentage scale, website loading speed is rated. Through evaluation using scoring, the efficiency of local government websites, like @dinas-pariwisatalampung @Keminfo, @waykambas. Score analysis uses indicators to evaluate policy changes developed in line with the developed objectives and, where appropriate, by weighting indicators deemed more relevant than other indicators [41]. A specific score will be given concerning the analysis of the evaluation scale in calculating the comparative rating [30] information quality, service quality, perceived website quality, user satisfaction with previous experiences, perceived website trust, and Viral marketing. If the information is available/available, will be given a score of 1 if the data is not available/not available, then a 0 will be given if the data is not available/not available. The formula for the estimated weight score is as follows (Table 1):

$$\text{Score Weight} = \frac{\text{Total website score} \times 100\%}{\text{Total score}}$$

**Table 1.** Model

| Score (%) | Information quality | Service quality | Perceived website quality | Satisfaction | Website trust | Viral marketing |
|---|---|---|---|---|---|---|
| 0–25 | | | | | | |
| 51–75 | | | | | | |
| 26–50 | | | | | | |
| 76–100 | | | | | | |

Local government website access speed is also calculated using Google Tools, namely Page Speed Insights. Website loading speed is rated on a percentage scale, slow (0%-49%), average (50%-89%), and fast (90%-100%). Meanwhile, Twitter content was analyzed using Nvivo 12 Plus with the NCapture feature to systematically document and analyze the data [42].

## 4   Result

Based on the assessment and research on local government websites, namely, the provinces of Lampung were assessed by 40 review units. Evaluation is done by giving a score on information quality, service quality, perceived website quality, user satisfaction with previous experiences, perceived website trust, and Viral marketing. The convenience of the public in accessing the website needs to be considered by the city and regional governments because the official website has a vital role in communicating or conveying information to the public. Data from Google shows that 53% of people in Indonesia will leave a website that takes more than 3 s to load. The website speed of the three account twitter in Lampung Timur can be seen in the following Fig. 1:

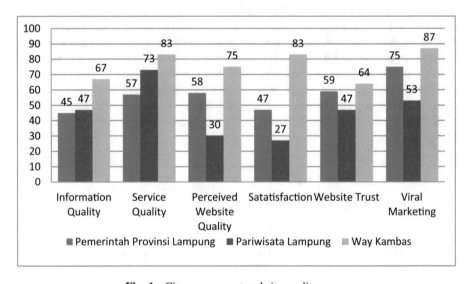

**Fig. 1.** City government website quality score

The pig above shows the way kambas website obtained the highest score with a percentage value of 89%. While the website of @waykambas, 75%, has the lowest score with a percentage value of 43% is @dinaspariwisatalampungprov. The account waykambas.org website is in the high-quality category, while the other two government websites are still in the low category.

### 4.1   Use of Local Government Social Media (Twitter Accounts)

Account waykambas.org must socialize project and work policies to the community in absolute terms and respond directly. Twitter has been used as a medium for contact and interaction with groups by Province Lampung area. The official Twitter accounts of website city governments in Lampung Timur Province are as follows (Table 2).

**Table 2.** Website government Twitter account

| No. | Website government | Twitter account |
|-----|--------------------|-----------------|
| 1. | Waykambas.org | @tnwaykambas |
| 2. | Waykambas | @w_kambaslampung |
| 3. | Dinaspariwisatalampungprov.go.id | @VISIT_LAMPUNG |

Only waykambas has a piece of much information shared or is checked by Twitter from the three official Twitter of the regional governments studied, namely, @tnwaykambas. Every Tweet from the provincial government provides various information related to the event and related problems in each of these areas. The results of the NVivo 12 Plus study using Word's cloud function define a publicly known issue. The issues that emerged on the provincial government's Twitter can be seen in this picture (Fig. 2).

**Fig. 2.** Wordcloud for website quality in Lampung area

The image above illustrates that the website government's Twitter discusses a lot about tourism provincial government is more about providing information to the public, namely issues of activities, work programs, and policies in running the wheels of the government tourism area. Issues that play a role include government-community cooperation, development, tour and travel marketing, education, and religion. Issues that are development marketing in tourism and travel also increase the provincial government's Twitter activity. This increase proves that the tourist is actively working on various kinds of activities and other activities. This has a terrible impact on tourism management and marketing development on the government in increasing public trust in performance and information disclosure in the realization of Good Governance.

## 5   Conclusion

Based on the results of data processing and analysis that has been carried out in this study, it can be concluded that the measurement of the quality of social media websites, namely the @WayKambas tourism Twitter account, Central Lampung district, can be measured from the quality of the website produced. Meanwhile, the research also has weaknesses like the limited knowledge of the researcher to analyze the detail in the source. These results can be seen where users have felt the ease of using the website, the existence of good links, the searching menu, loading speed, and the security in operating the website. Meanwhile, the measurement of information on the @WayKambas tourism website can be measured from the information output produced, where users feel that the information displayed on the website is considered useful, valuable, and credible. With these two factors, namely the quality of the website and the quality of information, it is proven that it can affect the level of user confidence in the website of the Lampung Tourism Office. Meanwhile, based on research results, service quality, user satisfaction, and user experience and capabilities have not been proven to increase user trust in tourism websites. User trust in tourism websites is proven to increase users' intention to adopt tourism information and encourage users to carry out viral marketing activities.

## References

1. Chung, J.: Effect of quality uncertainty, regulatory focus, and promotional strategies on perceived savings for sustainable marketing. Sustain. **12**(14) (2020). https://doi.org/10.3390/su12145653
2. Shabbir, M.S., Sulaiman, M.A.B.A., Al-Kumaim, N.H., Mahmood, A., Abbas, M.: Green marketing approaches and their impact on consumer behavior towards the environment—a study from the UAE. Sustain. **12**(21), 1–13 (2020). https://doi.org/10.3390/su12218977
3. way kambas: https://waykambas.org/
4. Yao, W., Qian, S.: From Twitter to traffic predictor: Next-day morning traffic prediction using social media data. Transp. Res. Part C Emerg. Technol. **124**(August 2020), 102938 (2021). https://doi.org/10.1016/j.trc.2020.102938
5. Katare, J., Banerjee, S.: Index for comparative assessment of municipal websites. Electron. Gov. **13**(1), 49–68 (2017). https://doi.org/10.1504/EG.2017.083942
6. Paulo, M.: The role of E-Governance in Europe's image of the Chinese communist party. Int. Commun. Gaz. **78**(1–2), 39–63 (2016). https://doi.org/10.1177/1748048515618105
7. Scott, D.M.: The New Rules of Marketing and PR: How to Use Social Media, Online Video, Mobile Applications, Blogs, News Releases, and Viral Marketing to Reach Buyers (2015)
8. Liu, F., Liu, C., Zhao, Q., He, C.: A hybrid teaching-learning-based optimization algorithm for the travel route optimization problem alongside the urban railway line. Sustain. **13**(3), 1–17 (2021). https://doi.org/10.3390/su13031408
9. Baumöl, U., Hollebeek, L., Jung, R.: Dynamics of customer interaction on social media platforms. Electron. Mark. (2016). https://doi.org/10.1007/s12525-016-0227-0
10. Oberoi, R.: Institutionalizing corporate social responsibility: a study of provisions and implications of Indian companies act 2013. In: Developments in Corporate Governance and

Responsibility, vol. 24, pp. 165–187. Emerald Group Publishing Ltd., University of Delhi, India (2018). https://doi.org/10.1108/S2043-052320180000014008

11. Khanra, S., Joseph, R.P.: E-Governance maturity models: a meta-ethnographic study. Int. Technol. Manag. Rev. **8**(1), 1 (2019). https://doi.org/10.2991/time.b.190417.001

12. Arfian, A.: Analisis Manfaat E-agricultural Mengunakan Metode Sempls (Study Kasus Limakilo. Id), INTI Nusa Mandiri (2020). http://ejournal.nusamandiri.ac.id/index.php/inti/article/view/1142

13. Gberevbie, D., Ayo, C., Iyoha, F., Ojeka, S., Abasilim, U.: E-Governance and accountability: towards achieving the goals of public agencies in Nigeria. In: Proceedings of the European Conference on e-Government, ECEG, vol. 2016-Janua, pp. 71–77 (2016). https://www.scopus.com/inward/record.uri?eid=2-s2.0-84979600224&partnerID=40&md5=e7d024a28a799680ea4b407e1ad9504e

14. Homburg, V.: ICT, E-Government and E-Governance: bits & bytes for public administration. In: Ongaro, E., Van Thiel, S. (eds.) The Palgrave Handbook of Public Administration and Management in Europe, pp. 347–361. Palgrave Macmillan, London (2018). https://doi.org/10.1057/978-1-137-55269-3_18

15. Suryanarayanan, P., et al.: AI-assisted tracking of worldwide non-pharmaceutical interventions for COVID-19. Sci. Data **8**(1), 1–14 (2021). https://doi.org/10.1038/s41597-021-00878-y

16. Ibragimov, I., Golubev, V., Balabanova, S., Filatova, O.: Open government data as a tool for cooperation between people and government: a case study of open data and e-governance resources in the Eurasian Economic Union. In: 2017 International Conference on Internet and Modern Society, pp. 195–199. IMS 2017 (2017). https://doi.org/10.1145/3143699.3143702

17. Naurin, E., Royed, T.J., Thomson, R.: Party Mandates and Democracy: Making, Breaking, and Keeping Election Pledges in Twelve Countries (2019)

18. Meijer, A.: E-Governance innovation: barriers and strategies. Gov. Inf. Q. **32**(2), 198–206 (2015). https://doi.org/10.1016/j.giq.2015.01.001

19. Santos, C., Ferreira, A., Marques, R.P., Azevedo, G.: EAGLE_Index: enhancement of an accountability guide for learning E-Government. In: Handbook of Research on Modernization and Accountability in Public Sector Management, pp. 103–129. University of Aveiro, IGI Global, Portugal (2018)

20. Sanmukhiya, C.: E-Governance dimensions in the republic of Mauritius. Humanit. Soc. Sci. Rev. **7**(5), 264–279 (2019). https://doi.org/10.18510/hssr.2019.7532

21. Kumar, V., Jenamani, M.: Identification and Prioritization of urban issues from Smart City data (2018). https://doi.org/10.1109/ICSCET.2018.8537360

22. Ariyanto, R., Rohadi, E., Lestari, V.A.: The effect of information quality, system quality, service quality on intention to use and user satisfaction, and their effect on net benefits primary care application at primary health facilities in Malang. IOP Conf. Ser. Mater. Sci. Eng. **732**(1) (2020). https://doi.org/10.1088/1757-899X/732/1/012084

23. Schivinski, B., Dabrowski, D.: The effect of social media communication on consumer perceptions of brands. J. Mark. Commun. (2016). https://www.tandfonline.com, https://doi.org/10.1080/13527266.2013.871323

24. Fajriansyah, A., Iskandarsyah, A., Puspitasari, I.M., Lestari, K.: Impact of pharmacist counseling on health-related quality of life of patients with type 2 diabetes mellitus: a cluster-randomized controlled study. J. Diabetes Metab. Disord. **19**(2), 675–682 (2020). https://doi.org/10.1007/s40200-020-00528-x

25. Karimi, S., Naghibi, H.S.: Social media marketing (SMM) strategies for small to medium enterprises (SMEs). Int. J. Inf. (2015). researchgate.net, https://www.researchgate.net/profile/Ahsan_Akbar/publication/346084844_The_Role_of_Online_Service_Quality_in_Enhancing_Customer_Satisfaction_An_Empirical_Investigation_of_Pakistani_Banks/links/5fbab558458515b797626bdd/The-Role-of-Online-Service-Quality-in

26. Tute, E., Scheffner, I., Marschollek, M.: A method for interoperable knowledge-based data quality assessment. BMC Med. Inform. Decis. Mak. **21**(1), 1–14 (2021). https://doi.org/10.1186/s12911-021-01458-1

27. Nunes, R.H., Ferreira, J.B., Freitas, A.S.D., Ramos, F.L.: The effects of social media opinion leaders' recommendations on followers' intention to buy. Revista Brasileira de …. core. ac. UK (2018). https://core.ac.uk/download/pdf/190596196.pdf

28. Toubal, E.B., Belkhir, A., Rahim, M., Kheldoun, A., Boudjebbour, K.: A web services-based secure platform for inter-institutions E-Governance (2019). https://doi.org/10.1109/ICASS.2018.8651963

29. Zraick, R.I., Azios, M., Handley, M.M., Bellon-Harn, M.L., Manchaiah, V.: Quality and readability of internet information about stuttering. J. Fluency Disord. **67**(October 2020), 105824 (2021). https://doi.org/10.1016/j.jfludis.2020.105824

30. Filieri, R., Alguezaui, S., McLeay, F.: Why do travelers trust TripAdvisor? Antecedents of trust towards consumer-generated media and its influence on recommendation adoption and word of mouth. Tour. Manag. **51**, 174–185 (2015). https://doi.org/10.1016/j.tourman.2015.05.007

31. Ayu, J.N.R.: Efektifitas viral marketing dalam meningkatkan niat dan keputusan pembelian konsumen di era digital. J. Manaj. Inov. **2**(1) (2020). https://doi.org/10.15642/manova.v2i1.353

32. Kimmons, R., Rosenberg, J., Allman, B.: Trends in educational technology: what facebook, twitter, and Scopus can tell us about current research and practice. TechTrends **65**(2), 125–136 (2021). https://doi.org/10.1007/s11528-021-00589-6

33. Kim, S.Y., Ganesan, K., Dickens, P., Panda, S.: Public sentiment toward solar energy—opinion mining of Twitter using a transformer-based language model. Sustain. **13**(5), 1–19 (2021). https://doi.org/10.3390/su13052673

34. Nunes-Da-Cunha, I., Martinez, F.M., Fernandez-Limos, F.: A global comparison of internationalization support characteristics available on the college of pharmacy websites. Am. J. Pharm. Educ. **83**(3), 393–400 (2019). https://doi.org/10.5688/ajpe6592

35. Lutzky, U.: You keep saying you are sorry. Exploring the use of sorry in customer communication on Twitter. Discourse Context Media **39** (2021). https://doi.org/10.1016/j.dcm.2020.100463

36. Masood, M.A., Abbasi, R.A.: Using graph embedding and machine learning to identify rebels on Twitter. J. Informetr. **15**(1), 101121 (2021). https://doi.org/10.1016/j.joi.2020.101121

37. Lee-Geiller, S., Lee, T.D.: Using government websites to enhance democratic E-Governance: a conceptual model for evaluation. Gov. Inf. Q. **36**(2), 208–225 (2019). https://doi.org/10.1016/j.giq.2019.01.003

38. Shahi, G.K., Dirkson, A., Majchrzak, T.A.: An exploratory study of COVID-19 misinformation on Twitter. *arXiv*, vol. 22, no. May 2020, p. 100104 (2020). https://doi.org/10.1016/j.osnem.2020.100104

39. Sinclair, A.J., Peirson-Smith, T.J., Borchers, M.: Environmental assessments in the Internet age: the role of E-Governance and social media in creating platforms for meaningful participation. Impact Assess. Proj. Apprais. **35**(2), 148–157 (2017). https://doi.org/10.1080/14615517.2016.1251697

40. Sumikawa, Y., Jatowt, A.: Analyzing history-related posts in twitter. Int. J. Digit. Libr. **22** (1), 105–134 (2020). https://doi.org/10.1007/s00799-020-00296-2
41. Alloghani, M.A.M.A., Hussain, A., Al-Jumeily, D., Aljaaf, A.J., Mustafina, J.: Gamification in E-Governance: development of an online gamified system to enhance government entities services delivery and promote public's awareness. In: ACM International Conference Proceeding Series, pp. 176–181 (2017). https://doi.org/10.1145/3029387.3029388
42. Athwal, N., Istanbulluoglu, D.: The allure of luxury brands' social media activities: uses and gratifications perspective. Inf. Technol. (2019). https://www.emerald.com/insight/content/, https://doi.org/10.1108/ITP-01-2018-0017/full/html

# Improving the Efficiency of the Pension System in the Digital Economy

Larisa Drobyshevskaya⬭, Elena Mamiy⬭, Alexander Chulkov⬭,
and Alina Timchenko^(✉) ⬭

Kuban State University, 149, Stavropolskaya st., 350040 Krasnodar, Russia

**Abstract.** The research focuses on the topical problem of the pension system development. The role of digitalization in economic development is revealed. The development of the digital economy and information space imposes new requirements on the level of transparency, accessibility and simplicity of the tax system. The study identifies the main problems of the development of the Russian Federation pension system. In the modern world, risks affecting the national pension system stability are exacerbated by the impact of global trends, problems and threats. COVID-19 has hit hard labor markets and pension mechanisms around the world. The essence of the risks for functioning of the pension system is revealed. In order to generalize the risks, affecting the Russian pension system, a route map of risks has been drawn up. A comparative assessment of the pension systems of Russia and the OECD countries performance was carried out. The analysis of the Melbourne Mercer Global Pension Index (MMGI) 2019 was made. In the course of the comparative analysis of MMGI and the author's research, the most effective national pension systems were identified. The directions of the entire pension system digital matrix formation and introduction of a digital platform, open and accessible, convenient and understandable for all its participants, have been substantiated.

**Keywords:** Pension system · Efficiency · Risks · Digital platform · Melbourne mercer global pension index

## 1 Introduction

**Goals and Objectives.** The purpose of the study is to form the main directions for the development of the pension system in the digital ecosystem, including the development of a system of measures, aimed at improving the efficiency of its functioning.

**Methodology.** In the course of the study the authors made use of dialectical and formal logic, systemic and structural-functional analysis, the method of expert assessments as well as the results of research by domestic and foreign economists.

**Results.** A risk map of the pension system of the Russian Federation has been developed. The key risks affecting the development of the pension system of the Russian Federation are highlighted. A system of measures aimed at increasing the effectiveness of the pension system of the Russian Federation is developed. **Conclusions.** A study of the effectiveness of the pension systems of the Russian Federation

T. Antipova (Ed.): DSIC 2021, LNNS 381, pp. 490–502, 2022.
https://doi.org/10.1007/978-3-030-93677-8_43

and the OECD countries has been carried out. As a result of a comparative analysis of the Melbourne Mercer Global Pension Index (MMGI) and the author's methodology, the most effective national pension systems have been identified. On the example of Denmark, Canada and the Netherlands, directions of increasing the value of MMGI are revealed.

## 2 Pension System: Problems and Efficiency

### 2.1 Main Problems and Risks of the Pension System

In the modern conditions, digitalization has a significant impact on the economies of all countries in the world. Big data, artificial intelligence, internet of things, machine learning, robotics, neural networks, blockchain, virtual reality and many others contribute to the integration of all data streams into the digital ecosystem. The fourth industrial revolution, based on the integration of physical and digital systems, will allow in the near future not only to ensure the autonomy of all production processes, but also make a breakthrough "from genetic engineering to nanotechnology, from programmable robots to artificial intelligence, from renewable energy sources to quantum computers" [1]. Industry 4.0 is developing exponentially, rapidly changing social and labor relations and labor markets, and has an impact on social protection systems.

The pension system stands as an important element of social protection of the population. The issue of maintaining a balance between social justice and economic efficiency is regularly raised within the world community, and each country is trying to find the most "correct", in terms of national characteristics, regulatory mechanisms. The pension system is a dynamic structure that must be flexible and adaptable to constantly changing conditions (both internal and external) in order to ensure sustainable development.

The development of pension systems is studied by such scientists as: Kravchenko E., Kuklin A. [2, 3]. The works of Yakushev E., Chervyakova A., Peris-Ortiz M., Alvarez-Garcia J., Dominguez-Fabian I., Devolder, P. [4, 5]. Features of the impact of demographic processes on the pension system are considered in the works of Nepp A., Okrah J. etc. [6]. Stewart F., Fabozzi F., Geddes T., Gratson J., Curguz I., Andreeva O., Kozlov P., Slepov V. are studying the risks of the pension system [7–11].

In the modern conditions, the fundamental problem of the Russian pension system is the long-term budget deficit of the Pension Fund of Russia (PFR) [12]. The gradual reduction of the deficit requires more attention from the state in order to maintain the balance of the system. Further prevalence of expenditures will reduce the effectiveness of the pension system, making it more and more burdensome for the state budget. Moreover, the extremely low replacement rate indicates the inefficiency of both, individual subsystems and the pension system as a whole, creating a threat to socio-economic security, the prevention and timely elimination of which is a key area of government activity.

In today's globalized world, risks affecting the stability of the national pension system are exacerbated by the impact of global trends, problems and threats. For example, COVID-19 has dealt an unprecedented blow to labor markets and pension mechanisms around the world. Even before the outbreak of the pandemic, pension savings and old-age pension systems faced serious problems all around the world, as well as in Russia. Population aging with increasing life expectancy to finance pensions and fewer people entering the labor market, as well as low economic and wage growth, low returns on traditional asset classes and low interest rates have already put serious pressure on accumulative and distributive private and public pension systems. COVID-19 exacerbates some of these problems and adds new ones. According to the results of research by O. Sinyavskaya, the number of insured persons will become equal to the number of recipients of insurance pensions already in 2022–2024, and starting from 2038–2041 the share of insurance pensioners in the total population of 18 years and older (potential voters) will exceed 40% [13].

In addition to the likelihood that economic growth, interest rates and yields will remain low for a long time into the future, the health and economic crisis increases the risk that people will not be able to accumulate enough funds for retirement. There will be an increase in pressure on public finances, already strained by demographic changes. Business disruptions due to teleworking, cyber-attacks, fraud and scams, as well as use of retirement assets to support economic recovery are all additional challenges that need to be addressed. Moreover, initiatives to provide short-term assistance by providing people with access to their retirement savings before they reach the retirement age are likely to have a negative impact on future retirement income, especially if access is widely available. The retirement savings mechanisms can be more resilient and address COVID-19-related early withdrawals if long-term savings mechanisms include both a retirement savings account and an emergency savings account.

We also note one of the tendencies of modern realities, aggravated during the pandemic - the ever-increasing number of people, involved in non-standard forms of employment. The sustainability and adequacy of pension systems includes ensuring that workers in precarious forms of employment are able to save for their retirement. This diverse population group, including part-time workers, temporary workers, self-employed and informal workers, has more limited access to public and private retirement programs and accumulates fewer pension entitlements, than permanent full-time workers. Retirement planners around the world need to consider targeted measures, including facilitating access to retirement plans, offering special or hybrid retirement savings products, enabling workers to maintain the same retirement plan when changing jobs, allowing flexible contributions, and promoting responsible attitude of citizens to their savings for retirement.

The Organization for Economic Community and Development (OECD) is currently revising the OECD Roadmap for the Proper Design of Defined Contribution Pension Plans to update its recommendations [14]. Such plans provide people with a choice. For example, they can choose the preferred investment strategy for placing their retirement savings. However it should be taken into account, when designing investment strategies, that some people may be unable or unwilling to make choices and choose the default investment strategies that protect them. People may end up using the default strategy or choosing a different investment strategy; however, it requires clear and consistent communication, which presents compromise solutions to people with trade-offs according to their risk profile and their level of risk tolerance, as well as their different mechanisms and goals for earning income in retirement.

People who save money for their retirement face the risk of longevity in addition to the investment risk. Sharing these risks among stakeholders increases the sustainability and viability of pension savings mechanisms [link]. For risk sharing to be sustainable, it is important to have a regulatory framework that supports the goal of equity in value transfer and the continuity of the pension arrangement through minimum funding requirements.

To identify the risks that may affect the Russian pension system, based on ISO 3100 and the works of Stewart F., Fabozzi F., Okhrin O., the authors compiled a risk map (see Fig. 1).

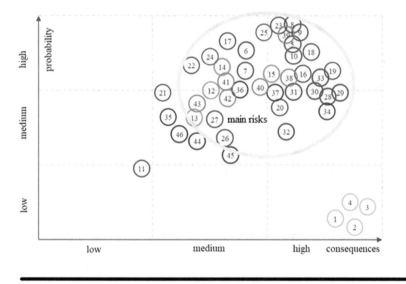

**Fig. 1.** Map of risks of the pension system of the Russian Federation.

All risks of the pension system can be divided into external (non-systemic) and internal (systemic). In its turn, external risks, or in other words, exogenous are subdivided into: political, demographic, social, legal, financial, macroeconomic and digitalization risks, and internal or endogenous into: PFR risks and labor risks. Political risks include: 1. Changes in the political course in the country; 2. Change of state priorities; 3. Coup d'état; 4. Military hostilities. Demographic: 5. Increase in the share of pensioners in the structure of the population; 6. Increasing the life expectancy of the population; 7. Decrease in the share of working pensioners; 8. Reducing the number of working-age population; 9. Decrease in the birth rate; 10. Negative natural population growth; 11. Violation of the sex and age structure. Legal risks include: 12. Changes in legislation; 13. The presence of gaps and contradictions in the laws. Social risks are: 14. Low level of financial literacy of the population and 15. Low level of trust in the existing pension system among the population. Macroeconomic risks include: 16. Decrease in GDP; 17. Decrease in labor productivity; 18. Decrease in incomes of the population; 19. Growth of unemployment; 20. Shadow economy; 21. Poor development of the real sector of the economy; 22. Changes in the economic environment; 23. Changes in oil prices; 24. Change in interest rates; 25. Changes in exchange rates; 26. Changes in the volume of investments; 27. Change in production volumes. Financial risks are subdivided into: 29. Inflation; 30. Growth of the federal budget deficit; 31. Reducing the size of the National Welfare Fund; 32. Limited supply of financial and savings instruments; 33. Decrease in the purchasing power of the national currency. The last group of external risks is digitalization risks, they include: 34. Structural changes in the economy (shift in capital and labor force); 35. Low level of information culture; 36. Risks when working with data; 37. Release of labor resources. Risks of the RF Pension Fund: 38. Low of collection of insurance premiums; 39. Growth of the PFR deficit; 40. Lack of balance in the budgets of pension funds; 41. Investment risks; 42. Information risks; 43. Risks of long-term forecasting errors. Labor risks include: 44. Risk of job loss (reduction); 45. Changes in pension legislation (increasing the retirement age, changing the method of calculating pensions, coefficients, points); 46. Risk of survival (death) until retirement age.

The identified highlighted risks can adversely affect the pension system of Russia and its financial security, causing a deterioration in indicators that determine its effectiveness, namely: the ratio of persons of retirement age to working age, replacement rate, demographic load, life expectancy after retirement, government spending on financing pensions etc.

In addition, it should be noted that a high proportion of "black" or "gray" wages in Russia has a negative impact on the pension system [15]. This is partly due to the population's distrust to the process of reforms, and especially to the possibility of receiving a decent pension.

On the other hand, this phenomenon can be caused by a decrease in the consumption of the population, and, consequently, in the profits of organizations. And the very essence of the pay-as-you-go pension system, which is that pensions are paid from the contributions of working citizens, depending on their size, is a path to its degradation if there are no additional well-tuned financing channels, since a change in the demographic climate will eventually lead to its unprofitableness, despite the increase in the retirement age. That is, the constantly decreasing number of workers will simply not be able to provide the growing number of the retired.

## 2.2  Comparative Analysis of Pension Systems Performance of the Countries of the World

The authors carried out a comparative analysis of the performance of the pension systems of Russia [16] and the OECD countries [17–20] based on the following indicators: government spending on pensions in relation to GDP, replacement rate, efficiency of the pension system, dependency ratio, coefficient substitutions per one percentage point of GDP.

As shown by the results of the conducted research, the pension systems of Iceland, New Zealand, Israel, Canada, Ireland, Luxembourg, USA and Denmark have the highest efficiency coefficient. The lowest are in Poland (0.67) and Turkey (0.89). As for the pension system in Russia, with such a high dependency ratio of 0.39, the replacement rate is catastrophically low (33%) in comparison with the average value of 56.5%.

Based on the results of the analysis, four groups of countries' distribution in the rating by the value of the efficiency coefficient of the pension system were identified:

- the first group includes countries with a low level of efficiency factor - up to 1.5. These are countries, such as Poland, Turkey, Greece, Slovenia, Chile and France.
- the second group is the most numerous and includes countries with an average efficiency factor of 1.5–2.0 (Finland, Italy, Russia, Austria, Germany, Belgium, etc.).
- the third group includes countries with a high efficiency factor of 2.0–3.0. This group of countries includes: Hungary, Estonia, Luxembourg, Japan, Great Britain, South Korea, Sweden, Denmark and the USA.
- the fourth group includes countries with the maximum values for this indicator - more than 3.0 - Israel, Canada, New Zealand, Ireland, the Netherlands.

Cross-country analysis of the effectiveness of pension systems is presented in the form of a thermal diagram (see Fig. 2).

A comparative analysis of the Melbourne Mercer Global Pension Index (MMGI) 2019 and the results of the study of the authors was carried out. The MMGI, developed by the Australian Center for Financial Research and consulting firm Mercer, uses three sub-indices - Adequacy, Sustainability and Integrity - to measure each pension system against more than 40 metrics.

The top includes the pension systems of Denmark and the Netherlands, the pension systems of Ireland, New Zealand and Canada were included in the top ten most effective systems [21].

Main characteristics of the most effective national pension systems, determined as a result of the study, as well as international The Melbourne Mercer Global Pension Index, namely Denmark, Canada and the Netherlands, are presented in the table (see Table 1).

The Danish pension system includes a government core pension scheme, cash-based supplementary pension benefits, a fully funded defined contribution plan and compulsory occupational schemes. The overall value of the index for the Danish system can be increased by raising the level of household savings and reducing their debt, as well as by increasing the level of participation in the labor force of older ages as life expectancy increases.

The Canadian pension system includes a universal fixed-rate pension, supported by an income-adjusted pension supplement; earnings-related pension, based on overvalued earnings throughout the working life; voluntary professional pension plans; and voluntary individual retirement savings plans. The overall value of the index for the Canadian system can be increased by:

| Country | Value | Group |
|---|---|---|
| Poland | 0,67 | Group I |
| Turkey | 0,89 | Group I |
| Greece | 0,92 | Group I |
| Chile | 1,07 | Group I |
| Slovenia | 1,22 | Group I |
| France | 1,47 | Group I |
| Finland | 1,60 | Group II |
| Portugal | 1,63 | Group II |
| Austria | 1,64 | Group II |
| Italy | 1,65 | Group II |
| Russia | 1,65 | Group II |
| Slovakia | 1,67 | Group II |
| Czech | 1,69 | Group II |
| Belgium | 1,81 | Group II |
| Germany | 1,83 | Group II |
| Spain | 1,88 | Group II |
| Norway | 1,89 | Group II |
| OECD | 1,93 | Group II |
| Australia | 1,98 | Group II |
| Hungary | 2,08 | Group III |
| Luxembourg | 2,08 | Group III |
| Estonia | 2,21 | Group III |
| Japan | 2,31 | Group III |
| South Korea | 2,49 | Group III |
| Great Britain | 2,54 | Group III |
| USA | 2,64 | Group III |
| Sweden | 2,64 | Group III |
| Denmark | 2,92 | Group III |
| Israel | 3,37 | Group IV |
| Canada | 3,71 | Group IV |
| New Zealand | 4,05 | Group IV |
| Irish | 4,9 | Group IV |
| Netherlands | 5,23 | Group IV |
| Iceland | 7,05 | Group IV |

**Fig. 2.** Cross-country analysis of the effectiveness of pension systems

**Table 1.** Comparative characteristics of the most efficient national pension systems.

| Index | Denmark | Canada | Netherlands |
|---|---|---|---|
| Gross replacement rate, % | 74,4 | 64,1 | 70,9 |
| Net replacement rate, % | 70,9 | 83,3 | 80,2 |
| Average monthly salary (PPP), $ | 60 578 | 43 022 | 64 662 |
| Average monthly pension, $ | 45 070 | 27 577 | 45 845 |
| Government spending on pensions % of GDP | 8,93 | 5,19 | 4,61 |
| Life expectancy at birth, years | 81,4 | 83 | 82,8 |
| Life expectancy at retirement, years | 20,3 | 20,9 | 21 |
| Retirement age (current), years | 65 | 65 | 65,8 |
| Retirement age pension (future), years | 74 | 65 | 71,3 |
| Dependency ratio (%) | 34,9 | 29,8 | 34,3 |

- growing coverage of workers with occupational retirement programs by developing an attractive product for those, who do not have an employer-sponsored retirement plan;
- growth of household savings and a decrease in household debt;
- a decrease in public debt as a percentage of GDP;
- an increase in the labor force participation rate at older ages, as life expectancy increases.

The Dutch pension system includes a fixed state pension and a quasi-mandatory corporate pension, linked to employment contracts. Most employees participate in these occupational plans, which are industry-wide defined benefit plans based on average lifetime earnings. The overall value of the index for the Dutch system can be increased by decreasing the level of household debt and, as in Denmark and Canada, increasing labor force participation in older ages as life expectancy increases.

As noted above, the value of the efficiency factor of the pension system of the Russian Federation is at an average level (1.65). Undoubtedly, the pension system of the Russian Federation needs a qualitatively different development. In this context, first of all, it is advisable to:

- encourage retirement later than the established age, which will increase savings and improve the ratio of the working population to pension recipients;
- encourage an increase in personal savings in order to reduce the likelihood that the state pension will become the only means of subsistence, while at the same time raising the pension expectations of many workers;
- increasing the insurance coverage for employees and/or self-employed in private provision for old age, because many people will not save money for the future if they are not automatically or forcibly included in the system;
- cancellation/reduction of the early retirement system, so that the accumulated funds can be used to generate pension income, which is possible with adequate tax support;

- revising the level of indexation of state pensions, since the method and frequency of increases are critical to ensure and maintain a balance in the real cost of a pension in terms of its long-term sustainability;
- improving the management of non-state pension funds and providing more transparency to increase the trust of the members of the foundation.

## 3   Results

Based on the conducted studies of the risks of the pension system of the Russian Federation, the authors have developed a system of measures aimed at increasing its effectiveness:

- optimizing costs and increasing the efficiency of using the resources of the pension system in general and the Pension Fund of Russia, in particular. This measure will include steps to digitize the pension system in order to reduce the burden on the budget from the maintenance of the pension fund buildings, personnel, etc. Also, the cost optimization will include limiting payments to working pensioners. In accordance with the latest changes in legislation, starting from 2016, working pensioners receive a pension without indexation (only for insurance pensions), and after the termination of employment, indexation begins to be taken into account, and for each year of later retirement, the insurance pension is increased by premium coefficients. The abolition of pensions for working pensioners will most likely simply lead to an even greater departure of workers to the shadow sector and, in principle, is not a fair measure. After all, a person who has completed the established length of service has the right to payment of funds from the solidarity system. Taking into account the use of the best world practices in the framework of incentives for later retirement, we consider it expedient to establish tax benefits for working pensioners, instead of using incentive coefficients. The pension system has been undergoing reform for 20 years, and during this time of constant "change of course", the population's confidence in the state apparatus has dropped significantly. In our opinion, those incentives that seem transparent and understandable to ordinary people can be favorably received, and show clearly where and in what amount the benefit will be obtained, which cannot be said about the coefficients system;
- the introduction of incentives for refusing to "gray" wages in envelopes, thereby increasing the income of the pension system. Now, according to various expert estimates, up to 25% of the workforce is employed in the illegal employment in Russia. This means that up to 20 million workers are excluded from the pension system. In addition, according to Rosstat, 20.6% of the workforce in 2019 was employed in the informal sector [22]. It can be noted, that the activities of enterprises in the informal sector may not be aimed at deliberate evasion of taxes and insurance premiums. Nevertheless, many informal sector companies choose to remain unregistered or unlicensed in order to avoid compliance with some or all of the regulations and thus reduce production costs, thereby depriving themselves of most of their pension entitlements;

- raising the retirement age is the most common reform measure as of 2019. The aim of this measure is to reduce the demographic burden on the existing pension system. According to the calculations of O. Sinyavskaya, E. Yakushev, A. Chervyakova, population aging will slow down GDP growth by 0.23% points on average annually from 2019 to 2025, despite the increase in the retirement age. However, it should be noted that the effect of this measure is short-lived - a maximum of 5 years - then the opposite effect will occur, which will lead to a new deficit of the pension system. And considering the fact, that the end of the current pension reform is planned for 2028, the meaning of such a measure is generally lost, the effect of which will be reduced to naught even before its full implementation;
- increasing financial literacy and creating responsibility of the population for their own future in the framework of teaching financial literacy. A person with a high level of financial culture will rationally use the resources at his disposal and save funds, forming a safety cushion in case of disability. Accordingly, with a high level of financial literacy of the population, the non-state pension system can receive a new impetus, which will lead to an increase in the replacement rate and an increase in the efficiency of the pension system as a whole;
- increasing the transparency of the pension mechanism in order to increase the level of trust among the population. There is a pension calculator in the Russian Federation, that allows one to only approximately calculate the future pension based on a number of entered metrics (gender, length of service, the expected period of further employment, etc.), and the State Services portal, the list of services of which is limited only by providing a certificate of pension savings;
- formation of a digital matrix of the entire pension system, which means the introduction of a digital platform that should be open and accessible, convenient and understandable for all participants. This will require the use of modern crypto protection tools, including blockchain technology, secure mobile platforms. A digital social security platform will make it possible to realistically assess the level of quality of life of a pensioner or a person in need of help, taking into account not only his income, but also his marital status, whether he has additional health insurance and participation in a non-state corporate pension system, whether or not he has benefits, disability. The need for the "physical" presence of the pension system institutions is decreasing, since today it is possible to carry out a whole range of financial transactions via mobile devices.

Moreover, the pandemic has accelerated the shift to the digital technology and online services. For example, the bodies of the Pension Fund of Russian provided state support to families with children during the coronavirus period, as quickly as possible, using the online public Portal of State Services. The concept of digital and functional transformation of the social sphere is designed until 2025. Currently, the Pension Fund, without a declaration, in other words, automatically provides two services based on the information already available in state registers and information systems: issuing certificates for maternity capital and SNILS for children born from 15.07.2020. Gradually, all processes for assigning pensions and social support measures should be transferred to this platform. In this regard, it is advisable to create a modern application integrated with other information resources and systems (part of the State Services portal), which will allow permanent and personalized tracking of the status of pension savings: both mandatory and voluntary, analyze them and conduct consultations using bots.

Digitalization transfers the economic and social responsibility of the employer to the employee. Self-control, personal responsibility, self-motivation of employees in the formation of pension rights (insurance contributions), wages and work rationing come out on top to ensure a decent level of income, both during the period of employment and in retirement. In general, the introduction of digital services is intended to optimize the process of interaction between the population and departments.

The success of creating an effective pension system in the Russian Federation largely depends on its transparency and accessibility to the majority of the population. In this regard, it is very important to ensure a balance of interests of the population in various regions of the country and sectors of the economy, as well as to strengthen the connection between the size of the pension and the labor contribution of each employee.

## 4   Conclusions

The purpose of the study is to form the main directions for the development of the pension system in the digital ecosystem, including the development of a system of measures aimed at improving the efficiency of its functioning. Based on the method of expert assessments, a risk matrix was compiled. The key risks affecting the development of the pension system of the Russian Federation are highlighted. The key risks include an increase in the share of pensioners in the structure of the population, an increase in the life expectancy of the population, a decrease in the birth rate, a low level of financial literacy of the population, financial crises, structural changes in the economy (a shift in capital and labor), a low level of collection of insurance premiums, an increase in the PFR deficit, a change in pension legislation (an increase in the retirement age, a change in the method of calculating pensions, coefficients, points). Comparative characteristics of the most effective national pension systems has been carried out. Measures have been developed to improve the efficiency of the pension system of the Russian Federation. Further research involves a detailed and quantitative assessment of the identified risks, as well as an assessment of the effectiveness of investment of pension savings.

Since the pension system is one of the most ambitious in terms of coverage and costly state social programs, any decisions to change the conditions of its functioning have a significant impact on various groups of the population and their economic behavior. This means, on the one hand, that all pension changes must be justified and their consequences calculated, and on the other hand, that it is necessary to ensure a balance between the economic and social goals of the functioning of the pension system.

In such a situation, it is necessary that all stakeholders have objective information that allows them to make reasoned and balanced decisions regarding the pension reform scenario.

# References

1. ISSA Report: 10 global challenges for social security. - Geneva, International Social Security Association (2019). https://ww1.issa.int/sites/default/files/documents/publications/2-10-challenges-Global-2019-WEB-263629.pdf. Accessed 23 July 2021
2. Kravchenko, E.V.: Modern pension systems: challenges, trends, prospects. State Munic. Adm. **2**, 83–91 (2018)
3. Kuklin, A.A., Shipitsyna, S.E.: From theory to practice of actuarial assessment of pension risks in Russia. Econ. Reg. **13**(3), 716–731 (2017)
4. Sinyavskaya, O.V., Yakushev, E.L., Chervyakova, A.A.: The Russian pension system in the context of long-term challenges and national development goals: Publishing House of the Higher School of Economics, Moscow (2021)
5. Peris-Ortiz, M., Alvarez-Garcia, J., Dominguez-Fabian, I., Devolder, P. (eds.): Economic Challenges of Pension Systems. A Sustainability and International Management Perspective. Springer Nature, Switzerland (2020)
6. Nepp, A., Larionova, V., Okhrin, O., Sesekin, A.: Optimal pension system: case study. Econ. Sociol. **11**(1), 267–292 (2018). https://www.researchgate.net/publication/324533202_Optimal_Pension_System_Case_Study. Accessed 09 Aug 2021
7. Stewart, F.: Pension funds' risk-management framework: regulation and supervisory oversight. In: OECD Working Papers on Insurance and Private Pensions, no. 40, OECD publishing, OECD (2010)
8. Fabozzi, F.J.: Measuring and explaining pension system risk. J. Pension Econ. Financ. **14**(2), 161–171 (2015). Special Issue on Assessing the U.S. Pension Insurance Modeling System (PIMS), https://www.cambridge.org/core/journals/journal-of-pension-economics-and-finance/article/abs/measuring-and-explaining-pension-system-risk/53AC5CE745E4807D8BEC3176ED368BBB. Accessed 09 Aug 2021
9. Geddes, T., Gratson, J., Curguz, I.: Pension Risk Transfer. Evaluating Impact and Barriers for De-Risking Strategies. Society of Actuaries, Schaumburg (June 2021). https://www.soa.org/globalassets/assets/files/resources/research-report/2021/2021-pension-risk-transfer.pdf. Accessed 05 Aug 2021
10. Andreeva, O.V., Kravchenko, E.V., Sukhoveeva, A.A.: Risks of financial stability of the domestic pension system in modern conditions. Financ. Econ. **11**, 109–111 (2020)
11. Kozlov, P.A., Slepov, V.A., Finogenova, Y.: Risk management of financial flows of the pension system of the Russian Federation. Insur. Bus. **9**(306), 26–34 (2018)
12. Central Bank of the Russian Federation Homepage. https://cbr.ru/. Accessed 03 June 2021
13. Sinyavskaya, O.V.: The Russian pension system in the context of demographic challenges and restrictions. Econ. J. High. Sch. Econ. **21**(4), 562–591 (2017)
14. OECD Pensions Outlook (2020). https://www.oecd.org/pensions/oecd-pensions-outlook-23137649.htm. Accessed 05 June 2021
15. Ministry of Finance of the Russian Federation Homepage. https://minfin.gov.ru/. Accessed 05 June 2021
16. Russian Pension Fund Homepage. https://pfr.gov.ru. Accessed 05 June 2021
17. Pension Markets in Focus (2020). https://www.oecd.org/pensions/private-pensions/pensionmarketsinfocus.htm. Accessed 05 July 2021
18. Wang, P., Zhang, M., Shand, R., Howell, K.: Retirement, Pension Systems and Models of Pension Systems. Economics Working Paper 1402, June 2014. https://www.plymouth.ac.uk/uploads/production/document/path/8/8845/models_of_pension_systems_wp.pdfsystems_wp.pdf. Accessed 06 Aug 2021

19. Pensions at a Glance 2019 OECD and G20 Indicators. https://www.oecd-ilibrary.org/sites/b6d3dcfc-en/index.html?itemId=/content/publication/b6d3dcfc-en. Accessed 02 July 2021
20. World Population Prospects 2019. United Nations, Department of Economic and Social Affairs. https://population.un.org/wpp/Publications/Files/WPP2019_Volume-I_Comprehensive-Tables.pdf. Accessed 02 July 2021
21. Melbourne Mercer Global Pension Index 2019 (MMGPI). https://www.mercer.com.au/our-thinking/mmgpi-2019.html. Accessed 01 July 2021
22. Federal State Statistics Service Homepage. https://rosstat.gov.ru. Accessed 03 June 2021

# Digital Technology and Applied Sciences

# Cyberattacks on Business Website: Case Study

Tatiana Antipova$^{(\boxtimes)}$ (ID)

Institute of Certified Specialists, Perm, Russia

**Abstract.** The main idea of this work based on real cyber attacks to our business website. In particular, our business website works world-wide for the collection via Submission System of scientific papers for publication. This study analyses a real-life situation with malicious actors to our Submission System by submitting empty form without attaching file contains paper body due to the bot used fake image/document to bypass the required file field. The problem in this case was defined as attacks by bot a little more "intelligent" than usual and qualified as Hacktivism with Ideological motivation by Disruption Goal to our Submission system and this problem has been solved. The best solution for this case was found as the identification attacker's IP address followed by ban malicious IP directly via including it into blacklist of anti-Spam plugin. The result of this work consists logical scheme how to stop cyberattacks to business website. Also, this work details how to identify malicious IP and find it location. The result of this study can be implemented by other business website users.

**Keywords:** Cyber security · Cyber-attacks · Website · Hacker · Malicious actors · Hacktivism · Plugin · Blacklist · Spam

## 1 Introduction

With the digitalisation of all spheres of human activity and businesses moving to online platform, the frequency of cyber-attacks on websites has increased enormously, raising the risk of online/IT business. Increasingly concerned, key voices are sounding the alarm. Two ongoing trends exacerbate this risk: (1) an unprecedented digital transformation, which is being related by the COVID-19 pandemic [1]. Meanwhile, the pandemic has heightened demand for online business/services and made work-from-home arrangements the norm. In this time of transformation, when an incident could easily undermine trust and derail such innovations, cybersecurity is more essential than ever. (2) malicious actors are taking advantage of this digital transformation and pose a growing threat to the financial stability, and confidence in the integrity of the system. The pandemic has even supplied fresh targets for hackers. Online business is experiencing the second-largest share of COVID-19–related cyberattacks [2].

According to [3], cyber attack is an attack, via cyberspace, targeting an enterprise's use of cyberspace for the purpose of disrupting, disabling, destroying, or maliciously controlling a computing environment/infrastructure; or destroying the integrity of the data or stealing controlled information.

Many research works address the problem of identifying the cyber attacks [4, 5]. According to the following Sources: ESRB [6], MI5 and Cambridge Centre for Risk Studies, cyber incidents classified as Nation-states, proxy groups; Cybercriminals; Terrorist groups, hacktivists, insider threats. The literature review demonstrated that the

© The Author(s), under exclusive license to Springer Nature Switzerland AG 2022
T. Antipova (Ed.): DSIC 2021, LNNS 381, pp. 505–512, 2022.
https://doi.org/10.1007/978-3-030-93677-8_44

approach and motivation for conducting cyberthreat differed from case to case. Among these, the ransomware was the most frequently used threat for disrupting the computer networks and servers [7].

Small and medium enterprises (SME) owners have a misconception that their legacy systems are not significant enough to be targeted by cybercriminals [8]. Additionally, cybersecurity personnel are currently available at excessively high rates, making it difficult for most SME's to afford, i.e. the most small organizations not having an in-house cybersecurity specialist to provide the needed guidance [8].

Like hacktivism, the cyber espionage is the illegal access to secret and delicate information (e.g. company strategy, private information, or intellectual capital). However, the cyber espionage aims to gain competitive advantages rather than create pressure and business disruption [6]. Thus, the consequences might be the loss of intellectual property, business profits and efficiency, and customer information, additional costs thanks to the interrupted business plan, and damage to company reputation [9]. Preventing cyberattacks can able to help develop an advanced, human-centric and human-consistent computer system for supporting consensus reaching which will use the strength of the human and the computer to solve a relevant problem [10, 11].

This study considers how we have found the way to deal with cyberattacks. To do this we have analysed simple case of preventing a real cyber-attacks on a business website.

## 2    Data and Methodology

Key methods of this paper by systemic approach are analysis, comparison and abstraction. This study analyses a real-life situation where existing problems have been solved. This case study consists of the following elements:

1. Identifying the problems.
2. Selecting the major problems in the case.
3. Suggesting solutions to these major problems.
4. Recommendation the best solution to be implemented.
5. Detalisation how this solution should be implemented.

## 3    Cyberattacks Analysis in Fact

Our website https://ics.events was created in 2018. As far as the research topic is concerned, our business website works world-wide for the collection via Submission System of scientific papers for publication. We have chosen WordPress platform because it is the platform of choice for over 42% of all sites across the web (https://wordpress.org/about/). Today, WordPress is built on Hypertext Preprocessor (PHP) and MySQL, and licensed under the General Public License (GPLv2).

In May-June 2021 we have received to our website via Submission System 44 empty submissions from different emails (approximately 2 times a day). Firstly, we could not understand how was it possible to submit empty form without attaching file contains paper body. This submission form intends for paper submission as attached.docx file for publication purpose. But in this case, it seems the bot could use fake image/document to bypass the required file field. A summary of this malicious actors is given in Table 1.

**Table 1.** List of cyberattacks

| Data | Time | Emails |
|------|------|--------|
| 10.05.2021 | 6:59 PM | jyrshnf@mail.ru |
| 11.05.2021 | 5:26 AM | kjhgf.lkjhg.2020@mail.ru |
| 12.05.2021 | 8:52 PM | htyugjh.esrgh@mail.ru |
| 13.05.2021 | 5:38 AM | geni2yevich11111123456@mail.ru |
| 13.05.2021 | 10:06 PM | geni2yevich111112@mail.ru |
| 14.05.2021 | 7:07 AM | geni2yevich111111@mail.ru |
| 17.05.2021 | 6:12 AM | geni2yevich111111234@mail.ru |
| 17.05.2021 | 3:08 PM | geni2yevich1111112345@mail.ru |
| 18.05.2021 | 8:33 PM | geni2yevich1111112345@mail.ru |
| 19.05.2021 | 5:19 AM | geni2yevich1111112345678@mail.ru |
| 19.05.2021 | 1:14 PM | geni2yevich111111234567@mail.ru |
| 19.05.2021 | 9:21 PM | ljhygfg@mail.ru |
| 20.05.2021 | 5:37 AM | jndgbs@mail.ru |
| 21.05.2021 | 2:13 PM | geni2yevich11111123456@mail.ru |
| 22.05.2021 | 1:25 AM | geni2yevich1111112345678@mail.ru |
| 22.05.2021 | 12:04 PM | geni2yevich111111234567@mail.ru |
| 23.05.2021 | 8:08 AM | geni2yevich11111123456@mail.ru |
| 23.05.2021 | 8:34 PM | geni2yevich11111123456@mail.ru |
| 24.05.2021 | 6:32 AM | geni2yevich1111112345678@mail.ru |
| 24.05.2021 | 5:12 PM | geni2yevich111111234567@mail.ru |
| 25.05.2021 | 9:52 PM | geni2yevich11111123456@mail.ru |
| 26.05.2021 | 6:43 AM | geni2yevich1111112345678@mail.ru |
| 26.05.2021 | 3:47 PM | geni2yevich111111234567@mail.ru |
| 29.05.2021 | 9:25 PM | ljhygfg@mail.ru |
| 30.05.2021 | 10:33 AM | leontev.valeriy111@mail.ru |
| 30.05.2021 | 10:50 PM | leontev.valeriy111@mail.ru |
| 31.05.2021 | 6:10 AM | ovatop@bk.ru |
| 31.05.2021 | 10:12 PM | ljhygfg@mail.ru |
| 01.06.2021 | 10:46 AM | alekxzs212234545@mail.ru |
| 01.06.2021 | 11:16 PM | leontev.valeriy111@mail.ru |
| 02.06.2021 | 10:56 PM | ujhnd.njtgd@mail.ru |
| 03.06.2021 | 12:30 PM | alekxzs2@mail.ru |
| 04.06.2021 | 1:20 AM | alekxzs219@mail.ru |
| 04.06.2021 | 12:06 PM | fd6rd5dryh@mail.ru |
| 05.06.2021 | 3:42 AM | gtkygjtyy@mail.ru |
| 05.06.2021 | 6:07 PM | alekxzs2198765434@mail.ru |
| 06.06.2021 | 5:35 AM | fgbvd.ngbfv.16521@mail.ru |
| 06.06.2021 | 5:23 PM | geni2yevich11111123456@mail.ru |
| 07.06.2021 | 4:51 AM | ryewrygfvx@mail.ru |
| 07.06.2021 | 4:39 PM | kjhgf.lkjhg.2020@mail.ru |
| 08.06.2021 | 4:36 AM | htyugjh.esrgh@mail.ru |
| 08.06.2021 | 4:19 PM | geni2yevich111112@mail.ru |
| 09.06.2021 | 5:23 AM | geni2yevich111111@mail.ru |

Source: author's compilation based on website data.

Analysing data in Table 1 we can see that cyber attacks came from different emails during one month one-two times a day predominantly before and after working hours. All of these submissions have not consisted attached file and were empty and we had ensured that it was the malicious actors. We could suggest that it was some of rejected authors or just joker. Almost each malicious submissions were sent from different email and only sometimes repeated the same emails like geni2yevich11111123456@mail.ru (highlighted in yellow in Table 1) that was used for six times. Apparently, attacker has not enough imagination to come up with totally new emails names.

All of malicious submissions were marked as Spam (unsolicited communications sent over the internet) in our Submission System but they are continued to arrive. One of suggestion to avoid this arriving from programmer was using the default input name [text* name-5464548] into contact form but it was not helpful and attacks were continued.

Thus, the problem in this case was identified as attacks by bots a little more "intelligent" than usual. According to [6] our case was qualified as Hacktivism with Ideological motivation with Disruption Goals of our Submission system. But it was still unknown how to solve this problem therefore we decided to start the correspondence with Contact Form designer/programmer.

Then programmer recommended us to install an antispam plugin because it seems the bot can use fake image/document to bypass the required file field. Since the plugin is a piece of software that adds new features or extends functionality on an existing application, we have chosen plugin "WordPress Zero Spam" for our WordPress website for antispam purposes. So, we installed plugin "WordPress Zero Spam". But attacks were keep going!

Then we requested programmer: "Is it possible to define an IP address from which these hackers always send us rubbish (empty form without paper)?".

Answer: "To get the IP of the user that is sending the email, place [_remote_ip] in your Submission System". We set option "to show IP address" ([_remote_ip] on PHP) and consequently we exactly stated that it was one person because his malicious submissions always came from the same IP address and we saw this address from our Submission System.

Then to identify attacker's IP location we used website https://whois.domaintools.com/ to get information about the attacker. We found IP location and physical address (see Fig. 1).

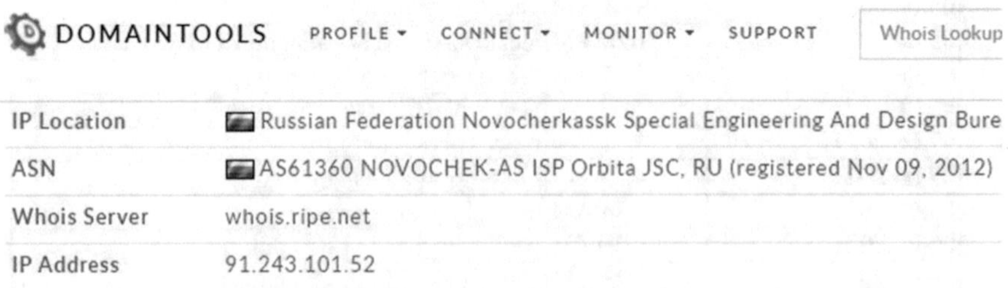

| DOMAINTOOLS | PROFILE ▾    CONNECT ▾    MONITOR ▾    SUPPORT | Whois Lookup |
|---|---|---|
| IP Location | 🏳 Russian Federation Novocherkassk Special Engineering And Design Bure | |
| ASN | 🏳 AS61360 NOVOCHEK-AS ISP Orbita JSC, RU (registered Nov 09, 2012) | |
| Whois Server | whois.ripe.net | |
| IP Address | 91.243.101.52 | |

**Fig. 1.** Identification of IP location.

We called to phone number revealed in above information, and there someone answered that can not define who exactly is owner of this IP address and we should send official letter to request defining this IP address owner with describing whole situation.

Bearing in mind that business correspondence lasts too long period of time we did not send any requirement to above office and tried to break attacks by other way.

In addition, programmer told us that it could also be someone who has taken possession of that server, without it necessarily being the IP of the hacker. So, we have reached to conclusion that having IP address location does not mean definitely that you have hacker's name and can stop attacks by requesting hacker to stop it.

And we have got another recommendation: "if the attack comes only from identified IP you can simply blacklist it to solve. You can ban the IP directly, by including it into blacklist!" Since we already installed Zero Spam plugin (see above) we included identified IP address into blacklist successfully (see Fig. 2).

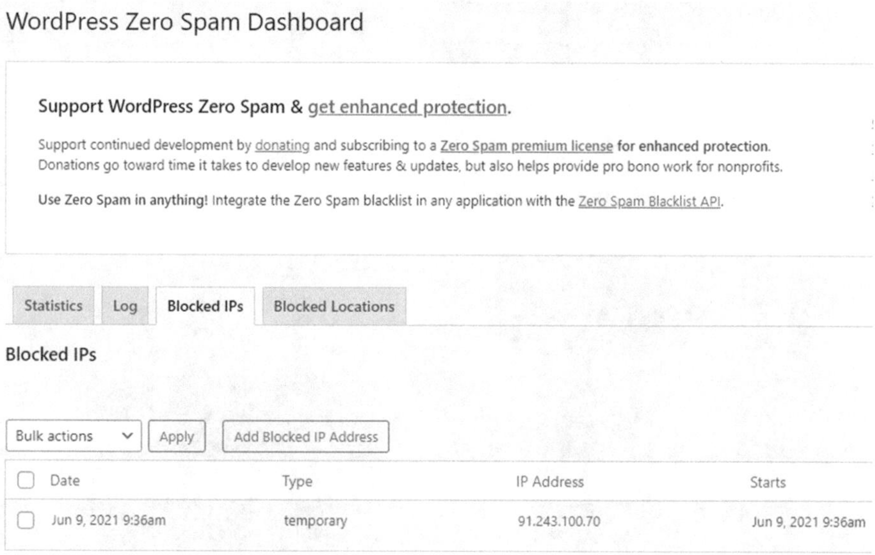

**Fig. 2.** Including identified IP address into Blocked IPs list of WordPress Zero Spam plugin.

So, cyber attacks have been stopped after including identified IP address of hacker into Blocked IPs list of "WordPress Zero Spam" plugin at 09.06.2021. And you can see in Table 1 that after 09.06.2021 we did not receive none of attacks from this IP after one month seeking the way. But you have to bear in mind that there is not enough just install antispam plugin, you should include identified malicious IP Address into black list of this installed plugin.

Actually, studied firm has not received significant consequences from these malicious actors excluding spending lots of working time, emotions and brain efforts to deal with such unpleasant things. But by tackling these attackers we gained new knowledge and confidence that we can cope with more serious attacks. Also, it is still important to understand information security investments in terms of compliance and risk, given that the cost of compliance is rising [12]. In doing so, it is particularly important to improve the skills of managers and IT personnel [13].

As a consequence, our recommendation for the best solution to be implemented is to ban malicious IP directly, by including it into blacklist of anti-spam plugin.

Having gained experience in stopping cyber-attacks, we have designed the following logical scheme/technique to stop cyber-attacks.

## 4   Logical Scheme/Technique to Stop Cyberattacks

To draw this logical scheme/technique we used the stages of procedures that described in paragraph 3 to stop cyberattacks (Fig. 3).

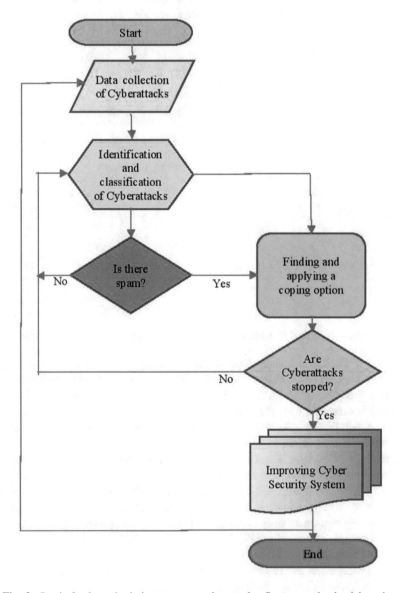

**Fig. 3.** Logical scheme/technique to stop cyberattacks. Source: author's elaboration.

This logical scheme/technique can help manager and IT-specialists to optimize Cyber Security System in their organization to avoid cyberattacks in the nearest future.

## 5   Conclusions

This study was considered some kind of cyber attacks - malicious submissions on business website. As a result of this case study we identified and selected the major problem as attacks by bots, and classified it as Hacktivism with Ideological motivation with Disruption Goals of our Submission system; suggested solutions to this major problem as including identified IP address of hacker into Blocked IPs list of anti-spam plugin; recommended the best solution to be implemented as the identification of malicious IP address; and detailed how this solution might be implemented given designed logical scheme/technique.

The limitation of this paper is that case study seems not significant at the first look. But our experience and technique could be useful for many others business websites and it is important since the pandemic has heightened demand for online services and made work-from-home arrangements the norm. Therefore, our logical scheme/technique can be implemented into any online services and work-from-home arrangements.

For future work we plan to research cyber attacks from country to country, from small to medium firm and bank/financial systems.

## References

1. Antipova, T.: Digital view on COVID-19 impact. In: Antipova, T. (ed.) ICCS 2020. LNNS, vol. 186, pp. 155–164. Springer, Cham (2021). https://doi.org/10.1007/978-3-030-66093-2_15
2. Maurer, T., Nelson, A.: The global Cyber Threat. Finance & Development, pp. 24-27, March 2021
3. Cyber Attack - Glossary | CSRC. https://csrc.nist.gov/glossary/term/cyber_attack
4. Dang, Q.-V.: Detecting the attacks to DNS. In: Antipova, T. (ed.) ICCS 2021. LNNS, vol. 315, pp. 173–179. Springer, Cham (2022). https://doi.org/10.1007/978-3-030-85799-8_15
5. Dang, Q.-V.: Predicting the signs of the links in a network. J. Digit. Sci. 2(2), 14–22 (2020). https://doi.org/10.33847/2686-8296.2.2_2
6. Systemic cyber risk. February 2020. https://www.esrb.europa.eu/pub/pdf/reports/esrb.report200219_systemiccyberrisk ~ 101a09685e.en.pdf?fdefe8436b08c6881d492960ffc7f3a9
7. Chalermpong, S.: Port cybersecurity and threat: a structural model for prevention and policy development. Asian J. Ship. Logist. 37(1), 20–36 (2021). https://doi.org/10.1016/j.ajsl.2020.05.001. ISSN 2092-5212
8. Eybers, S., Mvundla, Z.: Investigating cyber security awareness (CSA) amongst managers in small and medium enterprises (SMEs). In: Antipova, T. (ed.) ICCS 2021. LNNS, vol. 315, pp. 180–191. Springer, Cham (2022). https://doi.org/10.1007/978-3-030-85799-8_16
9. Platt, V.: Still the fire-proof house? An analysis of Canada's cyber security strategy. Int. J.: Canada's J. Glob. Pol. Anal. 67(1), 155–167 (2011)
10. Kacprzyk, J., Zadrożny, S.: Soft computing and Web intelligence for supporting consensus reaching. Soft Comput. 14, 833–846 (2010). https://doi.org/10.1007/s00500-009-0475-4

11. Tagarev, T., Atanassov, K.T., Kharchenko, V., Kacprzyk, J.: Digital Transformation Cyber Security and Resilience of Modern Societies. Springer, Heidelberg (2021). https://doi.org/10.1007/978-3-030-65722-2
12. Griffy-Brown, C., Chun, M., Miller, H., Lazarikos, D.: How do we optimize risk in enterprise architecture when deploying emerging technologies? J. Digit. Sci. **3**(1), 3–13 (2021). https://doi.org/10.33847/2686-8296.3.1_1
13. Danilina, E.I., Mingaleva, Z.A., Malikova, Y.I.: Strategic personnel management within innovational development of companies. J. Adv. Res. Law Econ. **7**(5), 1004–1013 (2016). https://doi.org/10.14505/jarle.v7.5(19).06

# Concepts of Improving Fuzzy and Neural Network Methods for Simulating Bankruptcy in Risk Management by the Bank's Loan Portfolio

Alexander Biryukov[1]([⊠]) (iD), Gulnaz Murzagalina[1]([⊠]) (iD),
Alina Kagarmanova[1]([⊠]) (iD), and Svetlana Kochetkova[2]([⊠]) (iD)

[1] Sterlitamak Branch of the Bashkir State University,
Lenin Avenue, 49, 453103 Sterlitamak, Bashkortostan, Russia
[2] Salavat Branch of Ufa State Petroleum Technical University,
Gubkin Str., 22 b, 453250 Salavat, Bashkortostan, Russia

**Abstract.** Fuzzy and neural network models for predicting bankruptcy are, first of all, relevant from the point of view of corporate governance in the real and financial and banking sectors. To achieve the efficiency of corporate governance, appropriate information and analytical support should be created, i.e. a system for monitoring the risk of bankruptcy of organizations in the context of the use of intelligent information technologies.

This article discusses the issues and presents the results of research on the management of a bank's loan portfolio using neural network models, which provide new opportunities to reduce risks at the stage of bankruptcy of organizations with different dynamics of changes in the financial and economic condition of borrowers.

Increase the efficiency of the bank's loan portfolio management are based on the neural network logistics iterative dynamic method. Attention is paid to the comparison of the developed dynamic model with many foreign and Russian static, expert and rating models.

A generalization of the proposed fuzzy model for assessing credit history and the neural network logistic iterative dynamic method in terms of increasing its predictive power in conditions of incomplete data on the intermediate values of the probability of bankruptcy of the borrowing organization is made. The proposed provisions were tested in computational experiments on real data and proposals were made for the generalized management of the bank's loan portfolio.

The conclusions obtained in the course of the study show that the neural network logistic iterative dynamic method allows one to build extrapolation models of bankruptcy of organizations when managing a bank's loan portfolio.

**Keywords:** Provisions · Bank · Creditworthiness of borrowers · Neural network model · Fuzzy model · Factor models of bankruptcy · Loan portfolio

T. Antipova (Ed.): DSIC 2021, LNNS 381, pp. 513–524, 2022.
https://doi.org/10.1007/978-3-030-93677-8_45

# 1  Introduction

The relevance of the research topic appeared in the conditions of the unstable state of the country's economy in 2020–2021, when the measures proposed by the state to support business created new economic instruments, which increased certain credit risks of the banking sector. The main prerequisites for the emergence of risks from banks are the low quality of assessing the creditworthiness of borrowers when deciding whether to issue a loan, as well as the lack of sufficiently accurate and adequate dynamic models for assessing the current financial and economic condition (FES) of borrowers by the bank in the process of servicing loans. Therefore, the issue of managing the restructuring of the organization's credit debt is very relevant. We emphasize that both parties to the credit transaction are interested in the effectiveness of the restructuring of credit debt: the bank and the organization - the borrower, unless, of course, cases of false (deliberate) bankruptcies are excluded from consideration.

In the process of monitoring the FES of borrowers when the bank serves its loan portfolio, it is important to catch the "point of no return", when already traditional methods of debt restructuring cannot save the borrower from bankruptcy, and radical instruments are required, for example, the issuance of corporate bonds, the re-crediting of the organization, the use of various models of public-private partnership based on an investment tax credit [1, 8].

To support the decisions made on the management of the bank's loan portfolio, the following adequate mathematical models with high predictive power in difficult modeling conditions must be created:

1. At the stage of incoming control (assessment of credit histories of borrowers), when there is usually no sufficiently complete data on quantitative financial and economic indicators for individuals and business organizations, it is effective to use fuzzy mathematical models [4].
2. At the stage of current monitoring for servicing the bank's loan portfolio and tracking the approach of the "point of no return", it is required to develop a dynamic neural network forecasting model with continuous time in conditions of incomplete data, when in some time slices for corporations, there are no "bankrupt - not bankrupt" marks [2].
3. At the stage after the "point of no return", when the onset of bankruptcy is inevitable, fuzzy Mamdani-type models based on fuzzy inference can be used to support decision-making on debt restructuring [4].
4. At stage 2), in order to increase the predictive power of the dynamic model of bankruptcy, it is advisable to introduce into it an aggregated quantitative indicator characterizing the "discipline" of the borrower in terms of the regularity of loan repayment.

Unfortunately, the methods and models for points 1), 3) and 4) have been practically little studied in relation to the problems of forecasting bankruptcies, since the issues of choosing a sufficiently complete system of factors for assessing the credit histories of potential borrowers, constructing membership functions in fuzzification of quantitative indicators have not been fully resolved and their subsequent aggregation, assessing the adequacy of the resulting fuzzy and neural network models. As for the proposed neural network logistic iterative dynamic method (NLIDM) for extrapolating the probability of bankruptcy risk [2], in preliminary experiments this method shows good predictive power (87.6% of correct identification of bankruptcies) and fast convergence of iterations to recover incomplete data in time slices.

## 1.1  The Degree of Study and Elaboration of the Problem

Using a specific example of data from 136 organizations of the construction industry in Russia from [2] (International database "Bereua Van Dijk"), an analysis of well-known domestic and foreign models of bankruptcies was carried out in terms of their adequacy and sensitivity, ie. the ability to differentiate the nature of the dynamic dependences of the probability of bankruptcy risk for various corporations.

Financial and legal tools for managing the restructuring of credit debt are closely related to the analysis of the corresponding stages of the developing process of the crisis in the organization. In different sources, these stages are distinguished in different ways.

It seems quite convenient to classify the stages from [5] with an approximate estimate of the probability of bankruptcy risk $P$ at each stage and the corresponding probabilities of bankruptcy risk (Table 1).

Table 1 stages of the crisis can be conditionally divided into three periods: pre-crisis, acute crisis, chronic crisis. The pre-crisis period includes stages 1, 2 and 3 and is characterized by the absence of a clear strategic management system in the organization (management crisis). The bank, by analyzing the borrower's reporting, can reveal a drop in production and sales volumes, and already at this stage, start negotiating with the borrower with a view to reorganizing his financial condition.

A period of acute crisis (financial crisis) begins with a liquidity crisis (stage 4). Credit officers need to pay attention to the increase in the enterprise's debt to all creditors and the deterioration of liquidity indicators. The insolvency of the organization can be temporary and caused by a lack of funds, for example, due to the seasonal nature of production or incomplete sale of receivables. Chronic debt can be caused by a lack of property in their assets. The bank should pay special attention to chronic insolvency, which can lead to bank losses.

In the period of an acute insurmountable crisis (stages 4 and 5), the organization no longer has liquid internal resources of its own, and in order to restore solvency, external sources of financing are needed. In this case, when the bank has identified the borrower's problems at this stage, resolving issues at the level of debt restructuring will not be effective, and the bank cannot return its asset without loss. One of the ways to resolve the issue may be debt refinancing, but such a step will cause an increase in the volume of risky assets, which means a decrease in equity capital adequacy.

After the stage of chronic insolvency, the organization enters the final period of the economic and legal crisis (stages 6 and 7). In this period, the organization is characterized by insolvency (absolute insolvency), which is a critical (bifurcation) point of development, after which it overcomes either the crisis process and continues to develop, or goes to stages 6 and 7 of bankruptcy with the launch of procedures for economic and legal regulation of insolvency.

**Table 1.** Stages of a crisis in an organization

| № | Crisis stage | Crisis type | Changes in the probability of bankruptcy of the organization | Linguistic characteristics of the bankruptcy risk |
|---|---|---|---|---|
| 1 | Strategic crisis | Management crisis | $0 \leq P < 0,15$ | Very low risk of bankruptcy |
| 2 | Structural crisis | Management crisis | $0,15 \leq P < 0,3$ | Low risk of bankruptcy |
| 3 | Operational crisis | Management crisis | $0,3 \leq P < 0,45$ | Average risk of bankruptcy |
| 4 | Liquidity crisis | Financial crisis | $0,45 \leq P < 0,6$ | Average risk of bankruptcy |
| 5 | Temporary insolvency | Financial crisis | $0,6 \leq P < 0,8$ | High risk of bankruptcy |
| 6 | Insolvency | Financial crisis | $0,8 \leq P < 1$ | Very high risk of bankruptcy |
| 7 | Bankruptcy | Financial and legal crisis | $0,8 \leq P < 1$ | Very high risk of bankruptcy |

From the above brief analysis, it can be seen that when managing a loan portfolio for a bank, it is very important at any current moment of time $t$ to know objective and reliable information about the stages of the developing bankruptcy of the borrower and, in particular, the dynamic curve of the probability of bankruptcy risk *P(t)*. For this, the dynamic model of bankruptcy with continuous time $t$, developed in [1, 3], is intended.

## 2   Data and Methodology

### 2.1   Provisions 1 of a Three-Stage Bank Loan Portfolio Management

According to the proposed provisions 1, the first stage is an analysis of the credit history of a potential borrower; the second stage is monitoring the financial and economic condition (FES) of the borrower in the process of servicing the bank's loan portfolio; the third stage is the adoption of a decision on the restructuring of credit debt [2].

From a systemic point of view, the proposed provisions is based on the well-known system-wide law of decreasing the entropy of the unified system $(S_1, S_2)$, formed by combining two isolated systems $S_1$ and $S_2$, if the systems $S_1$ and $S_2$ interact rationally. The meaning of the law is quite transparent: with the rational interaction of the systems $S_1$ and $S_2$ new connections appear in the unified system $(S_1, S_2)$, the degrees of freedom decrease and, accordingly, the number of admissible states decreases - entropy decreases.

In our problem of loan portfolio management, under the system $S_1$ we mean a fuzzy assessment model [4] of the credit history of a potential borrower of a credit institution, and under the system $S_2$ - a neural network logistic iterative dynamic model of bankruptcy (NLIDM) from [2].

First, we will formulate the proposed provisions, and then we will give detailed comments to it.

***Provisions 1*** for introducing qualitative information from the analysis of credit history into the dynamic bankruptcy model (DMB) is formed as follows: *in order to increase the predictive power of the DMB used to monitor the borrower's FES, it is proposed to create a unified information and analytical system $(S_1, S_2)$ by obtaining fuzzy credit indicators histories with their subsequent aggregation, de-diffusion and introduction into the NLIDM algorithm. At the same time, the de-differentiated quantitative aggregate $y^*$ is interpreted as a behavioral factor characterizing the borrower's "credit discipline".*

The idea of implementing the information and analytical system $S_1$ for *provisions 1* is as follows. In the lending system of most banks, the assessment of the credit history is carried out by a credit expert according to various kinds of indirect quantitative and qualitative characteristics contained in the credit history and allowing a wide range of interpretation. On the one hand, such an expert assessment method allows one to attract valuable experience and professional intuition of banking specialists in the lending system, which increases the quality of the assessment. But on the other hand, this method can lead to the introduction of subjective considerations that do not have sufficient grounds in the decisions on granting loans. As a result, in the assessment, an excessively large weight can acquire the subjective opinion of an expert up to the deliberate interpretation of information and the adoption of decisions detrimental to the bank.

It is especially difficult to describe the characteristics of the borrower's credit history. Setting strict (clear) restrictions on the values of its components can lead to the loss of potential borrowers and a decrease in the bank's profit. On the contrary, an excessive expansion of the permissible boundaries of characteristics is accompanied by a deterioration in the quality of the loan portfolio (errors of the second kind are growing when making a decision to issue a loan).

An effective solution to this contradictory situation, according to [7], consists in reducing the possibility of an expert's influence on the decision of the lending issue by formalizing the borrower's behavior on the basis of a fuzzy model.

As a result, we obtain an information-analytical fuzzy system $S_1$ for the proposed provisions 1.

Let us give a brief description of an illustrative example from [7]. Let the borrower's credit history contain quantitative indicators $\{u_k\}, k = \overline{1, n}$, included in the set of carriers $u \in U$. For example, $u_1$ is the total number of delinquencies; $u_2$- number of days overdue for a certain period; $u_3$- the ratio of the number of overdue payments to "successful" ones, etc. The specific content of the elements $\{u_k\} \in U$ should reflect the bank's policy in terms of credit risks.

The linguistic variable Zadeh [6] $X =$ "status of credit history", which has two terms as a set of meanings, was taken as the final assessment of credit history: $T_1 =$ "acceptable (several delays)"; $T_2 =$ "negative (many delays)".

The membership functions of the carrier values to fuzzy terms $\mu_T(u)$ in [4] are defined as the normalized components of the eigenvector for the pairwise comparison matrix $A_{ij}$ of the Saaty hierarchy analysis method [2]. For example, for four credit histories, the pairwise comparison matrix $A$ looks like Table 2:

**Table 2.** Pairwise comparison matrix for indicator $u_1$

|  | i | 3 | 5 | 7 | 8 |
|---|---|---|---|---|---|
| A = | j |  |  |  |  |
|  | 3 | 1 | 3 | 5 | 6 |
|  | 5 | 1/3 | 1 | 3 | 5 |
|  | 7 | 1/5 | 1/3 | 1 | 3 |
|  | 8 | 1/6 | 1/5 | 1/3 | 1 |

Here i, j = 3; 5; 7; 8 - the number of delinquencies for credit histories with numbers 1, 2, 3, 4, respectively, are arranged in ascending order from left to right. Elements of matrix $A$ are expert assessments (Saaty coefficients) of pairwise comparisons. For example, the number 6 in the cell (1; 4) means that the number of delinquencies $u_1 = 3$ is 6 times more preferable than the number of delays $u_1 = 8$. The number of such matrices of pairwise comparisons must be equal to the number of fuzzy quantitative indicators $u_k$. For Table 2, membership functions are calculated as normalized components of the eigenvectors of matrix $A$:

$$\mu_T(u) \in [0; 1] \tag{1}$$

$$\{\mu_{T1}(u)\} = \frac{\omega_{i1}}{\omega_{max1}}; i = 1, 2, 3, 4. \tag{2}$$

For Table 2 in [7], the values of the membership function of each i-th credit history to the fuzzy term $T_1$ = "several overdue payments" were obtained:

$$\mu_{1i} = 1; 0, 06; 0, 2170, 106.$$

If there are several quantitative indicators in the credit history, then in order to take into account their joint influence on the fuzzy logistic conclusion about the status of the credit history after the defasification of all $u_k$, it is necessary to build a knowledge base (system of production rules) of the form:

IF a "set of conditions", THEN «output".

In the Mamdani method [4], the aggregated fuzzy logistic inference is obtained by the formula:

$$\tilde{y} = \text{agg}\left(\sum_{j=\overline{1,m}} imp(\mu_{d,j}(\vec{X}), \mu_{d,j}(y)/y)\right), \tag{3}$$

where *imp* is a fuzzy implication operation, usually implemented as the operation of finding the minimum (intersection of sets); "agg" is a fuzzy aggregation operation, usually implemented as a maximum-finding operation (set union); $\overrightarrow{X}$ - vector of input variables in the knowledge base; $d_j$ - fuzzy inference term for the j-th production rule; $y_j$ - fuzzy classification of credit history according to the j-th rule; $\mu_{dj}(y)$ is the membership function of the fuzzy inference value to the term $d_j$.

Let us give the final result for the example under consideration. The crisp aggregated output value $y^*$ corresponding to the input vector $\overrightarrow{X}$ is determined as a result of defuzzification of the fuzzy set $\tilde{y}$. In [7], the center of gravity method was used:

$$y_i^* = \sum_{j=1,m} d_j \mu_{dj}(\overrightarrow{X}) / \left[ \sum_{j=1,m} \mu_{dj}(\overrightarrow{X}) \right] = 0,392; \tag{4}$$

The values $\{y_i^*\}$, $i = 1, 2, \ldots, l$- are calculated for each credit history and, according to the proposed *concept 1*, are introduced as an additional ("behavioral") factor into the neural network model (NNM) [10] for NLIDM forecasting diagnostics (extrapolation) bankruptcy risk.

*Comment.* The introduction into the neural network model (NNM) of the "behavioral" factor of the borrower, which characterizes his tendency to delays in the process of loan repayment, implements the connection between the first and second stages of the bank's loan portfolio management.

## 2.2 Provisions 2 Generalization of the Neural Network Logistic Iterative Dynamic Method

As shown by computational experiments on real data [2], the predictive power of the dynamic model of bankruptcies obtained by NLIDM is estimated at approximately 87.6% of correctly identified borrowers. A further increase in the accuracy of NLIDM comes up against a fundamentally unrecoverable incompleteness of the initial data in the training set of the neural network, which consists in setting discrete labels: "$P = 1$ (bankrupt)", "$P = 0$ (not bankrupt)". The latter is due to the legal aspects of declaring the borrower bankrupt. The dynamic model of bankruptcy obtained with the help of NLIDM would acquire a significantly higher quality if intermediate values of the probability $P$ between 0 and 1 appeared in the training set.

The article proposes the following approach to generalizing the neural network logistic iterative dynamic method (NLIDM):

*Provisions 2* or increasing the informativeness of the training set of a neural network, and, accordingly, generalizing NLIDM is formed as follows: *in order to increase the predictive power of NLIDM, it is proposed to introduce into the model a priori information about intermediate values of the probability of bankruptcy risk $P \in [0; 1]$, obtained on the basis of a Bayesian ensemble of static logistic models bankruptcies V. Yu. Zhdanov, Olson [5] and other types:*

$$P_q(\hat{y}(\overrightarrow{x})) = 1 / \left[ 1 + \exp(\hat{y}_q(\overrightarrow{x})) \right], \tag{5}$$

where $\widehat{y}_q = (\overrightarrow{x})$ is the regression dependence of the exponent on the vector of factors (financial coefficients) $\overrightarrow{x} = (x_1, x_2, \ldots, x_n)$ for the q-th model - the a priori hypothesis.

In our studies, we used 16 coefficients from four groups [2]:

Group 1 - *"profitability"*; group 2 - *"liquidity and solvency"*; group 3 - *"business activity"*; group 4 - *"financial stability"*.

Models - a priori hypotheses $\{h_q\}, q = \overline{1, Q}$ in the Bayesian ensemble differ in regression coefficients and the composition of factors:

$$\widehat{y}(\overrightarrow{x}) = b_0 + b_1 x_1, \ldots + b_j x_j, \ldots + b_n x_n \tag{6}$$

To increase the reliability of the estimates in (5), they were averaged over the filtered Bayesian ensemble:

$$\overline{P} = \left[\sum_{q=1}^{Q^*} P_q(\widehat{y}(\overrightarrow{x}))\right]/Q^* \tag{7}$$

where $Q^*$ is the number of models on the filtered ensemble.

## 3   Study Result

### 3.1   Ways to Restructure Credit Debt

The choice of debt restructuring methods mainly depends on two factors:

- on the stage of developing bankruptcy (stages 1… 7 in Table 1);
- on the bank's credit policy.

In this article there is no possibility of detailed analysis of the methods of restructuring. Therefore, we will simply list them, noting that for a multicriteria choice of a specific method from many alternatives, one can use the method of pairwise expert assessments, that is, the method of analysis of hierarchies (HAI), developed by T. Saati [2].

Methods of restructuring: 1) change in the maturity of the debt; 2) change in the amount of the payment; 3) change in the obligation with the termination of the original obligation; 4) issuance of a stabilization loan; 5) reduction of the interest rate; 6) increasing the terms of the loan; 7) suspension of interest accrual; 8) on-lending, etc.

### 3.2   Assessment of Sensitivity to Variation of Factors and Verification of a Neural Network Logistic Dynamic Model of Bankruptcies

Comparison of the neural network logistic iterative dynamic method (NLIDM) of bankruptcies in terms of sensitivity and detection of stages of the developing process of bankruptcy was carried out for 7 construction companies (Table 3). For comparison, 22 well-known bankruptcy models from the groups in Table 2 were used: linear models such as multiple discriminant analysis (MDA); modern "advanced" *logit*-models;

expert models; rating models; model according to the regulated methodology of the Government of the Russian Federation.

All these models and techniques are covered by the *QFinAnalysis* software product (version 1.9) developed by V.Yu. Zhdanov and I.Yu. Zhdanov [5]. The initial data was the standard accounting reporting. The results of the assessments are shown in Table 3, where the probability of bankruptcy risk is indicated in%, or as a qualitative indicator. In the regulated methodology of the Government of the Russian Federation, the letter "B" means that the enterprise has signs of bankruptcy.

**Table 3.** Summary table of the results of calculating the probability of bankruptcy risk using known models using the *QFinAnalysis* program

| Methods of analyzing the financial condition of the enterprise | Joint-stock company "$N_1$" | Joint-stock company "$N_2$" | Joint-stock company "$N_3$" | Joint-stock company "$N_4$" | Joint-stock company "$N_5$" | Join5t-stock company "$N_6$" | Joint-stock company "$N_7$" |
|---|---|---|---|---|---|---|---|
| Altman's two-factor model | Less than 50% | … | … | Less than 50% | … | … | Less than 50% |
| Fedotova's two-factor model | Very high | … | … | Very high | … | … | Very high |
| Altman's five-factor model | Untenable | … | … | Untenable | … | … | Untenable |
| Altman's modified five-factor model | Untenable | … | … | Untenable | … | … | Untenable |
| The four-factor Taffler model | Indefinite | … | … | Indefinite | … | … | Indefinite |
| The four-factor Fox model | Small | … | … | Small | … | … | Small |
| The four-factor Fox model is adjusted | Small | … | … | Small | … | … | Small |
| The Four-factor Springate model | Low | … | … | Low | … | … | Low |
| The four-factor IGEA model | 90–100% | … | … | 90–100% | … | … | 90–100% |
| The Sayfullin-Kadykov model | Low | … | … | Low | … | … | Low |

(*continued*)

**Table 3.** (*continued*)

| Methods of analyzing the financial condition of the enterprise | Joint-stock company "$N_1$" | Joint-stock company "$N_2$" | Joint-stock company "$N_3$" | Joint-stock company "$N_4$" | Joint-stock company "$N_5$" | Join5t-stock company "$N_6$" | Joint-stock company "$N_7$" |
|---|---|---|---|---|---|---|---|
| The Parenoy-Dolgolev model | Average | ... | ... | Average | ... | ... | Average |
| Model of the Republic of Belarus | Small risk | ... | ... | Small risk | ... | ... | Small risk |
| Savitskaya's model | Financially stable | ... | ... | Financially stable | ... | ... | Financially stable |
| The Altman-Sabato model, % | 100 | 98 | 98 | 0 | 0 | 98 | 0 |
| Model Lina-Play, % | 45 | ... | ... | 45 | ... | ... | 45 |
| The Juha-Tehong Model, % | 50 | ... | ... | 38 | ... | ... | 8 |
| The Georgian Model, % | 1 | ... | ... | 1 | ... | ... | 1 |
| Zhdanov's model, % | 1 | ... | ... | 1 | ... | ... | 1 |
| The Zaitseva model | High | ... | ... | High | ... | ... | High |
| KSTU model | Class 3-risk | ... | ... | Class 3-risk | ... | ... | Class 3-risk |
| Beaver Model | 2 years before bankruptcy | ... | ... | 2 years before bankruptcy | ... | ... | 2 years before bankruptcy |
| Regulated methodology The model of the Government of the Russian Federation in 1994 | "B" - a sign of bankruptcy | ... | ... | "B" - a sign of bankruptcy | ... | ... | "B" - a sign of bankruptcy |

*Evaluation Results.* If we fix corporations, i.e. columns of Table 3, then all the methods and the regulated methodology of the Government of the Russian Federation in 1994 give significantly contradictory results, which do not allow the development of effective practical recommendations for restructuring accounts payable.

If we fix the rows of Table 3, then all corporations are evaluated the same in the sense of closeness to bankruptcy, i.e. the fixed model does not distinguish between them. An exception is the Altman-Sabato model, which singled out three enterprises (joint-stock company "$N_4$", joint-stock company "$N_5$", and joint-stock company "$N_7$") as far from bankruptcy.

Thus, 21 out of 22 models do not have sufficient sensitivity to the signs and stages of the developing crisis process at enterprises. On the other hand, the original NLIDM from has the required sensitivity in dynamics, ie. differentiates organizations by the stages of bankruptcy observed at a given time $t$, and, most importantly, the method makes it possible to make a forecast for a given time horizon ($t = t_{\pi p}$). So for the corporation (joint-stock company "$N_2$") anti-crisis measures allowed to reduce the risk of bankruptcy from 0,95 to 0,4. At the same time, the corporation (joint-stock company "$N_3$") had a rather high risk of bankruptcy by 2020 ($P = 0,75$).

## 4  Conclusions

General conclusions on assessing the sensitivity of NLIDM to the dynamics of changes in factor signs:

1. For the original neural network logistic iterative dynamic method (NLIDM) for constructing a model for assessing bankruptcy risk taking into account continuous time t, which restores incomplete data in time "slices", a comprehensive verification of the sensitivity of the method to identifying signs and stages of an emerging corporate crisis was carried out.
2. By comparison with estimates for 22 known methods, including modern "advanced" logistic methods, it is shown that NLIDM has sufficient contrast, i.e. is able to differentiate the nature of the dynamic dependences $P(\overrightarrow{x}(t), t)$ of the bankruptcy risk probability. For various corporations, this allows the lender to track the dynamics of $P(\overrightarrow{x}(t), t)$ while servicing the loan portfolio and promptly start debt restructuring procedures. The mentioned 21 methods and techniques, which mainly cover the modern arsenal of bankruptcy risk models, do not have such a property of contrasting estimates. Consequently, NLIDM expands the capabilities of modern economic and mathematical tools and, most importantly, makes it possible to more effectively solve applied problems of financial management.
3. An analysis of their sensitivity was carried out on a wide range of well-known models of bankruptcies; the ability to differentiate the nature of the dynamic dependences of bankruptcy risk for different organizations. It is shown that out of 22 models, which mainly cover the modern tools for modeling bankruptcies, only the (Altman - Sabato) model has the ability to differentiate the probability of bankruptcy.
4. The neural network logistic iterative dynamic method (NLIDM) allows to build extrapolation models of bankruptcies with continuous time, to obtain a differentiated estimate of the dynamics of $P(t)$. The method has a predictive power of about 87% of correct corporations identification.Two provisions for improving the efficiency of the bank's loan portfolio management are proposed, based on the neural network logistics iterative dynamic method and the fuzzy Mamdani method.

# References

1. Bizyanov, E.E.: Implementation of fuzzy models in information systems of economic objects. Econ. Innov. Manag. № 4. P. 1 (2015). https://ekonomika.snauka.ru/en/2015/04/8351
2. Biryukov, A.N.: Bayesian regularization of neural network models for ranking and clustering economic objects. Ufa: Academy of Sciences of the Republic of Belarus, Publishing House "Gilem", 380 p. (2011)
3. Biryukov, A., Antonova, N.: Expert systems of real time as key tendency of artificial intelligence in tax administration. In: Antipova, T., Rocha, A. (eds.) DSIC18 2018. AISC, vol. 850, pp. 111–118. Springer, Cham (2019). https://doi.org/10.1007/978-3-030-02351-5_15
4. Borisov, V.V., Kruglov, V.V., Fedulov, A.S.: Fuzzy Models and Networks: Monograph. Hot line – Telecom (2007)
5. Zhdanov, V.Yu.: Diagnostics of the risk of bankruptcy of an enterprise in three-dimensional space. Manag. Econ. Syst. No. 8. 20 p. (2011)
6. Zadeh, L.A.: The concept of a linguistic variable and its application to approximate reasoning - I. Inf. Sci. **8**(3), 199–249 (1975). https://doi.org/10.1016/0020-0255(75)90036-5
7. Kuznetsov, L.A., Perevozchikov, A.V.: Assessment of the credit history of individuals based on fuzzy models. Manag. Large Syst.: Collect. Works. **21**, 84–106 (2008)
8. Lebedeva, M.: Fuzzy logic in economics – the formation of a new direction. Ideas and Ideals **11**(1–1), 197–212 (2019). https://doi.org/10.17212/2075-0862-2019-11.1.1-197-212
9. Holvoet, T., Valckenaers, P.: Exploiting the environment for coordinating agent intentions. In: Weyns, D., Parunak, H.V.D., Michel, F. (eds.) E4MAS 2006. LNCS (LNAI), vol. 4389, pp. 51–66. Springer, Heidelberg (2007). https://doi.org/10.1007/978-3-540-71103-2_3
10. Ni, X.: Research of data mining based on neural networks. World Acad. Sci. Eng. Technol. **39**, 381–384 (2008)

# Studying the Attack Detection Problem Using the Dataset CIDDS-001

Quang-Vinh Dang[(⊠)] [iD]

Industrial University of Ho Chi Minh City, Ho Chi Minh City, Vietnam
dangquangvinh@iuh.edu.vn

**Abstract.** Intrusion detection is an important problem in cybersecurity research. In recent years, researchers have leveraged different machine learning algorithms to empower intrusion detection systems (IDS). In this paper, we study the intrusion detection problem using the dataset CIDDS-001 released in 2017. The dataset is much different from the popular datasets using in the literature in that it is not equipped with a comprehensive feature list. We show empirically that we can effectively classify the attacks by using state-of-the-art machine learning algorithms.

**Keywords:** Intrusion detection system · Machine learning · DNS · Cyber-Security

## 1 Introduction

An intrusion detection system (IDS) [7] is an important component of cybersecurity today. Given the fact that almost every computer today is connected to the Internet, they pose a lot of vulnerabilities to attackers. Besides the anti-virus programs that are familiar to most of the end-users, the IDS is playing a crucial role to keep computers safe.

The task of the IDS is to prevent any malicious traffic from outside to enter the LAN network. Hence, it is natural to consider the IDS as a classifier: it classifies then allows the benign traffic to pass through and stop the suspicious traffic.

In recent years, researchers have studied various machine learning techniques to use in the intrusion detection problem. The researchers usually use some public datasets for training and evaluating their algorithms. To name a few, KDD'99 [10] and CICIDS [18] are two among the most used intrusion datasets. These datasets are equipped with a lot of calculated features that will ease the task of the classifiers.

However, these features might not be available or easy to be calculated in practice. It might limit the application of the established techniques in industrial settings.

In this paper, we study the intrusion detection problem using the dataset CIDDS [14]. Different from other datasets, the CIDDS dataset is provided in near-raw format. Hence we have a lot of freedom to analyze the dataset in its original form. We will show that the comprehensive feature set is not necessary to build a robust and powerful intrusion classifier.

© The Author(s), under exclusive license to Springer Nature Switzerland AG 2022
T. Antipova (Ed.): DSIC 2021, LNNS 381, pp. 525–532, 2022.
https://doi.org/10.1007/978-3-030-93677-8_46

The remaining of the paper is as follows. We review the related studies in Sect. 2. Then we present the dataset and the experimental results in Sects. 3 and 4. We conclude our paper in Sect. 5 and draw some further research directions.

## 2  Related Works

In this section, we quickly review some recent research studies that are closely related to the studying problem. We encourage the readers to refer to other review papers [15, 16, 19] for more detailed information.

In the work of [3], the researchers evaluated different outlier detection techniques [12, 20] to detect abnormal traffic networks. The technique does not require the researchers to explicitly define what benign and malicious traffic is, but relies on the assumption that the benign traffic will make up the majority of the traffic. Hence, the model assumes that if traffic is abnormal it is an attack. In general, the assumption might be violated [13], particularly during the event of an attack. However, the outlier detection algorithm family does not require the expensive labeled dataset. Other research works studied modern deep learning techniques like RNN-LSTM [21] to detect the anomaly.

In the work of [4], the authors performed a comprehensive comparison between many supervised algorithms, starting from Naive Bayes to ensemble methods. The authors claimed that the algorithm xgboost [2] achieved the highest predictive performance in various popular metrics.

The intrusion detection on IoT attracted a lot of attention recently [1]. Most intrusion detection techniques in literature are developed for generic computer systems that are not available for IoT systems.

In the work of [5], the authors argued that using the entire labeled training dataset is not needed, hence they propose to use active learning [17] to incrementally train the model. The researchers can achieve similar predictive performance to the best models of [4] but with less training data.

In the work of [6], the authors proposed a different tactic to approach the problem: they improve the models by trying to explain them. The work is extended in [7]. The main idea is that, by explaining a model, we might force the model to use robust features rather than non-robust ones, and in general we will improve the performance of the model.

The authors of [8] suggested using reinforcement learning algorithms. The authors achieved comparable performance to simple classification algorithms like Naive Bayes or logistic regression, but not yet reach the performance of xgboost.

Deep learning techniques are being studied actively [11]. However, given the fact that the network data are often formatted, the deep learning techniques do not significantly superb the traditional algorithms.

# 3  Datasets

We use the dataset CIDDS-001 [14]. The dataset contains data collected in four weeks divided into three classes: normal, victim, and attacker. We focus only on OpenStack data, that includes 28,033,635 normal instances, 1,656,605 attacker instances and 1,579,422 victim instances. There are three types of attacks: portscan, ping-scan, and brute-force. We display the number of each type of traffic flow in Table 1. The features of the datasets are presented in Table 2.

**Table 1.** Number of each type of traffic flows

| Flow type | Number |
|---|---|
| Normal | 28,033,635 |
| Attacker-dos | 1,480,217 |
| Victim-dos | 1,478,810 |
| Attacker-portScan | 168,177 |
| Attacker-pingScan | 4,134 |
| Attacker-bruteForce | 4,077 |
| Victim-portScan | 97,741 |
| Victim-pingScan | 1,956 |
| Victim-bruteForce | 915 |

**Table 2.** Features of the CIDDS-001 dataset

| Nr. | Name | Description |
|---|---|---|
| 1 | Src IP | Source IP Address |
| 2 | Src Port | Source Port |
| 3 | Dest IP | Destination IP Address |
| 4 | Dest Port | Destination Port |
| 5 | Proto | Transport Protocol (e.g. ICMP, TCP, or UDP) |
| 6 | Date first seen | Start time flow first seen |
| 7 | Duration | Duration of the flow |
| 8 | Bytes | Number of transmitted bytes |
| 9 | Packets | Number of transmitted packets |
| 10 | Flags | *OR* concatenation of all TCP Flags |
| 11 | Class | Class label (normal, attacker, victim, suspicious or unknown) |
| 12 | AttackType | Type of Attack (portScan, dos, bruteForce, —) |
| 13 | AttackID | Unique attack id. All flows which belong to the same attack carry the same attack id. |
| 14 | AttackDescription | Provides additional information about the set attack parameters (e.g. the number of attempted password guesses for SSH-Brute-Force attacks) |

There are four protocols presented in the dataset: TCP, UDP, ICMP, and IGMP. The distribution of protocols is displayed in Fig. 1. More than 80% of the traffic is transmitted using the TCP protocol.

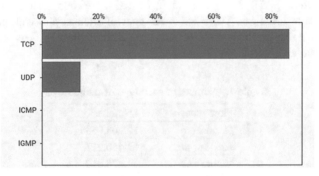

**Fig. 1.** Distribution of protocols

We note that the dataset itself is very large. Reading the data from the SSD disk to pandas took more than 20 min using our server. We can see that the CIDDS data is quite simple in terms of features, and most of the features such as IP addresses are not usable in classification machine learning models.

## 4    Experimental Results

All the experiments are done using our server with the configuration of CPU Intel E5-2683v3 14 cores 28 threads 2.00 GHz, 1TB of RAM. To keep our experiments be comparable with future research studies we do not use GPU for the computation. We used the language Python 3.8.10, scikit-learn 0.24.2, and catboost 0.26, which are the latest versions as of this writing.

### 4.1    Detecting Attacks

In this study, we focus on detecting the attacks in general but not the detailed type of attacks. So the task is to classify traffic into three classes: normal, attack, or victim.

**Table 3.** Comparison of different algorithms in detecting attacks

| Algorithm | Training time | Predicting time | Accuracy |
|---|---|---|---|
| Logistic regression | 22 m | 10 s | 89.7% |
| Random forest | 2 h 5 m | 1 m 43 s | 99.98% |
| CatBoost | 1 m 38 s | 26 s | 99.02% |

We evaluate three algorithms: Logistic Regression, Random Forest, and CatBoost [9]. We used the data of week 1 for training and the data of week 2 for testing. The data of week 3 and week 4 are not used because they do not contain any attacking traffic.

We display the results in Table 3. The Random Forest algorithm achieved the highest accuracy score, but require a very long training time (more than two hours) even with our relatively powerful server. On the other hand, the CatBoost algorithm achieved a comparable accuracy score with the Random Forest but with a much shorter training time.

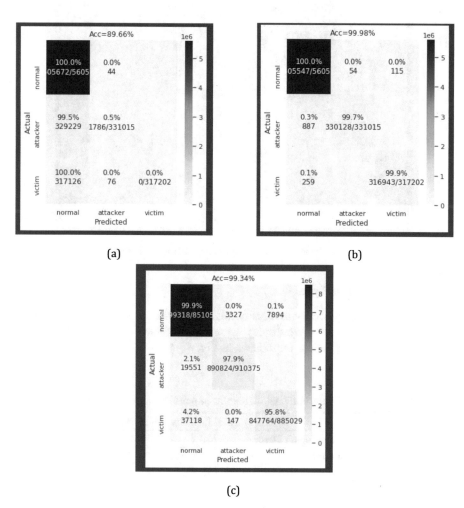

**Fig. 2.** Confusion matrix of three algorithms: a) logistic regression; b) random forest; and c) CatBoost

We display the confusion matrix of all algorithms in Fig. 2. The logistic regression algorithm does not learn anything but performs the majority prediction. The feature importance of the CatBoost algorithm is displayed in Fig. 3.

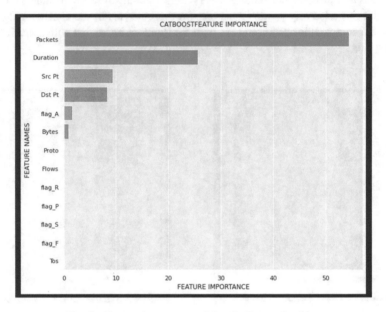

**Fig. 3.** Feature importance of the CatBoost algorithm

## 4.2 Detecting Attack Types

In this part of the study, we try to classify not only victims or attackers but also try to classify which attack type is. We display the results in Table 4. Similar to the previous section, we display the feature importance in Fig. 4.

**Table 4.** Comparison of different algorithms in detecting attacks and attack types

| Algorithm | Training time | Predicting time | Accuracy |
|---|---|---|---|
| Logistic Regression | 21 m | 10 s | 89.4% |
| Random Forest | 2 h 10 m | 1 m 56 s | 99.5% |
| CatBoost | 1 m 42 s | 25 s | 99.0% |

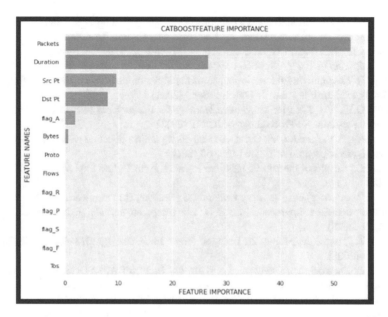

**Fig. 4.** Feature importance of the CatBoost algorithm in detecting attack types

## 5   Conclusions

In this paper, we study the intrusion detection problem using several classification algorithms. We evaluated the algorithms using the CIDDS-001 dataset. The experimental results showed that CatBoost is the suitable algorithm for the intrusion detection problem. However, the predictive accuracy of CatBoost can be improved further and that will be our research study topic.

## References

1. Benkhelifa, E., Welsh, T., Hamouda, W.: A critical review of practices and challenges in intrusion detection systems for iot: toward universal and resilient systems. IEEE Commun. Surv. Tutor. **20**(4), 3496–3509 (2018)
2. Chen, T., Guestrin, C.: Xgboost: a scalable tree boosting system. In: KDD, pp. 785–794. ACM (2016)
3. Dang, Q.V.: Outlier detection in network flow analysis. arXiv:1808.02024 (2018). 4
4. Dang, Q.V.: Studying machine learning techniques for intrusion detection systems. In: Dang, T., Küng, J., Takizawa, M., Bui, S. (eds.) Future Data and Security Engineering. FDSE 2019. LNCS, vol. 11814, pp. 411– 426. Springer, Cham (2019). https://doi.org/10.1007/978-3-030-35653-8_28
5. Dang, Q.V.: Active learning for intrusion detection systems. In: IEEE Research, Innovation and Vision for the Future (2020)

6. Dang, Q.V.: Understanding the Decision of Machine Learning Based Intrusion Detection Systems. In: Dang, T.K., Küng, J., Takizawa, M., Chung, T.M. (eds.) Future Data and Security Engineering. FDSE 2020. LNCS, vol. 12466, pp. 379–396. Springer, Cham (2020). https://doi.org/10.1007/978-3-030-63924-2_22

7. Dang, Q.V.: Improving the performance of the intrusion detection systems by the machine learning explainability. Int. J. Web Inf. Syst. (2021)

8. Dang, Q.V., Vo, T.H.: Reinforcement learning for the problem of detecting intrusion in a computer system. In: Proceedings of ICICT (2021)

9. Dorogush, A.V., Ershov, V., Gulin, A.: Catboost: gradient boosting with categorical features support. arXiv preprint arXiv:1810.11363 (2018)

10. Elkan, C.: Results of the kdd'99 classifier learning. Acm Sigkdd Explor. Newsl. **1**(2), 63–64 (2000)

11. Ferrag, M.A., Maglaras, L., Moschoyiannis, S., Janicke, H.: Deep learning for cybersecurity intrusion detection: approaches, datasets, and comparative study. J. Inf. Secur. Appl. **50**, 102419 (2020)

12. Liu, F.T., Ting, K.M., Zhou, Z.: Isolation forest. In: ICDM, pp. 413–422. IEEE Computer Society (2008)

13. MontazeriShatoori, M., Davidson, L., Kaur, G., Lashkari, A.H.: Detection of doh tunnels using time-series classification of encrypted traffic. In: 2020 IEEE International Conference on Dependable, Autonomic and Secure Computing, International Conference on Pervasive Intelligence and Computing, Intl Conf on Cloud and Big Data Computing, International Conference on Cyber Science and Technology Congress (DASC/PiCom/CBDCom/CyberSciTech), pp. 63–70. IEEE (2020)

14. Ring, M., Wunderlich, S., Grüdl, D., Landes, D., Hotho, A.: Flow-based benchmark data sets for intrusion detection. In: Proceedings of the 16th European Conference on Cyber Warfare and Security, pp. 361–369. ACPI (2017)

15. Salih, A.A., Abdulazeez, A.M.: Evaluation of classification algorithms for intrusion detection system: a review. J. Soft Comput. Data Min. **2**(1), 31–40 (2021)

16. Samrin, R., Vasumathi, D.: Review on anomaly based network intrusion detection system. In: 2017 International Conference on Electrical, Electronics, Communication, Computer, and Optimization Techniques (ICEECCOT), pp. 141–147. IEEE (2017)

17. Settles, B.: Active learning. Synthesis Lect. Artif. Intell. Mach. Learn. **6**(1), 1–114 (2012)

18. Sharafaldin, I., Lashkari, A.H., Ghorbani, A.A.: Toward generating a new intrusion detection dataset and intrusion traffic characterization. ICISSp **1**, 108–116 (2018)

19. Thakkar, A., Lohiya, R.: A review of the advancement in intrusion detection datasets. Procedia Comput. Sci. **167**, 636–645 (2020)

20. Wang, H., Bah, M.J., Hammad, M.: Progress in outlier detection techniques: a survey. IEEE Access **7**, 107964–108000 (2019)

21. Zhou, X., Hu, Y., Liang, W., Ma, J., Jin, Q.: Variational lstm enhanced anomaly detection for industrial big data. IEEE Trans. Ind. Inf. **17**(5), 3469–3477 (2020)

# Cognitive Screening Instruments for Community-Dwelling Older Adults: A Mapping Review

Rute Bastardo[1], João Pavão[2], Ana Isabel Martins[3] (ID),
Anabela G. Silva[3] (ID), and Nelson Pacheco Rocha[4(✉)] (ID)

[1] UNIDCOM, Science and Technology School, University of Trás-os-Montes
and Alto Douro, Vila Real, Portugal
[2] INESC-TEC, Science and Technology School, University of Trás-os-Montes
and Alto Douro, Vila Real, Portugal
jpavao@utad.pt
[3] CINTESIS.UA, Health Sciences School, University of Aveiro,
Aveiro, Portugal
{anaisabalmartins,asilva}@ua.pt
[4] IEETA, Department of Medical Sciences, University of Aveiro,
Aveiro, Portugal
npr@ua.pt

**Abstract.** This paper presents a systematic mapping review of the literature on innovative digital solutions to detect cognitive impairment of community-dwelling older adults. Seventy-six articles were included in this mapping review. Most of the included articles (i.e., 65 articles) reported the implementation and validation of computerized versions of paper-based neuropsychological tests. In turn, 11 studies are related to the application of emerging technologies to detect cognitive decline, including serious games, virtual reality, and data analytics (e.g., algorithms to analyse data from ubiquitous daily activity and interaction sensing) approaches. From these 11 studies, four include experimental setups to determine if the developed digital solutions can discriminate cognitive impairments. Based on the mapping review findings is possible to conclude that further research is required to develop cognitive screening approaches alternative to computerized versions of paper-based neuropsychological tests.

**Keywords:** Older adults · Cognitive screening · Digital solutions

## 1 Introduction

Neuropsychological assessment instruments, such as the Mini-Mental State Examination (MMSE) [1] or Montreal Cognitive Assessment (MoCA) [2], are used for extensive evaluation of multiple cognitive domains by trained practitioners. Since the application of these instruments is time and resource consuming, they are not a practical solution for cognitive screening in the general population.

Alternatively, there are computerized versions of neuropsychological tests. These computerized versions are being used to evaluate either multiple cognitive domains [3]

© The Author(s), under exclusive license to Springer Nature Switzerland AG 2022
T. Antipova (Ed.): DSIC 2021, LNNS 381, pp. 533–544, 2022.
https://doi.org/10.1007/978-3-030-93677-8_47

(e.g., Self-administered Gerocognitive Examination (SAGE) [4] or Cogstate [5]), and specific cognitive domains (e.g., attention, executive functions, or memory) [6]. However, these computerized instruments have been designed to mirror the neuropsychological assessment tests, applied by trained practitioners in a clinical setting and not for screening cognitive impairment in large populations.

Considering the importance of the early detection of mild cognitive impairment, it is envisaged the development of new instruments able to monitor the individuals in their residential environments [7]. In this respect, multiple technologies have the potential to support new methods to detect mild cognitive impairment and ultimately improve access to care [7, 8]. Smart devices (e.g., smartphones, smartwatches, or smart-home devices) may collect data that can be analysed (e.g., using machine learning technologies [9]) to detect subtle changes that may indicate decline in cognitive functioning [7]. Passive measures extracted from daily activities and interactions were proposed for a continuous neuropsychological monitoring, including sleep changes (e.g., increased number of awakenings and decreased deep or slow-wave sleep) [10], verbal fluency [11], behaviour [12], social interactions [13], gait [14] or fine motor control [15].

In addition, virtual reality might help empower individuals to monitor their own cognitive performance [6] and serious games allow for alternative approaches to cognitive tests. In fact, serious games might reduce feelings of test anxiety [16] and motivate the individuals to frequently complete the tests, which may be particularly relevant for the self-administration of tests at home where non-compliance can be an issue [17].

The mapping review presented in this article aimed to analyse current trends of innovative digital solutions for cognitive monitoring of older adults in their residential environments, and to provide systematic evidence of state-of-the-art solutions.

## 2   Methods

The mapping review was informed by the following research questions:

- RQ1 - What are the trends in developing digital solutions that might be used as a screening tool for age-related cognitive impairment of older adults in their residential environments?
- RQ2 - How emerging technologies such as serious games, virtual reality and data analytics are being used in the development of digital solutions that might be used as a screening tool for age related cognitive impairment?
- RQ3 - What type of experimental setups are being performed for the user-centred evaluation of the proposed digital solutions?

To perform the literature review, the authors defined a review protocol with explicit descriptions of the steps to be taken, which are described in this section.

The resources chosen for the review were three electronic databases (Scopus, Web of Science, and Medline). Queries were prepared to identify articles that have their titles, abstract or keywords conform with the following Boolean expression: ('cognitive screening' OR 'cognitive test' OR 'memory screening' OR 'memory test' OR 'attention screening' OR 'attention test') AND ('computer' OR 'game' OR 'gaming' OR 'virtual' OR 'online' OR 'internet' OR 'mobile' OR 'app' OR 'digital').

The electronic literature search was performed in January 2021 and included all the references published before December 31, 2020.

In terms of inclusion criteria, the articles were included if:

- Were full-text manuscripts.
- Reported on any independent digital solution that might be used as a generic self-administrated screening tool for age related cognitive impairment of community-dwelling older adults.
- Reported the evaluation of the proposed digital solution by end-users.
- Were written in English.

In turn, considering the exclusion criteria, articles were excluded if they reported on any digital solution that:

- Is intended to be used as a cognitive monitoring tool over time of patients diagnosed with any acute neurological condition.
- Was designed and evaluated considering patients with specific diseases.

The selection of the articles was performed in three steps:

- First step - the authors removed the duplicates, the articles without abstract and that were not written in English.
- Second step - the authors assessed all titles and abstracts for relevance and those clearly not meeting the inclusion and exclusion criteria were removed.
- Third step - the authors then assessed the full text of the remaining articles against the outlined inclusion and exclusion criteria and the final list of the articles for the review was created.

Throughout this entire process, all the articles were analysed by three authors and any disagreement was discussed and resolved by consensus.

Concerning data extraction, the following information was registered in a data sheet prepared by the authors for each of the articles included in the review: (i) the demographics of the published work (i.e., authors, year, or source of publication); (ii) the scope of the article; (iii) the purpose of the digital solution being reported; (iv) characterization of the participants; and (v) details of experimental setups for the evaluation by end-users.

Based on the demographic data of the included articles, a synthesis of studies' characteristics was prepared, which included: (i) the number of articles published in conference proceedings and the number of articles published in scientific journals; (ii) the distribution of the articles by year and the publication rate, which was calculated using RMS Least Square Fit; and (iii) distribution of the articles by country. Since some articles involved multidisciplinary teams, the institutional affiliation of the first author of each article was considered when determining the number of articles per nation.

The different digital solutions described by the included articles were coded in terms of technological support and development approaches. Concerning, the technological support, the articles were classified as reporting: (i) standalone applications;

and (ii) distributed applications, which were further divided as supported on personal computers, smartphones, tablets, wearables, and virtual reality setups. In turn, two classes were considered for the development approaches: (i) computerized versions of paper-based cognitive tests; and (ii) new approaches based on emerging technologies to assess cognitive functions, which was further divided into serious games, virtual reality, and data analytics.

Finally, considering the objectives of the experimental setups being reported, four different classes were identified: (i) studies aiming to assess the usability of the proposed digital solutions; (ii) studies aiming to evaluate the feasibility of the concepts being proposed; (iii) studies aiming to validate digital solutions to support tests to assess specific cognitive domains or test batteries to assess multiple cognitive domains; and (iv) studies aiming to evaluate the discrimination capability (i.e., whether the proposed digital solution is able to discriminate between individuals with and without cognitive impairment).

## 3    Results

### 3.1    Study Selection

A total of 7570 articles were retrieved from the initial search of the selected databases.

The first step yielded 3991 articles since 3579 articles were removed because: (i) were duplicated (i.e., 3283 articles); (ii) the abstracts were missing (i.e., 160 articles); or (iii) were not written in English (i.e., 136 articles).

During the second step, one article was excluded because it was retracted, and 3887 were removed because they did not meet the outlined inclusion and exclusion criteria.

Finally, also considering the inclusion and exclusion criteria, during the full-text analysis (i.e., the third step) 37 articles were removed due to the following reasons: (i) reported on solutions that are intended to be applied by clinicians and not to be self-administered (e.g. [18]); or (ii) were designed or evaluated for specific health conditions (e.g., [19]). Therefore, 76 articles were included for this systematic mapping review.

### 3.2    Demographics of the Included Articles

In terms of publication types, only four articles were published in conference proceedings. The remaining 72 studies were published in scientific journals.

Concerning the publication years, the oldest articles were published in 2010 (i.e., two articles). In turn, 16 articles were published in 2020. The diagram of Fig. 1 demonstrates that the publication trend is increasing, and more than 20% of articles (i.e., 16 articles) were published in the last year and more than 70% of the articles (i.e., 55 articles) were published in the last five years.

Figure 2 represents the distribution of the included articles by country. North America (28 articles) and Europe (28 articles) have the highest percentage of participation. Moreover, Asia contributed with 14 studies, while South America and Oceania together contributed with six articles.

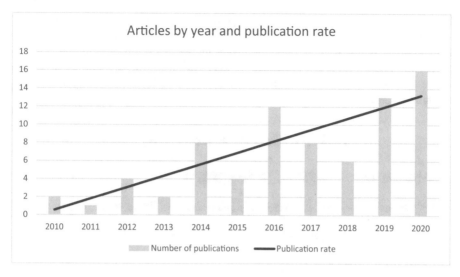

**Fig. 1.** Articles by year and publication rate (calculated using RMS Lest Square Fit).

**Fig. 2.** Distribution of the selected articles by country.

## 3.3 Technological Support

Figure 3 presents the nature of the applications (i.e., standalone or distributed) and the terminal equipment being used (i.e., personal computers, smartphones, tablets, wearable devices such as smartwatches, or virtual reality setups). A significant number of articles (i.e., 26 articles) reported the use of standalone applications, while the remaining articles (i.e., 50 articles) reported the use of web-based applications that can be accessed by different types of terminal equipment: personal computers (18 articles), smartphones (10 articles), tablets (20 articles) wearable devices (five articles) and virtual reality setups (two articles). The total number of occurrences in terms of the different terminal equipment is superior to the number of articles reporting distributed applications since some digital solutions support different types of terminals (e.g., smartphone and tablet).

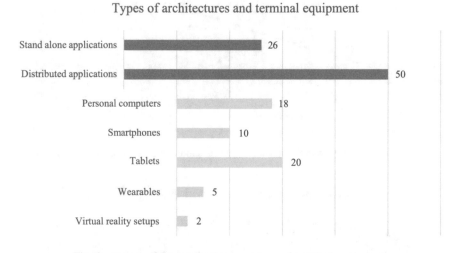

**Fig. 3.** Nature of the applications and type of terminal equipment.

## 3.4   Development Approaches

The included studies reported the use of various types of digital solutions to provide:

- Computerized versions of existing paper-based cognitive test batteries to assess multiple cognitive domains (e.g., MMSE [20], SAGE [21, 22], or Cogstate [23–27]), and cognitive tests to assess specific cognitive functions, namely attention (e.g., Trail Making Test [28, 29]), executive functions (e.g., Simple and Complex Reaction Time [30], Digit Span Backwards [23], or Stroop [29, 30]), and memory (e.g., Face-Name Associative Memory Exam [29, 31]).
- New approaches to assessing cognitive functions based on emerging technologies, including serious games, virtual reality, and data analytics.

Figure 4 represents the relative importance of the two approaches: 65 articles reported the development of computerized versions of paper-based cognitive tests, while 11 articles reported the implementation of new approaches to assess cognitive functions.

Looking specifically to these 11 articles [17, 32–41], six developed serious games [17, 34, 35, 37, 39, 41], two presented solutions based on virtual reality [33, 38], and three used data analytics [32, 36, 40]. In terms of serious games, one [34] presented a serious game based on a virtual supermarket, one [17] presented a gamification version of an executive function test (i.e., Go/No-Go Discrimination Task), another [35] provided a cognitive scoring using bimanual interaction with custom games, one [37] proposed new exercises to test executive functions, and two other [39, 41] proposed exercises to test multiple cognitive functions. In turn, concerning virtual reality, one [33] mimicked paper-based tests to assess specific cognitive domains, and another [38] assessed reactions to visual stimulus. Finally, looking for data analytics, one article [32] proposed a dementia scale classification based on ubiquitous daily activity and interaction sensing, one [36] assessed cognitive decline through quantitative analysis of parameters derived from body-worn inertial sensors, and another [40] examined whether passive aspects of responding to a remotely monitored weekly online questionnaire can be used to detect cognitive declines.

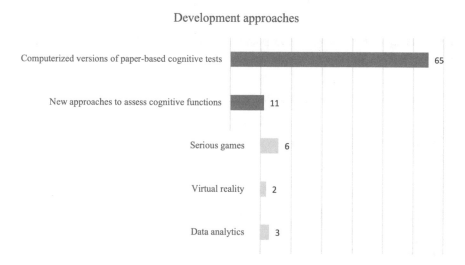

**Fig. 4.** Development approaches.

## 3.5 Objectives of the Experimental Setups

Figure 5 represents the distribution of the included articles by year, according to the reported experimental objectives (i.e., usability, feasibility, validation, or discrimination studies).

In turn, Fig. 6 presents the same distribution for the subset of articles reporting new approaches based on emerging technologies to access cognitive functions. In this respect, four articles reported experimental setups to evaluate the discrimination capabilities [33, 34, 37, 39], six aimed to evaluate the feasibility of the proposed solutions [17, 32, 35, 36, 38, 40], while one aimed to assess usability [41].

Number of articles by type of validation method and years

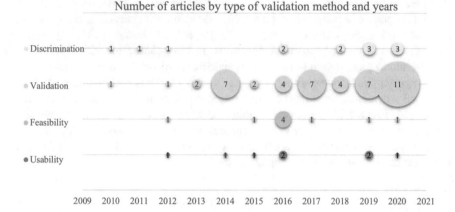

**Fig. 5.** Distribution of the number of studies according to the objectives of the experimental setups by year.

Number of articles by type of validation method and years

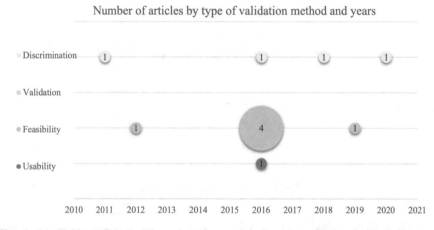

**Fig. 6.** Distribution of the number of studies reporting the use of emerging technologies to assess cognitive functions according to the objectives of the experimental setups by year.

## 4   Discussion and Conclusion

Concerning the first research question (i.e., what are the trends in developing digital solutions that might be used as a screening tool for age-related cognitive impairment of older adults in their residential environments?), during the last ten years, there was active research in the field and the review identified 76 studies. Moreover, looking for the distribution of the articles by year, it is possible to conclude that the publication rate is increasing.

The great majority of the included articles (i.e., 65 articles) aimed to propose digital solutions to support the application of existing paper-based instruments that are being used in clinical settings. In some studies, the aim was to develop digital versions of tests being used to assess specific cognitive functions (e.g., Trail Making Test, Simple and Complex Reaction Time, Digit Span Backwards, Stroop and Face-Name Associative Memory Exam), while in other studies the focus was to provide digital solutions to support the application of test batteries to assess multiple cognitive functions (e.g., MMSE, SAGE or Cogstate). In turn, a comparatively small number of studies (i.e., 11 studies) aimed to use emerging technologies, including serious games, virtual reality, and data analytics, to assess cognitive functions.

Concerning technological architectures, 26 studies reported standalone architectures and 50 studies reported distributed architectures. Moreover, as shown in Fig. 3 the distributed solutions are supported in a diversity of terminal equipment, including personal computers, smartphones, tablets, wearables, and virtual reality setups.

In what concerns the usage of emerging technologies such as serious games, virtual reality, and data analytics in the development of screening tools for age-related cognitive impairment of older adults in their environments (i.e., the second research question), six studies reported the use of serious games, two studies applied virtual reality and three studies developed different data analytics techniques.

The reported digital solutions are already endowed with a certain degree of development, as evidenced by the evaluations that were reported. In terms of the experimental objectives (i.e., the third research question), it was possible to identify four different classes: (i) usability evaluation; (ii) feasibility evaluation; (iii) validation of digital solutions based on paper-based scales to assess specific cognitive domains or to assess multiple cognitive domains; and (iv) evaluation of the discrimination capability.

Among the studies aiming to evaluate the discrimination capacity of the digital solutions, three studies were based on serious games [34, 37, 39] and one study was based on the usage of virtual reality [33]. In turn, in what concerns data analytics, none of the studies reported an assessment to determine the discrimination capability.

Based on the results of this review, it is possible to affirm that the potential of the emerging technologies, namely serious games, virtual reality, or data analytics, to support the cognitive screening of cognitive impairment of older adults in their residential environments seems to be not yet sufficiently explored. However, during the selection of the articles to include in this scoping review a significant number of articles that reported on the application of emerging technologies was excluded because they did not report on evaluations involving end-users (i.e., one of the inclusion criteria). This indicates that the ongoing research in the field is still not well consolidated to provide digital solutions to be assessed by end-users and, consequently, able to be translated to clinical settings.

The approach adopted in this work presents some limitations. The principal limitations are the chosen search keywords and the databases that were used for the identification of the articles. Although Scopus, Web of Science, and Medline are significant databases on scientific production, they still do not include the whole set of publications. Also, the reliance solely on the English language may reduce the number of studies considered in the analysis. Despite these limitations, the authors followed rigorous procedures for retrieving, selecting, and analyzing the articles, to guarantee the reliability of the results.

542    R. Bastardo et al.

# References

1. Cockrell, J., Folstein, M.: Mini-mental State Examination. In: Copeland, J., Abou-Saleh, M., Blazer, D. (eds.) Principles and practice of geriatric psychiatry, pp. 140–141. John Wiley & Sons, Hoboken, NJ, United States (2002)
2. Nasreddine, Z., et al.: The montreal cognitive assessment, MoCA: a brief screening tool for mild cognitive impairment. J. Am. Geriatrics Soc. **53**(4), 695–699 (2005)
3. Snyder, P., Jackson, C., Petersen, R., et al.: Assessment of cognition in mild cognitive impairment: a comparative study. Alzheimer's Dement. **7**(3), 338–355 (2011)
4. Scharre, D., Chang, S.-I., Murden, R., et al.: Self-administered gerocognitive examination (SAGE). Alzheimer Dis. Assoc. Disord. **4**, 64–71 (2010)
5. Fredrickson, J., Maruff, P., Woodward, M.: Evaluation of the usability of a brief computerized cognitive screening test in older people for epidemiological studies. Neuroepidemiology **34**, 65–75 (2010)
6. Diaz-Orueta, U., Blanco-Campal, A., Lamar, M., et al.: Marrying past and present neuropsychology: is the future of the process-based approach technology-based? Front. Psychol. **11**, 361 (2020)
7. Sabbagh, M., et al.: Early detection of mild cognitive impairment (MCI) in an at-home setting. J. Prev. Alzheimer's Dis. **7**, 171–178 (2020)
8. Gates, N., Kochan, N.: Computerized and on-line neuropsychological testing for late-life cognition and neurocognitive disorders: are we there yet? Curr. Opin. Psychiatry **28**(2), 165–172 (2015)
9. Pereira, C., Pereira, D., Weber, S., et al.: A survey on computer-assisted Parkinson's disease diagnosis. Artif. Intell. Med. **95**, 48–63 (2019)
10. Ibáñez, V., Silva, J., Cauli, O.: A survey on sleep assessment methods. PeerJ **6**, e4849 (2018)
11. Tröger, J., et al.: Telephone-based dementia screening I: automated semantic verbal fluency assessment. In: 12th EAI International Conference on Pervasive Computing Technologies for Healthcare, pp. 59–66. ACM, New York, NY, Unired States (2018)
12. Kourtis, L., Regele, O., Wright, J., Jones, G.: Digital biomarkers for Alzheimer's disease: the mobile/wearable devices opportunity. NPJ. Digit. Med. **2**(1), 1–9 (2019)
13. Marengo, D., Settanni, M.: Mining facebook data for personality prediction: an overview. In: Baumeister, H., Montag, C. (eds.) Digital Phenotyping and Mobile Sensing. Studies in Neuroscience, Psychology and Behavioral Economics, pp. 109–124. Springer, Cham (2019). https://doi.org/10.1007/978-3-030-31620-4_7
14. Grande, G., et al.: Measuring gait speed to better identify prodromal dementia. Exp. Gerontol. **124**, 110625 (2019)
15. Dagum, P.: Digital biomarkers of cognitive function. NPJ. Digit. Med. **1**, 1–3 (2018)
16. Lumsden, J., et al.: Gamification of cognitive assessment and cognitive training: a systematic review of applications and efficacy. JMIR Serious Games **4**(2), e11 (2016)
17. Tong, T., Chignell, M., Tierney, M., Lee, J.: A serious game for clinical assessment of cognitive status: validation study. JMIR Serious Games **4**, e7 (2016)
18. Bott, N., et al.: Device-embedded cameras for eye tracking–based cognitive assessment: validation with paper-pencil and computerized cognitive composites. J. Med.Internet Res. **20**(7), e11143 (2018)
19. Smith, A., III., Duffy, C., Goodman, A.: Novel computer-based testing shows multi-domain cognitive dysfunction in patients with multiple sclerosis. Mult. Scler. J. Exp. Transl. Clin. **4**(2), 2055217318767458 (2018)

20. Dharmasaroja, P., Ratanakorn, D., Nidhinandana, S., Charernboon, T.: Comparison of computerized and standard cognitive test in thai memory clinic. J. Neurosci. Rural Pract. **9** (1), 140–142 (2018)
21. Lauraitis, A., Maskeliunas, R., Damasevicius, R., et al.: A mobile application for smart computer-aided self-administered testing of cognition, speech, and motor impairment. Sensors **20**(11), 3236 (2020). https://doi.org/10.3390/s20113236
22. Scharre, D., Chang, S., Nagaraja, H., et al.: Digitally translated self-administered gerocognitive examination (eSAGE): relationship with its validated paper version, neuropsychological evaluations, and clinical assessments. Alzheimer's Res. Ther. **9**(1), 1–3 (2017)
23. Asgari, M., Gale, R., Wild, K., et al.: Automatic assessment of cognitive tests for differentiating mild cognitive impairment: a proof of concept study of the digit span task. Curr. Alzheimer Res. **17**(7), 658–666 (2020)
24. Koyama, A., Hagan, K., Okereke, O., et al.: Evaluation of a self-administered computerized cognitive battery in an older population. Neuroepidemiology **45**(4), 264–272 (2015)
25. Darby, D., Pietrzak, R., Fredrickson, J., et al.: Intraindividual cognitive decline using a brief computerized cognitive screening test. Alzheimer's Dement. **8**(2), 95–104 (2012)
26. Darby, D., Fredrickson, J., Pietrzak, R., et al.: Reliability and usability of an internet-based computerized cognitive testing battery in community-dwelling older people. Comput. Hum. Behav. **30**, 199–205 (2014)
27. Hammers, D., Spurgeon, E., Ryan, K., et al.: Validity of a brief computerized cognitive screening test in dementia. J. Geriatr. Psychiatry Neurol. **25**(2), 89–99 (2012)
28. Kokubo, N., Yokoi, Y., Saitoh, Y., et al.: A new device-aided cognitive function test, user experience-trail making test (UX-TMT), sensitively detects neuropsychological performance in patients with dementia and Parkinson's disease. BMC Psychiatry **18**(1), 220 (2018)
29. Troyer, A., Rowe, G., Murphy, K., et al.: Development and evaluation of a self-administered on-line test of memory and attention for middle-aged and older adults. Front. Aging Neurosci. **6**, 335 (2014)
30. Hafiz, P., Bardram, J.: The ubiquitous cognitive assessment tool for smartwatches: design, implementation, and evaluation study. JMIR mHealth uHealth **8**(6), e17506 (2020)
31. Alegret, M., Muñoz, N., Roberto, N., et al.: A computerized version of the short form of the face-name associative memory exam (FACEmemory®) for the early detection of Alzheimer's disease. Alzheimer's Res. Ther. **12**(1), 25 (2020)
32. Okada, S., et al.: Dementia scale classification based on ubiquitous daily activity and interaction sensing. In: 8th International Conference on Affective Computing and Intelligent Interaction (ACII), pp. 192–198. IEEE, New York, NY, United States (2019)
33. Wright, D., et al.: A novel technology to screen for cognitive impairment in the elderly. Am. J. Alzheimer's Dis. Dementias® **26**(6), 484–91 (2011)
34. Boz, H., Limoncu, H., Zygouris, S., et al.: A new tool to assess amnestic mild cognitive impairment in Turkish older adults: virtual supermarket (VSM). Aging Neuropsychol. Cogn. **27**(5), 639–653 (2020)
35. House, G., Burdea, G., Polistico, K., et al.: A serious gaming alternative to pen-and-paper cognitive scoring: a pilot study of BrightScreener™. J. Pain Manag. **9**(3), 255–264 (2016)
36. Greene, B., Kenny, R.: Assessment of cognitive decline through quantitative analysis of the timed up and go test. IEEE Trans. Biomed. Eng. **59**(4), 988–995 (2011)
37. Neto, H., Cerejeira, J., Roque, L.: Cognitive screening of older adults using serious games: an empirical study. Entertain. Comput. **28**, 11–20 (2018)
38. Oliveira, C., et al.: Development and feasibility of a virtual reality task for the cognitive assessment of older adults: the ECO-VR. Spanish J. Psychol. **19**, e95 (2016)

39. Ruano, L., Sousa, A., Severo, M., et al.: Development of a self-administered web-based test for longitudinal cognitive assessment. Sci. Rep. **6**(1), 19114 (2016)
40. Seelye, A., Mattek, N., Howieson, D., et al.: Embedded online questionnaire measures are sensitive to identifying mild cognitive impairment. Alzheimer Dis. Assoc. Disord. **30**(2), 152 (2016)
41. Boletsis, C., McCallum, S.: Evaluating a gaming system for cognitive screening and sleep duration assessment of elderly players: a pilot study. In: Bottino, R., Jeuring, J., Veltkamp, R. (eds.) Games and Learning Alliance. GALA 2016. LNCS, vol. 10056, pp. 107–119. Springer, Cham (2016). https://doi.org/10.1007/978-3-319-50182-6_10

# New Processor Architecture and Its Use in Mobile Application Development

Rostislav Fojtik$^{(\boxtimes)}$ (iD)

University of Ostrava, 30. dubna 22, Ostrava, Czech Republic
rostislav.fojtik@osu.cz

**Abstract.** The paper deals with the possibilities of new processor architectures. The most common processors are based on the x86 instruction set-based architecture among personal computers and many server platforms. On the other hand, mobile devices use processors based on the ARM architecture, which matches desktop processors' performance in many respects. It turns out that the new approach of creating a chip architecture can bring significantly higher processor performance and, at the same time, save energy consumption. One of the technologies is the creation of SoC. The paper focuses on a practical comparison of Intel and Apple processors' performance, representing two different architectures. It presents the results of functional tests, their comparison and demonstrations of the possibilities of using the new architecture in mobile computers and desktops and notebooks. The paper uses benchmarks, a comparison of compilation and execution of programs, and standard applications. The article shows the possibilities, advantages and disadvantages of ARM processor architecture in mobile applications.

**Keywords:** Architecture · ARM · Instruction set · Education · Processor

## 1 Introduction

The development of information and communication technologies affects all areas of human action. Despite the rapid growth of hardware, there is still a demand for higher performance, miniaturisation, and lower power consumption. Mobile technologies, wearables and the Internet of Things are playing an increasingly important role. New processors need to be developed to increase the quality of these devices. In the field of mobile devices, ARM (Advanced RISC Machine) processors are commonly used today. These processors initially outperformed i86 processors by consuming less power. But nowadays, their performance is also increasing. Their quality and computing power are beginning to compete with the processors used for desktops and laptops. These devices typically use x86 instruction set processors. ARM processors are currently used in other computing devices [8].

After introducing the new Apple Silicon M1 processor, based on the ARM architecture and the SoC (System on a Chip) system, comparative tests emerge between this new processor and the x86 instruction set processors. Test results show that the M1 processor can compete with x86 processors and outperform them in many ways [2, 7, 11]. There are many reasons why M1 processors match or outperform x86 processors.

© The Author(s), under exclusive license to Springer Nature Switzerland AG 2022
T. Antipova (Ed.): DSIC 2021, LNNS 381, pp. 545–556, 2022.
https://doi.org/10.1007/978-3-030-93677-8_48

For example, it is an SoC solution. M1 uses instructions of the same length, unlike Intel and AMD processors, which must have more complex decoders for instructions of varying sizes, which causes more complex and often slower instruction decoding. The older x86 instruction set causes problems in optimising and increasing computing power [15, 16].

The M1 contains a central processing unit (CPU), graphical processing unit (GPU), memory, input and output controllers, and many more things making up a whole computer. This is called a system on a chip (SoC). The M1 processor further includes, for example, Image Processing Unit (ISP), Digital Signal Processor (DSP), Neural Processing Unit (NPU), Video encoder/decoder, Secure Enclave [1, 3]. Apple M1 processors and other modern processors make efficient use of multiple cores and AI units [7, 17].

At the beginning of the comparison, a research question was defined. Can the Apple Silicon M1 compete with currently manufactured x86 desktop processors? This work aims to practically compare x86 processors with ARM processors, especially Apple Silicon M1 processors, using tests and practical activities. Based on the existing knowledge, the hypothesis was established: Apple Silicon M1 processors surpass current x86 processors for the same class of computers. The performance of the new M1 processors is also advantageous for the development of mobile applications.

## 2    Methodology

To compare different processors, it is advisable to use standard test applications, specially designed programs, and typical applications. ARM processors and processors were compared with the x86 instruction set using standard benchmarks Geekbench 5, Cinebech R23.200 and Novabench 4.0.2. These tests are often used and can be run on both processor platforms. The time required to execute programs created in the Swift and C++ programming languages was another comparison aspect. The article's author wrote those programs, and their goal was to test the time load of demanding algorithms working with a more significant amount of data. The programs were created using Xcode, which is optimised for both processor architectures. Because benchmarks and specialised tests may not always reflect the needs of regular users, a practical comparison of the performance of selected applications was also performed, monitoring the speed of application, CPU load and warm-up. Additional tests were performed to measure the time required to compile projects in development environments.

Comparative tests were performed on several devices with different processors. I completed most of the tests on two notebooks. The first was a 13″ MacBook Pro with an Intel Core i5-8259U processor, 4 Cores, GPU Iris Plus Graphics 655, 16 GB RAM, 512 GB SSD, macOS 11.2.1 (assembly 20D74). The second was a MacBook Air 13″ with an Apple M1 processor, 8 Cores (4 efficiency cores), 16 GB RAM, SSD 1 TB, macOS 11.2.1 (assembly 20D74). The tests were performed repeatedly, usually ten times. The computer with an Intel processor is offered as a medium-performance notebook. The MacBook Air with M1 processor is provided as a basic notebook model. In a practical comparison, other devices were used: iPhone 12 Pro Max, iPad Pro 11, MacBook Pro 13″ (i7), MacBook Pro 16″ (i9), Huawei P30, Samsung Galaxy Tab S4 10″.

The results were compared using the T-test: Two-Sample Assuming Equal or Unequal Variances. Two sample t-test is used to compare the difference between the two populations. The significance level was set at 0.05. This parametric test assumes that the variances are the same in both groups. The F-Test Two-Sample tested this assumption for Variances. The data variability measured was measured in each item, and the variation coefficient was used. A t-test was performed to interpret the results of the second-degree classification. The results were processed using MS Excel and statistical software Wizard for Mac OS X's operating system and statistical software Statistics Visualizer for iPad [12].

# 3 Results

## 3.1 Processor Performance Comparison

All tables and graphs were created based on a practical comparison of the mentioned devices. The author made all comparisons of the article. The first test was performed using the Cinebench tool (Table 1). In the test on one or more cores, a computer with an Apple Silicon M1 processor received more points and a better score. In both cases, the Apple Silicon M1 was almost 37% more powerful than the Intel Core i5. The results for M1 were better for calculations on one or more cores.

**Table 1.** Cinebench score.

| Cinebench | Multi-Core | Single Core |
|---|---|---|
| MacBook Pro 13, Intel C5, Iris Plus Graphics 655 | 4459 | 944 |
| MacBook Air 13, Apple M1 | 7026 | 1493 |

The second test was performed repeatedly using the Novabench tool. Table 2 contains the results, which show that a computer with an Apple Silicon M1 processor is more powerful than a laptop with an Intel Core i5 processor. More points mean better results. The data confirm that the difference in computer performance affects the CPU and GPU the most. The M1 processor and its integrated GPU score significantly more than the Intel Core i5 and Iris Plus Graphics 655 GPUs.

**Table 2.** Novabench score.

| Novabench score | Computer | CPU | GPU | RAM | Disk |
|---|---|---|---|---|---|
| MacBook Pro 13, Intel C5, Iris Plus Graphics 655 | 1555 | 850 | 267 | 274 | 164 |
| MacBook Air 13, Apple M1 | 2118 | 1071 | 567 | 309 | 171 |

Testing with Geekbench 5 was performed on eight devices with different processors (Fig. 1). This graph shows the number of points achieved. More points mean a better result. The scores executed by Intel desktop processors and ARM processors are not as diametrically different as they were a few years ago. The Intel Core i9 processor achieved the highest score when working with multiple cores but only slightly out-performed the Apple Silicon M1 processor. Let's consider that the Intel Core i9 processor is designed for powerful personal computers, and the M1 processor is calculated more for basic personal computer configurations. The difference seems negligible, and the question arises whether it is worth buying a more powerful processor. The computing power of multiprocessor processors is significant today. Many applications use multiple processor cores [5, 7].

The Apple Silicon M1 in the MacBook Air 13 "and the Apple A14 in the iPhone 12 Pro Max achieved the best results when testing a single processor core. Processors with ARM architecture thus surpassed the monitored Intel processors. The results show that processors initially designed primarily for mobile devices match or even beat x86 desktop processors. The Apple Silicon M1 processor is designed for basic computer configurations. Nevertheless, it achieved similar results in the test as the Intel i9 processor, designed for the most powerful personal computers.

Practical tests were performed with typical applications on both mentioned note-books. Applications optimised for the M1 processor and applications that required Rosetta 2 emulation to run were used. None of the applications ran slower on a computer with M1 processor. On the contrary, most of their operations were performed faster, even for some running in emulation in Rosetta 2 [4, 10]. The same procedures were always performed on both computers and monitored, where they were completed more quickly. The following were used as optimised applications: iMovie and Final Cut Pro editing video applications; development tools for programming Xcode, IntelliJ IDEA, PyCharm, CLion; Pixelmator Pro photo-editing application; MS Office applications; Chrome, Safari, Firefox, Edge browsers. Non-optimised applications were Visual Studio and Android Studio development tools, VLC video player, OBS screen recording application, MS Teams and Skype communication applications.

The difference between the monitored computers was noticeable even at higher loads. During an experiment with a video meeting at MS Teams and recording the screen with OBS, the MacBook Pro with an Intel Core i5 processor heated above 45 °C and active cooling made a noticeable noise and worked hard to cool the processor. During the same operation, the MacBook Air with the M1 processor did not exceed the computer's chassis temperature above 41 °C, even though the laptop lacked active cooling. The difference was also noted in the export of 4K video, which was 28 min long. The M1 computer managed 23% faster export than the Intel Core i5 computer. The computer with the Intel processor warmed up more and active cooling was working at total capacity.

It turns out that an architecture unloaded with the old x86 instruction set does not compromise so much and can provide high performance with relatively low power consumption. The processor can also decode M1 instructions faster.

The great advantage of the Apple Silicon M1 processor is the ability to run universal applications. That is, applications that can be run on desktop computers with the macOS operating system and at the same time on mobile devices with the iOS and iPad OS operating systems. This is possible because the Apple Silicon M1 processor continues the A12 Bionic mobile processors [14]. For example, the LumaFusion video editing application has been practically tested. Editing and subsequent export of the video went smoothly and faster than on the iPad.

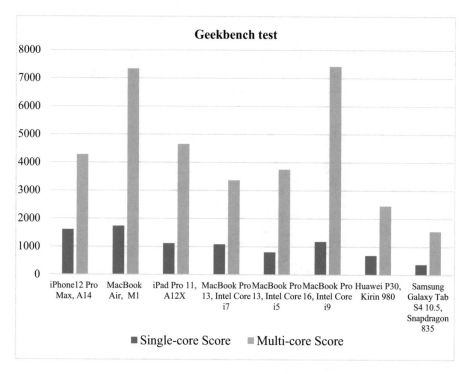

**Fig. 1.** Geekbench test.

The Geekbench also generated a Compute Score, in which a laptop with M1 processor regained better results by 62% (Table 3). Although both processor platforms are designed for equally equipped computers, the Apple Silicon M1 has outperformed the Intel i5 processor.

**Table 3.** Geekbench 5 - Compute Score.

| Geekbench 5 | Compute Score |
|---|---|
| MacBook Pro 13, Intel C5, Iris Plus Graphics 655 | 7425 |
| MacBook Air 13, Apple M1 | 19587 |

## 3.2   Development of Mobile Applications on Computers with M1 Processor

The results of measuring the performance of ARM M1 processors are excellent. But the performance of processors alone is not enough. Software support is required for practical application. Tests of program codes and the performance of development tools were performed.

Android Studio for Android applications and Xcode for iOS applications were used as test development tools. Both tools are recompiled for the Silicon M1 processor. Applications that are ready for M1 processors are listed at: https://isapplesiliconready.com.

From the performed measurements, it can be concluded that Android Studio is not yet sufficiently optimised for M1 processors. Also, not enough mobile device emulators are ready. Currently, there is only one virtual device available on GitHub [13]. Conversely, for processors with the x86 instruction set, there are enough emulators (Fig. 2).

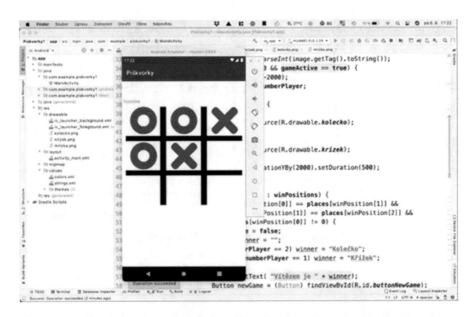

**Fig. 2.** Mobile phone emulator on a computer with M1 processor.

Android Studio was in version 2020.3.1 and was compared on two notebooks. The first was a 13″ MacBook Pro with an Intel Core i5-8259U processor, 4 Cores, 16 GB RAM, 512 GB SSD, macOS 11.2.1. The second was a MacBook Air 13″ with an Apple M1 processor, 8 Cores, 16 GB RAM, SSD 1 TB, macOS 11.2.1.

Figure 3 shows the results of a comparison of the time required to compile programs. Shows the ratio of compilation time on a laptop with M1/compilation time on a computer with i5 processor. Fifteen different projects were compared. In most cases, programs are compiled faster on an i5 computer. If multiple projects were opened, compilation on a laptop with M1 was more quickly.

Developing mobile applications on computers with M1 processor is possible, but it brings more complications. Compilation of projects is usually slower than on computers with x86 processors, and at the same time, a sufficient number of emulators is missing. Developers are forced to make more use of real mobile devices and test applications on them.

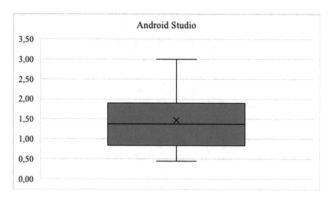

**Fig. 3.** Compilation time ratios - projects for the Android OS.

On the contrary, developing mobile applications with the operating system iOS, iPad OS, Watch OS on computers with the M1 processor is more advantageous. Development tools and necessary libraries are sufficiently optimised. Figure 4 shows the results of project compilation time ratios on an i5-based computer and M1-based computer. Fifteen different projects were compiled. The graph shows that compilation on M1 computer is faster. Simulators of all mobile devices are ready for computers with M1 processors.

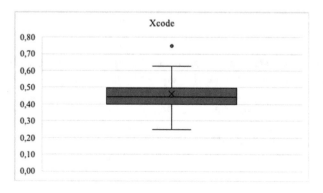

**Fig. 4.** Compilation time ratios - projects for iOS.

The following graphs (Fig. 5, 6, 7, 8, 9, 10) show the results of comparing the selected programs' execution on a MacBook Pro with an Intel Core i5 processor and a MacBook Air with M1 processor. The horizontal axis shows the time in seconds of the program being executed. The first five graphs show the results of the time required to complete programs written in the Swift programming language, compiled and run on these computers. The other four programs were written in the CPP programming language. Each test was performed ten times. Execution time was measured in each program. The measured data were processed using F-test and T-test. The null hypothesis cannot be rejected for only one program. In this case, the statistical difference was insignificant, and it cannot be stated that the program on the M1 processor was processed faster than on the Intel Core i5 processor. It was a program written using the Swift language, in which the elements of the array were deleted. The difference was always apparent in the remaining examples, exceeding the significance level, and the programs on the M1 processor were executed in a shorter time.

The graph in Fig. 5 compares the Fibonacci sequence calculation time for value 45. The program was created in the Swift programming language. It is clear from the chart and the statistical evaluation that the null hypothesis must be rejected in the comparison. The results are different for both computers, and the M1 processor managed to perform the calculation in less time than the Intel Core i5 processor. The graph shows the time in seconds of program execution on individual processors using a box graph.

**Fig. 5.** Fibonacci sequence – Swift language.

Figures 6, 7, 8, 9 show the comparison of programs in which eighteen thousand elements were added to the field. Subsequently, the fields were sorted in ascending order. Another algorithm performed a reverse operation with array elements. All programs were run ten times, and execution times were recorded. Repeated measurements of program execution speed have shown that the M1 processor executes these algorithms significantly faster than the Intel Core i5. The graphs show the execution time in seconds of the programs on the individual processors using box graphs.

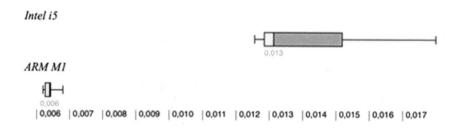

**Fig. 6.** Add elements to an array - Swift language.

**Fig. 7.** Sorting array elements - Swift language.

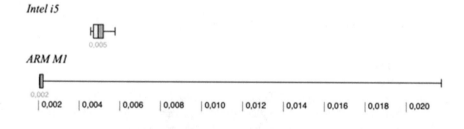

**Fig. 8.** Reverse - Swift language.

Figure 6 shows the only comparison for which the difference was not significant enough. It was a program written in the Swift programming language, in which six thousand elements in an array were taken.

```
for index in 0...6000{
    arrayCollection.remove(at: index)
}
```

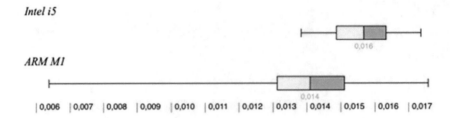

**Fig. 9.** Delete array elements - Swift language.

Similar results were measured for programs created in the C++ programming language. Figure 10 shows the results of time measurements when sorting array elements.

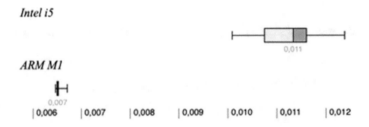

**Fig. 10.** Sorting array elements – C++ language.

## 4    Conclusion

The results prove that the Apple Silicon M1 processor, which uses the ARM architecture, can compete and essentially replace desktop processors with the x86 instruction set. Tools were benchmarking show that the computing power of M1 processors is better than that of the Intel Core i5 processor. The differences in the results were sufficiently convincing. The M1 processor also obtained better results by tracking the time required for programs created in Swift and C++. Using and testing typical applications has shown that the computing power of the processor is adequate even for applications that are not yet optimised for the ARM architecture, and therefore must run using Rosetta2 emulation.

The significant advantage is the ability to create universal applications that can be run on the desktop operating system macOS and at the same time on mobile devices such as iPhone or iPad. Economic and time requirements for the development of universal applications can be lower. Application distribution is also more comfortable.

The comparison results show that personal computers with ARM processors can increase companies' performance even in basic configurations. In many ways, the Apple Silicon M1 processor, designed for the base class of computers, rivals today's mid-range and high-end x86 processors. Businesses can also save on the energy costs of processors through the ARM architecture.

The M1 processor also has its shortcomings. The first drawback is optimising applications for the new architecture so that processor performance does not have to be devoted to the Rosetta 2 emulator. A more significant obstacle is the impossibility of virtualising other operating systems on macOS. Computers with M1 processor do not yet support dual boot for another operating system. This can be a significant barrier for users who need applications only for the MS Windows operating system on the x86 platform. The development and extension of MS Windows on the ARM operating system can provide hope in this direction.

Although some development environments, compilers, or libraries are not fully optimised for new processors, processors can be used to develop mobile applications.

Developing mobile applications for Android OS on computers with M1 processor is possible but less efficient with x86 processors. On the contrary, developing mobile applications with the operating system iOS, iPad OS, Watch OS on computers with the M1 processor is more advantageous.

**Acknowledgement.** The Ministry of Education supported this research, Youth and Sports Czech Republic within the Institutional Support for Long-term Development of a Research Organization in 2021.

# References

1. Becker, H., Hwang, V., Kannwischer, M.J., Yang, B., Yang, S.: Neon NTT: Faster Dilithium, Kyber, and Saber on Cortex-A72 and Apple M1. Cryptology ePrint Archive: Report 2021/986. IACR (2021)
2. Deakin, D.R.: Apple M1 vs Intel i7: Cherry-picked benchmarks and shifting Tiger Lake SKUs leave Intel looking revolutionary rather than evolutionary. https://www.notebookcheck.net/Apple-M1-vs-Intel-i7-Cherry-picked-benchmarks-and-shifting-Tiger-Lake-SKUs-leave-Intel-looking-revolutionary-rather-than-evolutionary.519176.0.html. Accessed 07 Feb 2021
3. Engheim, E.: Why Is Apple's M1 Chip So Fast? https://debugger.medium.com/why-is-apples-m1-chip-so-fast-3262b158cba2. Accessed 28 Nov 2020
4. Haihua, Q.: Research on key technology of embedded Linux system based on ARM processor. In: Proceedings of 2020 IEEE 3rd International Conference on Information Systems and Computer Aided Education (ICISCAE) (2020). ISBN 978-1-7281-8304-6
5. Joy, J.: Multi-core processors: concepts and implementations. Int. J. Comput. Sci. Inf. Technol. (IJCSIT) **10**(1), 1–10 (2018). ISSN 0975-3826
6. Forsell, M., Nikula, S., Roivainen, J., Leppanen, V., Traff, J.L.: Performance and programmability comparison of the thick control flow architecture and current multicore processors. J. Supercomput. (2021). ISSN0920-8542, https://doi.org/10.1007/s11227-021-03985-0

7. Guzide, O., Sloboda, S.: Is Apple's new M1 chip a gamechanger in computing? In: Proceedings of the West Virginia Academy of Science, vol. 93, no. 1 (2021). https://pwvas.org/index.php/pwvas/article/view/784

8. Groote, J.F., Morel, R., Schmaltz, J., Watkins, A.: The raspberry Pi and the ARM processor. In: Logic Gates, Circuits, Processors, Compilers and Computers, pp. 169–193. Springer, Cham (2021). https://doi.org/10.1007/978-3-030-68553-9_8

9. McMahon, J.: Apple M1 vs. Intel: Should You Trade in Your Old MacBook for a Slick New Processor? (2021). https://theinventory.com/apple-m1-vs-intel-i7-1846070507

10. Messaoudi, K., et al.: Hardware/software implementation of image processing systems based on ARM processor. In: Proceedings of the International Conference on Recent Advances in Electrical Systems, pp. 78–83. Tunisia (2019). ISBN 978-9938-9937-2-1

11. Rounak, J.: Comparing the Apple m1 MacBook vs the Intel MacBook. https://www.businessinsider.in/tech/news/comparing-the-apple-m1-macbook-vs-the-intel-macbook/articleshow/79733783.cms. Accessed 8 Dec 2020

12. Řehák, J., Brom, O.: SPSS, Praktická analýza dat. Computer Press (2015). ISBN 978-80-251-4609-5

13. Sai Balai Setting up Android Emulator in M1 Mac. https://medium.com/techiepedia/setting-up-android-emulator-in-m1-mac-fff58ec8bff8. Accessed 07 Feb 2021

14. Sarkar, S.: A survey of Apple A11 Bionic processor (2018). https://www.researchgate.net/publication/340580371_A_survey_of_Apple_A11_Bionic_processor

15. Sharma, K.: Apple M1 Vs Intel 11th Gen Vs AMD Ryzen 4000 (Laptops Processors Comparison). https://candytech.in/apple-vs-intel-vs-amd-ryzen/. Accessed 21 June 2021

16. Turley, J.: Apple M1 vs. Intel Core i7: The Benchmark Wars Continue. https://www.eejournal.com/article/apple-m1-vs-intel-core-i7-the-benchmark-wars-continue/. Accessed 15 Feb 2021

17. Venieris, S.I., Panopoulos, I., Leontiadis, I., Venieris, I.S.: How to reach real-time AI on consumer devices? Solutions for programmable and custom architectures. In: Proceedings 2021 IEEE 32nd International Conference on Application-specific Systems, Architectures and Processors (ASAP), pp. 93–100 (2021). https://doi.org/10.1109/ASAP52443.2021.00022

# Constructing a Model of the Main Highway of Cybersecurity Policy in a Digital Society Based on Multi-level Analysis

Dzhalladova Irada Agaverdi-kyzy[(✉)], Batechko Nina Grigoryevna,
Kaminsky Oleh Yevhenovych, and Gladka Yuliya Anatolievna

Kyiv National Economic University named after Vadym Hetman,
Kyiv 03057, Ukraine

**Abstract.** The article deals with the problem of formation of information culture of cyberspace by methods of system analysis. An analysis of modern approaches to the formation of the concept of "information culture" taking into account trends in the development of information and communication technologies is conducted. A methodology for forming information culture levels on the basis of the systematic approach is presented. The peculiarities of informational culture formation in conditions of uncertainty are specified.

**Keywords:** Cybersecurity · Cyberwar · System analysis · Digital society · Sociology of knowledge

## 1 Introduction

Cybersecurity covers the broad range of technical and social issues that must be considered to protect networked information systems. The importance of the concept has increased as so many government, business, and day-to-day activities globally have moved online. Cybersecurity is an important matter in (inter)national politics. Cybersecurity sensitivities arise primarily due to the continued proliferation of digital technologies in many aspects of human life. With the rise of online platforms where individuals could gather and spread information came the rise of online cybercrimes aimed at taking advantage of not just single individuals but collectives. In response, researchers and practitioners began trying to understand this digital playground and the way in which individuals who were socially and digitally embedded could be manipulated. The article deals with the problem of formation of information culture of cyberspace by methods of system analysis. An analysis of modern approaches to the formation of the concept of "information culture" taking into account trends in the development of information and communication technologies is conducted. A methodology for forming information culture levels on the basis of the systematic approach is presented. The peculiarities of informational culture formation in conditions of uncertainty are specified [1].

T. Antipova (Ed.): DSIC 2021, LNNS 381, pp. 557–566, 2022.
https://doi.org/10.1007/978-3-030-93677-8_49

## 2  Statement and Goal of Problem

The importance of cybersecurity has increased as so many government, business, and day-to-day activities around the world have moved online. But especially in emerging economies, "[m]any organizations digitizing their activities lack organizational, technological and human resources, and other fundamental ingredients needed to secure their system, which is the key for the long-term success" [2].

In research Linda Hantrais [3] assembles evidence from different areas of social science expertise about the impacts of Covid-19 in digitised societies and policy responses. At the micro-level, families are shown to have become 'digital by default', as children were exposed to online risks and opportunities. Globally, the spread of the pandemic provided a fertile ground for cybercrime, while digital disinformation and influencing risked becoming normalised and domesticated.

The more technically-focused information security is still in widespread use in computer science. But as these issues have become of much greater societal concern as "software is eating the world" [4], cybersecurity has become more frequently used, not only in the rhetorics of democratic governments as in the 2000s, but also in general academic literature.

The information policy of the governments of most countries is aimed at the transition to a digital society, building a new digital economy, preserving the multiculturalism of nations based on the digital environment. National programs for the development of digitalization prioritize government policy in the information sphere - the growth of national wealth through the creation of a competitive environment in the information highway, the appropriate regulatory framework and free access based on legal provisions to universal information services. Information policy is aimed at ensuring the leadership of states in the introduction and use of communication systems and electronic information, the creation of e-governments, which ensures the transparency of government structures and public access to government departments by electronic means. For example, in Canada, the national program "Building the Information Society: Moving Canada into the 21st Century" includes such strategic elements as the construction of information superhighways focused on public interests and market demand; election (creation) of a responsible government, which acts as a catalyst for the development of the information society and new technologies; transformation of the national economy, stimulation and encouragement of transnational projects in the financing of the information highway [5].

The work Ling Li and Wu He [6] suggested conceptual framework is tested using survey results from 579 business managers and professionals. The results show that when employees are aware of their company's information security policy and procedures, they are more competent to manage cybersecurity tasks than those who are not aware of their companies' cybersecurity policies. The study also indicates that an organizational information security environment positively influences employees' threat appraisal and coping appraisal abilities, which in turn, positively contribute to their cybersecurity compliance behavior.

Cybersecurity policy is implemented through specialized European institutions and national programs for the development of the digital society.

… The other side of modeling the cybersecurity policy of states… At the current stage of development of the geopolitical situation in the world to bring the state out of a difficult economic situation, which are planned or prepared by the target country in the "hot spots" of the planet with the use of military force and other international military activity of foreign countries in the world [7].

Many technical domains are of direct relevance to cybersecurity, but the field designed to synthesise technical knowledge in practical contexts has become known as security engineering: "building systems to remain dependable in the face of malice, error, or mischance" [8, p. 3]. It concerns the confluence of four aspects—policy (the security aim), mechanisms (technologies to implement the policy), assurance (the reliability of each mechanism) and incentives (of both attackers and defenders).

The article Štitilis D. and Pakutinskas P. [9] analyzed also revealed the importance of the global approach in cyber security phenomena. The majority of first strategies emphasized the cyber security capacity-building inside the country itself, while almost all the second strategies emphasize the building of capacities internationally.

The intelligence activities of the intelligence services to collect economic data are based, in accordance with strategic guidelines, on the tactics of "all azimuths": the collection of operational classified information not only in the CIS and former "socialist camp", but also in the United States and other NATO. In terms of counterintelligence, the main efforts are aimed at combating industrial espionage, terrorism, drug trafficking and disinformation.

Research focussed on the interaction between cybersecurity and society has also expanded the relevant set of risks and actors involved. While the term cybersecurity is often used interchangeably with information security (and thus in terms of the CIA triad), this only represents a subset of cybersecurity risks. Insofar as all security concerns the protection of certain assets from threats posed by attackers exploiting vulnerabilities, the assets at stake in a digital context need not just be information, but could, for example, be people. Moreover, traditional threat models in both information and cybersecurity can be limited. For example, domestic abusers are rarely considered as a threat actor [10] and systems are rarely designed to protect their intended users from the authenticated but adversarial users typical in intimate partner abuse [11].

Anderson suggests the continued integration of software into safety-critical systems will require a much greater emphasis on safety engineering, and protection of the security properties of systems like medical devices (even body implants) and automotive vehicles for decades—in turn further strengthening political interest in the subject [8, p. 2].

In addition to the above approaches, there exist several studies that focus on the security aspects of specific vertical areas. For example, [12] focuses on unoccupied aerial systems (such as drones) and identifies several research challenges including trustworthiness, monitoring, and resilience. Similarly, [13] addresses attacks on autonomous vehicles, and focuses on attacks related to machine learning. Nader et al. [14] focus on smart cities and argue that a data-driven approach would significantly improve the security posture of smart cities.

**The goal of our study** is a systematic analysis of the logical, structural relationships between different levels of socio-cyber security. Investigate the diversity of links between aspects of sociocybernetic security in the first stage, followed by the formulation of precise step-by-step actions to make optimal decisions in a pandemic and other threats to humans and society.

## 3   Research Method

The article will use theoretical-analytical and practical models, methods of analysis of complex systems, system-oriented, inductive and deductive approaches to the interpretation of research results, the study of mutual influence and stability of economic objects.

## 4   The Basic Tenets and Results

Applying a systems approach, we will consider cybersecurity as a set of interconnected structural elements with vertical (mega -, macro -, meso -, micro -, nano -) and horizontal (foreign economic, macroeconomic, production, energy, financial, investment, food, demographic, social and intellectual) security, levels (see Fig. 1).

| $g_5$ mega | | | | | | | | | | |
|---|---|---|---|---|---|---|---|---|---|---|
| $g_4$ macro | | | $Ch_3g_4$ | $Ch_4g_4$ | $Ch_5g_4$ | $Ch_6g_4$ | $Ch_7g_4$ | $Ch_8g_4$ | $Ch_9g_4$ | $Ch_{10}g_4$ |
| $g_3$ meso | | | | | | | | | | |
| $g_2$ micro | | | | | | | | | | |
| $g_1$ nano | | | | | | | | | | |
| h 0 | $h_1$ | $h_2$ | $h_3$ | $h_4$ | $h_5$ | $h_6$ | $h_7$ | $h_8$ | $h_9$ | $h_{10}$ |

**Fig. 1.**  A set of interconnected structural elements of cybersecurity.

Note that the vertical levels (along the Og axis), we associate with cybersecurity policy at the global - g5 (mega) level, state - g4 (macro), regional - g3 (meso), sectoral (corporate) - g2 (micro), and personality levels - g1 (nano).

Horizontal levels (along the Oh axis) will be associated with cybersecurity policy in foreign policy - h1; macroeconomics - h2; production sphere - h3; energy - h4; financial sphere - h5; investment sphere - h6; food sector - h7; demographic - h8; social - h9; and intelligent - h10 (see Fig. 1).

All these areas are in one way or another related to the life of society, and cybersecurity policy makes it possible to implement the process of digitalization of society. Let us trace some highways of digitalization of society with indication of the corresponding levels: gi (i = 1,2, ..., 5), and hj (j = 1, ..., 10).

The government's task is to create a competitive environment in which firms can increase national wealth. The federal government must ensure that the digitization highway creates jobs and promotes economic growth in every sector of the economy: **Ch3 g4 – Ch10 g4.**

Where market forces cannot provide equal access or preconditions for it, the government must act. A national strategy for providing access to basic services through legislation regulating access to information for all citizens is proposed.

In the new digital economy, success will be determined by the market, not the state. Thus, the main role of the state should be to establish rules, and it itself should be a kind of model (example, model). Public authorities themselves must also undergo a reengineering stage [15].

By liberalizing telecommunications regulation rules, the government should seek to remove outdated and unnecessary barriers to competition and introduce protection against anti-competitive practices. In addition, the state itself must become a leader in the implementation and use of electronic information and communication systems, which will allow all citizens to be able to communicate and interact with government departments and agencies by electronic means.

Thus, the **role of the state** is to ensure a balance between competition and regulation, the freedom to use cyberpolitics to protect privacy and personal communications, and the need to protect the public interest from terrorists, freedom of speech and protection of morals and the interests of minors. This balance must be established and reviewed by the state itself, because market forces cannot do so. These include education, telemedicine, the idea of universal access to network services and information, access to government information.

The highway of digitalization of society at the state level is presented in Fig. 2, where all sectors Ch3 g4 - Ch10 g4 are accumulated.

Note that the development of the digitalization highway of society will be effective in synergistic interaction with sectors at levels g3 (sectors Ch 3 g4 - Ch10 g4), g2 (sectors Ch g2 - Ch10 g2), g1 (sectors Ch3 g1 - Ch10 g1). For example, the inefficient functioning of the C1 h9 highway sector will negatively affect the functioning of the entire system, and over time will lead to an imbalance in its existence.

The implementation of cybersecurity policy in the sectors Ch1 g5 - Ch10 g5, (see Fig. 1), has certain features related to the activities of special services of individual states.

For example, the current trend in the external activities of intelligence officers is based on the principle of "civilized intelligence", ie "intelligence and diplomatic" activities, along with the need to organize joint counteraction to transnational threats, which include the above factors in addressing cooperation with representatives. foreign intelligence services party (by the way, as well as German, Israeli, American) pursues, first of all, a purely intelligence purpose - to obtain the necessary information and provide beneficial influence directly during business communication with the intelligence services of partner countries. Expansion of official business contacts with colleagues of partner countries is used to create conditions for beneficial information and psychological influence, control over the reaction to secret actions carried out by special services against a foreign state, and covert management of the behavior of foreign colleagues.

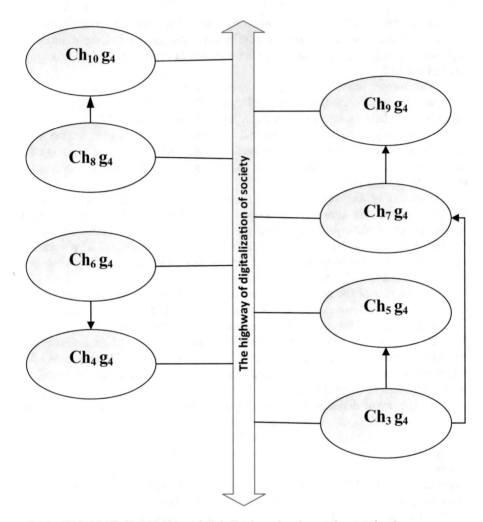

**Fig.2.** The highway of digitalization of society at the state level.

For example, France has developed and successfully implemented a targeted state program of French cultural (ideological) and French-speaking expansion for states with a "weak national elite" to create conditions for consistent economic and financial expansion and subordination of their economies to the interests of the French nation. A significant role in the implementation of such a program during the information and psychological activities is given to the cultural association "Alliance Française" and the Institute of Civil Service - "Cooperation Service" (alternative military), which reports to the French Ministry of Foreign Affairs represented by diplomatic missions abroad.

The main activity in modern conditions is the penetration of agents into higher state structures and the environment of senior officials of foreign states in order to carry out covert management of the behavior of the first persons of states during the adoption of responsible political decisions. To do this, the traditional source of important information is the leaders and influential officials of the opposition movements of the state-object of intelligence. Such work abroad is traditionally carried out by members of the political intelligence departments of the secret services, who operate under deep diplomatic cover. formations and groups of the state of accreditation.

It is a traditional practice to actively use the opportunities of the network of various non-governmental structures created in the host countries.

To stimulate and consolidate lasting relations with the selected persons, as a rule, their trips abroad are organized at the expense of the host party, where "guests" were given and are now given increased attention to creating particularly comfortable conditions and honors at receptions, meetings with representatives are organized. higher levels of the Western elite.

During the special operational work, the representatives of foreign states: first, have received and continue to receive information (especially closed or secret) about the processes taking place in the state; secondly, based on its analysis and development of its approaches, it is planned to provide moral and financial support through advice, recommendations and thus have a beneficial impact on the political, military, economic and moral situation, and thirdly, it gave and allows to focus on information and forecasts of dignitaries, more objectively assess the state and prospects of the situation in certain regions, in a particular field of economics and science, socio-political problem.

This multilevel model (see Fig. 1) can cover the interaction of different types of society, its digitization and cybersecurity policy of this process. However, each sector of interaction has its own characteristics, inherent in these levels, which in one way or another affect the overall process of digitalization, this stability and invulnerability. Note that the implementation of this process will be facilitated by the following *basic principles on which the levels for system analysis* are built [16]:

1. interaction and interconnection of networks;
2. cooperation in the field of public and private sector development;
3. protection of privacy;
4. network security;
5. lifelong learning.

The strategy for access to digital services and content is based on four principles:

1. universal, accessible and equal access;
2. consumer orientation and diversity of information;
3. competence and participation of citizens;
4. open and interactive networks

## 5 Recommendations

1) Develop an action plan to involve public authorities and make the transition to a digital society and economy through the digitalisation of the state highway by creating conditions for competition and regulation that would meet the public interest, promote innovation, investment, development of new services;

2) To achieve this goal requires coordinated action by the state, the private sector and public organizations.
   a) increase and develop national dialogue, create new jobs, give a new impetus to economic development;
   b) realize economic and social benefits for all citizens, which will allow them to personally participate in the formation of the information society;
   c) make the state more accessible and responsible.

   Also, as recommendations to the state:

1. The government must recognize the urgent need to address regulations and remove barriers to competition;
2. backbone networks and new infrastructure should be created by the private sector, and risk and rewards should be borne by shareholders;
3. the highway must move through the country in accordance with market requirements;
4. The development of the information highway should be "technology neutral" (ie the state should not support any of the technologies).

## 6   Conclusions

By applying a systematic approach to and breakdown at the level, the managed role of the state should be seen in the context of the role of the private sector, which invests and bears financial risk. its policy should be aimed at creating jobs and national wealth, stimulating competition, research and development. its purpose is to participate in the development of standards, ensure interaction, stimulate competition, accelerate the creation of new technologies and protect consumers.

The formation of a digital society in a state is not possible without taking into account the activities of other states. It is necessary to focus on the fact that legal intelligence is a type of organization of intelligence activities based on the legal penetration of intelligence into the state using the status of participants in international relations and their actions to gather intelligence and conduct actions of direct undermining on their own without establishing agency relations, in particular with representatives of the host state.

This geostrategic competition is forcing the military-political leadership of various states to show growing concern about the lag of national research in the field of information security from the leading United States in this field. Noting the emergence of a new kind of war - information and cyber warfare, it is necessary to accelerate and expand the relevant national research and development work.

In the digital society, it is not possible to build a cybersecurity policy without discussing the role and importance of information warfare in modern conditions. In their opinion, information weapons come out on top among the types of weapons used in modern warfare. This has led to the acceleration of a number of European programs in the field of communications, space control using satellites. At the same time, the steps of European states to increase their own potential of Information Weapons are met with serious opposition from the United States. In an effort to slow down Europeans' own developments, the United States is even selling dumped prices for space surveillance and line control systems and providing access to accumulated databases.

Experts from the **Institut de relations internationales et strategiques** believe that the United States' NATO allies are now heavily militarily dependent on the United States, as they use US software extensively in their weapons systems (including a missile deployment system) [17].

Analyzing new aspects of American military doctrine, aimed at achieving victory by bloodless methods with minimal losses, the main means is not control over territory, but domination in the "information sphere", which is a set of information systems and communications on the planet. At the same time, the war acquires a virtual and diplomatic character, which allows to achieve a decisive advantage without deploying hostilities.

THEREFORE, IT IS NECESSARY TO TAKE INTO ACCOUNT WHEN BUILDING CYBER SECURITY POLICY - means of control over information.

The need to create an information-analytical system to support decision-making of the top leadership of the state in any crisis situation is obvious. This is impossible without a mathematical basis [18, 19].

Cybersecurity policy is impossible for information development and financing of military production, where priority is given to information systems, including reconnaissance satellites, telecommunications networks, controls, and at the same time actively use civilian informatics, given the pace of electronics and informatics represented in all weapons systems. It is necessary to increase the service life of the material part, carrying out regular modernization at the expense of electronic information components. It is also necessary to significantly increase the share of costs for systems that allow modeling both at all stages of the creation of weapons systems and at the stage of their combat use, which will ensure a high level of technological research [20].

States should pay close attention to the protection of their information resources from unauthorized intrusion and manipulation.

The best protection of information in networks is provided only by encryption methods. In this regard, the active implementation of cryptographic software in the most important systems. In the world Work on the creation of appropriate devices is carried out by the companies "Thomson", "Sagem", "Mitra".

Analysis of expert assessments shows that the threat of information warfare is perceived as real, but this position is met with the inertia of the military spending system and lack of awareness of all the possibilities of information weapons, which go far beyond traditional means of information protection.

# References

1. Dzhalladova, I, Batechko, N., Gladka, Y.: System approach to the formation of the information culture levels in cyberspace. In: Antipova, T., Rocha, Á. (eds.) DSIC 2019. AISC, vol. 1114, pp. 275–283. Springer, Cham (2020). https://doi.org/10.1007/978-3-030-37737-3_25
2. Kshetri, N.: Cybersecurity and development. Markets. Globalization Dev. Rev. **1**(2) (2016)

3. Hantrais, L., et al.: Covid-19 and the digital revolution. Contemp. Soc. Sci. **16**(2), 256–270 (2021)
4. Andreessen, M.: Why software is eating the world. Wall Street J. (2011)
5. Angousturc, R.A., Pascal, V.: Diasporas et financement des conflits, t'conomie des guerres civiles. In: Economie des guerres civiles, pp. 495–642. Hachette, Paris (1996)
6. Ling Li, W., He, L.X., Ash, I., Anwar, M.: Xiaohong Yuan investigating the impact of cybersecurity policy awareness on employees' cybersecurity behavior. Int. J. Inf. Manage. **45**, 13–24 (2019)
7. Kuzmenko, A.: AInformatsiino-psykholohichna viina epokhy hlobalizatsii. Chastyna 10. Rozdil 2. Osoblyvosti chynnykiv vplyvu na formuvannia operatyvnoi obstanovky u sferi informatsiino-psykholohichnoi borotby Rosii (nimetski ta frantsuzki otsinky rosiiskoho chynnyka). Yurydychnyi zhurnal, **10** (2008)
8. Anderson, R.: Making security sustainable. Commun. ACM **61**(3), 24–26 (2018)
9. Štitilis, D., Pakutinskas, P., Malinauskaitė, I.: Preconditions of sustainable ecosystem: cyber security policy and strategies. Entrepreneurship Sustain. Issues **4**(2), 174–182 (2016)
10. Levy, K., Schneier, B.: Privacy threats in intimate relationships. J. Cybersecurity, **6**(1), tyaa006 (2020)
11. Freed, D., Palmer, J., Minchala, D., Levy, K., Ristenpart, T., Dell, N.: A Stalker's paradise: how intimate partner abusers exploit technology. In: Proceedings of the 2018 CHI Conference on Human Factors in Computing Systems, pp. 1–667 (2018)
12. Anisetti, M., Ardagna, C.A., Carminati, B., Ferrari, E., Perner, C.L:. Requirements and challenges for secure and trustworthy UAS collaboration. In: 2020 Second IEEE International Conference on Trust, Privacy and Security in Intelligent Systems and Applications (TPS-ISA), pp. 89–98. IEEE (2020)
13. Kyrkou, C., Papachristodoulou, A., Kloukiniotis, A., Papandreou, A., Lalos, A., Moustakas, K., et al.: Towards artificial-intelligence-based cybersecurity for robustifying automated driving systems against camera sensor attacks. In: 2020 IEEE Computer Society Annual Symposium on VLSI (ISVLSI), pp. 476–481. IEEE (2020)
14. Mohamed, N., Al-Jaroodi, J., Jawhar, I.: Opportunities and challenges of data-driven cybersecurity for smart cities. In: 2020 IEEE Systems Security Symposium (SSS), pp. 1–7. IEEE (2020)
15. Petryk, V.M., Kuzmenko, A.M., Ostroukhoe, V.V.: Sotsialno-pravovi osnovy informatsiinoi bezpeky: Navchalnyi posibnyk. Za red. In: Ostroukhova, V.V.K. (ed.) Rosava, p. 495 (2007)
16. Petryk, V.M., Ostorukhov, V.V., Shtokvysh, A.A.: Ynformatsyonno-psykholohycheskaia bezopasnost v эpokhu hlobalyzatsyy: Uchebnoe posobye/Pod red. In: Ostroukhova, V.V.K. (ed.) (2008)
17. Schmidt-Eenboom, E.: Nachrichtendienste in Nordamerika, Europa und Japan Länderportraits und Analysen und Analysen Weilheim STOPPEL-Verlag, pp. 3–70 (1995)
18. Dzhalladova, I.A., Kaminsky, O.Y.: Stabilizing steps the security of human and society in the Covid-19 pandemic. Econ. Bus. Organ. Res., pp. 295–308 (2020)
19. Dzhalladova, I.A.: Systemnyi analiz zahroz sotsiokibernetychnoi bezpeky v umovakh pandemii. Modeliuvannia ta informatsiini systemy v ekonomitsi. **100**, 50–58 (2020)
20. Kaminskyi, O.: Poliit D: Analiz dostovirnosti informatsii shchodo pandemii COVID-19 Ukraini (na prykladi svitovykh ahrehatoriv danykh). Modeliuvannia ta informatsiini systemy v ekonomitsi. **100**, 83–93 (2020)

# Covid-19 Pandemic and Disruptive Technologies Across Scientific Areas: A Bibliometric Review

Aleksander Aristovnik⬚, Dejan Ravšelj$^{(\boxtimes)}$⬚, and Lan Umek⬚

Faculty of Public Administration, University of Ljubljana, Ljubljana, Slovenia
{aleksander.aristovnik,dejan.ravselj,
lan.umek}@fu.uni-lj.si

**Abstract.** Disruptive technologies have been recognized as a key facilitator of the Covid-19 pandemic response and recovery efforts. Therefore, they have recently gained increased attention. However, due to the novelty of this research area, there is the lack of knowledge. Therefore, the main of the paper is to a provide bibliometric analysis on Covid-19 and disruptive technologies research with the focus on their application across different scientific areas. Bibliometric analysis is based on the Scopus database that contains 2353 documents published until July 2021. In this context, several innovative bibliometric approaches are applied. The results show that most of the research has been conducted within Physical Sciences, while Health Sciences are identified to have the most prominent scientific impact, as further confirmed of most relevant documents being more frequently cited compared to other scientific areas. Further, the highest authors' collaboration is observed for Life Sciences and Health Sciences, while Social Sciences exhibit the lowest authors' collaboration. Moreover, the *IEEE Access* from Physical Sciences is identified as the most relevant source, while *Health Informatics* from Multidisciplinary Sciences as the most relevant research field. Finally, *artificial intelligence (AI)* is by far the most applied disruptive technology across all scientific areas, especially in the context of Life Sciences. Contrary, *3D printing*, *augmented reality (AR)*, and *drones* seem to be not so relevant in the general Covid-19 related context, particularly in Health Sciences and Life Sciences. The findings of the paper add to the existing scientific knowledge and facilitate evidence-based policymaking.

**Keywords:** Covid-19 · Disruptive technologies · Bibliometric review

## 1 Introduction

The outbreak of the Covid-19 pandemic in early 2020 shocked the world and almost bringing it to an unprecedented stop. The pandemic is a typical public health emergency. Its high infection rate means it is a huge threat to global public health. However, despite the pandemic being primarily health crisis, its rapid proliferation has not only affected the lives of many people on the planet, but disrupted patterns of economic, social, and sustainable development (Aristovnik et al. 2020; Colavizza et al. 2021). In order to facilitate the Covid-19 pandemic response and recovery efforts, disruptive

© The Author(s), under exclusive license to Springer Nature Switzerland AG 2022
T. Antipova (Ed.): DSIC 2021, LNNS 381, pp. 567–580, 2022.
https://doi.org/10.1007/978-3-030-93677-8_50

technologies have been recognized as being important provider of effective and efficient support to day-to-day challenges related to the pandemic (Queiroz and Waba 2021). In practice, the following eight disruptive are often highlighted, namely: *artificial intelligence (AI), blockchain, internet of things (IoT), virtual reality (VR), 3D printing, augmented reality (AR), robotics* and *drones* (PwC 2021). Each of these types of disruptive technologies can facilitate addressing contemporary challenges in an innovative way.

Despite recognized benefits of disruptive technologies from the theoretical perspective (Love et al. 2020), recent scientific literature emphasize that their performance is not yet stable due to nonavailability of enough Covid-19 dataset, inconsistency in some of the dataset available, nonaggregation of the dataset due to contrasting data format, missing data, and noise. It is also emphasized that the security and privacy of information is not totally guaranteed (Mbunge et al. 2021). This may, however, seriously undermine practical application of disruptive technologies. Accordingly, knowledge of the applications, possibilities, and limitations of disruptive technologies requires specialization. Thus, using the appropriate knowledge is needed to leverage the full potential of disruptive technologies in the fight against the Covid-19 pandemic in a short time (Guggenberger et al. 2021).

There are some attempts in the existing scientific literature to provide systematic literature or bibliometric review, emphasizing that although the relationship between humans and digital technologies has been examined extensively over the last decades, there is still lack of the systematic and comprehensive review through the lens of the current Covid-19 pandemic (Vargo et al. 2021). A literature review by Golinelli et al. (2020) focused on healthcare and performed on 124 documents from PubMed and MedrXiv databases reveals that most of the documents examined the use of digital technologies for diagnosis, surveillance, and prevention. Moreover, the results show that digital solutions and innovative technologies have mainly been proposed for the diagnosis of Covid-19, whereby numerous suggestions on the use of artificial-intelligence-powered tools for the diagnosis and screening of Covid-19 have been identified. Finally, it is found out that digital technologies can be useful also for prevention and surveillance measures, for example through contact-tracing apps or monitoring of internet searches and social media usage. In another short literature review performed on 281 documents from Google Scholar, Web of Science, Scopus and PubMed, Vargo et al. (2021) have found out that various forms of technologies have been used (ranging from computers to artificial intelligence) several populations of users are using these technologies (primarily medical professionals), numerous generalized types of activities are involved (including providing health services remotely, analyzing data, and communicating) and various effects have been observed (such as improved patient outcomes, continued education, and decreased outbreak impact). Finally, in a bibliometric study by Queiroz and Waba (2021) based on 1297 documents from Web of Science database explores the interplay between disruptive technologies and Covid-19, considering the operations-related fields. Their findings show that cutting-edge AI technologies like machine learning and deep learning are the most popular approaches to support operations in this pandemic outbreak, but that other related technologies-like big data analytics, blockchain, simulation, AR/VR, 3d printing, etc. are also increasingly since the outbreak of the pandemic.

Most of the existing studies are focused on Health Sciences (Vaishya et al. 2020), while other scientific areas remain largely neglected. Accordingly, this bibliometric study tries to expand and extend the previous attempts in the scientific literature by providing bibliometric analysis on Covid-19 and disruptive technologies research with the focus on their application across different scientific areas. Specific objective of the analysis of Covid-19 and disruptive technologies research are the following: 1) to examine basic or descriptive indicators, including the most pertinent documents; 2) to find the most relevant and impacting journals and research fields; and 3) to examine the application of disruptive technologies across scientific areas. To pursue these objectives, a bibliometric analysis is used, allowing for an innovative literature review approach. The remainder of the paper is structured as follows. The second section presents the materials and methods applied in the paper. The third section presents in detail the results of the bibliometric analysis. The obtained results are concluded in the fourth section in which the main findings and implications are summarized.

## 2    Materials and Methods

Comprehensive bibliometric data were retrieved from Scopus, a world-leading bibliographic database of peer-reviewed literature. The choice of Scopus was based on the idea that it is a larger database than its competitors like Web of Science (Falagas et al. 2008). This was confirmed by the initial search in both databases since Scopus retrieved more documents for the intended search conditions than Web of Science. In addition, Web of Science has also been described as a database that significantly underrepresents scientific disciplines of the Social Sciences and Arts and Humanities compared to the Scopus database (Mongeon and Paul-Hus 2016). Accordingly, the Scopus database has larger coverage across different scientific areas and seems to be more relevant and tailored to the needs of bibliometric analysis. Bibliometric data were retrieved from Scopus in July 2021 using the advanced online search engine. The search strategy was based on title, abstract and keywords search with consideration of Covid-19 and eight essential disruptive technologies, having the most potential. Accordingly, the following search query was utilized: *(TITLE-ABS-KEY (covid-19) AND TITLE-ABS-KEY (artificial AND intelligence) OR TITLE-ABS-KEY (blockchain) OR TITLE-ABS-KEY (internet AND of AND things) OR TITLE-ABS-KEY (virtual AND reality) OR TITLE-ABS-KEY (3d AND printing) OR TITLE-ABS-KEY (augmented AND reality) OR TITLE-ABS-KEY (robot\*) OR TITLE-ABS-KEY (drone\*) OR TITLE-ABS-KEY (disruptive AND techn\*)).*

Based on this, a database of 4348 documents related to Covid-19 and disruptive technologies was obtained. Further, documents, which do not have the eight disruptive technologies listed in the authors' keywords were excluded, reducing the database to 2988 documents. In order to link the documents with the corresponding scientific or subject area, the fundamental Scopus database on document information was merged with Scopus CiteScore metrics that contain source-related information. The merging process revealed that some documents had no match, meaning that they were not considered in the bibliometric analysis. According to the presented multiphase process, the final dataset consists of 2353 documents.

According to the Scopus classification, the documents may be classified in five different subject areas: Health Sciences, Life Sciences, Physical Sciences, Social Sciences and Humanities and Multidisciplinary Sciences. However, these subject areas strongly intersect, meaning that an individual document can be classified in several subject areas at one time. In order to classify the documents into the corresponding subject areas, the following procedure was followed. In case that the document belongs to just one subject area (without intersecting with other subject areas), it was considered as a part of so-called pure science, classified in one of the four thematic subject areas, otherwise, it was considered as a part of Multidisciplinary Sciences. Based on this, the database was divided into 5 disjoint sets of documents, representing corresponding subject areas.

On this basis, the innovative bibliometric approaches were conducted with the following software tools. The calculations of bibliometric metrics were facilitated by using the Python data analysis libraries Pandas and Numpy (McKinney 2017) and visualized using Python's most powerful visualization libraries Pyvenn (VanderPlas 2016) and Matplotlib (Hunter 2007).

## 3  Results

### 3.1  Descriptive Overview

The documents may be classified in five different subject areas: Health Sciences, Life Sciences, Physical Sciences, Social Sciences and Humanities and Multidisciplinary Sciences, whereby each document can be classified in several subject areas at one time (see Fig. 1). Covid-19 and disruptive technologies research is spread across different subject areas, including their intersections. According to the number of documents related to the individual subject area (the number of documents on pure science is presented in italics), most of the research has been conducted within Physical Sciences (1398, *858*), followed by Health Sciences (901, *565*), Social Sciences and Humanities (516, *183*) and Life Sciences (328, *88*). The highest intersection across mentioned subject areas can be observed between Physical Sciences and Social Sciences and Humanities, as suggested by the highest number of documents (231) in the intersection between them. Some documents (44) are highly multidisciplinary as they are classified in all four subject areas at one time. However, to divide all documents into 5 disjoint sets of documents and made further bibliometric analysis possible, documents presented in the intersection were included in the so-called Multidisciplinary Sciences.

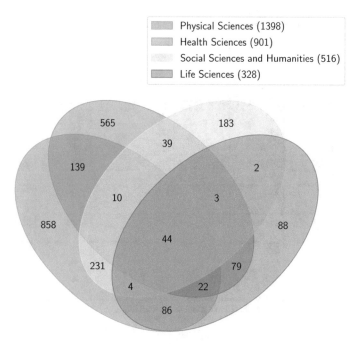

**Fig. 1.** Venn diagram by number of documents across subject areas.

An overview of the characteristics of scientific literature on Covid-19 and disruptive technologies research across pure scientific or subject areas is presented in Table 1. The bibliometric study relies on a total of 2353 documents with a total of 12429 citations written by 11430 distinct authors and published in 1069 sources (journals, books, etc.) in the last two years. Despite Physical Sciences having the largest share of total documents (36.46%), the largest share of citations can be observed for Health Sciences (36.04%) as further confirmed by the highest average number of citations (7.93), suggesting the most prominent scientific impact of this subject area. Moreover, the results show that Life Sciences have the highest average of cited references per document (61.05) followed by Social Sciences and Humanities and Multidisciplinary Sciences, meaning that the research conducted within these subject areas is grounded on a wider range of literature than within Physical Sciences and Health Sciences. Finally, the highest authors' collaboration can be observed for Life Sciences and Health Sciences, as confirmed by the highest average number of authors per document (6.56 and 6.41 respectively) and value of collaboration index (6.99 and 6.82 respectively), while Social Sciences exhibit the lowest values of both basic metrics (3.16 and 3.69 respectively).

**Table 1.** Overview of characteristics of scientific literature.

| Description | Total | Pure physical sciences | Pure health sciences | Pure social sciences and humanities | Pure life sciences | Multidisciplinary sciences |
|---|---|---|---|---|---|---|
| **Main information** | | | | | | |
| Timespan | 2020–2021 | 2020–2021 | 2020–2021 | 2020–2021 | 2020–2021 | 2020–2021 |
| Sources | 1069 | 263 | 321 | 129 | 59 | 297 |
| Documents | 2353 | 858 | 565 | 183 | 88 | 659 |
| Citations | 12429 | 2829 | 4479 | 622 | 671 | 3828 |
| Authors | 11430 | 3488 | 3623 | 578 | 577 | 3558 |
| **Basic metrics** | | | | | | |
| Average citations | 5.28 | 3.30 | 7.93 | 3.40 | 7.62 | 5.81 |
| Average references | 44.18 | 39.66 | 35.95 | 54.91 | 61.05 | 51.88 |
| Average authors | 4.86 | 4.07 | 6.41 | 3.16 | 6.56 | 5.40 |
| Collaboration Index | 5.38 | 4.66 | 6.82 | 3.69 | 6.99 | 5.84 |

Note: Documents published until July 2021 are considered in the bibliometric analysis.

Table 2 lists the most relevant (top 4) highly cited documents for each subject area. The results show that top cited documents from Health Sciences receive a much higher number of citations compared to other subject areas, especially to Social Sciences and Humanities, exhibiting the lowest number of citations of top cited documents. The top cited document on Covid-19 and disruptive technologies research is related to Health Sciences in which Peeri et al. (2021) provide an overview of the three major deadly coronaviruses and identify areas for improvement of future preparedness plans by recognizing IoT as one of the crucial technologies for mapping the spread of infection in the future. Within Life Sciences, the top cited document written by Mei et al. (2020) provide in-depth insights on how artificial intelligence can enable rapid diagnosis of patients with Covid-19. The top cited document related to Physical Sciences is by Chamola et al. (2020), providing a comprehensive review of the Covid-19 pandemic and the role of IoT, Drones, AI, Blockchain, and 5G in managing its impact. Within Social Sciences, the most cited document by Jiang et al. (2020) examines the effects of Covid-19 on hotel marketing and management with the focus on AI and robotics. Finally, top cited document related to Multidisciplinary Sciences by Ozturk et al. (2020) is about automated detection of Covid-19 cases using deep neural networks with X-ray images with the focus on application of advanced AI techniques.

**Table 2.** Most relevant documents by number of citations by subject area.

| Authors | Year | Document title | Country | Source title | Cited by |
|---|---|---|---|---|---|
| Pure Physical Sciences | | | | | |
| Chamola V. et al. | 2020 | A comprehensive review of the COVID-19 pandemic and the role of IoT, drones, AI, blockchain, and 5G in managing its impact | India | IEEE Access | 157 |
| Shi F. et al. | 2021 | Review of artificial intelligence techniques in imaging data acquisition, segmentation, and diagnosis for COVID-19 | China | IEEE Reviews in Biomedical Engineering | 144 |
| Yang G.-Z. et al. | 2020 | Combating COVID-19-The role of robotics in managing public health and infectious diseases | China | Science Robotics | 123 |
| Jiang X. et al. | 2020 | Towards an artificial intelligence framework for data-driven prediction of coronavirus clinical severity | United States | Computers, Materials and Continua | 119 |
| Pure Health Sciences | | | | | |
| Peeri N.C. et al. | 2021 | The SARS, MERS and novel coronavirus (COVID-19) epidemics, the newest and biggest global health threats: what lessons have we learned? | United States | International Journal of Epidemiology | 497 |
| Yang Z. et al. | 2020 | Modified SEIR and AI prediction of the epidemics trend of COVID-19 in China under public health interventions | China | Journal of Thoracic Disease | 408 |
| Li L. et al. | 2020 | Using Artificial Intelligence to Detect COVID-19 and Community-acquired Pneumonia Based on Pulmonary CT: Evaluation of the Diagnostic Accuracy | China | Radiology | 315 |
| Vaishya R. et al. | 2020 | Artificial Intelligence (AI) applications for COVID-19 pandemic | India | Diabetes and Metabolic Syndrome: Clinical Research and Reviews | 200 |
| Pure Social Sciences and Humanities | | | | | |
| Jiang Y. et al. | 2020 | Effects of COVID-19 on hotel marketing and management: a perspective article | Australia | International Journal of Contemporary Hospitality Management | 87 |

(*continued*)

**Table 2.**  (*continued*)

| Authors | Year | Document title | Country | Source title | Cited by |
|---|---|---|---|---|---|
| Zeng Z. et al. | 2020 | From high-touch to high-tech: COVID-19 drives robotics adoption | United States | Tourism Geographies | 83 |
| Kitchin R | 2020 | Civil liberties or public health, or civil liberties and public health? Using surveillance technologies to tackle the spread of COVID-19 | Ireland | Space and Polity | 27 |
| Abdel-Basset M. et al. | 2021 | An intelligent framework using disruptive technologies for COVID-19 analysis | United Kingdom | Technological Forecasting and Social Change | 24 |
| Pure Life Sciences | | | | | |
| Mei X. et al. | 2020 | Artificial intelligence–enabled rapid diagnosis of patients with COVID-19 | United States | Nature Medicine | 183 |
| Alimadadi A. et al. | 2020 | Artificial intelligence and machine learning to fight COVID-19 | United States | Physiological Genomics | 93 |
| Stebbing J. et al. | 2020 | Mechanism of baricitinib supports artificial intelligence-predicted testing in COVID-19 patients | United States | EMBO Molecular Medicine | 53 |
| Germain M. et al. | 2020 | Delivering the power of nanomedicine to patients today | France | Journal of Controlled Release | 40 |
| Multidisciplinary Sciences | | | | | |
| Ozturk T. et al. | 2020 | Automated detection of COVID-19 cases using deep neural networks with X-ray images | Turkey | Computers in Biology and Medicine | 388 |
| Ardakani A.A. et al. | 2020 | Application of deep learning technique to manage COVID-19 in routine clinical practice using CT images: Results of 10 convolutional neural networks | Iran | Computers in Biology and Medicine | 135 |
| Oh Y. et al. | 2020 | Deep learning COVID-19 features on CXR using limited training data sets | South Korea | IEEE Transactions on Medical Imaging | 111 |
| Toğaçar M. et al. | 2020 | COVID-19 detection using deep learning models to exploit Social Mimic Optimization and structured chest X-ray images using fuzzy color and stacking approaches | Turkey | Computers in Biology and Medicine | 109 |

## 3.2    Scientific Production

Scientific production varies greatly by subject area regardless of the observed dimension. Figure 2 presents most relevant (top 4) sources by number of documents and citations for each subject area. The *IEEE Access*, which is related to Physical Sciences, is found to be the most influential source in Covid-19 and disruptive technologies research, confirmed by the biggest number of documents (51) and the highest value of H-index (10).

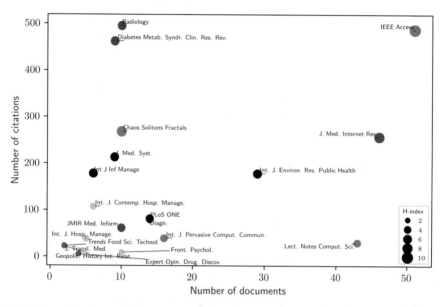

**Fig. 2.** Most relevant sources by number of documents and citations across subject areas. Note: Green – Physical Sciences; Blue – Health Sciences; Yellow – Social Sciences and Humanities; Red – Life Sciences; Black – Multidisciplinary Sciences

Despite a much smaller number of documents (10) compared to the previously mentioned source, *Radiology*, which is related to Health Sciences, is identified as a source with the biggest number of citations (495). On average, sources from Multidisciplinary Sciences rank somewhere in the middle among the most relevant sources with *International Journal of Environmental Research and Public Health* having the biggest number of documents (29) and the highest value of H-index (6) within this subject area. Further, *International Journal of Contemporary Hospitality Management* is found to be the most relevant source within Social Sciences and Humanities, as indicated by the biggest number of citations (107) and the highest value of h-index (3) within this subject area. Finally, sources related to Life Sciences are in general ranked on the bottom among the most relevant sources, however, as suggested by the biggest number of documents (14) and citations (78) and the highest value of h-index (3) within this subject area, *Diagnostics* is identified as the most relevant source in this subject area.

Moreover, Fig. 3 shows the most relevant (top 4) research fields by number of documents and citations for each subject area. The *Health Informatics*, which is a part of Multidisciplinary Sciences, is found to be the most relevant research field in Covid-

19 and disruptive technologies research, as suggested by the biggest number of citations (1226) and the highest value of h-index (14). Further, *Computer Science Applications* is identified as a research field with the biggest number of documents (185), thus being one of the most relevant within Physical Sciences.

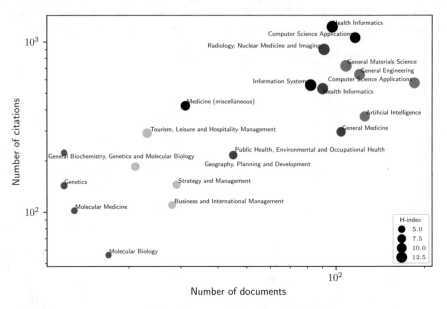

**Fig. 3.** Most relevant research fields by number of documents and citations across subject areas. *Note: Green – Physical Sciences; Blue – Health Sciences; Yellow – Social Sciences and Humanities; Red – Life Sciences; Black – Multidisciplinary Sciences*

Considering Health Sciences, *Radiology, Nuclear Medicine and Imaging*, is identified as the most influential research field as revealed by the biggest number of citations (902) and the highest value of H-index (14) within this subject area. On average, research fields related to Social Sciences and Humanities rank somewhere in the middle among the most relevant research fields with *Tourism, Leisure and Hospitality Management* having the biggest number of citations (291) and the highest value of H-index (9) within this research field. Finally, research fields related to Life Sciences are in general ranked on the bottom among the most relevant research fields, however, as suggested by the highest value of H-index (5) within this subject area, *Genetics* is identified as the most relevant research field in this subject area.

### 3.3    Disruptive Technologies Across Scientific Areas

In order to examine the application of eight essential disruptive technologies across scientific or subject areas, the shares of documents having selected disruptive technologies in their keywords has been calculated across corresponding subject areas (see Fig. 4). The results show that in general, *AI* is by far the most applied disruptive technology across all subject areas, since 28% of all documents list *AI* as a keyword, whereby it is particularly exposed in the context of Life Sciences (43%). Despite *robotics* being ranked on the second place, they are not applied within Life Sciences (0%), however, they seem to play an important role in Physical Sciences (12%) and Social Sciences and Humanities (11%) and Health Sciences (9%). Moreover, *IoT* is the most applied technology in the context of Physical Sciences (10%), while *VR* alongside with *blockchain* in the context of Social Sciences and Humanities (8%). The remaining three disruptive technologies (*3D printing, AR,* and *drones*) seem to be not so relevant in the general Covid-19 related context, as none of them exceed 5% of documents, describing them in keywords in general as well as across subject areas, particularly in Health Sciences and Life Sciences.

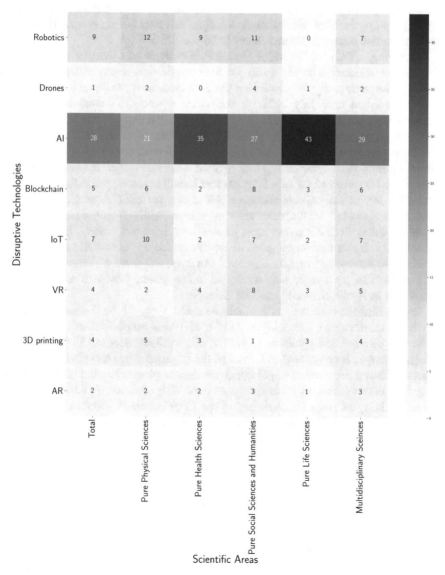

**Fig. 4.** Application of disruptive technologies across scientific areas (in %).

## 4   Conclusion

The outbreak of the Covid-19 is a typical public health emergency where the high infection rate poses a huge threat to not only global public health but economic, social, and sustainable development. In order to be able to solve such emergencies, it is vital to fully understand the potential of different solutions and applications of disruptive technologies across different scientific areas that may be effective and efficient in addressing potential devastating consequences. Therefore, scientific knowledge on Covid-19 and disruptive technologies research is essential because it leads to answers to real-life questions. However, the extent of the Covid-19 pandemic as well as unrecognized potential of disruptive technologies call for in-depth knowledge so as to allow numerous issues in different areas to be identified.

Accordingly, the paper provides bibliometric analysis on Covid-19 and disruptive technologies research with the focus on their application across different scientific areas. Bibliometric analysis is based on the Scopus database that contains 2353 documents published until July 2021. In this context, several innovative bibliometric approaches are applied. The results show that most of the research has been conducted within Physical Sciences, while Health Sciences are identified to have the most prominent scientific impact, as further confirmed of most relevant documents being more frequently cited compared to other scientific areas. Further, the highest authors' collaboration is observed for Life Sciences and Health Sciences, while Social Sciences exhibit the lowest authors' collaboration. Moreover, the *IEEE Access* from Physical Sciences is identified as the most relevant source, while *Health Informatics* from Multidisciplinary Sciences as the most relevant research field. Finally, *artificial intelligence (AI)* is by far the most applied disruptive technology across all scientific areas, especially in the context of Life Sciences. Contrary, *3D printing, augmented reality (AR),* and *drones* seem to be not so relevant in the general Covid-19 related context, particularly in Health Sciences and Life Sciences.

Several limitations of the study should be noted. First, the bibliometric analysis is only based on disruptive technologies and Covid-19 related documents retrieved from the Scopus database and published in journal with available Scopus CiteScore metrics. Although Scopus is considered to be one of the largest abstract and citation databases of peer-reviewed literature, it might not cover the complete collection of the research. Therefore, the inclusion of other databases e.g., Web of Science and especially the expanding body of preprints available in the Google Scholar database, could have provided additional insights not available in this study. Second, this study is based on a short time period (2020 and first half of 2021). Although this limitation cannot be solved at this stage, a repeated study with a longer period would yield further time-dimensional insights.

Notwithstanding the above limitations, the findings of the paper highlight the importance of a comprehensive and in-depth approach that considers different scientific disciplines in Covid-19 and disruptive technologies research. Namely, understanding of the evolution of emerging scientific knowledge on Covid-19 and disruptive technologies research is not only beneficial for the scientific community, but for evidence-based policymaking with a view to fully addressing the implications of the pandemic.

**Acknowledgements.** The authors acknowledge financial support from the Slovenian Research Agency (research core funding No. P5–0093 and project No. J5–2560).

# References

Aristovnik, A., Ravšelj, D., Umek, L.: A bibliometric analysis of COVID-19 across science and social science research landscape. Sustainability **12**(21), 9132 (2020)

Chamola, V., Hassija, V., Gupta, V., Guizani, M.: A comprehensive review of the COVID-19 pandemic and the role of IoT, drones, AI, blockchain, and 5G in managing its impact. IEEE Access **8**, 90225–90265 (2020)

Colavizza, G., Costas, R., Traag, V.A., van Eck, N.J., van Leeuwen, T., Waltman, L.: A scientometric overview of CORD-19. PLoS ONE **16**(1), e0244839 (2021)

Falagas, M.E., Pitsouni, E.I., Malietzis, G.A., Pappas, G.: Comparison of PubMed, Scopus, web of science, and Google scholar: strengths and weaknesses. FASEB J. **22**(2), 338–342 (2008)

Golinelli, D., et al.: Adoption of digital technologies in health care during the COVID-19 pandemic: systematic review of early scientific literature. J. Med. Internet Res. **22**(11), e22280 (2020)

Guggenberger, T., et al.: Emerging digital technologies to combat future crises: learnings from COVID-19 to be prepared for the future. Int. J. Innov. Technol. Manage. 2140002 (2021)

Hunter, J.D.: Matplotlib: a 2D graphics environment. Comput. Sci. Eng. **9**(3), 90–95 (2007)

Jiang, Y., Wen, J.: Effects of COVID-19 on hotel marketing and management: a perspective article. Int. J. Contemp. Hosp. Manag. **32**(8), 2563–2573 (2020)

Love, P.E., Matthews, J., Zhou, J.: Is it just too good to be true? unearthing the benefits of disruptive technology. Int. J. Inform. Manage. **52**, 102096 (2020)

Mbunge, E., Akinnuwesi, B., Fashoto, S.G., Metfula, A.S., Mashwama, P.: A critical review of emerging technologies for tackling COVID-19 pandemic. Hum. Behav. Emerg. Technol. **3**(1), 25–39 (2021)

McKinney, W.: Python for Data Analysis: Data Wrangling with Pandas, NumPy, and IPython, 2nd edn. O'Reilly Media Inc., Sebastopol (2017)

Mei, X., et al.: Artificial intelligence–enabled rapid diagnosis of patients with COVID-19. Nat. Med. **26**(8), 1224–1228 (2020)

Mongeon, P., Paul-Hus, A.: The journal coverage of web of science and scopus: a comparative analysis. Scientometrics **106**(1), 213–228 (2015). https://doi.org/10.1007/s11192-015-1765-5

Ozturk, T., Talo, M., Yildirim, E.A., Baloglu, U.B., Yildirim, O., Acharya, U.R.: Automated detection of COVID-19 cases using deep neural networks with X-ray images. Comput. Biol. Med. **121**, 103792 (2020)

Peeri, N.C., et al.: The SARS, MERS and novel coronavirus (COVID-19) epidemics, the newest and biggest global health threats: what lessons have we learned? Int. J. Epidemiol. **49**(3), 717–726 (2020)

PwC: Eight emerging technologies and six convergence themes you need to know about (2021). https://www.pwc.com/us/en/tech-effect/emerging-tech/essential-eight-technologies.htm. Accessed 01 Aug 2021

Queiroz, M.M., Wamba, S.F.: A structured literature review on the interplay between emerging technologies and COVID-19–insights and directions to operations fields. Ann. Oper. Res. 1–27 (2021). https://doi.org/10.1007/s10479-021-04107-y

Vaishya, R., Haleem, A., Vaish, A., Javaid, M.: Emerging technologies to combat the COVID-19 pandemic. J. Clin. Exp. Hepatol. **10**(4), 409–411 (2020)

VanderPlas, J.: Python Data Science Handbook: Essential Tools for Working with Data. O'Reilly Media Inc., Sebastopol (2016)

Vargo, D., Zhu, L., Benwell, B., Yan, Z.: Digital technology use during COVID-19 pandemic: a rapid review. Hum. Behav. Emerg. Technol. **3**(1), 13–24 (2021)

# Bibliometric Research in the Field of Mercantilism and Cameralism of Peter I

Ekaterina Zuga$^{(\boxtimes)}$ ⓘ and Svetlana Karelskaia ⓘ

St Petersburg State University, St Petersburg, Russian Federation
{e.zuga,s.karelskaya}@spbu.ru

**Abstract.** In this research, a bibliometric review was made on an issue related to the economic policy of Peter I. The following keywords were identified that characterize it: mercantilism, cameralism, monopoly, state-owned enterprise, and accounting. Scopus was selected as the database. It is one of the leading international abstract and citation databases. The research period is limited to 1990–2020. As a result of the study, a description of the current situation with research on these keywords was given, the time periods of publication activity, as well as its "peaks" - 2018–2020, were identified; the main types of publications were determined - articles and book chapters, as well as branches of knowledge, to which they relate - Social Sciences and Economics, Econometrics and Finance. In addition, the leading countries were highlighted, which works prevail in total - USA and UK. Also problem areas of research were specified, on which researchers interested in the development of this issue could focus. For instance, the importance of detailed studies of monopolies and state-owned enterprises in the Petrine era should be especially highlighted.

**Keywords:** Peter I · Mercantilism · Bibliometric review

## 1 Introduction

The coming era of digitalization presupposes global changes in many spheres of society, including the educational and scientific spheres. Educational institutions are equipped with modern software, which provides students and teachers with access to high-quality educational resources, allows to actively implementing online learning, etc. Researchers in their studies are increasingly using modern opportunities that have become available thanks to digitalization, for example, digitized archives, libraries, electronic catalogues, international databases with a huge number of books and articles in various languages (EBSCO, JSTOR, Scopus, Web of Science, etc.), which allow conducting high-quality research based on voluminous reviews of literary sources.

In addition, reviews are quite popular as a separate type of scientific research. Reviews are attracting close attention of the scientific community today, since they are synthesized knowledge that guides and informs researchers in the relevant field [1, p. 136].

Examples of such studies in Russia are: Vodyanitskaya (2021) - on the concept of scientific discourse using the ScienceDirect resource as of January 2021; Generalova and Popova (2015) - on accounting for goodwill in Russian and foreign publications

T. Antipova (Ed.): DSIC 2021, LNNS 381, pp. 581–590, 2022.
https://doi.org/10.1007/978-3-030-93677-8_51

for 25 years (1990–2014) using the Elibrary, it was continued by Ivanov (2016), but for the period 1990–2015 and using the Elibrary as well as Scopus; Muzyko (2020) - on assessing the effectiveness of innovative projects using bibliometric analysis carried out on the basis of Web of Science and Scopus and limited to 2019, etc. As you can see, authors use different databases for bibliometric analysis, since each of them has certain restrictions and their choice is conditioned, as a rule, by the purpose and objectives of the research.

Similar studies can be single out among foreign studies, for example, Anderson (2002), who analyzed 155 articles in the journal Accounting, Business and Financial History for the period 1990–2000 using the analysis of their content and citations; Carnegie and Potter (2000) reviewed 149 publications in English over the period 1996–1999 in three leading journals on accounting history; Esteve (2008) reviewed 683 accounting history articles in 62 of the world's leading journals over the period 2000–2007 using Blackwell, Emerald, Historical Abstracts, Ingenta, ScienceDirect, and Scopus; Matthews (2017) analyzed accounting publications for a record long period of 1899–2015 using the American Accounting Faculty Directory, British Accounting Review Research Register, etc. There are many examples of such studies; they are united by the use of the method of content analysis and analysis of citations.

Such studies are essential for the development of scientific research. The bibliometric review allows not only to determine the boundaries of a specific problem field and identify the main trends in its study, but also to localize gaps in existing knowledge within its framework. The use of bibliometric reviews allows you to get the most objective understanding of the level of prior studies of the analyzed issues at the international level, which means that it allows you to reduce information noise in the field of scientific communication [1, p. 136].

This study is the result of a bibliometric review on a historical topic, dedicated to the influence of the doctrines of mercantilism and cameralism on the economic policy of Peter the Great, with an emphasis on its implementation in terms of the development of state-owned enterprises and state monopolies in Russia as well as the influence of these processes on changing accounting practices.

## 2   Methods

A bibliometric review of the available literature in a particular scientific field allows us to identify new directions of research that are relevant for the study [2, p. 1057]. As it was shown in the introduction, this approach is very popular in the scientific community and often applicable.

The main characteristics of a bibliometric review are moderate coverage, insignificant depth, focus on databases, and the use of bibliometric analysis as a methodology [1, p. 134], which was also confirmed at the preliminary stage of the study. According to Lychagin et al. (2014) [3, p. 135], bibliometric analysis can be used for systematization in the subject area. Methods of bibliometric analysis make it possible to single out the research traditions of various disciplines, consider their development, find new scientific fields not expertly, but on the basis of algorithmic analysis [4, p. 9].

There are two main groups of methods used in modern bibliometrics: bibliographic (search, selection and systematization of information sources for the accurate formation of analyzed document collections and citations) and quantitative (frequency analysis of texts, correlation, factorial, cluster analysis, principal component analysis, etc.) [5, p. 22]. At the same time, the analysis of the citation statistics of publications and the analysis of citation communications are often carried out. According to Pavlova, citation statistics allow us to identify patterns in the development of science, the likely pace of development and "breakthroughs". Analysis of citation communications is a bibliometric tool that allows, on the one hand, to identify the disciplinary structure of science and to discover emerging research areas, on the other hand, to conduct a quantitative assessment of scientific research [5, p. 22–23]. In this study, the bibliographic method will be applied.

At the same time, the goal of bibliometric research, as a rule, is to "give an objective description of a scientific direction development, assess its relevance, potential, laws for the formation of information flows and the dissemination of scientific ideas" [5, p. 21]. The purpose of this study is to identify new areas of research or issues, requiring additional development, for more effective implementation of the research objectives dedicated to the economic policy of Peter I.

The main feature of the bibliometric review is the use of bibliometric tools in the analysis of sources. Most often, the literature for such a review is selected in databases based on the number of citations of publications, the impact factor or the Hirsch index of journals, etc. [1, p. 135].

This research is conducted using Scopus database. Scopus is the world largest abstract and citation database that includes peer-reviewed scientific literature, and applies tracking, analysis and data visualization tools. It contains more than 23.7 thousand publications from 5,000 international publishers, 166 thousand books, 8.3 million conference papers, etc. in the field of natural, social and human sciences, technology, medicine and art, and also it covers publications for a very long period, the earliest publication dates back to 1788.

For the designated main research issue on the influence of the doctrines of mercantilism and cameralism on the economic policy of Peter the Great, a list of the main keywords was formed for bibliometric analysis: "Peter I" (Peter the Great), "mercantilism", "cameralism", "state-owned enterprise" (state-owned company), "monopoly", which were searched for publications in the Scopus. Also all possible pair combinations of the above concepts have been investigated: "Peter I (Peter the Great)" and "mercantilism", "Peter I (Peter the Great)" and "cameralism"; "Peter I (Peter the Great)" and "state-owned enterprise (state-owned company)"; "Peter I (Peter the Great)" and "monopoly"; "Peter I (Peter the Great)" and "accounting"; "mercantilism" and "cameralism"; "mercantilism" and "state-owned enterprise (state-owned company)"; "mercantilism" and "monopoly"; "mercantilism" and "accounting"; "cameralism" and "state-owned enterprise (state-owned company)"; "cameralism" and "monopoly"; "cameralism" and "accounting"; "monopoly" and "state-owned enterprise (state-owned company)"; "monopoly" and "accounting"; "state-owned enterprise (state-owned company)" and "accounting". The concept of "accounting" was used only in combination with other keywords; it was not studied separately, since its very detailed bibliographic studies in the historical context have already been repeatedly conducted [6–9, etc.].

## 3   Results

As it was mentioned earlier, the research was conducted using Scopus. The period 1990–2020 was studied. The search was carried out using keywords ranked according to the degree of importance for the research subject. The search results are presented in a summary matrix (see Fig. 1).

| Keywords | Number of works | | | | | |
|---|---|---|---|---|---|---|
| | Search for keywords | Search for combinations of keywords | | | | |
| | | P | M | C | S | Mo |
| Peter the Great, Peter I (P) | 137 | | | | | |
| Mercantilism (M) | 557 | 0 | | | | |
| Cameralism (C) | 97 | 2 | 10 | | | |
| State-owned company, state-owned enterprise (S) | 8178 | 0 | 0 | 0 | | |
| Monopoly (Mo) | 3726 | 5 | 14 | 0 | 218 | |
| Accounting (A) | - | 8 | 8 | 3 | 275 | 209 |

**Fig. 1.** Search results for keywords and their intersections using Scopus.

As seen from Fig. 1 there are a lot of articles devoted to keywords separately, but pair combinations are quite rare. Moreover, 30% of combinations (5 out of 15) did not give any results at all. Combinations with the concept of "state-owned enterprise (state-owned company)" are especially rare, despite the rather impressive number of articles on this issue as a whole. The largest number of matches (503) was found when analyzing "accounting" with other keywords. It is especially noteworthy that it is the only one that has combinations with all other keywords. The second place is "state-owned enterprise (state-owned company)" (493), but there are matches only with two other "popular" keywords - "monopoly" and "accounting". Further, the third most popular combination with the concept of "monopoly" - 446 matches, but there are no matches with the concept of "cameralism". There are only 32 matches with the concept of "mercantilism", and 15 matches - with the concepts "cameralism" and "Peter I (Peter the Great)". Due to the results obtained, no research was conducted on the combination of three or more keywords.

The obtained results indicate a lack of research on the economic policy of Peter I - mercantilism and cameralism, and, accordingly, their importance and necessity. It confirms the relevance of the project implemented by the authors and the possible demand for its results for the scientific community. The need for the study of monopolies and, in particular, state-owned enterprises in the Peter the Great era are also quite obvious.

The most important of all the keywords is "Peter I (Peter the Great)". A search in Scopus for this word yielded 442 articles. The search here and further in relation to the rest of the selected keywords was carried out by the title of the article, keywords, and short description. The control of search results showed that due to the peculiarities of the search in Scopus, some of the 442 articles were extremely indirect or even had nothing to do with the Russian emperor. For example, some articles were devoted to the fauna around the island of Peter I[1]. As a result, it was necessary to screen out articles "manually", as a result of which 69% of them were excluded and only 137 of them remained.

All articles included in the sample were published during the stated research period 1990–2020 (see Fig. 2).

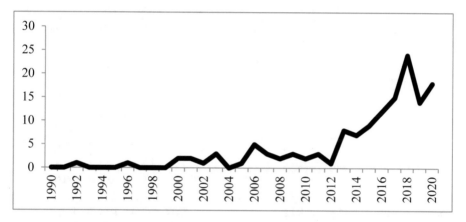

**Fig. 2.** Dynamics of publications by the keyword "Peter I (Peter the Great)" in Scopus for the period 1990–2020.

Analysis of the dynamics of publications in which Peter I is mentioned showed a sharp increase in interest in this issue, starting in 2013 - more than 78% of articles were published in the last eight years of the period. The maximum was reached in 2018 - 24 articles. After study by decades, the situation of the "publication boom" becomes even clearer: 1990–1999 - two articles, 2000–2009 - 10 articles, 2010–2019 - 95 articles.

Keyword "cameralism" demonstrates a similar dynamics, but with a slight shift: more than 59% of articles were published in recent years, starting from 2015, the maximum was reached in 2019 - 21 articles. But the distribution over decades is more even.

Most of the works that include the concept of "mercantilism", 83.5% were published in the second period of the study, starting in 2005. The maximum was reached twice - in 2016 and 2018 - 45 and 44 works, respectively.

---

[1] Peter I Island is an uninhabited island in the Antarctic part of Pacific Ocean, in the Bellingshausen Sea, 400 km from the Antarctic coast, in the area adjacent to Alexander I Land.

"State-owned enterprise (state-owned company)" is the most frequently encountered concept - 8178 works. Publication activity gradually increased during the study period, and a particularly sharp increase occurred in the last five years, starting in 2016, when their number doubled up to 848 articles in 2020.

The number of publications where the concept of "monopoly" is mentioned has grown by 7.9 times in 31 years up to 205 in 2020. However, there were no dramatic changes; growth, unlike other concepts, was gradual and very measured. The distribution of articles over decades is also quite even: in the first decade - 419 articles, in the second - 1072, in the third - 2030. The maximum number of publications was reached in 2018.

Comparative analysis by type of documents is presented in Table 1.

**Table 1.** Distribution of works by types of documents

| Type of documents | Number of works | | | | Share of works in total, % | | | |
|---|---|---|---|---|---|---|---|---|
| | (M) | (C) | (S) | (Mo) | (M) | (C) | (S) | (Mo) |
| Article[a] | 344 | 52 | 5833 | 2359 | 61.8 | 53.6 | 71.3 | 63.3 |
| Book chapter | 102 | 24 | 510 | 564 | 18.3 | 24.7 | 6.2 | 15.1 |
| Review[b] | 67 | 19 | 374 | 374 | 12.0 | 19.6 | 4.6 | 10.0 |
| Book | 31 | 1 | 106 | 138 | 5.6 | 1.0 | 1.3 | 3.7 |
| Conference paper | 6 | 0 | 917 | 194 | 1.1 | 0.0 | 11.2 | 5.2 |
| Business article | 1 | 0 | 209 | 29 | 0.2 | 0.0 | 2.6 | 0.8 |
| Editorial, note, Short survey and other | 6 | 1 | 229 | 68 | 1.1 | 1.0 | 2.8 | 1.8 |
| Total | 557 | 97 | 8178 | 3726 | 100.0 | 100.0 | 100.0 | 100.0 |

[a]Article - original research or opinion. Characteristics: Articles published in peer-reviewed journals, usually several pages long, consist of several parts: abstract, introduction, materials and methods, results, conclusion, discussion and references. However, reports, technical and research notes are also considered articles and can be up to one page in length. Articles in specialized journals are generally shorter than articles in peer-reviewed journals. They can be one page [10].
[b]Review - significant review of original research, including conference papers. Characteristics: Reviews usually have an extensive references. Educational publications that address specific issues are also considered reviews. The reviews are not original articles and therefore lack the sections that are characteristic of the original articles (e.g. methods and results) [10].

For all keywords, more than half of the documents (53.6–71.3%) are articles, the rest of the documents are mainly book chapters (15.1–24.7%) or, slightly less often, reviews (10.0–19.6%). The distribution by types of documents for the "monopoly" differs, since the second most important type of document is conference papers (11.2%).

Comparative analysis by branches of knowledge is presented in Table 2.

**Table 2.** Distribution of works by main branches of knowledge

| Branch of knowledge | Number of works | | | | Proportion of works in total, % | | | |
|---|---|---|---|---|---|---|---|---|
| | (M) | (C) | (S) | (Mo) | (M) | (C) | (S) | (Mo) |
| Arts and humanities | 269 | 64 | 368 | 711 | 48.3 | 66.0 | 4.5 | 19.1 |
| Business, management and accounting | 105 | 31 | 2776 | 489 | 18.9 | 32.0 | 33.9 | 13.1 |
| Economics, econometrics and finance | 224 | 43 | 2541 | 764 | 40.2 | 44.3 | 31.1 | 20.5 |
| Engineering | 26 | 1 | 1185 | 268 | 4.7 | 1.0 | 14.5 | 7.2 |
| Social sciences | 340 | 58 | 2746 | 2116 | 61.0 | 59.8 | 33.6 | 56.8 |

Analysis of the distribution of work by branches of knowledge showed that, in general, the key branches of knowledge are: for "mercantilism" and "monopoly" - Social Sciences, for "cameralism" - Art and Humanities, for "state-owned enterprise (state-owned company)" - Business, Management and Accounting. Despite the seeming inconsistency and difference in the branches of knowledge, they are close, and we can say that social and economic focus is predominating.

Table 3 shows the distribution of work by country.

**Table 3.** Comparative analysis of works by country

| Indicator | (M) | (C) | (S) | (Mo) |
|---|---|---|---|---|
| Number of countries | 64 | 23 | 129 | 110 |
| Top-3 countries in terms of the number of works | USA<br>UK<br>France | UK<br>Germany<br>USA | China<br>USA<br>UK | USA<br>UK<br>Germany |
| Number of works from Russian Federation (in %) | 8 (1.3) | 9 (8.9) | 129 (1.3) | 151 (3.6) |

As it can be seen from the Table 3, the issues of monopoly and state-owned enterprises are studied more than in a hundred countries, which is quite natural due to the rather large total number of works. The smallest country diversity is presented for "cameralism", however, if we analyzed it in relative terms (taking into account the total number of works - only 97), then, on the contrary, it turns out - the best indicator out of four keywords.

For all keywords, two countries - USA and UK - were in the top-3 in terms of the number of works, which is quite consistent with the global tendencies of their dominance in terms of the number of research conducted in various scientific fields, including the economy. Also, Germany was twice in the top-3, which is quite understandable if you pay attention to the fact that cameralism is often called the German version of mercantilism.

As to the contribution of Russian scientists to the development of these issues, Russia twice falls into the top-5 countries with the largest number of works: in "cameralism" - fourth place and in "monopoly" - fifth. Moreover, its proportion in the total number of works is quite large - 8.9 and 3.6%, respectively.

The analysis carried out confirmed the earlier conclusions that it is necessary to more actively study mercantilism and cameralism in various aspects.

## 4   Discussion

It is possible to make some generalizations based on the results of the study:

– in relation to the dynamics of publication activity

All keywords can be divided into two groups: first – "Peter I (Peter the Great)" and "cameralism"; the second – "mercantilism", "state-owned enterprise (state-owned company)", and "monopoly". The first group is characterized by extremely low publication activity during most of the analyzed period with a rather rapid growth in the last 3–5 years; the second one - a gradual increase in publication activity during the entire study period.

A generalization of the results by decades is presented in Table 4.

As it can be seen from Table 4, four keywords have similar distributions over the period of publication activity – "mercantilism", "cameralism", "state-owned enterprise

**Table 4.**  Distribution of works by research period

| Research period | (P) | (M) | (C) | (S) | (Mo) |
|---|---|---|---|---|---|
| 1990–1999 | 1.7 | 9.9 | 11.4 | 9.6 | 11.9 |
| 2000–2009 | 18.5 | 24.6 | 21.6 | 31.0 | 30.4 |
| 2010–2019 | 79.8 | 65.5 | 67.0 | 59.4 | 57.7 |
| Total | 100.0 | 100.0 | 100.0 | 100.0 | 100.0 |

(state-owned company)", and "monopoly". At the same time, scientific research in which Peter I is mentioned or is the "protagonist" began to be actively carried out only in the last decade.

– in relation to the distribution of works by types of documents.
   About 90% of works are articles, book chapters and reviews. The rest of the types of documents are sporadic. "State-owned enterprise (state-owned company)" is the exception: conference papers occupy a significant proportion of 11.2% instead of reviews; it may indicate the presence of controversial issues requiring discussion.
– in relation to the distribution of works by branches of knowledge.
   As previously shown, the bulk of the work is presented in four fairly similar directions, which in general can be called social and economic. A more detailed analysis reveals that there are much more directions - more than 20, but the number of works on them is usually sporadic. Keyword "cameralism" is the exception - only nine branches of knowledge.
– in relation to the distribution of works by country.

In general, we can say that the study of the research area, despite the first glance noticeable country diversity is concentrated in the top-2 countries (Table 5).

**Table 5.** The proportion of works from top-2 countries

| Keywords | Countries | Proportion of works, % |
|---|---|---|
| Mercantilism | USA+UK | 36.8 |
| Cameralism | UK+Germany | 36.6 |
| State-owned company, state-owned enterprise | China+USA | 36.9 |
| Monopoly | USA+UK | 32.0 |

As it can be seen from the Table 5, scientists from the leading countries (top-2) published about 1/3 of all works (32–36%). At the same time, ¼ of works falls on the first country out of the top-2 for all keywords, except for "cameralism" - for which the proportion of works is approximately equal in both countries.

The contribution of Russian researchers is quite noticeable only on the subject of "cameralism" (8.9%).

At the same time, the analysis of pair combinations of keywords showed the need for research at the junction of the concepts being studied.

## 5   Conclusions

Conducting a bibliometric review is an important and integral part of any qualitative research. Such reviews make it possible to objectively assess the degree of knowledge of the subject under study, depending on the choice of the database - at the Russian and/or international level. In addition, the bibliometric review reveals little-studied areas that may require closer study and attention from researchers.

The results obtained indicate an increase in research interest in the study of the Petrine era, as well as mercantilism and cameralism in recent years, especially in the last five years. Nevertheless, attention should be paid to areas of research that would combine these concepts, perhaps this would give a synergistic effect and reveal a number of interesting research areas.

**Acknowledgments.** The reported study was funded by RFBR, project number 20–010–42004.

## References

1. Raitskaya, L.K., Tikhonova, E.V.: Reviews as a promising kind of scholarly publication, its types and characteristics. Nauchnyi Redaktor i Izdatel '= Science Editor and Publisher **4**(3–4), 131–139 (2019). https://doi.org/10.24069/2542-0267-2019-3-4-131-139 (In Russ.)
2. Muzyko, E.I.: Bibliometric analysis of world economic literature in the field of evaluating the effectiveness of innovative projects (based on Web of Science and Scopus). Econ. Entrepreneurship **8**, 1057–1065 (2020). (In Russ.)

3. Lychagin, M.V., Mkrtchyan, G.M., Suslov, VI: The Concept of System-Innovative Bibliometric Analysis and Mapping of Economic Literature. Vestn. Novosib. State University. Series: Socio-Economic Sciences 14, 127–141 (2014). (In Russ.)

4. Moiseev, S.P., Maltseva, D.V.: Selection of sources for a systematic literature review: comparison of expert and algorithmic approaches. Sociol. Methodol. Methods, Math. Model. **47**, 7–43 (2018). (In Russ.)

5. Pavlova, A.S.: Bibliometric methods: types, tasks, problems (analytical review). Proc. State Public Libr. Sci. Technol. SB RAS **9**, 20–29 (2015). (In Russ.)

6. Anderson, M.: An analysis of the first ten volumes of research in accounting, business and financial history. Acc. Bus. Finan. Hist. **12**(1), 1–24 (2002). https://doi.org/10.1080/09585200110107939

7. Carnegie, G.D., Potter, B.N.: Publishing patterns in specialist accounting history journals in the English language, 1996–1999. Account. Hist. J. **27**(2), 177–198 (2000)

8. Esteve, E.H.: Some reflections on the orientations and volume of accounting history research in the 21st century. DE Comput. **9**, 97–135 (2008)

9. Matthews, D.: Publications in accounting history: a long-run statistical survey. Acc. Historians J. **44**(2), 69–98 (2017)

10. Scopus. Content Coverage Guidelines. https://elsevierscience.ru/files/Scopus_Content_Guide_Rus_2017.pdf. Accessed 02 Aug 2021

11. Vodyanitskaya, A.A.: Bibliometric analysis as a methodological principle for the study of academic discourse. Bull. S.-W. State Univ. Ser.: Linguist. Pedagogy **11**(1), 71–81 (2021). (In Russ.)

12. Generalova, N.V.: Popova EYu: analysis of scientific publications on accounting for goodwill in Russian and international periodicals for 1990–2014. Audit Statements **2**, 63–72 (2015). (In Russ.)

13. Ivanov, A.E.: Who and in what aspects researches goodwill abroad: publication analysis. Audit Statements **12**, 73–86 (2016). (In Russ.)

# In Search of Insight from Unstructured Text Data: Towards an Identification of Text Mining Techniques

Sunet Eybers[(⊠)] and Helgard Kahts

Department of Informatics, University of Pretoria,
Hatfield, Pretoria, South Africa
Sunet.eybers@up.ac.za

**Abstract.** The availability of massive sets of unstructured data has opened up new opportunities for businesses to gain meaningful insights into what is currently happening in their organization and market place. Unfortunately, unstructured data is often challenging to work with, as it requires specialized toolsets, techniques, knowledge and skills to engage with the data. This study used a systematic review (SLR) process to explore the current and conversant state of affairs in the field of text mining (TM) techniques that exist for the processing of natural language in unstructured text datasets across a broad expanse of applications and domains. A comprehensive literature search yielded 1022 eligible articles from five prominent bibliographic databases, narrowed down to 89 articles for review. Information related to TM techniques, the TM process, applications, challenges and recommendations for the improvement of TM results were extracted and synthesized. Eighteen TM techniques were identified and used to complete a variety of tasks such as data pre-processing, information retrieval, information extraction, text classification, clustering, topic modeling, text summarization and sentiment analysis or opinion mining. No single TM technique may be suitable for all text representation and extraction requirements. Therefore, context and applicability are key factors for TM technique selection.

**Keywords:** Text mining · Text mining techniques · Unstructured data · Knowledge discovery

## 1 Introduction

Data analytics takes on an unprecedented denotation in the presence of the 'Big Data' phenomenon as analysis of these high voluminous and variety-rich datasets can no longer be performed using traditional analytical techniques and tools [1]. It's no longer 'business as usual' in the field of text data analytics. However, the exponential increase in the amount of available data, when properly leveraged, provides numerous opportunities for businesses to create value through knowledge discovery [2]. Organizations expend significant resources collecting and storing large numbers of natural language text documents. This can represent a huge strategic resource, however, many fail to take advantage of, or materially benefiting from the intrinsic value of the information,

T. Antipova (Ed.): DSIC 2021, LNNS 381, pp. 591–603, 2022.
https://doi.org/10.1007/978-3-030-93677-8_52

accessible through big data analytics (BDA) [3]. This information, if effectively repurposed, may assist in improving operational efficiencies, enhance decision-making or drive new revenue streams to gain competitive advantage [4, 5].

As more organizations start adopting and deploying BDA, meaningful insights may be extracted from the information [4]. These diverse sets of data comprising structured, semi-structured and unstructured data may potentially yield huge economic benefits to organizations who choose to exploit big data (BD) strategies [2]. Unfortunately, a large percentage of this data, 75–80% according to [6] even as high as 95% according to [2], is in the form of unstructured text data that superimposes some profound limitations with regard to the mining techniques that can effectively be used to extract meaningful insights [1].

Although traditional information retrieval systems enabled the retrieval of large amounts of data, it still required enormous amounts of effort from domain specialists in order to obtain information that could be classified as 'useful' from which to derive meaningful insight. Thus, a re-look at information retrieval strategies is required to enable users to gain a more holistic view of information contained within databases and therefore process information beyond the purpose of retrieval only. It is perceived that text mining (TM), also known as text data mining (TDM), could provide this dexterity to effectively and efficiently retrieve only useful information [7].

Over and above the requirement of more dexterous TM techniques, cognizance should be taken that different data types require different techniques for processing. The mining of normal structured data usually requires some sort of pre-processing techniques like data reduction, data integration and data cleansing, even more so with unstructured and semi-structured text data, however, these pre-processing techniques are much more complex and difficult to execute successfully [8].

In addition to normal information retrieval strategies, contemporary TM techniques, to be effective and efficient, require the application of a formal systematic process of mining as proposed in [1, 9–13]. The TM process generally encompasses the obtaining of targeted text data, pre-processing of the text data, applying an appropriate TM technique and lastly analyzing the TM results to determine the effectiveness of the TM technique used before the extracted information can be analyzed to gain meaningful and actionable insight.

How effectively a TM technique could be used to extract actionable new knowledge from unstructured text data depends by and large on the correct application of an appropriately selected TM technique and the associated challenges with using certain TM techniques on certain types of unstructured text datasets. Here, a conversant perception of the domains of TM and TM techniques may proof invaluable. The aim of this systematic literature review (SLR) is to identify existing TM techniques and to provide a synopsis of different TM techniques available to extract useful information from unstructured text data, with the primary research question: *What TM techniques exist for analyzing unstructured text data?*

The paper starts with a description of the research approach and literature review process. The various text mining techniques identified in the literature are discussed followed by a short conclusion.

## 2  Research Approach

[14] contend that the most compelling reasons for conducting a SLR are to; summarize existing empirical evidence; identify gaps in current research; and lastly to provide a background for new research activities. Thus, in terms of the scope of this study a SLR is seen as an appropriate methodology to apply in order to answer the primary research question for this study. Moreover, this study may provide scholars and practitioners with a conversant state of affairs on TM techniques, providing a critical account of a small subset of the research field of TM [15].

Although various authors have different views on the number and order of activities comprising the SLR, most however, agree on the concept of a three-phased approach consisting of a planning, conducting the actual review followed by a reporting phase [14, 16, 17]. The following activities represent the complete SLR process, across the mentioned three phases, followed in this study. The steps were:

Phase 1: Planning: This phase include the research problem formalization, and refer to the identification of the research question(s), search strategy, inclusion and exclusion criteria, screening procedures, quality assessment, data extraction strategy, data analysis, synthesis and reporting as elements of the SLR).

Phase 2: Conducting the Review (Sect. 3): Performing the actual literature review, which include tasks such as, identifying primary studies for inclusion (based on pre-defined criteria); performing quality assurance on included sources; extracting data; and analyzing the content of the final literature review article pool.

Phase 3: Report Findings (Sect. 4): The findings are discussed and reported on.

## 3  Literature Review Process (Conducting the Review)

For the purpose of this study, journal articles and reviews previously conducted on TM and TM techniques, form the basis for identifying which academic journals and databases to consider for literature searching. However, reference lists from pertinent primary studies and previous reviews were also scrutinized to obtain additional pertinent sources that may otherwise not have been obtained and that may manually be included [14]. The following table provides a summary of the main considerations during the literature review process (Table 1):

**Table 1.**  Literature review tasks

| Task | Description |
|------|-------------|
| Database search | SCOPUS, ACM Digital Library and Web of Science, IEEE Xplore and Springer Link |
| Search terms | "and", "or", "text mining'; "text analytics' "knowledge discovery"; "web-based analytics' "information retrieval"; "natural language processing"; "information extraction"; "text summarization"; "machine learning"; "clustering"; "visualization"; "web mining"; "questions answering"; "topic identification"; "sentiment analysis"; "text classification"; "text categorization"; "unstructured data"; "unstructured text"; "semi-structured data"; "semi-structured text" |
| Inclusion/Exclusion criteria | English articles<br>Online libraries available through the institutional library portal<br>Peer reviewed |
| Data extraction | 1022 potential articles<br>780 after title screening<br>390 after focus on articles were considered<br>195 after considering abstract and conclusions<br>89 was identified as final article pool |
| Quality assessment | Considered study design, bias, internal and external validity, and the overall generalizability and replicability of study results<br>Distribution of articles across publications: IEEE Access platform (8% of articles), followed by Sustainability (Switzerland) (4.6%), International Journal of Computer Applications (3.5%), and the ACM Transactions on Knowledge Discovery from Data (2.2%) |

The final step in the process was to analyse and synthesize the content of the final article pool. Thematic analysis was used to identify the main TM techniques, which can be defined as "the process of identifying patterns or themes within qualitative data." [18]. A good thematic analysis encompasses more than merely summarizing data, it also provides interpretation and an explanation of the data. [19] provides a six-step framework for conducting thematic analysis that were applied in this study. The steps were becoming familiar with the data; generating initial codes; searching for themes; reviewing themes; defining the themes and lastly writing-up the findings.

## 4   Discussion of Findings

The main objective of this study was to identify TM techniques that exist for mining large datasets of unstructured textual data in order to derive valuable and meaningful insight. A total of eighteen techniques were identified and discussed below.

**Text Mining (TM):** According to [20, 21] and [8] TM is the process in which useful insight or knowledge could be discovered from analyzing large unstructured and dissimilar text datasets. Besides, most data encountered on a daily basis, such as text documents, social media text and other web-based or electronic text are primarily composed of unstructured data and it is foreseen that TM will become progressively more important as digitization of text documents and the storing of increasingly more and larger text datasets become more prevalent. TM techniques are useful to analyses language constructs ubiquitous in large text datasets due to insufficiencies that exist with classic data mining approaches. Moreover, it allows for the processing of large text datasets according to specific needs and by focusing on different aspects in the text such as textual similarities, textual semantic and implied logic. The knowledge gained through such analysis may then be used to promote informed decision-making and trend predictions [22]. TM has rapidly evolved into an interdisciplinary research field and thus there are multiple TM techniques that may be applied across domains in various application such as NLP, IR, IE, text categorization, text clustering, summarization, semantic analysis and sentiment analysis [21–24].

**Natural Language Processing (NLP)** refers to the use of computers to process and analyses large amounts of text using ML algorithms designed to analyses natural language, extract associated semantics and interpret the text [25–28]. What differentiates NLP from classical TM techniques is the method in which lexical analysis is performed i.e. the underlying metadata, context and patterns in text are analyzed as opposed to only looking at the words themselves [22, 29, 30]. [26] further argue that large sets of sample data make NLP a good contender for ML and deep learning algorithms. Thus, NLP comprise a wide array of syntactic, semantic and other rule-based or statistical ML techniques used to analyze, categorize or extract text from large text datasets [29]. However, no single TM or ML approach may be suitable for all text representation and extraction requirements and a suitable technique(s) is therefore selected based on the intended application. Depending on what type of ML approach (supervised or unsupervised) is applied will also dictate the type of technique to use. For most applications, traditional unsupervised ML techniques still achieve unparalleled performance, however, deep learning approaches hitherto were more appropriate using supervised approaches [30]. Another domain in which ML approach has become indispensable is that of Web mining and opinion mining from for example, social media platforms. Short text is pervasive in different formats throughout the Web and likely to continue growing over time. [31] claim that it is important to be able to successfully categorize short text datasets, firstly because of its pervasiveness and the sheer quantity thereof, but also for its importance in many TM applications such as sentiment analysis and identification of new trends.

**Text Categorization or Text Classification** is described as the dividing of documents in a text dataset into two or more classes or categories (sequence of texts). This categorization may be performed using supervised or unsupervised document classification techniques. A supervised ML technique is used to assign pre-defined category labels or classes, derived from the content, to text documents based on probabilities suggested by a classification ML algorithm using a set of labelled training text. With supervised ML techniques these predefined category labels or classes are provided by an external mechanism i.e. human intervention. Examples of ML classification algorithms include decision trees (DT), naïve Bayes (NB) text classifier, Support Vector Machines (SVM), and the k-nearest neighbor model (kNN).

Some of these categorization techniques are based on a binary system, considering only two categories as in the case with IR to determine the relevance/non-relevance of documents or sentiment classification into positive/negative categories. Other techniques are multi-class categorization techniques such as topic categorization and email routing or even hierarchical type categorizations [32]. [33] describes text categorization as a filtering feature for categorizing text into positive and negative classes based on frequency. However, the authors argue that using this method favors the terms used more frequently in a larger class while ignoring relative document frequency across the classes. In contrast, [34], propose the use of a multi-channel convolutional neural network (CNN) approach for text classification. This approach use two sets of word vectors, namely a static- and fine-tuned vector. For capturing better semantic features, however, the authors propose a dynamic CNN enabling the extraction of both local semantic features as well as long-range relations from text datasets.

**Clustering:** Many scholars defined clustering as a predominantly unsupervised learning technique for organizing distinct text and features from various unstructured text datasets into different clusters to enhance the retrievability thereof and also facilitate better browsing of the content [35–38]. [35] and [20] argue that the quality of the results from text clustering depend on three aspects; the text representation model, the similarity metrics and the ML algorithm applied for clustering the text dataset. [45] reference two main types of clustering approaches namely hierarchical and partition clustering models that remain pre-eminent for data clustering. In the context of TM, clustering may be applied to documents, paragraphs or sentences depending on the requirement. The similarity of words and phrases comprising a cluster is computed using cluster algorithms such as Vector Space Model (VSM), Cosine similarity, Manhattan distance, and Euclidean distance [22, 39]. Best suited for text clustering activities are usually unsupervised ML techniques that are applied to unlabeled data, leaving it to the algorithms to discover potential patterns in large-scale text datasets [26, 39, 40]. [39] acknowledge a gap that exist between supervised and unsupervised clustering approaches and propose performing the clustering process using a semi-supervised algorithm model to bridge the gap. Document clustering is a TM technique used to assign text documents into classes or categories, for example letters, newspapers, articles and blog posts. In terms of document-level clustering, techniques include, among other; partitioning clustering algorithms, density-based clustering algorithms and agglomerative hierarchical clustering algorithms. Different techniques have different applications and their use merely depend on the specific scenario or requirement for example, topic extraction or IR [36, 39]. [41] Propose a 'soft' clustering TM approach that can allow features to be assigned across different semantic classes but further propose that scholars should continue to search and develop new approaches and ML algorithms to address its limitations.

**Text Summarization (TS):** According to [22] and [42] summarization is a means of representing the meaning, words and phrases found in documents in a succinct manner without losing the main information or context. [42] differentiate between extractive and abstractive summarization while [43] refer to concept-guided document summarisation. The former is described as an extraction of key sentences from text and arranging it in close relation to what they appear in the original text, whereas the latter arrange those sentences in a meaningful order based on certain linguistic rules. These rules may only be derived using NLP techniques enabling the identification of the inherent syntax, semantics and context in the original text. [44] argue that most summarization approaches currently cater for traditional text documents and not necessarily focus on short texts such as social media content.

**Sentiment Analysis or Opinion Mining** is a field in NLP with the purpose of analyzing and understanding human emotions, opinions, and sentiments. It may be used to analyze and understand human opinion or sentiment. Generally, the main goal of sentiment analysis is to automatically classify documents or other types of text by representing its polarity in a binary fashion based on the subjective sentiments expressed in the text, for example; positive/negative, like/dislike or good/bad. Sentiment analysis, mostly performed using supervised ML algorithm, is stated to heavily rely on the integrity and the completeness of the labelled classification-feature training dataset [22, 26, 35, 39, 42]. Sentiment analysis is generally performed on three levels i.e. document, sentence and aspect or feature-based level. Currently, literature points to sentiment analysis most prominently being used in the analysis of short text extracted from social media platforms. In addition, rapid advancement in the field of deep language models led to the development and optimization of far superior sentiment classifiers based on recurrent neural networks (RNN) and convolutional neural networks (CNN) [45]. In [39] sentiment analysis based on short social network texts are described using a two-layer CNN for training characters, words, and sentences.

**Semantic Analysis** goes beyond the normal syntax-based extracting of textual content, grammar and logic in relation to topics by incorporating the discovery of semantic knowledge based on statistical techniques, past history and its learning through establishing semantic networks in large textual datasets. This allows, for example, search engines to provide more relevant and contextually accurate search results and facts when mining a text dataset. There are several approaches to building types of semantic representations of knowledge contained within text [22, 26, 29, 46]. [46] propose a 'Subject-Action-Target'-structured model to achieve a relatively high degree of expressiveness in a domain-agnostic representation of text datasets. The authors conclude by stating that domain-specific relations could, through the use of semantic networks, be linked to domain-agnostic representations of text datasets.

[30] describe Latent Semantic Analysis (LSA) as a model able to link the semantic relations that exist between features in text using the Bag of Words (BoW) approach. However, it has proven to be more effective than vector space model (VSM) based on its use of a unique dimensionality reduction algorithm. The success of LSA is further ascribed to its capability for discovering hidden relations between words that may never have co-occurred in the text dataset.

**Machine Learning (ML)** refers to the ability of ML algorithms to automatically learn and improvise, based on previous experiences, without human intervention or explicit programming. Thus, the focus of ML is to develop algorithms that can identify scenarios or problems and device and implement a solution to such problems. ML approaches mainly employ supervised and unsupervised ML algorithms [13, 20]. In supervised learning the algorithm is 'trained' using labelled training data, previously compiled from valid and expected data outputs. Therefore, the learning depends on human intervention necessary to first 'train' the machine [47]. In unsupervised the machine is required to draw inferences and find hidden patterns and structures from datasets after being trained with input data, however, without the labelled responses as was the case with supervised learning algorithms [47]. [23] argue that the focus of ML in recent years was mainly centered around automating the text classification procedure. As a result of extensive research, rapid progress was made especially with regards to various ML approaches such as Naïve Bayes and Decision Trees.

**Topic Extraction/Modelling** is a classic problem in NLP and ML. Topic modelling refers to the extraction of unknown topics from large text datasets while at the same time discovering the distribution of topics for each document comprising a text dataset. Topic modelling is widely used text mining technique for the discovering of latent topics from large unstructured text datasets [39]. According to [39] keyword extraction play an important role in topic modelling. Keyword extraction per se, is a TM technique that is generally performed using unsupervised ML algorithms such as TF-IDF, Kullback–Leibler divergence and chi-square test, all statistical word feature or keyword extraction algorithms based on probabilistic topic models with the most prominent here being LDA, or network graph algorithms such as TextRank, Rapid Automatic Keyword Extraction (RAKE) and TopicRank. [48] support the notion of LDA being the most popular and widely used topic modelling technique.

**Information Extraction (IE):** [49] describe IE as the extraction of linguistic patterns and domain-specific terminology and concepts from unstructured text. Domain experts establish the relationships between the concepts and attributes depending on the domain. [1, 50] suggest that domain experts are required to possess in-depth knowledge of the relevant field in order to obtain the best results from this technique. [51] argue that in the context of question answering systems, the main focus of IE is in finding key relations between named entities in order to construct the knowledge bases that represent the main resource in cognitive processes.

**n-Gram Model:** [51] and [33] define an n-gram or noun phrase based model as a purely statistical approach of identifying or extracting features in text datasets irrespective of the language, lexical or semantic properties. These features typically encompass single words, keywords or phrases.

**Named Entity Recognition (NER):** [8] claim that NER comprise two steps namely, the identification of entity boundaries and the determination of entity classes. These steps may be extremely challenging in domains where natural language texts are marred with grammatical errors, polysemy, synonymy and other word ambiguities such as abbreviations used in text. Moreover, entities may consist of phrases or compounds. All these challenges may affect the effectiveness of the entity recognition approach in some text domain datasets. Notwithstanding, [8] suggest ML-based NER techniques are widely used in the field of clinical texts in spite the fact that clinical text datasets contain many of the mentioned issues. Named Entity Recognition (NER) or concept extraction/recognition is a TM technique that is highly dependent on a well-defined dictionary and stop-word lists [52, 53]. However, [8] argue that the algorithm applied and the features in text datasets are the two main determining factors of success in NER approaches. [51] describes NER as part of the pre-processing phase to improve the relation extraction process and argue that NER is a statistical-based method and highly adaptable across various domains and languages.

**Vector Space Model (VSM)** is a ML technique that provides an efficient numerical method for analyzing and representing large text datasets. With VSM each word is assigned a weighted score, based its importance using a numeric value. In spite it being a simple and fast method, VSM may proof problematic as it suffers the inherent problem of dimensionality. It also uses a BoW-vectorisation to represent word counts in text with a total disregard of all the semantic, syntactical, or lexical relations between words comprising the text dataset [30, 33, 35]. The resultant vectorised matrix, based on the occurrence representation of words, provides a more informative view on the frequency of words in a document. When several documents are compared, a weighted approach is followed to show the occurrence of words across all documents.

The **Bag of Words (BoW)** representation is employed in many TM clustering techniques for the converting of features in text datasets to a structured format of vectors. With this approach a text dataset is represented as a matrix with each row representing a document in the dataset and each column representing a distinct term comprising the document. The intersect between the document and term contains the information regarding the frequency the term appears in the document. This value is usually in the TF-IDF format i.e. a function of both the frequency of a single distinct term in a document of the dataset and the inverse value of the number of documents comprising the text dataset [22, 30, 36, 54] argue that the BoW approach may not be suitable to use when semantic relations between words are important as it completely ignores word-order and grammar in representing a text dataset.

**Latent Dirichlet Allocation (LDA)** is one of the most successful generative latent topic models developed and referred to as the state-of-the-art unsupervised technique for unlabeled text document topic extraction. LDA is a useful method for the statistical analysis of text datasets. The topic model is a probability distribution technique where each document is modelled based on the latent topics found therein. However, the model requires widespread experimental analysis with various configurations before optimal results may be achieved [55, 56]. The main premise of LDA is based on the assumption that a document contains a discrete distribution of topics and the individual topics, in turn, are defined by a restricted lexis found in a text dataset. The representation of topics in the documents are based on the percentage value of the topic occurring in each document. The LDA algorithm therefore estimates the distribution of the latent topic and term sequences based on the occurrence of topics and term sequences in documents comprising a text dataset using a sampling or optimization approach [55, 56].

**TF-IDF:** [20] define TF-IDF as a statistical-based ML approach used to determine the distinction of a term or feature in a text document or across various documents comprising a text dataset. The method is based on ranking term importance based on its occurrence in a document. Thus, the higher the frequency a term appears in a document, the higher the importance of the term in the classification of the document. However,

[30] states that TF-IDF and other pure statistical-based methods are insufficient when analyzing short text datasets, as it might indicate terms of equal importance.

**Naïve Bayes:** [13] describe naïve Bayes as simple and effective non-iterative probabilistic classifier based on the idea that a single extracted feature from a document is not in any way related to any other feature in that document. However, it is stated that naïve Bayes is not a single algorithm but rather a combination of algorithms using a common principle for the task of classifying text documents. The Naive Bayes classifier considers each feature to contribute independently to the probability of its occurrence in a text document regardless of any possible correlations that may exist with other features in a text document.

**Word2vec:** [57] and [30] define Word2vec as a shallow, two-layer neural network widely used with great success in the ambit of NLP. The approach is primarily used for the embedding of words, especially in the context of sparse short text datasets to reconstruct contextual information. In addition, the approach is highly efficient in learning embedded words in unstructured text. Internally, Word2vec uses the CBOW or Skip-Gram models, which are algorithmically similar for learning word vectors that may be efficiently trained on large text datasets.

## 5   Conclusion

A comprehensive, current and conversant list of TM and ML techniques that exist for the processing of natural language in unstructured text datasets across a broad selection of applications and domains was identified and described. The main TM techniques are centered around a small number of main tasks in TM namely; pre-processing, IR, IE, text classification or text categorization, clustering, topic modelling, text summarization and sentiment analysis or opining mining. ML techniques are critical to NLP to analyses, categories or extract text from large text datasets. As TM will become progressively more important due to digitization of text documents and large datasets, TM will continue to play a pivotal role in the data analysis process. However, users should take cognizance of the fact that no single TM or ML approach may be suitable for all text representation and extraction requirements and a suitable technique(s) should therefore be selected based on the intended application and business context.

## References

1. Talib, R., Hanif, M., Ayesha, S., Fatima, F.: Text mining: techniques, applications and issues. Int. J. Adv. Comput. Sci. Appl. **7**, 414–418 (2016)
2. Tanwar, M., Duggal, R., Khatri, S.K.: Unravelling unstructured data: a wealth of information in big data. Presented at the 2015 4th International Conference on Reliability, Noida (2015)
3. Karl, A.T., Wisnowski, J., Rushing, W.H.: A practical guide to text mining with topic extraction (2015). https://doi.org/10.1002/WICS.1361
4. Sivarajah, U., Kamal, M.M., Irani, Z., Weerakkody, V.: Critical analysis of big data challenges and analytical methods. J. Bus. Res. **70**, 263–286 (2016)

5. Castellanos, A., Castillo, A., Lukyanenko, R., Tremblay, M.C.: Understanding benefits and limitations of unstructured data collection for repurposing organizational data. In: Wrycza, S., Maślankowski, J. (eds.) SIGSAND/PLAIS 2017. LNBIP, vol. 300, pp. 13–24. Springer, Cham (2017). https://doi.org/10.1007/978-3-319-66996-0_2

6. Cogburn, D., Hine, M.: Introduction to text mining in big data analytics Minitrack. Presented at the Hawaii International Conference on System Sciences (2017). https://doi.org/10.24251/HICSS.2017.110

7. Alwidian, S., Bani-Salameh, H., Alslaity, A.: Text data mining: a proposed framework and future perspectives. Int. J. Bus. Inf. Syst. **18**, 127–140 (2015). https://doi.org/10.1504/IJBIS.2015.067261

8. Sun, W., Cai, Z., Li, Y., Liu, F., Fang, S., Wang, G.: Data processing and text mining technologies on electronic medical records: a review. J. Healthc. Eng. **2018**, 4302425 (2018). https://doi.org/10.1155/2018/4302425

9. Gupta, V., Lehal, G.: A survey of text mining techniques and applications. J. Emerg. Technol. Web Intell. **1**, 60–76 (2009). https://doi.org/10.4304/jetwi.1.1.60-76

10. Gaikwad, S.V., et al.: Text mining methods and techniques. Int. J. Comput. Appl. **85**(17) (2014)

11. Shrihari, C., Desai, A.: A review on knowledge discovery using text classification techniques in text mining. Int. J. Comput. Appl. **111**(6) (2015)

12. Kaushik, A., Naithani, S.: A Comprehensive Study of Text Mining Approach, p. 8 (2016)

13. Aggarwal, A., Singh, J., Gupta, D.K.: A review of different text categorization techniques. Int. J. Eng. Technol. **7**(3.8) (2018). https://doi.org/10.14419/ijet.v7i3.8.15210

14. Kitchenham, B., Charters, S.: Guidelines for performing systematic literature reviews in software engineering. Technical Report EBSE 2007-001, Keele University and Durham University Joint Report (2007)

15. Hofstee, E.: Constructing a good dissertation - a practical guide to finishing a masters, MBA or Ph.D on schedule. Exactica (2006). https://www.loot.co.za/product/erik-hofstee-constructing-a-good-dissertation/ggld-123-g790?referrer=googlemerchant&gclid=CjwKCAjwndCKBhAkEiwAgSDKQVAM3xZ1SA4J-0N-RifK2_cR67cUQVjRCK01ZSsB4p-xXuQiAB1aoRoC9GgQAvD_BwE&gclsrc=aw.ds.    Accessed 29 Sep 2021

16. Okoli, C., Schabram, K.: A guide to conducting a systematic literature review of information systems research. Social Science Research Network, Rochester, NY, SSRN Scholarly Paper ID 1954824, May 2010. https://doi.org/10.2139/ssrn.1954824

17. Xiao, Y., Watson, M.: Guidance on conducting a systematic literature review. J. Plan. Educ. Res. **39**(1), 93–112 (2019). https://doi.org/10.1177/0739456X17723971

18. Maguire, M., Delahunt, B.: Doing a thematic analysis: a practical, step-by-step guide for learning and teaching scholars. Irel. J. High. Educ., **9**(3) (2017). https://ojs.aishe.org/index.php/aishe-j/article/view/335. Accessed 29 Sep 2021

19. Braun, V., Clarke, V.: Using thematic analysis in psychology. Qual. Res. Psychol. **3**(2), 77–101 (2006). https://doi.org/10.1191/1478088706qp063oa

20. Zhang, L., et al.: Assessment of career adaptability: combining text mining and item response theory method. IEEE Access, 1 (2019). https://doi.org/10.1109/ACCESS.2019.2938777

21. Elakiya, E., Rajkumar, N.: In text mining: detection of topic and sub-topic using multiple spider hunting model. J. Ambient. Intell. Humaniz. Comput. **12**(3), 3571–3580 (2019). https://doi.org/10.1007/s12652-019-01588-5

22. Yuksel, M.E., Fidan, H.: A decision support system using text mining based grey relational method for the evaluation of written exams. Symmetry **11**(11), 1426 (2019). https://doi.org/10.3390/sym11111426

23. Alonso-Abad, J.M., López-Nozal, C., Maudes-Raedo, J.M., Marticorena-Sánchez, R.: Label prediction on issue tracking systems using text mining. Prog. Artif. Intell. **8**(3), 325–342 (2019). https://doi.org/10.1007/s13748-019-00182-2

24. Kim, E.-G., Chun, S.-H.: Analyzing online car reviews using text mining. Sustainability **11** (6), 1611 (2019). https://doi.org/10.3390/su11061611

25. Yüksel, A.S., Tan, F.G.: A real-time social network-based knowledge discovery system for decision making. Automatika **59**(3–4), 261–273 (2018). https://doi.org/10.1080/00051144.2018.1531214

26. Nahili, W., Rezega, K., Kazar, O.: A new corpus-based convolutional neural network for big data text analytics. J. Intell. Stud. Bus. **9** (2019). https://doi.org/10.37380/jisib.v9i2.469

27. Yao, J.: Automated sentiment analysis of text data with NLTK. J. Phys. Conf. Ser. **1187**(5), 052020 (2019). https://doi.org/10.1088/1742-6596/1187/5/052020

28. Eskici, H., Koçak, N.A.: A text mining application on monthly price developments reports. Cent. Bank Rev. **18**, 51–60 (2018). https://doi.org/10.1016/j.cbrev.2018.05.001

29. Dreisbach, C., Koleck, T.A., Bourne, P.E., Bakken, S.: A systematic review of natural language processing and text mining of symptoms from electronic patient-authored text data. Int. J. Med. Inf. **125**, 37–46 (2019). https://doi.org/10.1016/j.ijmedinf.2019.02.008

30. Pal, T., Kumari, M., Singh, T., Ahsan, M.: Semantic representations in text data. Int. J. Grid Distrib. Comput. **11**, 65–80 (2018). https://doi.org/10.14257/ijgdc.2018.11.9.06

31. Bollegala, D., Atanasov, V., Maehara, T., Kawarabayashi, K.-I.: ClassiNet – predicting missing features for short-text classification. ACM Trans. Knowl. Discov. Data, **12**(5), 55:1–55:29 (2018). https://doi.org/10.1145/3201578

32. Ghawi, R., Pfeffer, J.: Efficient hyperparameter tuning with grid search for text categorization using kNN approach with BM25 similarity. Open Comput. Sci. **9**(1), 160–180 (2019). https://doi.org/10.1515/comp-2019-0011

33. Kim, K., Lee, S.-K., Park, H., Chae, J.: Academic conference analysis for understanding country-level research topics using text mining. Int. J. Comput. Inf. Syst. Ind. Manage. Appl. **11**, 001–016 (2019)

34. Ma, Q., Yu, L., Tian, S., Chen, E., Ng, W.W.Y.: Global-local mutual attention model for text classification. IEEEACM Trans. Audio Speech Lang. Process. **27**(12), 2127–2139 (2019). https://doi.org/10.1109/TASLP.2019.2942160

35. Grida, M., Soliman, H., Hassan, M.: Short text mining: state of the art and research opportunities. J. Comput. Sci. **15**(10), 1450–1460 (2019). https://doi.org/10.3844/jcssp.2019.1450.1460

36. Somasekar, H., Naveen, K.: RNS Institute of Technology, Text Categorization and graphical representation using Improved Markov Clustering. Int. J. Intell. Eng. Syst. **11**(4), 107–116 (2018). https://doi.org/10.22266/ijies2018.0831.11

37. Mustafi, D., Sahoo, G.: A hybrid approach using genetic algorithm and the differential evolution heuristic for enhanced initialization of the k-means algorithm with applications in text clustering. Soft. Comput. **23**(15), 6361–6378 (2018). https://doi.org/10.1007/s00500-018-3289-4

38. Sangaiah, A.K., Fakhry, A.E., Abdel-Basset, M., El-henawy, I.: Arabic text clustering using improved clustering algorithms with dimensionality reduction. Clust. Comput. **22**(2), 4535–4549 (2018). https://doi.org/10.1007/s10586-018-2084-4

39. Qin, L., et al.: A review of text corpus-based tourism big data mining. Appl. Sci. Web **9**(16), 3300 (2019)

40. Madhusudhanan, S., Jaganathan, S.J.L.S.: Incremental learning for classification of unstructured data using extreme learning machine. Algorithms, **11**(10) (2018). https://doi.org/10.3390/a11100158

41. Zhukov, D., Andrianova, E., Otradnov, K., Istratov, L.: Soft clustering method for text mining, with an opportunity to attribute them to different semantic groups. ITM Web Conf. **18**, 03004 (2018). https://doi.org/10.1051/itmconf/20181803004

42. Gupta, S., Gupta, S.K.: Natural language processing in mining unstructured data from software repositories: a review. Sādhanā **44**(12), 1–17 (2019). https://doi.org/10.1007/s12046-019-1223-9

43. Anoop, V.S., Asharaf, S.: Extracting conceptual relationships and inducing concept lattices from unstructured text. J. Intell. Syst. **28**(4), 669–681 (2019). https://doi.org/10.1515/jisys-2017-0225

44. Qiang, J., Chen, P., Ding, W., Wang, T., Xie, F., Wu, X.: Heterogeneous-length text topic modeling for reader-aware multi-document summarization. ACM Trans. Knowl. Discov. Data, **13**(4), 42:1–42:21 (2019). https://doi.org/10.1145/3333030

45. Luo, L.-X.: Network text sentiment analysis method combining LDA text representation and GRU-CNN. Pers. Ubiquit. Comput. **23**(3–4), 405–412 (2018). https://doi.org/10.1007/s00779-018-1183-9

46. Piad-Morffis, A., Gutiérrez, Y., Muñoz, R.: A corpus to support eHealth knowledge discovery technologies. J. Biomed. Inform. **94**, 103172 (2019). https://doi.org/10.1016/j.jbi.2019.103172

47. Padhi, B.K., Nayak, D.S.S., Biswal, D.B.N.: Machine learning for big data processing: a literature review. Int. J. Innov. Res. Technol. **5**(7), 359–368 (2018)

48. Cortez, P., Moro, S., Rita, P., King, D., Hall, J.: Insights from a text mining survey on expert systems research from 2000 to 2016. Expert Systems (2018). https://onlinelibrary.wiley.com/doi/10.1111/exsy.12280. Accessed 30 Sep 2021

49. Intarapaiboon, P., Theeramunkong, T.: An application of intuitionistic fuzzy sets to improve information extraction from thai unstructured text. IEICE Trans. Inf. Syst. **E101.D**(9), 2334–2345 (2018). https://doi.org/10.1587/transinf.2017EDP7423

50. Sharma, K., Sharma, A., Joshi, D., Vyas, N., Bapna, A.: A review of text mining techniques and applications. Int. J. Comput. IJC **24**(1), 170–176 (2017)

51. Momtazi, S., Moradiannasab, O.: A statistical approach to knowledge discovery: bootstrap analysis of language models for knowledge base population from unstructured text. Sci. Iran. 26, no. Special Issue on: Socio-Cognitive Engineering, 26–39 (2019). https://doi.org/10.24200/sci.2018.20198

52. Westergaard, D., Staerfeldt, H.H., Tonsberg, C., Jensen, L.J., Brunak, S.: A comprehensive and quantitative comparison of text-mining in 15 million full-text articles versus their corresponding abstracts. PLoS Comput. Biol. **14** (2018). https://journals.plos.org/ploscompbiol/article?id=10.1371/journal.pcbi.1005962. Accessed 30 Sep 2021

53. Manimaran, J.V.T.: Evaluation of named entity recognition algorithms using clinical text data. Int. J. Eng. Technol. **7**, 295–302 (2018). https://doi.org/10.14419/ijet.v7i4.5.20093

54. Soares, S.: Data Governance Tools: Evaluation Criteria, Big Data Governance, and Alignment with Enterprise Data Management., 1st edition. MC Press (2014)

55. Allahyari, M., et al.: A Brief Survey of Text Mining: Classification, Clustering and Extraction Techniques. *ArXiv170702919 Cs*, July 2017. http://arxiv.org/abs/1707.02919. Accessed 30 Sep 2021

56. Nagwani, N., Verma, S.: A comparative study of bug classification algorithms. Int. J. Softw. Eng. Knowl. Eng. **24**, 111–138 (2014). https://doi.org/10.1142/S0218194014500053

57. Yuan, X., Chang, W., Zhou, S., Cheng, Y.: Sequential pattern mining algorithm based on text data: taking the fault text records as an example. Sustainability, **10**(11) (2018). https://doi.org/10.3390/su10114330

# Author Index

T. Antipova (Ed.): DSIC 2021, LNNS 381, pp. 605–607, 2022.
https://doi.org/10.1007/978-3-030-93677-8